T0331119

Terrestrial Biosphere-Atmosphere Fluxes

Fluxes of trace gases, water, and energy between the terrestrial biosphere and the atmosphere govern the state and fate of these two coupled systems. This "breathing of the biosphere" is controlled by a large number of interacting physical, chemical, biological, and ecological processes. In this integrated and interdisciplinary book, the authors provide the tools to understand and quantitatively analyze fluxes of energy, complex organic compounds such as terpenes, and trace gases including carbon dioxide, water vapor, and methane.

The book first introduces the fundamental principles that affect the supply and demand for energy and trace gas exchange at the leaf and soil scales: thermodynamics, diffusion, turbulence, and physiology. It then builds on these principles to model the exchange of energy, water, carbon dioxide, terpenes, and stable isotopes at the ecosystem scale. Detailed mathematical derivations of commonly used relations in biosphere-atmosphere interactions are provided for reference in appendices.

An accessible introduction for graduate students to this essential component of Earth system science, this book is also a key resource for researchers in many related fields such as atmospheric science, hydrology, meteorology, climate science, biogeochemistry, and ecosystem ecology.

Online resources at www.cambridge.org/monson:
- A short online mathematical supplement guides students through basic mathematical principles, from calculus rules of derivation and integration, to statistical moments and coordinate rotation.

Russell Monson is Louise Foucar Marshall Professor at the University of Arizona, Tucson and Professor Emeritus at the University of Colorado, Boulder. His research focuses on photosynthetic metabolism, the production of biogenic volatile organic compounds and plant water relations from the scale of chloroplasts to the globe. He has received numerous awards, including the Alexander von Humboldt Fellowship, the John Simon Guggenheim Fellowship, and the Fulbright Senior Fellowship, and was also appointed Professor of Distinction in the Department of Ecology and Evolutionary Biology at the University of Colorado. Professor Monson is a Fellow of the American Geophysical Union and has served on advisory boards for numerous national and international organizations and projects. He is Editor-in-Chief of the journal *Oecologia* and has over 200 peer-reviewed publications.

Dennis Baldocchi is Professor of Biometeorology at the University of California, Berkeley. His research focuses on physical, biological, and chemical processes that control trace gas and energy exchange between vegetation and the atmosphere and the micrometeorology of plant canopies. Awards received include the Award for Outstanding Achievement in Biometeorology from the American Meteorological Society (2009), and the Faculty Award for Excellence in Postdoctoral Mentoring (2011). Professor Baldocchi is a Fellow of the American Geophysical Union and is a member of advisory boards for national and international organizations and projects. He is Editor-in-Chief of the *Journal of Geophysical Research: Biogeosciences* and has over 200 peer-reviewed publications.

Terrestrial Biosphere-Atmosphere Fluxes

RUSSELL MONSON

University of Arizona

DENNIS BALDOCCHI

University of California, Berkeley

Shaftesbury Road, Cambridge CB2 8EA, United Kingdom

One Liberty Plaza, 20th Floor, New York, NY 10006, USA

477 Williamstown Road, Port Melbourne, VIC 3207, Australia

314–321, 3rd Floor, Plot 3, Splendor Forum, Jasola District Centre, New Delhi – 110025, India

103 Penang Road, #05–06/07, Visioncrest Commercial, Singapore 238467

Cambridge University Press is part of Cambridge University Press & Assessment, a department of the University of Cambridge.

We share the University's mission to contribute to society through the pursuit of education, learning and research at the highest international levels of excellence.

www.cambridge.org
Information on this title: www.cambridge.org/9781107040656

First published 2014

A catalogue record for this publication is available from the British Library

Library of Congress Cataloging-in-Publication data
Monson, R. K. (Russell K.), 1954–
Terrestrial biosphere-atmosphere fluxes / Russell Monson, Dennis Baldocchi.
pages   cm
ISBN 978-1-107-04065-6 (hardback)
1. Atmospheric circulation.   2. Atmospheric turbulence.   3. Biosphere.
I. Baldocchi, Dennis D.   II. Title.
QC880.4.A8M658   2013
551.51–dc23
2013024741

ISBN   978-1-107-04065-6   Hardback

Additional resources for this publication at www.cambridge.org/monson

Contents

Color plate section is found between pages 202 and 203

Supplement available at **www.cambridge.org/monson:**
- Supplement 1: Mathematical concepts

Preface

This book is about *interactions* – those that occur between the terrestrial biosphere and the atmosphere. Understanding biosphere-atmosphere interactions is a core activity within the discipline of *earth system sciences*. Many of the most pressing environmental challenges that face society (e.g., the anthropogenic forcing of climate change, urban pollution, the production of sustainable energy sources, and stratospheric ozone depletion), and their remedies, can be traced to biosphere-atmosphere interactions within the earth system. Traditionally, biosphere-atmosphere interactions have been studied within a broad range of conventional disciplines, including biology, the atmospheric and geological sciences, and engineering. In this book we take an integrated, interdisciplinary perspective; one that weaves together concepts and theory from all of the traditional disciplines, and organizes them into a framework that we hope will catalyze a new, synergistic approach to teaching university courses in the earth system sciences.

As we wrote the initial outline for the book, we recognized that the interdisciplinary perspective we sought, in a subtle way, had already emerged; it simply had not been formally collated into a synthetic format. For the past several years, biologists have been attending meetings and workshops traditionally associated with meteorology and geochemistry and conversely meteorologists and geochemists have been attending biology meetings. As a result, newly defined and integrative disciplines have already appeared with names such as "biometeorology," "bioclimatology," and "ecohydrology." Thus, the foundations for the book had already been laid. We simply needed to find the common elements and concepts that permeated these emerging disciplines and pull them together into a single treatment.

We have written the book as two colleagues who have migrated from different ends of the biology-meteorology spectrum – one (Monson) from formal training in biology and one (Baldocchi) from formal training in meteorology – but who also have struggled throughout their careers to grasp concepts at these disciplinary interfaces. In many ways this book is autobiographical; it reflects the challenges that both of us faced as we developed collaborations across these disciplines. We actually met for the first time at a conference in Asilomar, California in 1990, which was dedicated to bridging the gaps among biologists, meteorologists, and atmospheric chemists. Thus, the interdisciplinary foundation for the book has deep roots that were initiated over two decades ago. From that initial friendship we developed a collaboration in which we began to compile and combine materials that we extracted from our respective course lectures.

This book is intended to be used as both a textbook and reference book. As a textbook it is intended to support courses for advanced undergraduate students or beginning graduate students. As a reference book it is intended to provide detailed mathematical derivations of some of the most commonly used relations in biosphere-atmosphere interactions. In order to address both aims, we have written the primary text of the chapters to provide what we consider to be *the rudiments*; those concepts essential to an introductory understanding of process interactions and fundamental theory. Detailed mathematical derivations are presented as "appendices" at the end of many chapters. These derivations are intended mostly as reference material; however, in our own experiences we discovered that formal derivations, such as these, also served as an important resource to students. In fact a well-received feature of some of our classes was the "Derivation Derby" held as an evening session in which students were required to use the chalk board to present, in their own words, the foundations of some of the more classic biophysical relations; of course with good food and drink as accompaniment. We have used a second tool to develop advanced topics of more conceptual, rather than quantitative, nature – the "boxes" that are embedded in many chapters. In the boxes we have tried to bring out current topics and issues that appear to have captured the attention of the field at the moment, or we have described studies that have used the concepts under discussion in unique ways. Once again, the boxes will be most effectively used to provide supplementary material that embellishes the rudimentary topics presented in the main text of the chapters. We have tried to use a modest frequency of citations in most chapters. Much of the material we cover is of an elementary nature, and in order to sustain continuity in those discussions we have not interrupted the text with frequent citations. In those cases where we thought that a citation might be useful for further explorations of a topic, especially where a review article or an article of historical significance might be useful, we have provided citations. In the sections that cover contemporary concepts, especially those still being defined through active debate in the literature, we have provided a more complete record of citations. Furthermore, many of the figures were adopted from past studies, and we have provided citations in the figure legends, which will be useful in directing students to primary sources in the literature.

One of the initial decisions we made as we organized material for the book involved the strategy for topical organization. We considered two possible frameworks: chapters that focused on single environmental factors (e.g., a chapter on water, a chapter on light, a chapter on temperature, and so on), or chapters that build in spatiotemporal scale, from processes at smaller scales to those at larger scales (e.g., a chapter on cells and metabolism, a chapter on leaves and diffusion, a chapter on canopies and turbulent transport, and so on). Conventional treatments, especially in texts that deal with environmental physics, have followed the former model, and they have done so with good success. However, we recognized that many of the observations and much of the theory that has emerged in recent years has been framed around hierarchical scaling, and we wanted to develop a treatment that could be used within this framework. After much discussion and deliberation, we decided to follow the second model, though with a bit of introgression from the first model. Thus, the chapters build in scale, beginning with chloroplasts, progressing to leaves and canopies, and culminating with the planetary boundary layer. Each of these scaled chapters is preceded with one or more chapters on the nature of relevant environmental factors as drivers of processes. Thus, the chapter on leaf scale transport is preceded with a chapter on diffusion, and the chapter on turbulent transport is preceded with a chapter on stability in the planetary boundary layer. Exceptions to these patterns are the initial three chapters, which deal with broad topics in thermodynamics and chemical rate theory, and the final three chapters, which deal respectively with soil carbon and nitrogen fluxes, fluxes of volatile reactive compounds and atmospheric chemistry, and fluxes related to stable isotope fractionation. These chapters are intended to provide a framework for understanding the relations among fluxes, sources/sinks, and gradients, in the case of the earliest chapters, and to elaborate on some important recent directions in earth system sciences research, in the case of the latest chapters.

The overall emphasis of the book is on understanding processes that control fluxes. Less emphasis is placed on descriptions of biogeochemical pools and reservoirs. We also pay less attention to instrumentation and experimental protocols. Most of the chapters focus on CO_2, H_2O, and energy fluxes, although we also take up the topic of other trace gases in briefer format. Finally, we note that our book focuses exclusively on terrestrial ecosystems. Our decision not to wade into the oceans was determined by recognition of our strengths and weaknesses as scientists and authors, and this decision does not reflect a bias against the importance of ocean processes to earth system dynamics.

We appreciate the many discussions we have had with generous colleagues as we wrote the book and sought critical feedback. Reviews and discussions of several of the chapters in early form were provided by Dave Bowling, Tom Sharkey, John Finnigan, Rowan Sage, Ray Leuning, Laura Scott-Denton, Peter Harley, Tony Delany, Dan Yakir, Jielun Sun, Mike Weintraub, Dave Moore, Paul Stoy, Dave Schimel, and Keith Mott. Many thanks to all of you! While these colleagues provided many useful insights and suggestions, responsibility for the book's final form belongs with us.

Symbols

In writing a book with as broad a set of mathematical relations as that presented here we had to make decisions as to whether to create new symbols for cases of duplicated usage, or retain those most often used, by convention, in the scientific literature. We tried to use conventional symbols as often as was possible, and we allowed for some overlap in designation, especially when duplicated symbols were used in different chapters.

Uppercase, non-italicized Latin

A	CO_2 assimilation rate (μmol m^{-2} s^{-1})
A_c	canopy net CO_2 assimilation rate
A_n	net CO_2 assimilation rate
A_g	gross CO_2 assimilation rate
E	energy (J) or energy content (J mol^{-1})
E_a	energy of activation (J mol^{-1})
E	surface evaporation or leaf transpiration flux density (mol m^{-2} s^{-1})
E_t	total enzyme protein content (mol l^{-1})
E^o	standard reduction potential (J coulomb^{-1})
F	flux density (mol m^{-2} s^{-1})
F_c	flux density of CO_2
F_w	flux density of H_2O
F_j	flux density of constituent j
F_J	photosynthetic electron transport flux density
F_{vm}	vertical atmospheric mean flux density
F_{vt}	vertical atmospheric turbulent flux density
F	Faraday's constant (coulomb mol^{-1})
G	conduction flux density of heat (J m^{-2} s^{-1})
G	free energy (J) or molar free energy content (J mol^{-1})
G^0	standard free energy (J) or molar free energy content (J mol^{-1})
G	rate of biomass increase (g s^{-1})
GPP	gross primary productivity (mol m^{-2} s^{-1} or mol m^{-2} yr^{-1})
H	enthalpy (J) or enthalpy content (J mol^{-1})
H	conduction of heat (W m^{-2})

H_{se} conduction of sensible heat (from the surface to the atmosphere) (W m^{-2})

H_G conduction of sensible heat (from the atmosphere to the ground surface) (W m^{-2})

I   photon flux density (mol photons m^{-2} s^{-1})

I_D direct photon flux density

I_d diffuse photon flux density

I_s isoprene emission flux density (nmol m^{-2} s^{-1})

J   joule unit of energy (kg m^2 s^{-1})

LAI   leaf area index (m^2 leaf area m^{-2} ground area)

L   leaf area index (used in equations)

L_e effective LAI

N   newton unit of force (kg m s^{-1})

N_a Avogadro's number

NDVI   normalized difference of vegetation index (dimensionless)

NPP   net primary productivity (mol m^{-2} s^{-1} or mol m^{-2} yr^{-1})

P   total atmospheric pressure (N m^{-2}, Pa)

P   statistical probability

P_0 probability of photon penetration to a canopy layer

P_{sf} probability of a sunfleck in a canopy layer

Q   thermal energy (J) or molar thermal energy content (J mol^{-1})

Q_{10} respiratory quotient (ratio of R_d at two temperatures separated by 10 °C)

R   radiant energy flux density (J m^{-2} s^{-1} or W m^{-2})

R_S shortwave radiant energy flux density (J m^{-2} s^{-1} or W m^{-2})

R_L longwave radiant energy flux density (J m^{-2} s^{-1} or W m^{-2})

R_n net radiation flux density (J m^{-2} s^{-1} or W m^{-2})

R   isotope abundance ratio

R_d "dark" (mitochondrial) respiration (μmol m^{-2} s^{-1})

R_e ecosystem respiration

R_g growth mitochondrial respiration

R_m maintenance mitochondrial respiration

S   molar entropy content (J mol^{-1} K^{-1})

S   amount of substrate (moles)

S   sink or source "strength," as a flux density (mol m^{-2} s^{-1})

S_{rel} relative specificity of Rubisco (unitless)

[S]   enzyme substrate concentration (mol l^{-1} or mol m^{-3})

T   temperature (K or °C)

TKE   turbulence kinetic energy (J)

TPU   triose phosphate utilization flux density (μmol m^{-2} s^{-1})

U   internal energy (J) or molar internal energy content (J mol^{-1})

V   volume (m^3)

V_{max} Michaelis–Menten velocity coefficient (mol s^{-1})

V_{cmax} Michaelis–Menten velocity coefficient for Rubisco carboxylation

V_{omax} Michaelis–Menten velocity coefficient for Rubisco oxygenation

W   work (J) or molar work content (J mol^{-1})

W_p total plant biomass (g)

Y_g growth yield (fraction of substrate converted to biomass)

Uppercase, italicized Latin

A surface area (m^2)

A_G ground area

A_L leaf area

B feedback multiplier (unitless)

B_k permeability coefficient for viscous flow (m^2)

C_D drag coefficient (dimensionless)

C_{Ex} flux control coefficient (unitless)

E_x radiative transfer extinction function (fraction of total PPFD)

F force (N)

F_d molar diffusive force (N mol^{-1})

F_D drag force (g m s^{-2})

G fraction of leaf area oriented normal to I_D in radiative transfer models

G gain of feedback loop (unitless)

G_c closed-loop feedback gain

G_o open-loop feedback gain

K_d molecular diffusion coefficient (m^2 s^{-1})

^kK_d Knudsen diffusion coefficient (m^2 s^{-1})

K_{dh} diffusion coefficient for heat (m^2 s^{-1})

K_{dw} diffusion coefficient for H$_2$O

K_{dc} diffusion coefficient for CO$_2$

K_D eddy diffusion coefficient (m^2 s^{-1})

K_e equilibrium constant (unitless)

K_I canopy PPFD extinction coefficient ($K_I = G/\cos\theta$)

K_m Michaelis–Menten coefficient (mol l^{-1} or mol m^{-3})

K_c Michaelis–Menten coefficient for dissolved CO$_2$

K_o Michaelis–Menten coefficient for dissolved O$_2$

K_s steady state constant (mol^{-1})

Kn Knudsen number (dimensionless)

L turbulent length scale (m) (generally used)

L Obukhov length scale (m) (specifically used)

Nu Nusselt number (dimensionless)

R	universal gas constant ($J\ K^{-1}\ mol^{-1}$)
Re	Reynolds number (dimensionless)
Ri	Richardson number (dimensionless)
Ri_c	critical Richardson number
Ri_b	bulk Richardson number
S_c	radiative transfer scattering function (fraction of total PPFD)
$S(\kappa)$	spectral density as a function of wavenumber
V	specific volume ($m^3\ kg^{-1}$)
$\overline{V}_w$	partial molal volume of H_2O ($m^3\ mol^{-1}$)

Lowercase, non-italicized Latin

a	radiant or photon absorptance (fractional)
aPAR	fraction of absorbed photosynthetically active radiation
c	concentration as mole fraction
c_{ac}	atmospheric CO_2 mole fraction
c_{aw}	atmospheric H_2O mole fraction
c_{aw}*	atmospheric H_2O mole fraction at saturation
c_{cc}	chloroplast CO_2 mole fraction
c_{co}	chloroplast O_2 mole fraction
c_{ic}	intercellular CO_2 mole fraction in the leaf air spaces
c_{iw}	intercellular H_2O mole fraction in the leaf air spaces
c_{sc}	CO_2 mole fraction at leaf surface
c_{Ex}	mole fraction concentration of enzyme x
fPAR	fraction of absorbed photosynthetically active radiation
g	conductance ($m\ s^{-1}$ or $mol\ m^{-2}\ s^{-1}$)
g_b	boundary layer conductance ($m\ s^{-1}$ or $mol\ m^{-2}\ s^{-1}$)
g_{bw}	boundary layer conductance to H_2O diffusion ($m\ s^{-1}$ or $mol\ m^{-2}\ s^{-1}$)
g_{bc}	boundary layer conductance to CO_2 diffusion ($m\ s^{-1}$ or $mol\ m^{-2}\ s^{-1}$)
g_s	stomatal conductance ($m\ s^{-1}$ or $mol\ m^{-2}\ s^{-1}$)
g_{sw}	stomatal conductance to H_2O vapor diffusion ($m\ s^{-1}$ or $mol\ m^{-2}\ s^{-1}$)
g_{sc}	stomatal conductance to CO_2 diffusion ($m\ s^{-1}$ or $mol\ m^{-2}\ s^{-1}$)
g_{ic}	internal leaf conductance to CO_2 diffusion ($m\ s^{-1}$ or $mol\ m^{-2}\ s^{-1}$)
g_{tw}	total leaf conductance to H_2O vapor diffusion ($m\ s^{-1}$ or $mol\ m^{-2}\ s^{-1}$)
h	height (m)
m	mass (g)
n	molar quantity (mol)
p	pressure or partial pressure of a gas constituent ($N\ m^{-2}$, Pa)
p_r	probability of recollision (secondary collision) of a photon

r	radius (m)
r	reflectance of incident PPFD (fractional)
r	resistance (s m^{-1})
r_a	aerodynamic resistance (s m^{-1})
r_{bl}	boundary layer diffusive resistance (s m^{-1})
r_i	internal leaf diffusive resistance (s m^{-1})
r_s	stomatal diffusive resistance (s m^{-1})
t	transmittance of incident PPFD (fractional)
v	speed or velocity (mol l^{-1} s^{-1} or m s^{-1})
v_c	Rubisco carboxylation rate on leaf area basis (μmol m^{-2} s^{-1})

Lowercase, italicized Latin

a	acceleration (m s^{-2})
c	speed of "light" (m s^{-1})
c	specific heat (J kg K^{-1})
c_p	specific heat of dry air at constant pressure (J kg^{-1} K^{-1})
c_v	specific heat of dry air at a constant volume (J kg^{-1} K^{-1})
d	boundary layer length scale (m)
d_H	canopy displacement height (m)
f	frequency (s^{-1})
f_a	fraction of canopy woody surface area
g	gravitational acceleration ($\sim$ 9.8 m s^{-2})
h	Planck's constant (J s)
h_c	heat transfer coefficient (J m^{-2} s^{-1} K^{-1})
k	reaction rate constant (s^{-1} or mol^{-1} s^{-1})
k_{cat}	enzyme catalytic rate constant
k	von Karman's constant (dimensionless)
k_B	Boltzmann constant (J K^{-1})
k_H	Henry's Law partitioning coefficient (kPa liter mol^{-1})
k_N	canopy nitrogen allocation coefficient (dimensionless)
l	length (m)
$\hat{m}$	mechanical advantage of the epidermis (dimensionless)
p	porosity of a soil or leaf volume (fractional)
r_p	radial width of penumbra (cm)
t	time (s)
t_E	Eulerian time scale (s)
t_L	Lagrangian time scale (s)
u	molar flow rate (mol s^{-1})

xix   List of symbols

u	longitudinal wind velocity (m s^{-1})
u'	turbulent longitudinal wind velocity (m s^{-1})
$\bar{u}$	mean longitudinal wind velocity (m s^{-1})
u_j	Einstein–Smoluchowski mobility of constituent j (s kg^{-1})
u_*	friction velocity (m s^{-1})
v	cross-stream wind velocity (m s^{-1})
w	vertical wind velocity (m s^{-1})
w'	turbulent vertical wind velocity (m s^{-1})
$\bar{w}$	mean vertical wind velocity (m s^{-1})
z	electrical charge
z	vertical length (m)
z_{bl}	vertical depth of boundary layer (m)
z_0	aerodynamic roughness length (m)
z_p	depth of pore (mm)

Lowercase, non-italicized Greek

α	isotope effect (unitless)
γ	foliar clumping (fraction of LAI)
δ	isotope abundance ratio (delta notation) (‰)
ε	TKE dissipation rate (s)
ε	radiation-use efficiency in remote sensing modeling (g C MJ^{-1})
κ	wavenumber (m^{-1})
λ	canopy clumping index (dimensionless)
λ_a	mean free path of diffusion in air (m)
λ_w	latent heat of vaporization for H$_2$O (J mol^{-1})
$\lambda_w E$	latent heat flux density (J m^{-2} s^{-1})
μ	molar chemical potential (J mol^{-1})
μ^*	standard molar chemical potential (J mol^{-1})
ν	kinematic viscosity (m^2 s^{-1})
ρ	density (g m^{-3})
ρ_a	mass density of air (g m^{-3})
ρ_m	molar density (mol m^{-3})
ρ_{mw}	molar density of water (typically of air; mol m^{-3})
ρ_w	mass density of water (g m^{-3})
σ	standard deviation
τ	atmospheric lifetime (s)
τ	momentum flux density (N m^{-2})
ϕ	fractional leakage of mass from a metabolic pathway
ϕ	ratio of the rates of oxygenation and carboxylation for Rubisco

ϕ molar quantum yield of photosynthesis (mole fraction)

ψ_w total water potential (Pa)

ψ_g gravitational component of water potential (Pa)

ψ_m matric component of water potential (Pa)

ψ_p pressure component of water potential (Pa)

ψ_π osmotic component of water potential (Pa)

$\psi_{\pi g}$ osmotic potential of guard cell (Pa)

$\psi_{\pi s}$ osmotic potential of subsidiary cell (Pa)

Lowercase, italicized Greek

α Kolmogorov constant for turbulent inertial subrange (dimensionless)

α surface albedo (percentage of incident solar flux density)

ε radiant emittance (fractional)

ε_L leaf emittance of longwave radiation

$\varepsilon_j^{v_x}$ elasticity coefficient of reaction x with respect to metabolite j (unitless)

θ solar zenith angle (degrees or radians)

θ_t potential temperature (K)

θ_{vt} virtual potential temperature (K)

κ thermal conductivity (J s^{-1} m^{-1} K^{-1})

κ_E Eyring transmission coefficient (fractional)

λ wavelength (m)

μ dynamic viscosity (kg m^{-1} s^{-1})

ν frequency of electromagnic wave

σ Stefan–Boltzmann constant (5.673×10^{-8} J s^{-1} m^{-2} K^{-4})

τ tortuosity of a pore system (dimensionless)

ϕ Monin–Obukhov scaling coefficient (dimensionless)

ϕ solar azimuth angle (degrees or radians)

ϕ Bunsen solubility coefficient for gases (m^3 gas m^{-3} solution)

φ_E electrical potential (J coloumb^{-1})

χ stomatal mechanical coefficient (mmol H$_2$O m^{-2} s^{-1} MPa^{-1})

ω photon scatter coefficient (dimensionless)

Uppercase, non-italicized Greek

Γ CO$_2$ compensation point (μmol mol^{-1})

Γ_* CO$_2$ photocompensation point (μmol mol^{-1})

Δ isotope discrimination (‰)

Δc_j finite difference in mole fraction of chemical species j

Λ_E Eulerian length scale (m)

Λ_L Lagrangian length scale (m)

Ω angle of solar photon interactions with a surface (degrees or radians)

Ω_L angle of leaf surface orientation

A Note on the Parenthetical Formatting of Function Relations and Collected Sums or Differences

Conventional algebraic notation indicates that a dependent variable is a 'function of' an independent variable through use of parenthetical formatting. Thus, dependent variable y is related to independent variable x according to $y = f(x)$. However, other symbols can be used to designate dependent and independent variables using parenthetical notation. Take the example of atmospheric vapor pressure (often designated as e_s) determined as a function of air temperature (often designated as T_a). We can write an equation with e_s expressed as a function of T_a, and related to surface temperature (T_s), and a linear slope (s), as: $e_s [T_a] \approx e_s [T_s] + s (T_a - T_s)$. This relation is read as 'e_s' evaluated as a function of 'T_a' is approximated by 'e_s' as a function of 'T_s' plus the product between a linear slope 's' and the difference between T_a and T_s. The terms containing e_s on the left and right sides of the equation *should not* be read as "e_s multiplied by T_a or T_s"; rather, the reader should be aware from the context of the equation that the notation is referring to e_s as a function of T_a or T_s. The mathematical difference between T_a and T_s on the right side of the equation is gathered as a "collected difference" within parentheses. Similar parenthetical nomenclature is used to indicate "collected sums". Both *collected differences* and *collected sums*, unlike the terms indicated as *parenthetical functions*, are indeed active variables of the relation. We have tried to assist the reader in making these distinctions by using squared brackets around those terms intended as functional relations (e.g., $[T_a]$), and rounded parentheses around those terms intended as collected sums or differences (e.g., $(T_a - T_s)$).

1 The general nature of biosphere-atmosphere fluxes

The atmosphere and its manifold changes have held fascination for men and women ever since human beings have trod this Earth. Its study played an integral role in the evolution of natural philosophy from which all of our present sciences have sprung.

F. Sherwood Rowland, Nobel Prize Banquet Speech, 1995

It is widely believed that the abundance of the principal gases N_2 and O_2 is determined by equilibrium chemistry. One of the larger problems in the atmospheric sciences is that of reconciling that belief with the uncomfortable fact that these same gases are cycled by the Biosphere with a geometric mean residence time in thousands of years.

James Lovelock and Lynn Margulis (1974)

Sherwood Rowland's comment at the banquet held to honor receipt, along with Paul Crutzen and Mario Molina, of the 1995 Nobel Prize in Chemistry, places the atmosphere at the center of some of the most influential scientific discoveries to have been made during human history. Within Rowland's comment we can recognize Thales of Miletus who in the sixth century BC struggled to understand the different states of water and the process of evaporation, Lavoisier in the late eighteenth century discerning the exchange of oxygen between organisms and the atmosphere, and Arrhenius in the early part of the twentieth century calculating the relation between the carbon dioxide content of the atmosphere and the earth's surface temperature. The importance of the atmosphere in the history of natural philosophy is clearly underscored by these seminal studies. Within all of these studies, however, is the undeniable influence of the earth's surface and in particular the earth's biosphere, on the chemical composition and dynamics of the atmosphere. The two are linked in a type of "co-dependency" in which processes and change can only be understood through studies that include both biotic and abiotic systems. The requisite nature of the nexus between the biotic and abiotic domains of the earth system is recognizable, albeit in extreme form, in the controversial concept of "Gaian homeostasis" laid out by James Lovelock and Lynn Margulis in 1974. While we (the authors) do not, in its entirety, endorse the tenets of a Gaian earth, we do recognize the value of this concept in defining the biosphere and atmosphere as coupled and interdependent systems. It is this interdependency, and the processes that maintain it, that will be the focus of this book.

In this opening chapter, we present the concept of biosphere-atmosphere interactions through some relatively general aspects of biogeochemical fluxes. We begin by establishing the biosphere and atmosphere as components of an earth system connected through coupled biogeochemical cycles. We then establish the concept of flux, as a unifying principle in the transport of mass and energy, and explore the fundamental driving forces and constraints that govern the directions and magnitudes of fluxes. Finally, we consider some of the general cybernetic features of biosphere-atmosphere exchange, noting in particular the tendencies for non-linear relations and feedbacks, and we present a hierarchical framework within which to understand biosphere-atmosphere interactions. It is important at the outset to make the point that in this book mass and energy will be discussed with regard to *changes in state and exchanges among components of the earth system*. We will not focus on pool sizes or biogeochemical budgeting. This is a challenge because we are much more confident in our ability to observe states of the biosphere and atmosphere, compared to changes in them. It is the interactions among biogeochemical pools and earth system components, however, that drives dynamics in the state variables and thus forces time-dependent changes in those variables. It is the role of biosphere-atmosphere exchange as an agent of change in the earth system that we want to keep as a principal point of focus as we launch into our initial discussions.

1.1 Biosphere-atmosphere exchange as a biogeochemical process

Biogeochemistry is the study of changes in the earth system due to the combined activities of chemical, geological, and biological processes and reactions. Many of the trace gases that are exchanged between the biosphere and atmosphere are produced from the reduction-oxidation (redox) reactions that drive the biogeochemical cycling of elements, principally C, O, N, and S. Biogeochemical cycling is powered by the flow of solar photons through the atmosphere and biosphere, where a part of it is absorbed and used to energize electrons, which in turn drive redox reactions. This coupling between the solar energy flux and redox chemistry is exemplified in the biological processes of photosynthesis and nitrogen fixation. In both processes the energy state of electrons is increased at the expense of solar energy, and those energized electrons are used in the construction of organic compounds. The energy from the electrons is captured within the molecular structure of organic compounds, providing them with a higher energy state than their inorganic precursors. This is biogeochemical cycling in its most basic form, and it is the basis by which energy is used to construct and maintain the biotic component of the earth system.

The fundamental biogeochemical unit within the earth system is the *ecosystem*. Formally, an ecosystem is the sum of the organisms and their environment within a three-dimensional space. However, it is the *synergistic interactions* among biotic and abiotic processes that create interdependencies between organisms and their environment, and which determines the dynamic nature of the ecosystem. Thus, we will treat the ecosystem as *both* the "theater" and the "play" within which organisms transform energy into the redox reactions that drive biogeochemical cycling. Although we will spend most of our time on biogeochemical exchanges within ecosystems, it is important to recognize that ecosystems aren't truly discrete – they have porous, diffuse boundaries through which exchange occurs with the surroundings. Interactions at those boundaries tie ecosystems into broader regional and global processes.

1.2 Flux – a unifying concept in biosphere-atmosphere interactions

Reciprocity is obvious in biosphere-atmosphere interactions. Reciprocity underlies the interdependencies that mutually bind components of the earth system, and it is reflected in the flows of mass and energy among those components. Within the earth system, changes in the size of biogeochemical pools are forced by unequal reciprocities. Reciprocity in biogeochemical flows is formally described through an analysis of *fluxes*. Proper terminology for flux according to the Système International is *flux density*, which refers to flow (of mass, momentum, or heat) per unit of surface area per unit of time. For brevity and convenience, the terms *flux density* and *flux* will be used interchangeably in this book. In some cases, we will refer to *total flux* within the context of an integral quantity with respect to finite space and time. For example, the total flux for global CO_2 uptake is ~ 123 Pg C yr^{-1}.

Flux forms the basis of the biosphere-atmosphere interactions we will consider in this book; whether it be the exchange of shortwave radiant energy between the sun and earth (a flux of photons), or the movement of water between leaves and the atmosphere (a flux of compound mass). We will be most concerned with the fluxes of mass, momentum, and heat. These entities are described by scalars, in the case of mass and heat, or vectors in the case of momentum. A *scalar* is any quantity that can be described by a quantitative scale, but has no specified direction (e.g., mass or time). A *vector* can be described with a quantitative scale and has an associated direction (e.g., wind velocity). Fluxes, themselves, are vector quantities (they have direction); in accordance with vector mathematics, gross fluxes can be summed to provide net fluxes. Gradients in scalar or vector density are also vector quantities, as they have direction.

Formally, flux is velocity multiplied by density. The nature of a flux as the product between velocity and density can be appreciated through an examination of units. Velocity is expressed in units of m s^{-1} and density (at least for the case of scalars with mass) is expressed in units of mol m^{-3} (or mass m^{-3}). In reconciling the product between velocity and density, the units emerge as mol m^{-2} s^{-1} or mass m^{-2} s^{-1}, which are both examples of flux density.

Following fundamental thermodynamic tendencies, net fluxes of mass and energy will occur in the direction that opposes their associated gradients in density; i.e., *net flux* will occur from higher density toward lower density. The net flux works to diminish the gradient. The magnitude of a flux is dependent on, and proportional to, the magnitude of its associated density gradient. In most biogeochemical systems, density gradients are determined by the location and magnitude of *sources* and *sinks*. The terms "source" and "sink" are used in different ways within the biogeochemistry research community; some usages refer to locations (e.g., "the oceanic sink or terrestrial sink") whereas others refer to processes (e.g., "photosynthesis as a CO_2 sink or respiration as a CO_2 source"). We will tend to use these terms within the latter context – as processes that determine density gradients and thus drive fluxes. The sources and sinks that drive biogeochemical fluxes of mass are those chemical or biochemical processes that consume or produce compounds. The sources and sinks that determine the direction and magnitude of atmospheric momentum are inertia and viscosity, respectively.

One prominent example of the relations among fluxes, density gradients, and the location of sources and sinks occurs at the hemispheric scale in the record of CO_2 concentration that has been deduced for the past two millennia. In pre-industrial times (prior to the late 1800s), the global carbon budget was roughly in balance; on average, the net CO_2 *source from the biosphere to the atmosphere* almost exactly compensated for the net CO_2 *sink from the atmosphere to the biosphere* (i.e., global respiration was, on average, equal to global gross primary productivity). This balance is deduced from observations of the CO_2 content in ancient ice recovered from deep inside polar ice sheets; air samples taken from the cores reflect a nearly constant atmospheric CO_2 concentration over the millennium prior to the late nineteenth century (Figure 1.1A). Since the Industrial Revolution in the late 1800s, human activities have emitted progressively more CO_2 to the atmosphere, thus creating an additional CO_2 source and unbalancing the natural carbon cycle. The anthropogenic CO_2 added to the atmosphere is partly absorbed by biological processes in ocean and terrestrial ecosystems. However, not all the additional CO_2 can be absorbed, leading to an overall imbalance in global sources and sinks. This imbalance has caused an increase in the earth's atmospheric CO_2 concentration. The pre-industrial balance in the global carbon budget, and subsequent imbalance, provide a convenient lesson in the concept of *continuity*, which we will take up in more detail in a later chapter. Continuity requires that in the presence of equal exchanges of conserved quantities, such as those for mass and energy, within the bounds of a controlled volume, the concentration of the quantity must remain constant. If an imbalance appears in the sources and sinks, and thus in the associated fluxes, it must be reflected as a change in concentration. We can observe the principle of continuity in action in the near-continuous record of atmospheric CO_2 concentration that has been collected since the 1950s from an observatory at Mauna Loa, Hawaii and since the 1970s from the South Pole, and is familiar to most students of the earth system sciences (Figure 1.1B). Oscillations in the atmospheric CO_2 mole fraction reflect seasonal changes in the shifting dominance between photosynthesis and respiration, superimposed on a nearly constant emission of anthropogenic CO_2 pollution. Thus, with the onset of the growing season, the photosynthetic flux of CO_2 from the atmosphere to terrestrial ecosystems (hemispheric sink) is greater than the respiratory flux from ecosystems to the atmosphere (hemispheric source), and the CO_2 mole fraction of the atmosphere decreases as the growing season progresses. This pattern is reversed during the winter. Here, in an example of global proportions, we can fully appreciate the coupling between CO_2 fluxes to density gradients, and the determination of density gradients by the location and magnitude of sources and sinks.

It is important to recognize that a gradient in the density of a scalar or vector is only one determinant of a flux. State variables that characterize the system, such as temperature and pressure, also influence the flux. Temperature, pressure, and density are related through the equation of state, which will be discussed in a future chapter. Furthermore, the flux of one scalar or vector can be coupled to the flux of a different scalar or vector. For example, the flux of solar photons into a leaf is coupled to the flux of CO_2, through the processes of photosynthesis. In general, we will refer to the principal influences on flux magnitude as *driving variables*.

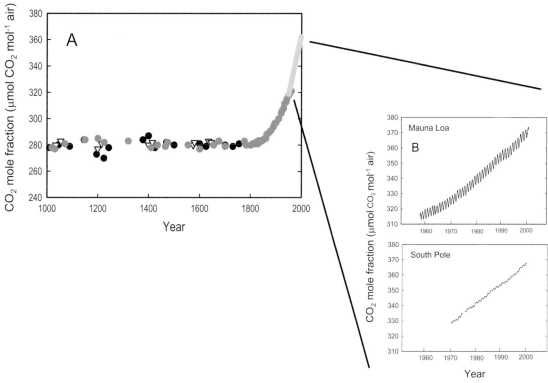

Figure 1.1 **A**. Atmospheric CO_2 mole fraction obtained from measurements using ice cores (prior to the twentieth century) and direct measurements of the atmosphere (since the late twentieth century). The stable CO_2 mole fraction through most of the past millennium is indicative of balanced CO_2 fluxes into and out of the atmosphere, on average, including those associated with the biosphere. The increase in CO_2 mole fraction over the past century is due to higher fluxes of CO_2 from the earth's biosphere to the atmosphere, compared to those from the atmosphere to the biosphere. The different symbols are indicative of different data sets. Redrawn from the Intergovernmental Panel on Climate Change, Policy Makers Report, Working Group 1 (IPCC 2001, p. 6). **B**. Details of the seasonal variations in CO_2 mole fraction measured at Mauna Loa, Hawaii or the South Pole. Data from the US National Oceanic and Atmospheric Administration.

1.3 Non-linear tendencies in biosphere-atmosphere exchange

One of the challenges that we face in describing biosphere-atmosphere interactions is the non-linear form of mathematical relations between fluxes and their associated driving variables (Table 1.1). Non-linear relations originate at the smallest spatial and temporal scales in ecosystems, and are amplified as process relations are transferred to progressively larger and longer scales. As examples of non-linear tendencies, we will briefly consider three processes; two at the sub-cellular scale and one at the leaf-to-landscape scale. In the case of enzymes, the nature of enzyme-substrate interactions changes as the availability of

Table 1.1 Causes and effects of non-linearities in biosphere-atmosphere exchanges

Causes
1. Processes tend to be highly dependent on initial conditions
2. Processes are controlled simultaneously by multiple rate-limiting, forcing variables
3. Processes are influenced by interacting scales of space and time
4. Processes are subject to positive and negative feedbacks

Effects
1. Emergent (synergistic) interactions among processes
2. Power-law scaling between process rates and driving variables
3. Time- and space-dependent patterns of abrupt (amplified or muted) change

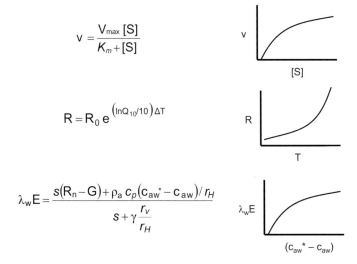

$$v = \frac{V_{max}\,[S]}{K_m + [S]}$$

$$R = R_0\, e^{\left(\ln Q_{10}/10\right)\Delta T}$$

$$\lambda_w E = \frac{s(R_n - G) + \rho_a\, c_p\left(c_{aw}{}^* - c_{aw}\right)/r_H}{s + \gamma \dfrac{r_v}{r_H}}$$

Figure 1.2 General non-linear forms of mathematical relations between some of the important processes and environmental drivers involved in plant-atmosphere exchanges. **Top**. Relation between the enzyme-catalyzed reaction velocity (v) and substrate concentration [S], as determined by biochemical parameters of enzymes (represented in this particular model as V_{max} and K_m). **Middle**. Relation between respiration rate (R) and temperature (T) in plant mitochondria. **Bottom**. Relation between leaf or canopy latent heat exchange and the difference between the saturation concentration of water vapor in air ($c_{aw}{}^*$) and the actual concentration of water vapor in air (c_{aw}) at a constant temperature.

substrate changes; thus, forcing the relation between velocity and substrate concentration toward non-linearity. We can represent this non-linearity with the Michaelis–Menten model presented in the uppermost equation of Figure 1.2. At low substrate concentration the rate of the reaction is determined by the affinity of the enzyme for the substrate (reflected in the K_m term), the rate by which the enzyme can convert the substrate into product (reflected in the V_{max} term), and the frequency by which enzyme molecules interact with substrate molecules (which is dependent on substrate concentration, [S]). As substrate concentration increases, the rate of the catalyzed reaction is determined less by K_m, and more by V_{max}. Mathematically, the dependence of reaction rate on [S] is resolved as a rectangular hyperbola, reflecting a shift from approximately first-order dependence at low substrate concentration toward approximately zero-order dependence at high substrate concentration.

As a different example, we see that the response of respiration (a flux of CO_2) to temperature is also non-linear, but in a manner that reflects increased capacity for metabolism as temperature increases. In this case, increased temperature forces enzyme and substrate molecules to collide more frequently. Additionally, due to flexibility in protein molecules, changes in temperature can cause changes in the shape of enzymes, which in turn can change the nature of their catalytic interactions with substrates. The net result of these temperature-dependent effects is an amplification, or acceleration, of respiratory reactions as temperature increases. We can describe this acceleration with an increasing power-law function (as shown in the second example equation in Figure 1.2).

In a third example, the transfer of latent heat through evaporation from a leaf to the atmosphere can be described by the Penman–Monteith model (as shown in the third example equation in Figure 1.2). The flux of water from the wet, cellular surfaces inside a leaf is dependent on the atmospheric water vapor concentration. The air in the intercellular spaces of a leaf is typically saturated with water vapor; so it is the air in the ambient atmosphere that determines the overall gradient in water vapor density, and thus the evaporative flux. However, the flux of water from leaves is mitigated by a diffusive resistance, because the evaporation stream is channeled through narrow pores, known as stomata. The diameters of those pores, and thus their diffusive resistances, are sensitive to changes in atmospheric humidity. Furthermore, changes in the loss of latent heat from the leaf causes leaf temperature to decrease, which in turn causes the water vapor concentration inside the leaf to decrease (cooler air holds less moisture at vapor saturation). These multiple effects interact in complex ways as atmospheric humidity changes, forcing the relation between leaf latent heat loss and atmospheric humidity to reflect the non-linear form represented in the Penman–Monteith model.

Non-linear relations provide challenges to the modeling of ecosystem-atmosphere exchanges. Non-linear responses mean that flux densities exhibit large changes (either increase or decrease) in response to small changes in the value of a driving variable. Thus, small errors in defining the values of driving variables can produce large errors in predicted flux. Additionally, non-linear relations do not lend themselves easily to linear averaging; failure to recognize non-linear relations between an averaged dependent variable estimated from an averaged independent variable can also cause significant error. We will consider these errors further when we discuss the details of non-linear flux relations in regard to leaf CO_2 and H_2O exchange.

1.3.1 Feedback – a frequent source of flux non-linearities

The redistribution of mass, momentum, or heat caused by a flux can cause feedback that modifies the flux (see Box 1.1). *Feedback* is defined as mitigation (negative feedback) or amplification (positive feedback) of a flux caused by the flux itself. Feedback is one of the

Box 1.1 **The mathematical concept of feedback**

We will discuss several examples of feedback as we progress through this book. Feedback will be an important component of metabolic flux control, the coupled responses of stomata to light and CO_2, and the interactions between surface evapotranspiration and atmospheric humidity. Thus, it is important at the outset to get a sense for the nature of feedback. Here, we present a quantitative treatment of feedback as borrowed from the discipline of electrical engineering.

Feedback modifies the degree to which for external forcing variables to affect a flux. Feedback can mute or amplify a flux in the case of negative or positive feedback, respectively. A flux that is subjected to negative feedback will be altered in iterative fashion over time until a new, reduced, but stable flux is achieved (i.e., altered flux causes altered feedback which causes altered flux and so on to a stable point). The new stable flux is often called the set-point (or attractor) and in biology the process involving the negative feedback is often called *homeostasis*. A flux subjected to positive feedback is incapable of reaching a set-point unless the feedback itself is susceptible to negative feedback. In the case of positive feedback, a flux will continue to increase away from its original set-point, eventually reaching an explosive state; an example can be appreciated as the outcome of nuclear fission. The influence of feedback on a flux occurs through a *feedback loop* (Figure B1.1). Thus, we can describe a flux at an arbitrary initial time point (we will call this the input flux, or F_1), which is altered through sensitivity to an external forcing variable, to produce a new flux at a later point in time (we will call this the output flux, or F_2). Returning to our example of negative feedback on forest evapotranspiration, the forcing variable is the amount of solar radiation incident on the ecosystem (in this case influenced by clouds), and F_1 and F_2 would represent the evapotranspiration flux in the presence of solar radiation without clouds or with clouds, respectively. If we start by assuming no feedback, then we can describe the system in the *open-loop mode*. Sensitivity of the flux to the forcing variable is defined by the open-loop gain (G_o), such that: $F_2 = G_o F_1$ and therefore we can define the open-loop gain as: $G_o = F_2/F_1$. If we now engage a closed feedback loop, such that

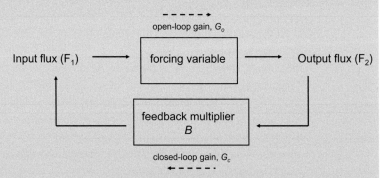

Figure B1.1   Scheme showing information flow in open and closed feedback loops. The open-loop gain (G_o) describes sensitivity of a flux to a forcing variable when the feedback loop is not engaged. The closed-loop gain occurs when the output flux "feeds back" to influence the original flux through a feedback multiplier (B). The closed-loop gain (G_c) describes the sensitivity of the original flux to the feedback multiplier scaled to the output flux.

information (in this case due to increasing clouds) is transferred back to the original flux, modifying it, we can write:

$$F_2 = F_1 G_o + F_2 G_o B, \qquad \text{(B1.1.1)}$$

where B is the feedback multiplier, being positive in sign for positive feedback and negative in sign for negative feedback. Returning once again to our example of evapotranspiration, B represents the composite of atmospheric processes that convert the evapotranspiration flux into clouds and determine the attenuation of solar radiation by those clouds. We can use algebra to remove F_2 from the right-hand side of Eq. (B1.1.1), resulting in the following equation:

$$F_2 = \frac{F_1 G_o}{1 - G_o B}. \qquad \text{(B1.1.2)}$$

Equation (B1.1.2) defines the resultant flux (F_2) in *closed-loop mode*. We can define the closed-loop gain (G_c) as F_2/F_1, and using Eq. (B1.1.2) to substitute for F_2, we can write:

$$G_c = \frac{G_o}{1 - G_o B}. \qquad \text{(B1.1.3)}$$

principal causes of non-linearities in the mathematical relations that describe the earth system. The difficulties in understanding feedback arise because we not only have to understand the quantitative nature of primary interactions between driving variables and fluxes, but also the secondary interactions that have the potential to modify the primary interactions. This creates higher-order interactions and non-linearities in mathematical relations that evolve over time as a flux proceeds. To illustrate the concept of feedback, let's return to the process of latent heat transport, in this case from a forest landscape to the atmosphere. As solar energy is absorbed by the forest, water evaporates from leaves and the soil. Turbulent wind eddies transport the water vertically through the atmosphere to a critical height where some fraction is likely to condense on suspended particles and form clouds. Clouds reflect solar radiation, causing a reduction in the energy flux received by the forest, and reducing the rate of evapotranspiration (Figure 1.3). If there is adequate moisture in the forest and a strong thermal gradient, vertical wind eddies can take on high velocities, creating convective updrafts that form dark, cumulus clouds and trigger precipitation. When taken together, these processes exemplify negative feedback in two ways. First, the flux of energy to the forest has caused the formation of clouds, a condition that reduces the further flux of energy to the forest and thus reduces surface warming. Second, in the condition of adequate moisture, the flux of water vapor from the forest has caused a condition that triggers rain and increased atmospheric humidity, both of which tend to reduce the evaporative flux from the forest.

Feedbacks have come to the forefront of discussions on interactions between the biosphere and the climate system (Field *et al.* 2007). In this case, many of the feedbacks occur at the regional-to-global scale. Fundamental questions that have emerged from this analysis include: (1) whether biospheric carbon fluxes will change sign, from net uptake to net loss, switching from a negative to positive feedback on climate change, respectively; (2) whether changes in the earth's albedo due to melting ice at high latitudes will trigger a

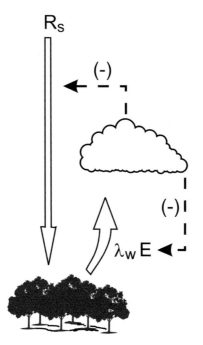

Figure 1.3   Diagrammatic representation of negative feedback with respect to latent heat flux ($\lambda_w E$) from a vegetated surface. The absorption of shortwave radiant (solar) energy (R_S) and consequent evapotranspiration of water vapor results in cloud formation which decreases the further flux of solar energy (a negative feedback), and decreases the rate of further latent heat flux from vegetation (also a negative feedback).

positive feedback by enhancing solar energy absorption; and (3) whether human-caused land-use change will trigger positive or negative feedbacks by altering the distribution of carbon-sequestering ecosystems from the landscape or altering surface albedo. Beyond questions as to the mathematical sign of potential feedbacks, future models of ecosystem-atmosphere interactions will have to be capable of resolving the magnitude of regional and global feedbacks, and the relation among feedbacks with different time constants. Understanding the rate at which feedbacks operate is crucial to predicting the rates by which fluxes can cause acceleration or deceleration of compound turnover in important biogeochemical pools and cause imbalance in the global energy budget in the face of future climate change.

1.4 Modeling – a tool for prognosis and diagnosis in ecosystem-atmosphere interactions

If there is one dominant theme that emerges from this book we hope it is the utility of organizing observations within a framework of quantitative models. Mathematical models provide the means to convert observations into predictions. Models are the tools for prognosis, allowing us to ask "what if" questions about how processes operate in conditions that transcend those associated with observations. Models also provide a means of organizing observations into explicit mathematical expressions from which mechanistic insight can be extracted. Please recognize, however, that both models and observations are imperfect representations of the true state of the earth system. Because of these imperfections, we are forced to accommodate errors and uncertainty in both observations and models.

Models are generated in one of two ways: (1) observations are organized into statistical correlations, which are then used as the means to generate values of dependent (unknown) variables from values of independent (known) variables; and (2) knowledge of process theory is used to relate a dependent (unknown) variable to an independent (known) variable. Models of the first type are referred to as *empirically based models*, and those of the second type are referred to as *mechanistic*, or *first-principle models*. Empirically based models contain the implicit assumption that multivariate correlations are conserved in conditions different than those of the original observations. First-principle models are often limited by available theory; in the absence of adequate theory, assumptions are often made to fill gaps in knowledge. Conventional wisdom states that mechanistic models are more accurate at predicting unobserved states because empirically based models are based on inadequate observations that don't necessarily overlap with future states of the system. It is important to realize, however, that gaps in theory can introduce just as much, or more, error into model predictions as inadequate observations.

Models are developed and used with either a "bottom-up" or "top-down" approach. In bottom-up modeling, processes and parameters are defined at the smallest scales of space or time and used as dependent variables that permit prediction for larger scales. The mechanistic forms of these models are often designed to include transport mechanisms and concentration gradients. Bottom-up modeling is often referred to as *forward modeling*, because it reflects the traditional "forward" logic of defining effects (at higher scales) from causes (at lower scales). In top-down modeling, parameters are measured at the highest scale in order to constrain the upper bound (or output) of a nested mathematical scheme. Working backwards through the nested scheme of equations, the highest layer is forced to define the output for the next layer down, and so on until an output is derived for the lowest scale of consideration. Top-down modeling is often referred to as *inverse modeling*. In inverse modeling we aim to estimate causes from effects (Box 1.2).

Inverse modeling has become increasingly favored in the earth system sciences. This trend has been driven by the increased availability of broadly distributed, large sets of observational data, from which model outputs can be well-constrained. For example, using eddy flux tower networks, inverse modeling is increasingly used to infer the values of parameters that control processes at the leaf and plot scales, working backwards from observations at the landscape scale. Inverse modeling at the global scale has become especially useful for identifying the spatial distribution of sources and sinks given known gradients (Dewar 1992). At the global scale, researchers use a global transport model to "work backward" from a global network of observed gradients to resolve the spatial distribution of sources and sinks that most likely caused the gradients. Global inverse

Box 1.2   **The mathematical concept of forward and inverse modeling**

Modeling is a core activity in the earth system sciences. Modeling allows us to organize observations and generate hypotheses about the causes that underlie trends in the observations. Observations are integrated with models in one of two ways: either the model is parameterized by observations and run in *forward mode* to make predictions about the outcome of a process, or observations on the outcome of a process are assimilated into a model and the model is run in *inverse mode* to generate predictions about underlying parameter states. Forward modeling can be thought of as predicting effects from causes. Inverse modeling can be thought of as predicting causes from effects. The relationship between forward and inverse modeling can be expressed mathematically as: $dx/dt = s[t, x, \theta]$, where s is a mathematical model (incorporating variables t, x, and θ), x_0 at time t_0 is defined, and θ is a causative forcing variable. In the forward projection of the model, θ is defined and the model (represented by s) is used to find $x[t]$ for some $t > t_0$. In the inverse projection of the model, $x[t]$ is defined for $t > t_0$, and the model (s) is used to find θ. To make this logic more relevant to biosphere-atmosphere interactions, imagine that x is the CO_2 concentration in the atmosphere at some specific point in time and space and θ is the surface CO_2 flux that influences x. The model, s, then relates surface CO_2 flux (the cause) to a change in atmospheric CO_2 concentration (the effect). The forward projection of s will allow us to predict the change in CO_2 concentration downwind of the flux. The inverse projection will permit us to work backward to define the surface flux (θ) if the change in CO_2 concentration (x) is observed. (In this simple example, θ would also have to include a component describing the dispersion of CO_2 as it travels downwind.)

models have limitations caused by inadequate coverage of the observed concentration distribution, errors and uncertainties in the transport model, and high sensitivity of the inversion process to these limitations.

1.5 A hierarchy of processes in surface-atmosphere exchange

> Knowledge of processes in plants and vegetation is largely at small scales. The transfer of this knowledge up to larger spatial and longer temporal scales is an open-ended process with potential errors arising from heterogeneity and patchiness in the distribution of processes and non-linearities in the functional relationships between processes and environmental variables. (P. G. Jarvis 1995)

Scaling processes according to space, time, and biomass has become a fundamental focus in the earth system sciences, principally as a means for nesting the perspectives of reductionism into higher levels of biogeochemical and ecological organization, and for identifying universal scaling functions that simplify the organization of reductionist principles into simple quantitative relations at larger scales (Brown and West 2000). The biosphere interacts with the atmosphere through fluxes that occur across a broad range of spatial and temporal scales. At the smallest scales, organisms interact with the atmosphere through diffusive fluxes. The diffusion of molecules across a leaf surface, for example, can be described at the scale of micrometers or less. These small scale processes, however, can accumulate to produce global ramifications – an amplification that crosses spatial scales of 10^{18} or greater. The mechanisms that drive fluxes at larger scales are fundamentally different from those that drive fluxes at the smallest scales. As scale increases, mass transport takes over as a dominant flux mechanism, and diffusion becomes less important. As a general rule, as scale increases, flux mechanisms become more efficient at moving scalars and vectors across space (Figure 1.4). For example, at the global scale, geographic gradients in atmospheric pressure drive material around the circumference of the earth (on the order of 10^6 m) within a couple of weeks (on the order of 10^6 seconds). This translates to transport efficiency on the order of 10 m s^{-1}. At the sub-cellular scale, molecular concentration gradients

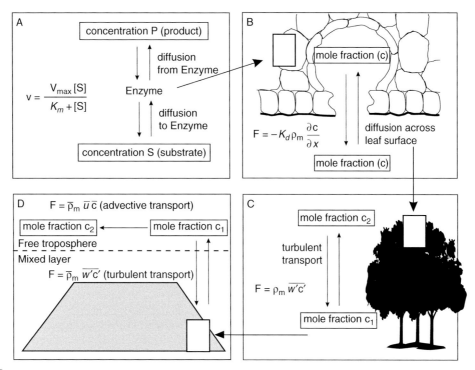

Figure 1.4 Scalar fluxes across spatiotemporal scales. **A.** At the scale of biochemistry, the flux of metabolites through metabolic pathways is determined by the kinetic features of protein enzymes. Flux (or in this case, reaction velocity, v) can be determined by the Michaelis–Menten model, which relates the potential for an enzyme to catalyze a specific chemical reaction to substrate concentration (S). **B.** At the scale of the leaf, fluxes are driven by diffusion through air. Flux density (F) can be modeled by an expression of Fick's First Law of diffusion in which the diffusion coefficient (K_d) relates flux to mole fraction concentration gradients. **C.** At the scale of a canopy, flux is driven by the turbulent motions of wind eddies. Vertical transport can be calculated from the time averaged statistical covariance in the vertical turbulent wind speed (w') and the turbulent scalar mole fraction (c'). **D.** At the scale of the landscape, flux is driven by turbulent and advective wind motions. Horizontal advective flux in the free troposphere can be determined from the time averaged product of the mean longitudinal wind velocity ($\overline{u}$) and the mean scalar mole fraction ($\overline{c}$).

drive diffusion across organelle membranes (on the order of 10^{-6} m) within fractions of a second (on the order of 10^0 seconds). This translates to transport efficiency on the order of 10^{-6} m s^{-1}. The process of diffusion, which is effective as a transport mechanism at the smallest scales, is negligible at the largest scales.

We have structured the book to reflect the hierarchy of processes presented in Figure 1.4. Processes that drive fluxes at the biochemical scale are considered first, progressing to processes at the leaf, canopy, and planetary boundary layer scales. We have tried to construct the discussions such that smaller scales are nested within larger scales through the progression of chapters. There is logic to this design in that higher scale processes reflect lower scale drivers, at least to some degree; with appropriate allowance for novel relationships that emerge.

2 Thermodynamics, work, and energy

A mathematician may say anything he pleases, but a physicist must be at least partially sane.

Josiah Willard Gibbs, *The Scientific Monthly*, December 1944

The playful statement written by Josiah Gibbs, who claimed identity as both a mathematician and physicist, provides some truth in jest. Mathematical treatments carry an elegance and beauty that can be appreciated within the abstract world of pure contemplation. However, descriptions of physical processes must be anchored within the allowable states and transitions defined by thermodynamics. As we begin to focus on the specific processes that drive biosphere-atmosphere fluxes we will return to the concept of flux that was established in the last chapter, but now we will construct a physiochemical foundation beneath that concept. Fluxes of scalars and vectors are driven by states of thermodynamic disequilibrium. A flux of mass or energy represents work that is done at the expense of internal energy that is derived from a state of thermodynamic disequilibrium. Thus, as we open this chapter we will develop the thermodynamic context for *energy* and *work*, and we will discuss their roles as underlying drivers of biogeochemical fluxes. This will require us to spend some time on the fundamental laws of thermodynamics, the concept of equilibrium, and the various forms of energy that drive biogeochemical processes. One of the concepts we will develop in some detail is the biogeochemical context of potential energy. Potential energy is how we define the capacity for components in a natural system to do work. Potential energy exists in a thermodynamic system that is in a state of disequilibrium; in contrast, a system that is at equilibrium lacks potential energy, and therefore lacks the capacity to generate work on its surroundings or on other systems. Following an initial introduction to energy and work we will move to the thermodynamic basis for descriptions of temperature and pressure, emphasizing these concepts with regard to the earth system. Gradients in temperature and pressure represent two of the driving forces that determine transport processes in the atmosphere, and ultimately determine the direction and magnitude of energy exchange between the earth's surface and the atmosphere. We will use the concepts of thermal equilibrium and disequilibrium as paths to a discussion of sensible heat flux; the first biogeochemical flux we will consider in some detail. Finally, we will take up the topic of energy transfer through electromagnetic radiation. Electromagnetic radiation from the sun ultimately powers most of the biogeochemical transformations we will discuss in this book, both in terms of their chemical states and kinetic motions.

2.1 Thermodynamic systems and fluxes as thermodynamic processes

Within the realm of biophysics, energy relations are often discussed with regard to a *thermodynamic system*, which has "boundaries" that separate it from the *surroundings*. In thermodynamic terms a system that can exchange heat and work with its surroundings, but not mass, is a *closed system*. It is important to note that a closed system is different than an *isolated system*. An isolated system does not exchange energy, work, or mass with its surroundings – it is indeed completely isolated! A system that exchanges heat, work, and mass with its surroundings is an *open system*. From a biogeochemical perspective the biosphere is an open system because it absorbs energy from its surroundings (e.g., from the sun), does work on its surroundings (e.g., organizes inorganic compounds into organic compounds), and exchanges mass with its surroundings (e.g., exchanges of CO_2 and H_2O between the biosphere and atmosphere) (Figure 2.1).

The potential for a thermodynamic system to do work is determined in relation to the state of *equilibrium*. At equilibrium the system is not capable of a change in the partitioning of its internal energy. (*Internal energy* is the total energy contained within a thermodynamic system and it is partitioned into components, the two main ones being kinetic energy and potential energy.) Equilibrium is the most stable state that a thermodynamic system can attain. A system at equilibrium cannot do work and its state properties are constant. (In our considerations the relevant *state properties* of a biogeochemical system will be temperature, pressure, volume, and mass.) A system at equilibrium, however, can enter a state of *disequilibrium* when disturbed by energy transfer to the system from the surroundings. Transition to a state of disequilibrium will be reflected as a change in the state properties of

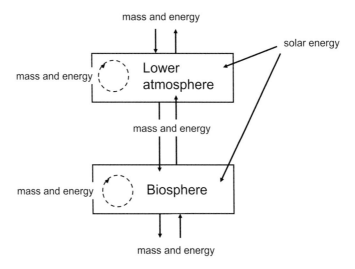

Figure 2.1   Representation of biogeochemical exchange between the biosphere and lower atmosphere as two coupled, open thermodynamic systems. Interconnecting fluxes are ultimately driven by solar energy.

the system. A system at disequilibrium can do work on its surroundings or it can redistribute its internal energy (e.g., between its kinetic energy and potential energy components) as it undergoes *spontaneous* transition back toward a state of equilibrium.

Within the context of biogeochemical systems the concept of disequilibrium refers to one of many states in which the system may exist, with at least some potential energy; that is, some capacity to conduct biological, geological, or chemical work. The potential energy of biogeochemical systems is typically due to gradients in the distributions of mass, momentum, or heat. Biogeochemical fluxes represent work that is done as the system moves from the state of disequilibrium toward the state of equilibrium; that is, toward a state without the existence of gradients. For the atmosphere and ecosystems of the earth system the continual flow of solar energy powers biogeochemical sources and sinks that in turn sustain gradients in the distributions of mass, momentum, and heat, and thus sustain thermodynamic disequilibrium.

2.2 Energy and work

Two of the most fundamental thermodynamic concepts that we will consider are *energy* and *work*; these concepts are inherently related. Energy is a scalar quantity that describes *the amount of work that can be done in response to a specific force*. Within the context of Newtonian physics, work is defined as *a force (mass times acceleration) applied across distance (m)*. The relation between energy and work can be appreciated by consideration of the standard measure given to energy – a *joule*. A joule of energy is capable of supporting a "newton-meter" of work. A *newton* (N) is a unit of force defined as $kg\ m\ s^{-2}$, which in turn can be understood as the product between mass (kg) and acceleration ($m\ s^{-2}$). Thus, a joule defines the work that can be done with a unit of energy within the context of position and momentum, two conjugate variables. (*Conjugate variables* are pairs of variables that interact mathematically in ways that define the thermodynamic potential of a system. Together, they form a type of time-dependent duality capable of relating incremental changes in the internal energy of a system to work.) The relation between work and energy can also be understood within the context of pressure and volume, two additional conjugate variables. The unit measure of pressure is $N\ m^{-2}$, and that of volume is m^3. Thus, expansion of a system, which is defined as the product between pressure and volume, carries the same unit measure as that of work, the newton-meter, which is in turn equivalent to a joule of energy. Through consideration of units of measure alone, we can begin to appreciate the conceptual relations of the terms, energy and work.

A number of theoretical treatments have attempted to categorize all components of the internal energy of a system into two fundamental types – potential energy and kinetic energy. *Potential energy* is the internal energy stored within a system that has the potential to do work. Potential energy can be converted to kinetic energy as work is done. *Kinetic energy* is the work required to accelerate a unit of mass to a given speed. In the following paragraphs of this section we will discuss the various "laws" of thermodynamics that govern the conversions between energy and work in natural systems. We will then progress to a discussion of the thermodynamics of gases, a topic especially relevant to our future discussions of biosphere-atmosphere trace gas exchanges.

The *Zeroth Law of Thermodynamics* (or the Law of Thermal Equilibrium) states that "two systems in thermal equilibrium with a third system are also in equilibrium with each other." It is the Zeroth Law that explains the measurement of temperature using a mercury-bulb thermometer. The Zeroth Law establishes the concept of thermal equilibrium between conjoined systems. While the earth and atmosphere are indeed conjoined systems they never reach thermal equilibrium because of the continual flow of solar energy to the earth's surface and the storage of some of that energy in the earth system. The "Zeroth Law" was so-named because at the time of its original formulation and statement, in the 1920s, the higher-numbered thermodynamic laws had already been named; yet, this law was even more fundamental in its formulation and so it was named according to the lowest-available ranked law – zero.

The *First Law of Thermodynamics* states: "the change in internal energy of a system (dU) is a function of the amount of thermal energy absorbed (δQ) and the amount of work done (δW) by the system." Stated mathematically:

$$d\text{U} = \delta\text{Q} - \delta\text{W}, \tag{2.1}$$

where δ notation is used for *inexact* differentials in the right-hand terms because Q and W cannot be defined *exactly* by differentiable functions. The First Law is based on the fact that a closed system can absorb thermal energy from the surroundings, which will result in an increase in the system's internal energy content. Furthermore, as a system gains internal energy, it can use that energy to do work on the surroundings; by the First Law, as work is done, the energy originally absorbed by the system must be returned to the surroundings. The First Law establishes the basis for *energy balance* in a thermodynamic system; any gain or loss of energy to work must be reflected in a net change in the system's internal energy. Two inferences that emerge from the First Law are: (1) energy cannot be created nor destroyed but it can be transferred from one form to another; and (2) in the absence of exchanges between a system and its surroundings, the total amount of energy in a system must remain constant as the system changes from one state to another. Energy is a conserved quantity. It is important to note that the First Law says nothing about how much internal energy exists; *only how much it changes* as energy is gained from the surroundings and work is done.

Within the context of the First Law the capacity to do work is measured in terms of the change in *free energy* (G; in J mol^{-1}):

$$d\text{G} = d\text{H} - \text{T}d\text{S}, \tag{2.2}$$

where H is enthalpy (in J mol^{-1}), T is temperature (in K), and S is entropy (in J mol^{-1} K^{-1}). *Enthalpy* defines the energy per mole of mass required to synthesize the system, along with the work required to "make space within the universe in order for the system to exist." Thus, enthalpy includes the kinetic energy present in the system, the potential work that can be done by the system, and the change in pressure and volume that had to be imposed on the surroundings to accommodate the system. One point of confusion that often arises concerns the difference between thermal energy (Q) and enthalpy (H). In the truest sense of thermodynamics, thermal energy is one component of enthalpy. Students are typically exposed to these terms in their introductory chemistry classes. In the case of a chemical reaction, at constant pressure and volume, changes in the internal energy of the reaction will be entirely due to the exchange of thermal energy between the reactants and their surroundings. Thus, within the realm of chemistry, the change in the thermal properties of the reaction is taken as the change in enthalpy. This can lead to confusion when the term "enthalpy" is then applied in a different context to systems in which work is performed and/or changes in the pressure and volume of the system occur.

Entropy refers to the degree of disorder in the system. A system with higher entropy has more disorder. Beyond associating entropy with disorder, it is difficult to obtain a meaningful measure of the term. There is no firm physical basis for entropy and it is normally treated as a statistical entity. (In fact, it is widely cited that in 1948 a discussion between John von Newman and Claude Shannon, two scientists who devoted their lives to understanding the concept of entropy, resulted in von Newman stating "nobody knows what entropy is in reality; that is why in the debate you will always have an advantage.") Despite some ambiguity, entropy is used in thermodynamic theory to define one of the tendencies that determine a system's potential to change *spontaneously*; by "spontaneously" we mean "without the addition of energy from the surroundings." Entropy can be linked to the transfer of thermal energy and work, in that it provides one means by which energy is converted into work. Thus, we can state $dS = \delta Q/T$. From Eq. (2.2) we begin to understand that the capacity to drive fluxes, or conduct biogeochemical work, within the earth system is linked to the total energy and state of disorder that exist within the system. Earlier, we used the term "driving forces" to describe those processes that enable biogeochemical fluxes to occur. As you can see from Eq. (2.2) these driving forces are not truly "forces" in the Newtonian context of the term. However, they lead to the same outcome as Newtonian forces in that biogeochemical work is enabled by changes in enthalpy and entropy, and these changes can be predicted on the basis of changes in the free energy state of the system.

The *Second Law of Thermodynamics* takes up the issue of irreversibility in *isolated systems* and states: "the entropy of an isolated system that is not in a state of equilibrium will increase over time and eventually reach a maximum at equilibrium." Once again, an isolated system differs from a closed system in that no heat or work is exchanged with the surroundings in an isolated system. The Second Law states that when left without energy input from the surroundings, and thus isolated, a transition toward greater disorder is inevitable. Stated mathematically:

$$\frac{dS}{dt} \geq 0. \tag{2.3}$$

Given what we learned from the First Law, the conditions of the Second Law make clear that free energy in an isolated system will decrease as entropy increases. The only way to reverse this tendency is for the system to lose its isolated condition, and gain energy from the surroundings. Work will have to be done at the expense of free energy from the surroundings in order to resist the inevitable transition toward greater entropy.

One common misconception is that the existence of life on earth violates the Second Law because living organisms are highly ordered "systems" with low entropy compared to their surroundings. In fact, as stated above, the Second Law is only valid with regard to isolated systems in their entirety. It does not pertain to changes in the entropy of individual components *within a system*. In considering the universe as an isolated system its overall entropy must increase over time. However, one component of the universe, such as the sun, can exhibit an increase in entropy while transferring energy to another component, such as the biosphere. The internal energy of the system must be conserved, but it can change states in different components. This is what happens through the process of photosynthesis. Solar energy is transferred to photosynthetic organisms which are able to capture a small portion and use it to synthesize biomass, thus preserving solar energy in the form of potential energy in the chemical bonds of the biomass. Heterotrophic organisms must be connected through food webs back to the photosynthetic primary producers (except for those groups of chemoautotrophs capable of extracting energy from reduced inorganic compounds); otherwise, the energy required to decrease entropy and drive growth and biological maintenance is not available. When an organism dies and ceases assimilating the products of photosynthesis, the inevitable transition toward greater entropy will ensue, resulting in bodily decay – a harsh reality, indeed, but one based on physics! The requirements of the Second Law will be further evident as we talk about redox reactions that involve the transfer of electrons along free energy gradients.

2.3 Free energy and chemical potential

Returning to Eq. (2.2), the minimum amount of energy required to change a thermodynamic system from one state to another is *free energy*. (Free energy is often referred to as Gibbs energy, in honor of Josiah Gibbs, a mathematician who developed the concept in 1873.) If the work required to change the system is capable of being done at the expense of the system's internal energy, then we refer to the change as an *exergonic process*. Exergonic processes can occur spontaneously. An example of an exergonic process is combustion. If the change in the system requires a sustained energy input from the surroundings we refer to the change as an *endergonic process*. An example of an endergonic process is the photosynthetic production of biomass.

We define *chemical potential* as free energy per mole of a substance. Thus, chemical potential is a measure of the capacity for the chemical constituents of a system to do work. The chemical potential of a substance depends in part on properties of chemical composition such as atomic structure and bond strength. However, other properties also contribute to chemical free energy as evident in the following relation:

$$\mu = \mu^* + RT \ \ln \ c + \bar{V}p + zF\varphi_E + mgh$$
$$ \text{I} \qquad \text{II} \qquad \text{III} \quad \text{IV} \quad \text{V} \qquad \text{Terms} \qquad (2.4)$$

where μ is chemical potential (J mol^{-1}), μ^* is standard chemical potential (J mol^{-1}), R is the universal gas constant (J K^{-1} mol^{-1}), T is temperature (K), c is concentration in mole fraction, $\bar{V}$ is partial molal volume ($m^3\ mol^{-1}$), p is pressure (Pa, where $1\ Pa = 1\ kg\ m^{-1}\ s^{-2}$), z is electrical charge (if any), F is Faraday's constant (coulomb mol^{-1}), φ_E is electrical potential (J coloumb^{-1}), m is mass (kg), g is gravitational acceleration ($\sim 9.8\ m\ s^{-2}$), and h is height (m). When grouped separately, Term I defines the reference potential determined under standard conditions of temperature and pressure (298 K, 101.3 kPa). Term II defines the influence of constituent concentration, and from this relation we see that chemical potential is exponentially and positively related to concentration. Term III defines the contributions of pressure and volume. Term IV defines influences due to the charged, or polar nature of reaction constituents. Term V defines the influence of height, and associated gravitational force. Thus, properties such as electrical charge, concentration, and position can affect chemical potential in addition to the inherent properties of chemical composition.

The additional factors in Eq. (2.4) (beyond the reference potential, μ^*) allow us to expand the use of chemical potential to evaluate the free energy change associated with fluxes and the coupling of chemical free energy to cellular and organismic processes. For example, through Term II the work associated with diffusive transport of a constituent within cells or across leaf surfaces can be represented thermodynamically within the context of a concentration gradient. Term III can be applied to the means by which plants use the chemical potential of water to create internal turgor pressure in cells and achieve growth. Term IV can be used to account for the chemical work of ion transport across membranes. Term V explains the limited ability for plants to transport water to great heights above the surface. In all of these cases the use of chemical constituents (such as water or charged ions) to drive biological work, or to extract mass and energy from the surroundings, is driven by the thermodynamic potential of the constituent or chemical system to provide free energy, which in turn is expressed in the system's chemical potential.

2.3.1 The chemical potential of water

Water moves through the earth system according to free-energy gradients. Formally, *water potential* (ψ_w) is the thermodynamic capacity for a unit of water to do work. Water potential can be expressed on the basis of molar, mass, or volumetric units. We will use the convention of molar chemical potential (J mol^{-1}). Water flows spontaneously from regions of higher ψ_w (higher free energy) to regions of lower ψ_w (lower free energy). By convention, the ψ_w of pure water at standard temperature and pressure is taken as zero. Also by convention, ψ_w is often expressed in units of pressure (Pa), rather than energy (J). (The units of pressure and energy can be reconciled if we consider that $\psi_w = p_w \bar{V}_w$, where p_w is water pressure (in $N\ m^{-2}$ or $m\ kg\ s^{-2}\ m^{-2}$) and $\bar{V}_w$ is the partial molal volume of water ($m^3\ mol^{-1}$); thus, the unit of pressure that defines the conjugate $p_w \bar{V}_w$ is equivalent to J mol^{-1} ($m^2\ kg\ s^{-2}\ mol^{-1}$), the unit of energy that defines ψ_w.) When you see ψ_w expressed as a unit of pressure alone (e.g., MPa), you should recognize the implicit assumption that this is water potential per unit of water volume (i.e., $\psi_w/\bar{V}_w$), which is required by the relation $p_w = \psi_w/\bar{V}_w$.

The molar chemical potential of water is influenced by four factors: solutes, pressure, gravity, and polar matrices. Three of the four factors – solutes, gravity, and matrices – reduce the chemical potential of water relative to its pure state (or, in the case of gravity, relative to pure water at the earth's surface). In other words, they reduce the potential for water to do work. (Note that the negative water potential produced by addition of a solute or matrix should not be equated with the negative ΔG of an exergonic reaction. In the case of water potential, as values get more negative, they reflect *lower free energy relative to pure water.*) Pressure can either increase or decrease the water potential of a solution depending on whether it exists as positive pressure or tension, respectively. Tension can exist in a column of water because of the polarity of water molecules and the associated properties of cohesion and adhesion. As the chemical potential of water is decreased due to these factors, the water system is capable of driving less biophysical work (e.g., less evaporative cooling, less pressure-driven cell expansion).

The addition of solutes to water will reduce the free energy of the resulting solution by reducing the mobility of polar water molecules as they are attracted to the charged or polar surfaces of the solutes; thus, the addition of solutes results in an *osmotic potential* (ψ_π) which is negative in value (i.e., $\psi_\pi < 0$). The molar osmotic potential (J mol^{-1}) can be represented as $\psi_\pi = (c_s RT) \bar{V}_w$, where c_s is the concentration of solutes (mol m^{-3}), R is the universal gas constant (8.31 m^3 Pa mol^{-1} K^{-1}), and T is temperature (K).

The addition of a positive pressure to the solution will increase its free energy by increasing the mobility of water molecules. Negative pressures (tensions) often develop in transpiring plants as water moves from the soil (with higher ψ_w) to the atmosphere (with lower ψ_w). Tension in the xylem of plants, for example, reduces the capacity for the water to do work, such as drive cellular growth. The molar pressure potential (ψ_p) scales as the absolute pressure of the water system (p_w) minus atmospheric pressure ($\psi_p = (p_w - P) \bar{V}_w$), with pressure defined in Pa and ψ_p defined in J mol^{-1}.

As water moves up a tree it comes under the influence of gravitational forces that resist its ascent. The gravitational resistance reduces the free energy of water producing a gravitational component to water potential (ψ_g), which is negative in sign. The molar gravitational potential is defined as $\psi_g = (\rho_w g h) \bar{V}_w$, where ρ_w is the mass density of the water (kg m^{-3}), g is gravitational acceleration (9.81 m s^{-2}), and h is height above the ground (m).

Matric potential (ψ_m) refers to the effect of a large polar or charged surface (matrix) in the water system. In biophysics this is most relevant in soil where soil particles and organic matter surfaces can interact with water molecules. The matric effect reduces the free energy of water molecules through attractive forces that reduce their mobility; thus, it is negative in sign. It is not possible to derive one theoretical relationship to define ψ_m, but rather it is dependent on the specific characteristics of the matrix causing the effect. We will discuss matric potential in more detail in Chapter 9, when we consider water flows through soil.

The total molar water potential of a system is reflected in the sum of its component potentials: $\psi_w = \psi_\pi + \psi_p + \psi_g + \psi_m$. Water potential can be expressed on a volumetric basis as: $\psi_{w(v)} = \psi_{w(m)} \rho_{mw}$, where $\psi_{w(m)}$ is the molar water potential and ρ_{mw} is the molar density of water (mol m^{-3}). Gradients in ψ_w provide the framework for understanding water fluxes. Evaporation occurs because of the large difference in ψ_w between atmospheric water vapor and liquid soil water. Water is driven from the soil through plants and into the atmosphere according to progressively more negative water potentials along the so-called soil-plant-atmosphere continuum. In living plant cells, water fluxes between the solution outside the cell (the apoplast) and inside the cell (the symplast) underlie several important ecophysiological processes, such as stomatal opening and closing, which in turn exert a major control over surface-atmosphere gas exchange.

The chemical potential of water and its linkage to work can be viewed in a different way when we consider the process of atmospheric humidification and dehumidification. The earth's water cycle is driven by the evaporation of water from ocean and terrestrial surfaces which combine to humidify the atmosphere. These evaporative processes are driven by the absorption of solar radiation. If not for precipitation, and in the presence of sufficient atmospheric mixing, evaporation from the surface would continue until the earth's atmospheric moisture content were to reach a state of saturation – or equilibrium. Disequilibrium is maintained when water molecules condense within clouds, and fall to the earth's surface as rain drops under the force of gravity. The process of condensation involves a loss of chemical potential in the water, which can be sensed as a transfer of heat to the surrounding atmosphere. Chemical work has been done during condensation; water potential has been reduced and free energy has been released to the atmosphere. Thus, the work of atmospheric dehumidification prevents a state of equilibrium and, along with the evaporative power of solar energy, drives the cycling of water in the earth system.

Osmosis is a special case of water transport in which water diffuses across a selectively permeable membrane from a solution with low solute concentration to a solution with high solute concentration. A selectively permeable membrane permits certain, but not all, solute or solvent molecules to cross. Selectivity is imposed by the nature of the membrane. In biological systems, osmosis is most relevant to the flux of water across the phospholipid membranes of cells and organelles. The phospholipid matrix of the membrane is hydrophobic, and is thus impermeable (though not ideally so) to most charged or polar molecules. Hydrophilic protein pores, known as aquaporins, traverse the phospholipid bilayers of cells, providing a path for the diffusion of water molecules.

Osmosis occurs across membranes from one reservoir of solution to another. The solution that contains the higher solute concentration is defined as *hypertonic* relative to the solution with lower solute concentration. Conversely, the solution with lower solute concentration is defined as *hypotonic*. Osmosis can also be defined in thermodynamic terms. A free-energy gradient exists from the hypotonic solution, where collisions among water molecules are more frequent, to the hypertonic solution, where collisions among water molecules are less frequent. On this basis alone it is clear that water molecules in the hypotonic solution have more potential to do work; i.e., they have higher chemical potential. Osmosis in response to gradients in ions is an important process driving the opening and closing dynamics of plant stomata. We will return to this topic in Chapter 7.

2.4 Heat and temperature

Heat is a form of energy that is transferred from one body to another. Heat is not a state property of a system, but rather it is most appropriately discussed in relation to the *process* of thermal energy transfer. The thermal energy of a system reflects the sum of the kinetic energy of the atomic and subatomic particles that constitute mass. The kinetic state of a system is defined in part by temperature, not heat. Heat will cross the boundary from a system with greater heat to a system with lesser heat. In thermodynamic terms the boundary between the systems defines a zone of disequilibrium. If the systems are allowed to interact, thermal energy flow will act to eliminate the zone of thermal disequilibrium. An eventual state of thermal equilibrium between the systems is predicted from the Zeroth Law. The flux of sensible heat (H_{se}) in response to a thermal gradient is defined by *Fourier's Law*, which for the one-dimensional condition (e.g., along the *x*-coordinate) and for the case of dry air can be written as:

$$H_{se}[x] = -K_{dh}c_p\rho_m\frac{\partial T}{\partial x},\tag{2.5}$$

where K_{dh} is the diffusion coefficient for heat in $m^2\ s^{-1}$, c_p is the specific heat of dry air in $J\ mol^{-1}\ K^{-1}$, ρ_m is the molar density of dry air in $mol\ m^{-3}$, T is absolute temperature in K, and *x* is in m, such that the relation reconciles to a sensible heat flux density, $H_{se}[x]$, in $J\ m^{-2}\ s^{-1}$ (or $W\ m^{-2}$). The combined term ($K_{dh}c_p\rho_m$) is often referred to as *thermal conductivity* (represented here as κ), with units $J\ s^{-1}\ m^{-1}\ K^{-1}$. The surface area implied in the unit m^{-2} of the heat flux density is assumed perpendicular to the coordinate defining the flux (in this case imagined as a vertically oriented, infinitely thin plate with surface area defined by the *y*- and *z*-coordinates). The *differential form* of Fourier's Law is used to define the local flux of sensible heat. The total amount of sensible heat transported in or out of a control volume can be calculated by the *integral form* of Fourier's Law:

$$\frac{\partial Q}{\partial t} = -K_{dh}c_p\rho_m\int_0^A\frac{\partial T}{\partial x}dA,\tag{2.6}$$

where Q is thermal energy (in J) that is flowing in one face of the volume and out the opposite face, and *A* is the total surface (m^2) of the perpendicularly oriented faces of the control volume through which heat is flowing. It is assumed that there is no time-dependent storage or depletion of heat within the volume. Equation (2.6) resolves to a form defined in finite space as:

$$\frac{\Delta Q}{\Delta t\ A} = -K_{dh}c_p\rho_m\frac{\Delta T}{\Delta x} = -\kappa\frac{\Delta T}{\Delta x}.\tag{2.7}$$

Temperature is a measure of the *average* kinetic energy of the molecules or atoms in a system. Temperature is an *intensive property*, which, in the physical sciences, means that while it reflects a measure of the state of matter (thermal energy content), its value is independent of the amount of matter within the system. Temperature is also a state variable, meaning that it is used with other state variables, such as pressure and volume, to define the thermodynamic state of a system. As a thermodynamic metric, temperature emerges from the Zeroth Law; it is required to explain the mechanism by which thermal equilibrium is achieved between conjoining systems of mass. In the example described above, as heat is transferred through molecular collision at the boundary of two systems, the temperature of the system that initially had a lower thermal energy content, will increase. Temperature is measured on various scales – e.g., Celsius, Fahrenheit. In this book we will use the Celsius scale when justified by tradition, or more frequently the Kelvin scale (symbolized as K), which is the most thermodynamically consistent measure of temperature. The Kelvin scale is referenced to the thermal state of a system at which there is no internal energy (taken as 0 K = –273.2 °C and –459.7 °F).

2.4.1 Fourier's Law and estimation of the soil heat flux

Quantification of the downward transport of heat from the surface of the soil to deeper layers is often required to balance the ecosystem energy budget. In certain circumstances, heat can be carried downward, through the soil, as air is advected from the surface and through air-filled soil pore channels. However, most heat is conducted downward through direct molecular contact in the solid or liquid (water) phases of the soil. Fourier's Law is used as the basis for measuring the vertical flux of heat at the ground surface. The soil heat flux is observed using *heat flux plates*, which are composed of thermocouples connected in series (forming what is called a *thermopile*), with the thermocouple junctions alternating at depths z_1 or z_2, where Δz replaces Δx in Eq. (2.7) and $\Delta z = z_1 + z_2$ (Figure 2.2). Thus, the voltage output generated from the entire thermopile reflects the temperature gradient across Δz. The thermopile is embedded in a plastic material of known thermal conductivity to form the heat flux plate. When the heat flux plate is buried in soil, with its surfaces parallel to the soil surface, and the value $\Delta T / \Delta z$ is observed, then $H_{se}[z]$ can be estimated from Fourier's Law.

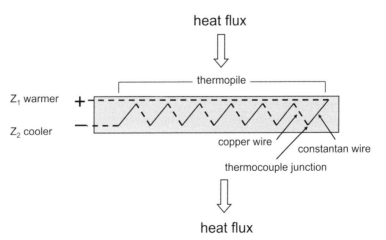

Figure 2.2 Schematic of a heat flux plate showing the series of copper-constantan thermocouples used to measure the voltage difference between the positive and negative poles of the thermopile, which in turn is proportional to the temperature difference between z_1 and z_2. The thermopile is embedded in a material of known thermal conductivity, so that the heat flux density can be calculated according to Fourier's Law.

2.5 Pressure, volume, and the ideal gas law

Pressure results from mass with kinetic energy and the resultant collisions that occur as particles of mass collide with each other and with any surfaces they contact, thus transferring momentum. In the case of gases, pressure and volume are described as conjugates in relation to one another according to the *ideal gas law*:

$$PV = nRT, \tag{2.8}$$

where P is pressure (Pa), V is volume (m^3), n is the number of moles of gas constituents, R is the universal gas constant (8.314 m^3 Pa mol^{-1} K^{-1}), and T is temperature (K). Most gases approximate an "ideal gas" meaning that transitions in their properties can be adequately described by Eq. (2.8). However, at extremely low temperatures or extremely high pressures, collisions among gas molecules may not be completely "elastic"; an often unstated condition of the ideal gas law. In those conditions, the molecular masses and inter-molecular interactions involving atomic charge (e.g., charge attraction or repulsion) can influence the thermodynamic behavior of the gas and Eq. (2.8) may be inadequate to quantitatively describe all behaviors. In most cases, the behavior of gases in the atmosphere, which is the topic of this book, can be adequately described by the ideal gas law. The ideal gas law is an example of an *equation of state*, which provides quantitative relations among the state variables used to describe matter. The ideal gas law governs the distribution of internal energy in a system of gas. For example, Eq. (2.8) makes it clear that any decrease in the pressure of a gas not accompanied by a proportional decrease in volume or increase in molar abundance must be accompanied by a proportional decrease in temperature. In another application, Eq. (2.8) can be rearranged to show that molar density (n/V) increases in direct proportion with pressure and decreases in inverse proportion with temperature.

In expressing the atmospheric concentrations of constituents (c) in this book we have used *mole fraction* because it is a unit of measure that is not influenced by pressure and temperature. Mole fraction is the amount of a constituent expressed as a fraction of all constituents in a mixture. From the ideal gas law, when the pressure and temperature of a gas-phase system are changed, the concentration density (mass per unit volume) will change for all constituents. However, the mole fraction of a single constituent will not change. While not formally expressed as a concentration or density, the measure of mole fraction is an expression of constituent abundance. A measure of abundance that is also not dependent on temperature or pressure and is commonly used in the atmospheric sciences is *mixing ratio*. Mixing ratio is defined as the mass or volume of a single constituent per unit mass or volume of total constituents (i.e., g g^{-1} or m^3 m^{-3}); mixing ratio is analogous to mole fraction, though expressed in different units. The most widely used measure of constituent concentration is *mass or molar density* (mass or moles per unit volume). For gases, mass and molar densities are dependent on pressure and temperature according to the ideal gas law.

In cases where the atmospheric concentration of a constituent influences chemical reactivity, such as the dependence of plant photosynthetic rates on the CO_2 concentration of the atmosphere, it is the *chemical potential* or "activity" of the constituent, not the concentration, that is most relevant. The chemical potential of ideal gases is approximated by *fugacity*. Fugacity is the pressure of an ideal gas that has the same chemical potential as the gas under consideration; in other words, fugacity allows "the ideal to converge to the reality." Thus, for the case of photosynthetic CO_2 uptake from the atmosphere, which is dependent on the fugacity of CO_2 at the active site of the enzyme ribulose 1,5-bisphosphate carboxylase (Rubisco), a measure of the *partial pressure* of CO_2 in the atmosphere is often used, rather than concentration. Partial pressure is defined as the pressure (with units Pa) created by a single gas constituent when present in a mixture of several constituents. The partial pressure of an atmospheric constituent is equal to the product between mole fraction and total atmospheric pressure. Under the assumption that CO_2 in the atmosphere behaves similarly to an ideal gas, its partial pressure is proportional to its fugacity.

Calculations that use expressions of concentration on a mass density or partial pressure basis must take their dependencies on pressure and temperature into account. The mixing ratio or mole fraction (represented as c(m)), and partial pressure (represented as p), can be related to mass density (ρ) according to:

$$\rho = c(m)\left(\frac{P}{RT}\right) = \frac{p}{RT}, \tag{2.9}$$

where P refers to the total atmospheric pressure (Pa). In reading the research literature associated with the atmospheric sciences, students may also come across the unit "parts per million" (or ppm), which is usually referenced to a unit volume (i.e., ppmv). This is also a ratio of constituent abundance and so is analogous to the mixing ratio or mole fraction. Also like mixing ratio and mole fraction, the measure of ppmv is not dependent on temperature and pressure.

In this book we use the mole fraction as a measure of atmospheric concentration, even when discussing photosynthetic reactions where a measure of partial pressure would be fundamentally more correct. We have chosen to do so in the interest of consistency across chapters. We have also tried to adhere to the convention of expressing gas concentration in terms of mole fractions relative to *dry air*. The reason for using dry air as our standard is that the constituent composition of dry air is conserved, whereas the mole fraction of a constituent in moist air is variable depending on humidity. If we start from a sample of dry air, and begin introducing H_2O molecules, then the mole fraction of other constituents must decrease in compensation. The mole fraction of constituent j relative to dry air can be related to that for moist air according to:

$$c_j = \frac{c_j'}{1 - c_w}, \tag{2.10}$$

where c_j is the mole fraction of constituent j in dry air, c_j' is the mole fraction of j in moist air, and c_w is the mole fraction of water vapor in the moist air sample. The mole fraction of water vapor in moist air tends to be small (< 0.02 mol H_2O mol^{-1} dry air at air temperatures less than 40 °C), so the correction of mole fraction from a dry air basis to a moist air basis is also small.

2.6 Adiabatic and diabatic processes

Two of the most important thermodynamic processes driving motions in the atmosphere are referred to as adiabatic or diabatic. An *adiabatic process* is one in which work is done without exchange of energy between the system and its surroundings. A *diabatic process* is one in which work is accompanied by the exchange of energy between the system and the surroundings. In order to illustrate the nature of adiabatic and diabatic processes and their potential to drive atmospheric motions we will consider a classic experiment conducted by James Joule in 1845. Joule constructed an apparatus with two glass bulbs connected by a closed stopcock. One of the bulbs was filled with gas and the other was evacuated. Joule insulated the experimental apparatus such that it could not exchange thermal energy with the surroundings; in other words $\delta Q = 0$. By definition, he had constructed an apparatus with the potential for observing adiabatic processes. Upon opening the stopcock, gas from the filled bulb expanded into the evacuated bulb, but because there was no pressure (or other force) to resist the expansion, no work was done; in other words $\delta W = 0$, but $dV > 0$ and $dP < 0$. Given that $dU = \delta Q - \delta W$, the internal energy of the system was constant during the experiment; despite an increase in V and decrease in P. Internal energy was redistributed within the system, but none was lost to work. Recall from our discussion above that the ideal gas law "governs the distribution of internal energy in a system of ideal gas." Based on this relation, Joule was able to conclude that the internal energy of a gas is not a function of V or P, and therefore must be only a function of T. We will return to this conclusion shortly.

Now, let's consider the case whereby an adiabatic system is allowed to do work on its surroundings. Imagine a parcel of air sealed off from the surrounding atmosphere by a boundary across which no heat can be exchanged, but which also offers no resistance to volumetric expansion; i.e., an ideal balloon. Once again, we have described an experimental condition governed by adiabatic processes. If the parcel is allowed to rise in the atmosphere, accompanied by a decrease in the external pressure of its surroundings, the air parcel will expand (Figure 2.3). In this case, unlike Joule's experiment, an external force exists that resists the parcel's expansion. Work must be done by the parcel to the surrounding atmosphere in order to drive expansion; i.e., $\delta W \neq 0$, but $\delta Q = 0$. The internal energy of the parcel will decrease, and returning to Joule's conclusion, the temperature of the parcel must also decrease. This explains why buoyant air parcels cool as they rise. By definition, all of the change in internal energy that occurs in an adiabatic system must be accounted for in the work that is done as the parcel expands. The degree of cooling as dry air rises adiabatically in the atmosphere is known as the dry adiabatic lapse rate and has a value of -9.8 K km^{-1}.

A diabatic process is one involving energy exchange with the surroundings. An example of a diabatic process is the evaporation of water from a wet surface, which is driven by the absorption of solar energy (indicative of an exergonic process). In this case, we can consider the wet surface to be an open thermodynamic system. As the surface absorbs energy from the sun, some of that energy increases the kinetic energy of liquid-state water molecules. With increased kinetic energy a fraction of the liquid water molecules will break free of the cohesive electrostatic interactions that hold them to other water molecules, and enter the vapor phase. Not all of the

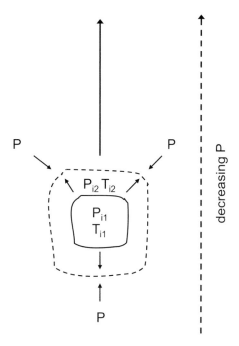

Figure 2.3 Schematic of a rising air parcel exhibiting adiabatic expansion in the face of decreasing atmospheric pressure (P). The internal pressure of the parcel (P_i) is greater than P. Two states are presented for the parcel – state 1 and state 2, with P_{i1} and T_{i1} representing the internal pressure and temperature of state 1 and P_{i2} and T_{i2} representing the internal pressure and temperature of state 2. Because work is done as the parcel expands and works against the resistance of external pressure, the internal energy of the parcel (U) decreases, and because the mass contained within the parcel remains constant the internal pressure must decrease as the volume increases. In the adiabatic condition the decrease in U must be accompanied by a decrease in T.

molecules that escape will remain in the vapor phase; some will lose energy through multiple collisions and rejoin the liquid phase. Thus, there are gross evaporative and condensation fluxes to and from the surface, but overall there exists a net evaporative flux away from the surface. It is important to note that in changing state, from a liquid to a gas, the gas does not change temperature. By definition, this means that the evaporating water molecules have not gained kinetic energy during the phase change; rather they have gained potential energy. We will come back to this principle – that a change in phase is not accompanied by a change in temperature – in a later chapter where we will consider latent heat exchange in more detail.

2.7 The Navier–Stokes equations

Among the most fundamental mathematical relations used to derive predictions about fluid flows are the Navier–Stokes equations. These relations follow from Newton's Second Law of motion, $F = ma$, where the acceleration of a fluid is described in terms of applied pressure forces. Unlike many relations that are defined in the earth system with regard to spatial coordinates, the Navier–Stokes equations are cast in terms of velocity, rather than position. The equations define temporal and spatial divergence in velocity within the constraint of conservation of momentum and energy. In their most fundamental forms the Navier–Stokes equations for the motions of incompressible fluids (fluids with constant density) are written as:

$$\rho_a \left(\frac{\partial u}{\partial t} + u\frac{\partial u}{\partial x} + v\frac{\partial u}{\partial y} + w\frac{\partial u}{\partial z} \right) = -\frac{\partial P}{\partial x} + \mu \left(\frac{\partial^2 u}{\partial x^2} + \frac{\partial^2 u}{\partial y^2} + \frac{\partial^2 u}{\partial z^2} \right) + F_x, \qquad (2.11)$$

$$\rho_a \left(\frac{\partial v}{\partial t} + u\frac{\partial v}{\partial x} + v\frac{\partial v}{\partial y} + w\frac{\partial v}{\partial z} \right) = -\frac{\partial P}{\partial y} + \mu \left(\frac{\partial^2 v}{\partial x^2} + \frac{\partial^2 v}{\partial y^2} + \frac{\partial^2 v}{\partial z^2} \right) + F_y, \qquad (2.12)$$

$$\rho_a \left(\frac{\partial w}{\partial t} + u\frac{\partial w}{\partial x} + v\frac{\partial w}{\partial y} + w\frac{\partial w}{\partial z} \right) = -\frac{\partial P}{\partial z} + \mu \left(\frac{\partial^2 w}{\partial x^2} + \frac{\partial^2 w}{\partial y^2} + \frac{\partial^2 w}{\partial z^2} \right) + F_z, \qquad (2.13)$$

where ρ_a is the mass density of air (kg m^{-3}). Combined, the left-hand sides of the equations describe the three-dimensional spatial divergences (accelerations or decelerations) in terms of u, v, and w, the wind speeds (m s^{-1}) in the x-, y-, and z-coordinates, respectively, plus the time-dependent divergence of each. *Divergence* in a scalar or vector quantity is defined in terms of a differential equation and reflects a gradient in the value of the scalar or vector. Respecting the requirement for conservation of mass and energy, a spatial gradient that decreases (or increases) the value of a scalar or vector must be balanced by a time-dependent increase (or decrease), reflecting local storage of that scalar or vector. This coupling between space- and time-dependent divergences is often referred to as *continuity*, a concept that will be discussed further in Chapter 6. The right-hand sides of the equations define the conservation condition that these divergences must be balanced by the divergent sum of the forces and resistances that influence fluid velocity: pressure (P in Pa), dynamic viscosity (μ in kg m^{-1} s^{-1}), and external "body forces" expressed per unit volume (F in kg m^{-2} s^{-2}). (A "body force" is a type of force density that is applied to an entire volume, as opposed to surface forces that are applied to the faces of a parcel or packet. Gravity is an example of a body force, whereas the shear between fluid "particles" that produces drag is an example of a surface force.) The second derivative terms on the right-hand side of the Navier–Stokes equations represent frictional "forces" that oppose the pressure force (thus, the difference in mathematical sign between the first and second terms on the right-hand side). The Navier–Stokes equations make clear that fluid motions, including the wind, are best described within the context of vector sums in a three-dimensional framework, and that divergence in fluid velocity must be causally coupled to those forces and resistances that impart momentum to the fluid (e.g., pressure) and create stress and strain within the fluid (e.g., friction and drag). The Navier–Stokes equations have not been resolved with analytical solutions, but their form as differential equations within a three-dimensional framework will provide the foundation for our studies of turbulent transport in future chapters.

2.8 Electromagnetic radiation

> All these fifty years of conscious brooding have brought me no nearer to the answer to the question: "What are light quanta?" Nowadays every Tom, Dick and Harry thinks he knows it, but he is mistaken. (Albert Einstein, *The Born–Einstein Letters*, 1954 (translated by Irene Born, 1971))

A physical description of light has challenged physicists since Max Planck first took up the issue in 1894 and Albert Einstein tried to clarify the matter with his Photoelectric Theory in 1905. Light is a form of electromagnetic radiation that exhibits properties described by both wave and particle physics. It is the dichotomous nature of electromagnetic radiation that eludes quantitative description. *Electromagnetic radiation* is defined as the energy emitted from mass when the vibrational motions of negatively charged electrons create alternating electric and magnetic fields. These fields give rise to self-propagating waves that travel through space. The energy of the waves is called electromagnetic energy or *radiant energy*. Radiant energy is quantized; that is, it is delivered to a surface in discrete amounts depending on the wavelength of the propagated wave. These discrete amounts of energy are defined in terms of quanta, or *photons*. The quantized nature of electromagnetic radiation originates in the quantized vibrational motions of electrons in the emitting matter. The energy of a photon can be defined as:

$$E_p = \frac{hc}{\lambda}, \tag{2.14}$$

where E_p is photon energy (J), h is Planck's constant (J s), c is the speed of electromagnetic radiation propagation (or speed of "light") (m s^{-1}), and λ is wavelength (m).

Any system of matter that exists above 0 K will have vibrational motion in its subatomic particles and will thus emit electromagnetic radiation, converting thermal energy to electromagnetic energy. A *black-body* is defined as a system of mass that ideally exchanges electromagnetic radiation with its surroundings, meaning that it absorbs all radiant energy it receives and re-emits all energy it absorbs. By definition, a black-body emitter will emit radiant energy as a function of wavelength according to *Planck's Radiation Distribution Law*:

$$\frac{d\mathrm{R}[\mathrm{T},\lambda]}{d\lambda} = k_1 \frac{\lambda^{-5}}{(e^{k_2/\lambda \mathrm{T}} - 1)}, \tag{2.15}$$

where R is the radiant energy flux emitted from a black-body surface into a unit solid angle (J m^{-2} s^{-1} sr^{-1}, where sr represents steradian, a unitless measure of solid angle), λ is wavelength (m), k_1 is a constant with units kg m^4 s^{-3}, k_2 is a constant with units m K, and T is surface temperature (K). Written in its differentiated form, as in Eq. (2.15), Planck's Law is defined in the limit as $\Delta\lambda \rightarrow 0$. If we set the derivative to 0 we can evaluate the fundamental relation between radiative flux density and each wavelength. In other words we can ask: what is the wavelength of maximum radiative flux? Mathematically, we state the condition as differentiating Planck's Law with respect to wavelength, and setting the derivative equal to zero, we obtain:

$$\frac{\partial R[T, \lambda]}{\partial \lambda} = 0. \tag{2.16}$$

and the solution to this condition yields:

$$\lambda_{max} = \frac{2897}{T}, \tag{2.17}$$

where T is the black-body temperature (K). Equation (2.17) is known as *Wien's Displacement Law*. Using Wien's Displacement Law we can calculate the wavelength of maximum radiant emission from the sun. The surface temperature of the sun is estimated to be 5500 K, which translates to a wavelength of maximum emission at 527 nm (Figure 2.4A). This wavelength also happens to be in the middle of the range of wavelengths most active in driving the photosynthetic systems of plants and in the utilization of "light" for vision processes in many animals. Thus, biological systems have evolved in a way that matches the wavelengths of light required to drive their energy transduction systems with the wavelength domain of greatest energy transfer from sun to earth. In this book we will refer to photons in the wavelength band from 400–700 nm as *photosynthetically active radiation* (PAR) and the photon flux in this waveband will be referred to as *photosynthetic photon flux density* (PPFD).

The total amount of radiant energy emitted by a surface can be calculated by integrating Eq. (2.15) across all wavelengths; in other words, establishing the other limit on Eq. (2.15) as:

$$dR[T, \lambda] = \int_{\lambda \to 0}^{\infty} k_1 \frac{\lambda^{-5}}{(e^{k_2/\lambda T} - 1)} d\lambda. \tag{2.18}$$

When this is done we obtain the *Stefan–Boltzmann Law*, which relates radiant flux density from a black-body to the fourth power of its temperature, i.e, $R = \sigma T^4$, where σ is the Stefan–Boltzmann constant (5.673×10^{-8} J s^{-1} m^{-2} K^{-4}), and takes account of the integration of constants k_1 and k_2 across all emitted wavelengths. For objects other than black-body emitters, the Stefan–Boltzmann Law is modified according to the object's *emittance* (ε_L; the fractional emission of longwave radiation from the surface of a non-black-body relative to a black-body). The modified Stefan–Boltzmann Law is represented as $R = \varepsilon_L \sigma T^4$.

Solar energy occurs as short, higher-energy wavelengths that are referred to generically as shortwave radiation. (By convention, radiant energy of wavelengths less than 2.5 μm is defined as *shortwave radiation* and that greater than 2.5 μm is defined as *longwave radiation*.) Often, the term *near-infrared radiation* is used for that part of the solar spectrum incident on the earth at wavelengths greater than 0.7 μm. The earth's surface emits radiant energy with an emission maximum near 9.6 μm, fully within the longwave portion of the spectrum (Figure 2.4B). The energy emitted by the earth's surface is also referred to as "re-radiated" energy since it represents shortwave energy that was originally emitted by the sun, absorbed by the earth's surface and re-radiated back to space as

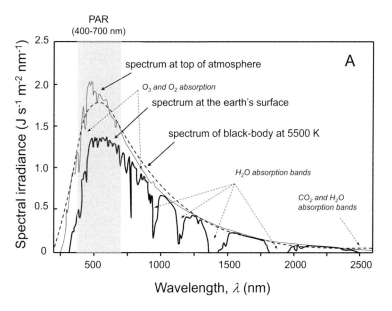

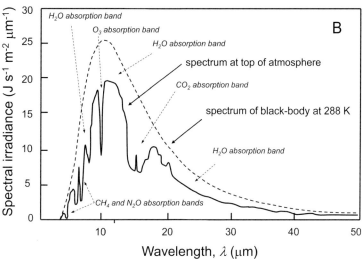

Figure 2.4 **A**. The solar spectrum at the top of the atmosphere and at the earth's surface showing peak energy emission from the sun in the domain of photosynthetically active radiation (PAR) and attenuation due to atmospheric constituents. The spectrum of an ideal 5500 K (estimated temperature of sun) black-body is shown for comparison. Redrawn from a figure produced by Robert A. Rohde, for the Global Warming Art Project and distributed under the GNU Free Documentation License. **B**. The longwave radiation spectrum at the top of the atmosphere, after radiant emission from the earth's surface, and the spectrum of an ideal 288 K (mean earth surface temperature) black-body. Redrawn from data originally presented in Gates (1980).

longwave energy. The *net radiation* (R_n) absorbed by a surface is the difference between absorbed and emitted radiant energy. Net radiation is the sum of the absorbed shortwave (R_S) and longwave (R_L) radiation minus re-radiation ($\varepsilon_L \sigma T_S^4$):

$$R_n = R_S (1 - \alpha) + R_L a_L - \varepsilon_L \sigma T_S^4, \qquad (2.19)$$

where α and a_L are the fractional albedo and absorptance, respectively, of the surface for short- and longwave radiation (Figure 2.5). Albedo is the ratio of reflected shortwave flux density to incident shortwave flux density.

As a surface absorbs energy from its surroundings its *energy budget* is altered; energy absorbed by the surface must be channeled to an increase in temperature and/or dissipated to balance the energy budget. Absorbed energy will be dissipated through a variety of mechanisms, including conduction of heat to depths beneath the surface, conduction of heat to a cooler fluid (such as air or water) that is flowing across the surface, and loss of latent heat through evaporation. At steady state, the fluxes that move energy away from the surface will equal the fluxes that bring it to the surface, the temperature of the surface will remain constant, and the surface energy budget will be balanced. Perturbation to the flux of incoming or outgoing energy will disrupt the energy balance, and cause the temperature of the surface to increase or decrease accordingly. Ultimately, surface temperature is the transducer that connects energy gain and energy loss and allows the surface to maintain energy balance in the face of variable energy loads. In future chapters we will consider the energy balance of various surfaces in the earth system, including leaves, canopies, and the planet.

2.9 Beer's Law and photon transfer through a medium

The absorption of electromagnetic radiation as it penetrates a medium (such as the atmosphere or a solution of molecules) is described quantitatively according to the

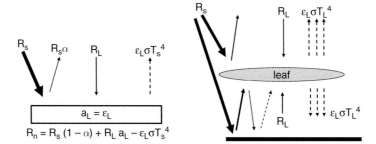

Figure 2.5 Net radiation with respect to two surfaces. **Left**. The solid ground surface absorbs shortwave radiation (R_S) from the sun and longwave radiation from the sky R_L, and emits longwave radiation according to the Stefan–Boltzmann relation. In some treatments, $R_S (1-\alpha)$ is replaced by α_s, the surface absorptance to shortwave radiation (i.e., $\alpha_s = R_S (1-\alpha/R_S)$. **Right**. A leaf that is elevated above the ground absorbs radiation inputs from both above and below its surfaces, and emits radiation from both surfaces.

Beer–Lambert Law (or *Beer's Law*). Beer's Law relates the absorption of a photon flux to the quantity and nature of photon-absorbing mass suspended in a medium. Photon absorption will be dependent on the total cross-sectional area of the suspended particles of mass, perpendicular to the flux, and to the inherent capacity of each molecule to absorb photons at the wavelengths represented in the flux:

$$I = I_0 e^{-\varepsilon l c}, \tag{2.20}$$

where I is the photon flux density exiting the medium (μmol photons m^{-2} s^{-1}), I_0 is the photon flux density entering the medium (μmol photons m^{-2} s^{-1}), ε is the molar absorption coefficient of the suspended particles (m^2 mol^{-1}), l is the length crossed by the photon beam (m), and c is molar concentration density of the particles (mol m^{-3}). The normalized molar absorption coefficient and length are often combined as the extinction coefficient (K_l, where $K_l = \varepsilon l$, with units m^3 mol^{-1}). The natural exponent (e) that scales exponential extinction of the photon flux emerges from the mathematical derivation of the Beer–Lambert Law. The Beer–Lambert Law takes a central role in many aspects of earth system sciences, including the foundation for radiative transfer models in the atmosphere and through vegetation canopies, and the basis for quantifying the concentration of molecules in many trace gas detector instruments.

As radiant energy is transferred from the sun, through the earth's atmosphere, it is partially absorbed by suspended molecules. Absorption occurs within specific bands of wavelengths depending on the oscillation frequency of atoms within the molecule (Box 2.1). In the atmosphere, O_3 and O_2 attenuate solar energy at relatively short wavelengths, whereas H_2O and CO_2 attenuate energy at longer wavelengths (Figure 2.4). Solar absorption by CO_2 is relatively minor compared to O_3, O_2, and H_2O. The principal CO_2 absorption bands are at wavelengths longer than 2000 nm, placing them well within the spectrum of earth-emitted radiation, but beyond most of the solar spectrum. Thus, recent warming of the atmosphere, which has been attributed to anthropogenic perturbations to the trace gas composition of the atmosphere, is not due to attenuation of solar energy, but rather absorption of re-radiated energy emitted by earth.

Box 2.1	What makes a greenhouse gas radiatively active?

The so-called *greenhouse gases* in the earth's atmosphere trap radiant energy emitted from the earth in the longwave (far infrared) part of the electromagnetic spectrum; part of the trapped energy is transmitted back to earth, creating the "greenhouse effect." Not all atmospheric molecules are radiatively active in the far infrared part of the spectrum. For example, N_2 and O_2, the two most abundant gases in the atmosphere, do not contribute to the greenhouse effect. Here, we first consider the properties of a molecule that permit it to absorb radiant energy in a particular waveband. Then, we will consider the properties of greenhouse gases that determine their radiative activity in the atmosphere.

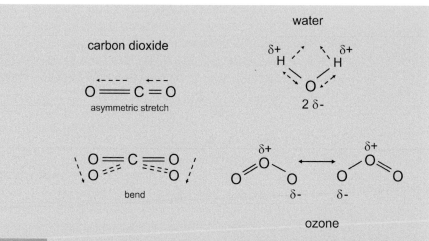

Figure B2.1   Molecular structures and oscillations in carbon dioxide, water, and ozone showing the potential for a dipole moment, and thus absorption of electromagnetic radiation. The δ symbol is used to indicate partial charges resulting from the asymmetric distribution of electron clouds. The interconversion of O_3 between two molecular forms represents the "resonance hybrid structure" of compounds in which no single theory of bonding conformation can account for stability in the molecular structure.

In order for a molecule to absorb radiant energy the frequency of oscillation in the electromagnetic field of emitted radiation must match the frequency of atomic oscillation within the molecule. Atomic oscillations occur within covalent bonds. The covalent bonds that allow two neighboring atoms to share electrons do not exist at a static length and angle. Rather the length of the bond can extend and shorten, and the angle of the bond can oscillate between alternative states as centers of positive charge (nuclei) and negative charge (electron clouds) are alternately repelled and attracted by electrostatic forces (Figure B2.1). As charged domains of a molecule oscillate asymmetrically a dipole moment is created. *Dipole moment* refers to the degree of polarity that exists within a molecule as a result of asymmetric charge distribution. An oscillating dipole moment creates an electromagnetic field capable of matching the resonant frequency in the electromagnetic field of emitted radiation, thus permitting energy absorption.

In the case of the CO_2 molecule a linear structure is predicted on the basis of optimal separation of shared electron clouds, with two pairs of shared electrons in each double bond between carbon and oxygen (Figure B2.1). Asymmetric extension and contraction of the double bonds on each side of the molecule creates an oscillating dipole; symmetric extension and contraction leads to no dipole and no potential for absorption of electromagnetic energy. A water molecule is predicted to take the shape of a tetrahedron. In this case, angular oscillation occurs as the bonded hydrogen atoms move back and forth in relation to each other, creating a dynamic dipole capable of energy absorption. There is a range of possible oscillating frequencies in molecules with dipoles, meaning that a range of wavelengths (within a waveband) can be absorbed. The most probable absorption frequencies lie in the center of the waveband, with lower absorption probabilities at the "wings." One interesting example to demonstrate the concepts of molecular asymmetry and dipole involves the radiatively active compound, ozone (O_3). This compound is composed of atoms of the same element, but they are bonded asymmetrically with respect to one another (Figure B2.1); thus, a polar dipole is created as the component atoms move about their bonds and as electrons move from atom to atom to form transient double bonds (forming a so-called resonance hybrid molecule).

Different types of greenhouse gases cause different amounts of global warming when added to the earth's atmosphere. One metric used to characterize warming is the *global warming potential* (GWP). The GWP provides an estimate of the warming caused by the emission of a unit mass of a trace gas compared to CO_2. For example, CH_4 is estimated to have a GWP of 72 for a 20-year time horizon and 25 for a 100-year time horizon; meaning that it is 72 and 25 times more potent as a greenhouse gas compared to CO_2 when calculated over 20-year and 100-year spans, respectively. Differences in the global warming potentials of CO_2 and CH_4 can be traced in large part to existing atmospheric concentrations of these gases and the degree to which each absorbs emitted energy. For example, CO_2 exists at a concentration several orders of magnitude higher than CH_4 ($\sim$ 380 μmol mol^{-1} for CO_2 versus $\sim$ 1.7 μmol mol^{-1} for CH_4). The concentration of CO_2 is high enough that the absorption of longwave radiation is saturated near the center of its absorption bands. This means that additional CO_2 will only cause warming through energy absorption near the waveband wings. Thus, as CO_2 is progressively added to the atmosphere its potential to induce warming, on a per mole basis, decreases non-linearly. Methane also exists at an atmospheric concentration that causes energy absorption to approach saturation near the center of its absorption bands. However, more of the overall absorption spectrum is available for enhanced energy absorption in the case of CH_4 compared to CO_2; this causes CH_4 to have a greater GWP. The reason that the GWP for CH_4 decreases between the 20-year and 100-year estimates, is that the effective atmospheric lifetime of CH_4 is shorter than that of CO_2 (8–10 years for CH_4 versus > 100 years for CO_2). Thus, relative to CO_2, the GWP of a pulse of CH_4 added to the atmosphere will decline over the time scale of a decade or more due to its oxidation and removal from the atmosphere. (For the convenience of calculating GWPs, the reference GWP accorded to CO_2 is kept constant at 1 for all considered time spans.)

3 Chemical reactions, enzyme catalysts, and stable isotopes

A corollary to this law shows how the chemical equilibrium varies with temperature – namely, how, as the temperature increases, more of the one compound is formed at the expense of the other, or vice versa. This corollary can be stated as follows: At low temperature the greater yield is always of that product whose formation is accompanied by evolution of heat.

Jacobus H. van't Hoff, Nobel Prize Lecture, 1901

In explaining the relation between temperature and chemical reactions in his *Etude de Dynamique Chemique* in 1884, Jacobus van't Hoff built a conceptual bridge between the concepts of thermodynamics, which had emerged in cogent fashion during the second half of the nineteenth century, and chemical equilibrium, which had emerged from studies on reversible reactions earlier in that century. In the quote above from his Nobel Prize lecture, van't Hoff refers to the concept of exothermic reactions as being the "favored" outcome of a chemical interaction, which we now understand to be predicted by thermodynamic laws. Chemical transformations within the earth system must follow the same thermodynamic laws that were presented in the last chapter and to which van't Hoff alluded in the statement above. In this chapter, we will build on those thermodynamic laws in order to establish a foundation for the more detailed considerations of biochemistry and metabolism. More specifically, we will delve into the theory underlying chemical reactions, their sensitivity to temperature, and their biochemical association with protein catalysts.

Chemistry requires interactions among particles of mass, either as individual atoms or as atomic components of molecules. As units of mass interact, subatomic particles will rearrange themselves into relations with one another that satisfy the most stable thermodynamic state. In some cases this results in the actual exchange of one or more electrons. In other cases electrons may move to a more stable, shared association with two nuclei, forming a covalent bond. The result of these chemical interactions is often reversible, and the probability of reversal can be predicted from knowledge about free energy dynamics during the interactions. We will begin this chapter with a discussion of chemical equilibrium, reaction kinetics, and reversibility. From this discussion we will move to the topic of reduction-oxidation reactions; those reactions involving the exchange of electrons. We will emphasize examples of reduction-oxidation reactions that are relevant to biogeochemical cycling. We will next take up the topic of enzyme catalysis. Most biochemical reactions would not occur at rates sufficient to sustain life at the temperatures that exist on earth, if not for the existence of biological catalysts known as enzymes. Knowledge about enzymes, and their kinetic effect on reaction rates, is crucial to understanding control over the metabolic fluxes that we will consider in the next chapter. Finally, we will begin a discussion about stable isotopes and their role in biogeochemical studies. In a future chapter (Chapter 18), we will return to the topic of changes in stable isotope ratio as a key metric by which to evaluate the kinetics of biogeochemical processes.

3.1 Reaction kinetics, equilibrium, and steady state

The *kinetic properties* of a reaction are defined as those factors that affect reaction rate. Consider the elementary bimolecular and reversible reaction whereby reactants A + B are transformed into products C + D, and through reverse reaction, reactants C + D are transformed into products A + B:

$$A + B \underset{k_2}{\overset{k_1}{\rightleftharpoons}} C + D, \tag{3.1}$$

where k_1 and k_2 are *reaction rate constants*. By defining reaction (3.1) as "elementary" we mean that A and B in the forward direction or C and D in the reverse direction react in a single-step collision, forming an intermediate (transition) state, before proceeding to products. Now, let's begin to define the kinetic constraints on reaction (3.1). The rate of an elementary reaction is controlled by the quantity of reactants and their inherent reactivity as they collide with one another. The *law of mass action* states that at a fixed temperature, the rate of a chemical reaction will be directly proportional to the concentration of reactants. The reaction rate constant (k) is a measure of *reactivity*; it is the constant of proportionality between reaction rate and reactant concentration, and reflects the energetic threshold required for each reactant to create product. The reaction rate constant is specific to each type of reaction. The kinetics of reaction (3.1) can be modeled as a *rate law*:

$$v_1 = k_1[A]^1[B]^1, \tag{3.2}$$

where v_1 refers to the rate of the reaction and the exponents attached to each reactant concentration are called *orders* and define the *stoichiometry* of the reaction. Stoichiometry refers to the relative proportions of reactants used and/or products formed. In the rate law described by Eq. (3.2) one molar unit of reactant A and one molar unit of reactant B combine in a simple bimolecular collision to produce a product. The stoichiometry of a reaction is not derived naturally from theory. Rather, it is determined through experimentation. For simplicity, we have set the exponents as 1.0 for both reactants in Eq. (3.2); most reactions are more complex in their stoichiometries.

As written, v_1 in Eq. (3.2) is dependent equally on the concentrations of A and B and this is referred to as a *second-order rate equation*. (Order exponents are added when the overall reaction rate is expressed as a function of all reactants. In this case, $1 + 1 = 2$, so the overall reaction is second order with respect to both reactants A and B.) In some cases, v_1 is dependent on the concentration of only one reactant. For example, if reactant B is present at high enough concentrations, v_1 will only be dependent on the concentration of reactant A; the so-called *rate-limiting reactant*. In that case we would write:

$$v_1 = k_1 [A]^1 [B]^0 = k_1 [A]. \tag{3.3}$$

Equation (3.3) is referred to as a *first-order rate equation*. Given that $[B]^0 = 1$, the rate equation for reaction (3.3) can be simplified to $v_1 = k_1 [A]^1$. Conversely, when concentration of A is at rate-saturation, $[A]^0 = 1$, and the rate equation simplifies to $v_1 = k_1 [B]^1$. In a first-order elementary reaction, a doubling of the concentration of the single rate-limiting reactant will cause a doubling of v. In a second-order elementary reaction, a doubling of the concentrations of both reactants will cause a quadrupling of v. The units carried by k depend on reaction order. In the first order, k carries units of reactions per second, or simply s^{-1}. Thus, v carries units of $mol \, l^{-1} \, s^{-1}$ (or alternatively $mol \, m^{-3} \, s^{-1}$). In the second order k carries units of $mol^{-1} \, s^{-1}$ and v, once again, carries units of $mol \, l^{-1} \, s^{-1}$.

The order of a reaction will be discussed once again in Chapter 5 where we will consider models of photosynthetic CO_2 assimilation. As you will see in that discussion, one of the most widely used models of CO_2 assimilation focuses on the reaction catalyzed by the enzyme ribulose 1,5-bisphosphate carboxylase (Rubisco); the initial photosynthetic reaction in which CO_2 is assimilated from the atmosphere and metabolically fixed to produce sugars. The model that we will discuss assumes that the reaction catalyzed by Rubisco is first order with respect to CO_2 or ribulose 1,5-bisphosphate (RuBP), its two substrates (or reactants). In some conditions the availability of RuBP is high and the reaction can be modeled as zero order with respect to RuBP and first order with respect to CO_2. In other conditions the reaction can be modeled as zero order with respect to CO_2 and first order with respect to RuBP.

Returning to reaction (3.1) the rate (v_2) of the reverse reaction, whereby reactants C + D are converted to products A + B, can be expressed as:

$$
\begin{aligned}
v_2 &= k_2 [C] \quad \text{or conversely} \quad v_2 = k_2 [D] \quad &\text{(first order)} \\
v_2 &= k_2 [C] [D] \quad &\text{(second order)}
\end{aligned} \tag{3.4}
$$

where C and D refer to mole units. The state of *chemical equilibrium* is defined as $v_1 = v_2$. Chemical equilibrium can only occur in a *closed reaction system*, in which new reactants are not added and product is not removed. For an elementary, second-order reaction at equilibrium we can write:

$$k_1 A \, B = k_2 C \, D, \tag{3.5}$$

and after rearrangement:

$$\frac{k_1}{k_2} = \frac{CD}{AB} = K_e, \tag{3.6}$$

where K_e is the *equilibrium constant*, and is unitless.

In an *open reaction system* equilibrium is not possible, but the condition of *steady state* is possible. Consider the following reaction:

$$A + B \underset{k_2}{\overset{k_1}{\rightleftharpoons}} AB \xrightarrow{k_3} C + D. \tag{3.7}$$

The reactions defined by k_1 and k_2 cannot exist at equilibrium because the conversion of AB to C + D is irreversible. However, if reactants A and B are added to the reaction system at the same rate that C and D are removed (thus producing an open reaction system), then the concentration of AB must remain constant as a function of time. For the case in which [AB] is constant, we refer to the reaction as being at steady state. In the condition of steady state we can write:

$$v_1 = k_1[A][B], \ v_2 = k_2[AB], \ v_3 = k_3[AB]. \tag{3.8}$$

In this case, k_1 carries units of $mol^{-1} \, s^{-1}$, whereas k_2 and k_3 carry units of s^{-1}. Taking the necessary condition that at steady state the rates of formation and breakdown of AB must be equal we can write:

$$k_1[A][B] = (k_2 + k_3)[AB]. \tag{3.9}$$

Rearranging to isolate [AB] we obtain:

$$[AB] = \frac{k_1}{k_2 + k_3}[A][B] = K_s[A][B], \tag{3.10}$$

where K_s is the *steady-state constant* (with units mol^{-1}) and represents the combined effects of all three rate constants.

3.2 The energetics of chemical reactions

In addition to mass action, reaction rate depends on the energetics underlying reactant collisions. This is one of the most important determinants of the value of k. Ultimately, the rate of a reaction depends on the number of reactant molecules that carry sufficient energy to exceed an energy threshold during collision. The energy threshold is referred to as the *energy of activation* (E_a; see Figure 3.1). If the collision energy of reactants exceeds E_a they will be forced into a *transition state*, an unstable state that leads spontaneously to the formation of product. Given two elementary reactions at equal temperature and reactant concentrations, the reaction with greater E_a will be slower because a smaller fraction of collisions will result in formation of the transition state, and ultimately product. In that case, k will be smaller in value. The presence of a *catalyst* has the potential to "speed up" a reaction by facilitating more reactant interactions that exceed E_a.

Let's take a closer look at the relation between E_a and k. Any given pool of reactant molecules will exist across a range of different kinetic energies; some molecules will exist at lesser energy and some at greater energy. It is not possible to know the exact number of molecules at each energy level, and thus it is not possible to predict the number of collisions that will exceed E_a. Instead, we rely on statistical models to predict the probability of energy distribution. The most commonly used model is known as the *Maxwell–Boltzmann energy*

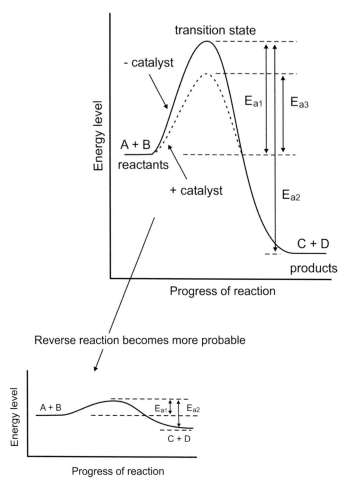

Figure 3.1   **Upper figure**. Energetics of a hypothetical chemical reaction in which reactants A + B react to form products C + D in the absence or presence of a catalyst. The energy of activation for the reverse reaction is represented as E_{a2}. The presence of a catalyst will reduce the energy of activation (represented as E_{a3}), allowing a greater fraction of reactions between A and B to form the transition state and proceed to products. **Lower figure**. A reaction in which the difference between E_{a1} and E_{a2} is small, and the overall difference in free energy between the reactants and products is small.

distribution. At a given temperature the theoretical distribution of kinetic energy across the pool of reactant molecules participating in a reaction can be described as:

$$n_E = n_T A\, e^{-E/RT}, \tag{3.11}$$

where n_E is the moles of reactant molecules that have energy that exceeds a given energy level (E, with units J mol^{-1}), n_T is the total moles of reactants, A is a synthetic coefficient that scales proportionally with the square root of E, R is the universal gas constant (with units J mol^{-1} K^{-1}), and T is Kelvin temperature. In this form of the Maxwell–Boltzmann distribution, a rather complex function $(2\pi(1/\pi RT)^{3/2}\sqrt{E})$ has been condensed to provide the synthetic coefficient, A. In this case we are using E to represent an arbitrarily chosen energy threshold. The chosen energy threshold could be the energy of activation, but not necessarily. It is also worth noting that R, the universal gas constant, appears in a lot of chemical contexts, not just those involving pressure–volume relations. This is because R has a fundamental relation to the Boltzmann constant, k_B, which is defined as $k_B = R/N_a$, where N_a is Avogadro's number.

It is clear from Eq. (3.11) that an increase in temperature shifts the distribution of kinetic energies toward the higher end of the distribution. As the temperature of a reaction is increased, the average kinetic energy per mole of reactants will also increase. This is the fundamental reason that reaction rate (and thus k) is dependent on temperature. The average kinetic energy of a reaction is related to the acceleration of reactants (and therefore, indirectly, to more frequent and more powerful collisions) according to $E = 0.5\,m\bar{v}^2$, where m is mass per mole and v is molecular speed in m s^{-1}, leaving E to be resolved in units of J mol^{-1}.

Using Eq. (3.11), we can define a probability density distribution of the fraction n_E/n_T as a function of E (Figure 3.2). In the study of chemical kinetics we are often more interested in the fraction of molecules that exceeds a specific value of E, rather than defining the entire distribution across all possible values of E. Thus, we can define the activation energy E_a as that threshold value for E above which reactants will attain the transition state and proceed to product. In reactions with a greater E_a the reaction rate will increase less in response to an increase in reactant concentration than in reactions with a lesser E_a. This is because a smaller fraction of collisions will have the energy capable of forming the transition state in reactions with greater E_a. Thus, in reactions with a high E_a, k will be correspondingly small.

A model that relates k to E_a and temperature, and is also derived from the Maxwell–Boltzmann distribution, is known as the *Arrhenius equation*:

$$k = Ce^{-E_a/RT}, \tag{3.12}$$

where k is the rate constant with units s^{-1}, C is a reaction-specific constant that accounts for components of the reaction that do not respond to temperature in exponential fashion (also called the "frequency factor") with units s^{-1}, E_a is the energy of activation with units

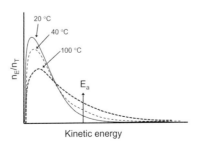

Kinetic energy

Figure 3.2   The Maxwell–Boltzmann energy distribution as a function of temperature. The molecular fraction (n_E/n_T) at the greater end of the energy distribution increases with an increase in temperature, causing more molecules to exist at kinetic energies above the energy of activation (E_a).

J mol^{-1}, T is absolute temperature with units K, and R is the universal gas constant (8.314 J mol^{-1} K^{-1}). On theoretical grounds, and referring to Eq. (3.12), a positive relation exists between E_a and the temperature sensitivity of k. As E_a is increased, the effect of each unit change in temperature on k will be amplified. The Arrhenius equation is one of the most widely used relations for predicting the response of biochemical reaction rate to temperature. As an empirical tool, the Arrhenius equation can be used to derive E_a from the relation between ln k/C and 1/T at a fixed reactant concentration. The linear slope of such a relationship will equal $-E_a/R$.

In the biological sciences, an alternative model is often used to describe the temperature dependence of process rates: the Q_{10} *model*. In the Q_{10} model the ratio of reaction rate constants defined at two temperatures ($k[T+\Delta T]/k[T]$) is constant and defined by a variable called the Q_{10}. The Q_{10} model obtains its name from the fact that $\Delta T = 10$ K. The Q_{10} model is derived from the same fundamental relations between temperature and reaction rate that were formulated by van't Hoff in the late 1800s. According to van't Hoff's theoretical work an approximation can be stated that the entropy and enthalpy changes of a reaction will remain constant as a function of changing temperature. Arrhenius built his model on this approximation and reasoned that the thermodynamic constancy assumed by van't Hoff could be expressed as E_a, which is assumed to be constant as a function of changing temperature (Arrhenius 1908). Application of the Arrhenius model to various chemical reactions resulted in the generalization that for every 10 K increase in reaction temperature, reaction rates were approximately doubled; this generalization, in turn, led to development of the Q_{10} model. The Arrhenius and Q_{10} models are not congruent with regard to formal mathematical derivation. However, they are related in form and provide similar means of predicting the temperature dependence of chemical processes. Formal derivations of the Arrhenius and Q_{10} models are provided in Appendix 3.1.

We caution readers in the use of Q_{10}. In the past, Q_{10} has been misused with the assumption that it is a constant property of a reaction. Many biochemical treatments have relied on the assumption of $Q_{10} \approx 2.0$ within the temperature range experienced by most organisms. This assumption is probably valid in homeothermic animals where the temperature does not vary significantly. However, the Q_{10} itself is sensitive to temperature because of changes in the rate by which the fraction of molecules that exceed E_a increases or decreases as a function of temperature (see discussion of the Maxwell–Boltzmann Law earlier in this section). The Maxwell–Boltzmann effect causes the Q_{10} to decrease as temperature increases.

3.2.1 Endergonic and exergonic reactions

Now that we have developed the conceptual foundations of chemical reactions and their relation to energetics, temperature, and collision theory, let's return to a discussion of thermodynamics and work. The potential for a chemical system to do work is determined by its free energy. Recalling our definition of free energy in Eq. (2.2) in Chapter 2, and working with finite forms of the algebra, we can write:

$$\Delta G = \Delta H - T\Delta S, \tag{3.13}$$

where the change in free energy (G) of a system at constant temperature is determined by the relative changes in enthalpy (H) and entropy (S). By convention, the mathematical sign of a change in H is negative when a chemical reaction is capable of occurring spontaneously (without energy from the surroundings) and positive when the reaction is not capable of occurring spontaneously. The opposite sign convention is used for changes in S. A negative change in the entropy of a chemical system undergoing reaction will accompany a non-spontaneous reaction and a positive change will accompany a spontaneous reaction. The changes in free energy that accompany a reaction should, in theory, be related to mass action. Thus, we can write:

$$\Delta G = \Delta G^0 + RT \ln \frac{[C][D]}{[A][B]}, \tag{3.14}$$

where ΔG^0 is the *standard free energy change of a reaction* and carries units J mol^{-1}. The term ΔG^0 reflects the inherent thermodynamic potential in a set of reactants to force changes in H and S as a reaction proceeds; it is determined for a standard set of conditions (298 K, 101.3 kPa, and all reactants and products set to concentrations of 1 mol l^{-1}). The value of ΔG^0 can be positive or negative in sign. A negative value for ΔG^0 means that under standard conditions the reaction will proceed spontaneously from reactant to product.

Let's consider these relations graphically (Figure 3.3). Consider a closed reaction system in which ΔG^0 is negative. Starting at the left-hand end of the reaction trend line, the ratio of reactants to products is high, ΔG is large and negative, the reaction from reactants to products is thermodynamically favored, and the reaction will proceed spontaneously. As the reaction proceeds and reactants are converted to products a change in free energy occurs in the positive direction, meaning that the reaction is moving from a state of greater displacement from

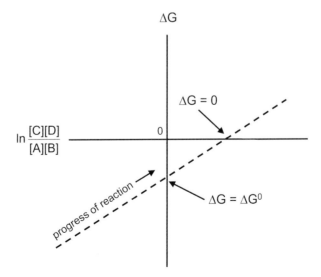

Figure 3.3 Graph showing relation between the mass action ratio of products to reactants and the change in free energy of a reaction as it progresses.

equilibrium toward a state of lesser displacement from equilibrium. Eventually, as the ratio of product to reactant approaches 1, ΔG will converge to ΔG^0, and at the y-intercept, $\Delta G = \Delta G^0$. Continuing up the trend line, at the x-intercept, $\Delta G = 0$. At this point the reaction has reached equilibrium. Continuing upward, beyond the x-intercept, the reaction will shift to the endergonic state and further conversion of reactant to product would require the extraction of free energy from the surroundings. The energy of activation of the reaction contributes to defining the slope of the trend line in Figure 3.3. A slightly greater value for E_a will be reflected in a slightly higher slope to the trend line, causing a greater (less negative) y-intercept and lesser (less positive) x-intercept for this particular example.

At equilibrium we can rewrite Eq. (3.14) to state:

$$\Delta G^0 = -2.3\, R\mathrm{T} \log \frac{[\mathrm{C}][\mathrm{D}]}{[\mathrm{A}][\mathrm{B}]} = -2.3 R\mathrm{T} \log K_e. \tag{3.15}$$

The form of Eq. (3.15) shows that there is a fundamental relation between ΔG^0 and the equilibrium constant (K_e) for a reaction. The relation shows that a large and negative ΔG^0 reflects a reaction that is highly displaced from equilibrium when at standard conditions, and will shift spontaneously toward that equilibrium, if permitted. Conversely, a reaction with a large and positive value for ΔG^0 is also displaced from equilibrium at the standard condition, but will require a large amount of energy from the surroundings in order to shift toward equilibrium. From Eq. (3.15), the utility of ΔG^0 is made clear; it provides a means of comparing reactions under a standard set of conditions and with regard to their inherent thermodynamic tendencies.

Exergonic reactions occur spontaneously; endergonic reactions must be supported through the addition of energy. The additional energy required to drive endergonic reactions can come through coupling to an exergonic reaction or through the removal of product by "downstream" exergonic reactions. Coupling to an exergonic reaction often occurs in biological metabolism through the transfer of "high energy" phosphate bonds from an intermediate compound, such as adenosine triphosphate (ATP), to a reactant in an otherwise endergonic reaction. The energy of the transferred phosphate bond can then be added to the energetics of the reaction forcing the overall change in free energy to be negative. In the case of product removal, an otherwise endergonic reaction can be sustained by the removal of product and maintenance of a negative ΔG through mass action effects. Endergonic reactions occur frequently in living cells, and are especially represented in anabolic metabolism, in which macromolecules are synthesized from simpler precursors. Exergonic reactions tend to be most frequently represented in catabolic metabolism in which chemical energy is extracted from the oxidation of energy-rich molecules.

3.3 Reduction-oxidation coupling

Coupled reactions in which the oxidation states of two elements or compounds are changed are called reduction-oxidation reactions, or *redox reactions*. The *oxidation state* of an element or compound defines the degree to which it can gain or lose electrons. Stated another way, oxidation state describes the electrical charge that an atom would have if it only participated in ionic bonds with other atoms. The oxidation state is a term of convenience; it does not describe a conserved property. The oxidation state of an element changes depending on the other elements with which it forms bonds. For example, in their pure elemental forms hydrogen and oxygen have oxidation states of zero (as do all elements in pure form). However, when two atoms of hydrogen are bonded to an atom of oxygen to form a water molecule, the hydrogen atoms take on an oxidation state of $+1$ and the oxygen atom takes on an oxidation state of -2. This means that the oxygen atom is more *electronegative* than the hydrogen atoms and electrons that are shared in the covalent bonds of the molecule spend more time associated with the oxygen nucleus than with either hydrogen nucleus. If, hypothetically, the hydrogen atoms were to give up their shared electrons to the oxygen atom and exist as ionically bonded entities, they would carry a charge of $+1$ and the oxygen would carry a charge of -2.

The concept of oxidation state is most easily applied in those cases where electrons are completely transferred from the atoms of one element to those of another. The removal of electrons from an atom through chemical reaction will increase its oxidation state, and we refer to such reactions as *oxidation reactions*; the atom has been chemically *oxidized*. The addition of electrons tends to *reduce* an atom's oxidation state, and we refer to such reactions as *reduction reactions*; the atom has been chemically *reduced*. Oxidation and reduction reactions must, by nature, be coupled; one atom gains the electron lost from another. In biological systems reduction is usually associated with reactions that not only add electrons, but also potential energy. Potential energy content increases when an electron is moved to an orbital at a greater distance from its associated nucleus (thus moving to an atomic zone of less charge attraction). In this energized state electrons have greater chemical potential.

An example of energy transfer during redox reactions is found in the process of *photosynthesis*. During photosynthesis solar photons are captured when an electron in a specialized photosynthetic molecule (chlorophyll) is excited and "jumps" to an atomic orbital at a greater (potential) energy level. In this excited state the electron is more susceptible to being lost from its original atom during a redox reaction. There are other possible fates for the excited electron, including falling back to its original ground state in a single step and emitting some of its potential energy as electromagnetic radiation (fluorescence photon emission), or falling back in multiple steps and emitting its energy as heat (radiationless transfer). When the electron is lost through a redox reaction, however, it moves its association to a different nucleus, in this case at the photosystem reaction center. The primary electron acceptor in plant chloroplasts is a chlorophyll derivative known as pheophytin. The reaction center molecule will replenish its lost electron through another set of redox reactions as it extracts an electron from a donor compound (e.g., H_2O, H_2S, or NH_4^+). The extracted electron will exist at the ground state energy level until photons of light are channeled once again to the reaction center.

Overall, photosynthesis exemplifies a "reductive" process in which electrons are removed from inorganic donors in a state of lesser chemical potential and transferred along with energy to organic acceptors with greater chemical potential. Respiration, in contrast, exemplifies an "oxidative" process in which electrons are removed from organic substrates

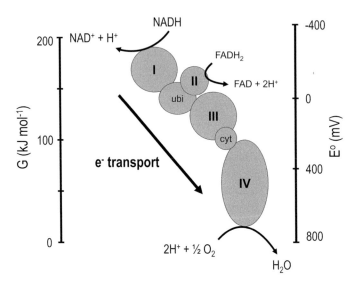

Figure 3.4 Electron transport in oxidative respiration illustrating successive redox reactions as electrons are passed along a chain of four primary protein complexes. The electrons are passed down a gradient in free energy (expressed here as free energy per electron, G) and reduction potential (E^o), where the most positive E^o indicates the greatest potential for oxidation. Oxygen is the ultimate electron acceptor with an E^o value of ~ 820 mV. The free energy of electron transport is eventually transferred to energy-rich organic phosphate compounds.

in a state of greater chemical potential and transferred to inorganic electron acceptors at a state of lesser chemical potential. Respiration can occur in the presence or absence of O_2, referred to as aerobic and anaerobic respiration, respectively. During *aerobic respiration* electron transfers occur in a series of steps, each catalyzed by a different protein (often a metalloprotein specialized for electron transfers). The proteins are collectively organized into electron transfer complexes embedded in the membranes of mitochondria. Ultimately, the energy from the sequential redox reactions is channeled into phosphate-containing organic compounds such as adenosine triphosphate (ATP), the "energy currency" of cellular metabolism. The channeling of electrons through the series of respiratory redox reactions is maintained by progressive irreversibility of the chain of reactions. Each step in the electron transport chain exists at a slightly lesser chemical potential than the previous, so according to thermodynamic constraints the redox sequence can only proceed in one direction (Figure 3.4).

The *reduction potential* (sometimes called redox potential) describes the maximum amount of energy per unit of charge that is available to do work when the charge is transferred in an electrochemical reaction. The reduction potential is a measure of thermo-dynamic potential to do electrical work; one component of the work associated with redox reactions. Energy is stored in the form of an electrical potential when unlike charges are separated. In order to convince yourself of this fact imagine the electrical work that can be done when particles of unlike charge are attracted and move spontaneously toward each other, releasing free energy. The *standard reduction potential* (E^o) of a compound can be related to the change in free energy of an oxidation/reduction reaction (ΔG) as:

$$E^{\circ} = -\Delta G / nF, \tag{3.16}$$

where n is the molar equivalent of electrons involved in the reduction and F is Faraday's constant expressed in units of coulombs mol^{-1}. Thus, E° carries units of volts (V, where 1 V = 1 joule per coulomb). Similar to the value for ΔG^{0}, the measure of E° is determined for a standard set of conditions: concentration of redox components at $1 \ mol \ l^{-1}$, pressure at 101.3 kPa, temperature at 298 K, and referenced to the standard hydrogen electrode. (The change in free energy required to add one electron to a proton (H^{+}) to produce the electrically neutral atom, H, is arbitrarily defined as a reduction potential of 0.) The negative sign in Eq. (3.16) forces the reduction potential to become positive for exergonic reactions. Those compounds that have the highest affinity for electrons and are most readily reduced in redox reactions will have the most positive E° values.

Reduction potential forms a central concept in the oxidation of organic and inorganic matter by microorganisms in soils. Soil organic matter represents the debris from living organisms, and thus the carbon contained in the organic matter exists in a chemically reduced state with high potential energy. Microorganisms use soil organic matter as a source of respiratory substrates, progressively oxidizing the carbon, and passing electrons through chains of metalloprotein electron carriers while simultaneously extracting free energy to do the work of ATP synthesis. The ultimate electron acceptor is typically a metal compound or highly electronegative compound (i.e., with a high affinity for electrons). It is clear that O_2 is the preferred final electron acceptor as it has a high reduction potential. When O_2 is present at low concentrations, such as in water-saturated soils, then soil microorganisms will use alternative electron acceptors, favoring the acceptor at the highest reduction potential. The overall metabolic sustenance of soil microorganisms is thus constrained by the available supply of electron donors and acceptors and the energy yield available from coupled redox reactions.

The reduction potential only defines the electrical component of the electrochemical potential. The full electrochemical potential (E, with units of volts) can be calculated using the *Nernst equation*:

$$E = E^{\circ} - \frac{RT}{nF} \ln \frac{[red]}{[ox]}, \tag{3.17}$$

$$\text{Term I} \qquad \text{Term II}$$

where "red" and "ox" refer to the molar concentrations of "reduced" and "oxidized" compounds, respectively. The Nernst equation combines the concepts of electrical and chemical disequilibrium and thus defines the full thermodynamic potential for a redox reaction to do chemical work. An examination of Eq. (3.17) shows that the electrochemical potential represents the summed thermodynamic potential due to reduction potential (Term I) and chemical disequilibrium (Term II). The highest electrochemical potential will occur for a reaction in which the oxidized member of a redox pair has a high reduction potential *and* the oxidized member is present in high concentration relative to the reduced member. It is also clear from Eq. (3.17) that redox pairs characterized by a negative E° (e.g., the $CO_2 \rightarrow CH_4$ redox pair in soils, see Figure 3.5) will only react spontaneously when the concentration ratio of oxidized (in this case CO_2) to reduced (in this case CH_4) compounds in the pair is relatively low and highly displaced from chemical equilibrium.

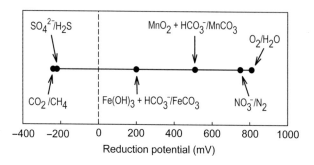

Figure 3.5 Reduction potential for significant redox couplings in soils. The left member of each pair is the oxidized member and the right member is the reduced member. Redox values taken from Liesack *et al.* (2000).

Biological redox reactions are responsible for producing some of the most important constituents of the atmosphere. For example, most of the atmospheric reservoir of O_2 represents the product of photosynthetic oxidation of H_2O. Photosynthetic O_2 began to accumulate in the atmosphere 2–3 billion years ago following some early epochs during which it was drawn out of the atmosphere by geochemical sinks, such as during the production of iron oxides (Nisbet *et al.* 2007). Soil redox reactions are also important for the production of atmospheric constituents, especially those that are radiatively active. For example, methane (CH_4) and nitrous oxide (N_2O) are both produced from soil redox chemistry. Methane and nitrous oxide have atmospheric lifetimes of 8–9 years and 120 years, respectively (Monson and Holland 2001). Both of these compounds have greater global warming potentials than CO_2 and have been increasing in their atmospheric concentrations over the past century. As for CO_2, microbial and root respiration rates in global soils are high enough to annually offset most of the mass of CO_2 taken out of the atmosphere by global gross primary production (approximately 80 Pg C yr^{-1} released in soil respiration and approximately 120 Pg C yr^{-1} taken up through gross primary production).

3.4 Enzyme catalysis

Enzymes are protein catalysts. As catalysts, enzymes are capable of increasing the rate of metabolic reactions by 10^{10}–10^{23}-fold over non-catalyzed rates at equivalent temperatures (Kraut *et al.* 2003). This represents an astonishing capacity for catalysis and it is crucial to sustain the energy and biomass transformations of organisms. The metabolic reactions catalyzed by an enzyme are determined by the enzyme's structure and distribution of electrical charges on the protein surface, which is ultimately determined by an organism's genetic code. Protein enzymes are constructed as chains of amino acid molecules that interact with each other through various hydrophobic and hydrophilic forces and form a three-dimensional shape with various pockets, clefts, and tunnels distributed across the protein surface. Enzymatic catalysis occurs within these pockets, clefts, or tunnels, and these foci of catalysis are referred to as *active sites*. Although the entire enzyme may contain hundreds or thousands of amino acids, typically only 10–20 are found within the active site. Interactions between the substrate and active site are highly specific and are determined by complementary relationships involving shape, hydrophobicity, and/or electrical charge. Research into the structural attributes of proteins have revealed that a relatively small number of characteristic folds and shapes exist at the active sites of enzymes across the eukaryotic and prokaryotic kingdoms, and these have been conserved, co-opted, and recycled during the evolution of a broad diversity of protein functions (Hrmova and Fincher 2001).

Various hypotheses have been proposed to explain the catalytic mechanism of enzymes, all of which focus on how the active site facilitates transformation of the substrate(s) to the transition state (Figure 3.6). In the *strain hypothesis*, catalysis results when the structural integrity of the substrate is strained after binding to the active site. Strain occurs because the active site complements the structural and electrochemical nature of the transition state more than the original form of the substrate. Strain allows for the rearrangement of electron orbitals, forcing the substrate through its transition state and on to product formation. In the *induced-fit hypothesis*, the active site complements the structural and electrochemical nature of the substrate, but loosely so. When the substrate binds to the active site, polarity and charge interactions between the substrate and enzyme force the enzyme to "flex" and more tightly "envelope" the substrate. Flexing of the enzyme aligns critical catalytic groups that complement the transition state, forcing the form of the substrate to distort toward that of the transition state. In both hypotheses the binding energy that goes into the enzyme-substrate interaction is channeled into distortion of the substrate(s) to the point that the E_a of reaction

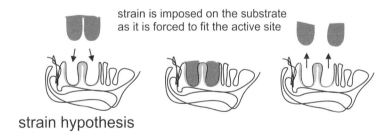

strain is imposed on the substrate
as it is forced to fit the active site

strain hypothesis

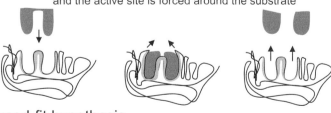

the transition state is formed as the protein flexes
and the active site is forced around the substrate

induced-fit hypothesis

Figure 3.6   Two hypotheses to explain catalysis by enzymes as they promote the progressive conversion of substrate through the transition state to product.

is exceeded, allowing the substrate(s) to pass through the transition state and, spontaneously, to product formation. In energetic terms, *the free energy released during binding of the substrate to the active site is channeled in a way that reduces the activation energy of the reaction*. Free energy must be released during substrate-enzyme binding for the reaction to occur spontaneously. Thus, enzyme-catalyzed reactions do not change an otherwise endergonic reaction into an exergonic reaction – they only reduce E_a.

3.4.1 Models of enzyme kinetics

The rate at which an enzyme-catalyzed reaction occurs is determined by the combined effects of how tightly the enzyme's active site binds the substrate and how rapidly the active site releases product. Consider the simplest of the enzyme-catalyzed reactions – one in which a single substrate (S) is catalytically converted with the aid of an enzyme (E) to a single product (P). The reaction and its relevant reaction constants can be represented as:

$$E + S \underset{k_2}{\overset{k_1}{\rightleftharpoons}} ES \xrightarrow{k_3} E + P, \tag{3.18}$$

where ES represents the enzyme-substrate complex which is formed according to rate constant k_1, and can either produce the transition state followed by dissociation to form product and free enzyme (according to rate constant k_3), or decompose back into the original substrate and free enzyme (according to rate constant k_2). The *Michaelis–Menten model* of enzyme kinetics is built on the assumption that the concentrations of E, S, and ES reach equilibrium at a rate that is rapid relative to the steady-state rate at which ES is converted to E + P. Thus, the overall velocity of the reaction is limited by k_3 and [ES]. The Michaelis–Menten model is often referred to as the *rapid equilibrium model*, and has been derived to provide a term that reflects the product between k_3 and the total enzyme concentration (reflected mathematically in the term V_{max}), and a combination of terms that reflects the equilibrium that ultimately determines [ES] (expressed mathematically as $[S]/K_m + [S]$, where K_m is an equilibrium enzyme-substrate dissociation constant). The Michaelis–Menten model model takes the form:

$$v = \frac{V_{max} \ [S]}{K_m + [S]}, \tag{3.19}$$

where v is the velocity of the reaction (mol l^{-1} s^{-1}), V_{max} is the maximum velocity that the reaction is capable of reaching (i.e., at a saturating concentration of substrate) (mol l^{-1} s^{-1}), [S] is substrate concentration (mol l^{-1}), and K_m is the Michaelis–Menten constant (mol l^{-1}). A derivation of Eq. (3.19) is presented in Appendix 3.2.

Graphically, Eq. (3.19) describes a rectangular hyperbola (Figure 3.7). Under conditions when $[S] \ll K_m$, we can ignore [S] in the denominator on the right-hand side of Eq. (3.19), and $v \approx (V_{max}/K_m)$ [S]. Thus, at low [S], the relation between v and [S] approaches first-order kinetics, with a first-order reaction coefficient equal to V_{max}/K_m. As [S] increases to the point where $[S] \gg K_m$, the K_m can be ignored and Eq. (3.19) reduces to $v \approx V_{max}$. Thus, at high [S], v approaches zero-order dependence on [S] and is approximated by an asymptotic limit represented by V_{max}.

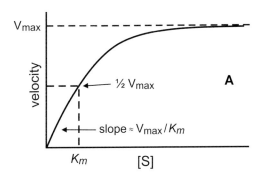

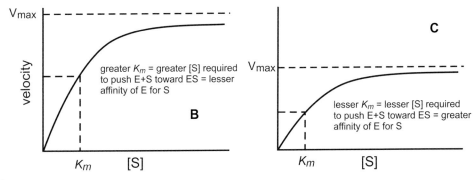

Figure 3.7   **A**. Graphical representation of the Michaelis–Menten model describing the dependence of reaction velocity on substrate concentration [S]. **B**. Graphical representation showing that a relatively high K_m for an enzyme requires that [S] be relatively high before 0.5 V_{max} is reached. Recall that as v → V_{max} catalysis is determined to a progressively greater extent by [ES]. Thus, a greater K_m reflects lesser potential to form ES as [S] is increased; i.e., lesser affinity of the active site for S. **C**. Graphical representation showing the opposite influence of a lesser K_m, which reflects a greater potential to form ES as [S] is increased and thus greater affinity of the active site for S.

We can gain a deeper perspective on the V_{max} by placing Eq. (3.19) within the context of the reaction sequence depicted in Eq. (3.18). At saturating substrate concentration, the concentration of free enzyme molecules (E) will be negligible since nearly all of the enzyme should be bound to substrate as ES. With this condition, [ES] ≈ [E$_t$], where [E$_t$] is the total concentration of enzyme. Recognizing that by definition V_{max} occurs at saturating [S], V_{max} can be approximated as the product between [E$_t$] and k_3. The rate constant k_3 is also known as k_{cat}, the *catalytic rate constant* or the *enzyme turnover number*. Under the assumption of rapid equilibrium among E, S, and ES, k_3 and the total concentration of enzyme proteins [E$_t$] represent the principal kinetic constraints to v as [S] increases toward zero-order influence. The enzyme turnover number can be viewed as the rate at which the active site is capable of converting a unit quantity of ES, through the transition state, to E + P; the enzyme turnover number carries the unit s^{-1}. As the reaction approaches the condition of zero-order dependence on [S], catalytic increases in reaction rate are only achieved by increases in [E$_t$] or k_{cat}.

An examination of Eq. (3.19) also reveals further insight into the nature of K_m. Resolving the equation for the condition $K_m = [S]$, yields the relation, $v = 0.5\ V_{max}$. Thus, by definition, the K_m is the substrate concentration that yields half V_{max}. An enzyme with a lesser K_m for a specific substrate is capable of catalyzing a reaction at half its V_{max} at a lesser substrate concentration, compared to an enzyme with greater K_m for that substrate. From that conclusion, we can state that the lesser the K_m, the less the tendency for ES to dissociate back to E and S. Using this logic the K_m for an enzyme reflects the apparent affinity of the enzyme for its substrate. The *lesser* the value of the K_m, the tighter is the association between enzyme and substrate, and the *greater* is the affinity of the active site for its substrate (Figure 3.7). From Eq. (3.19), and under the assumption of rapid equilibrium, it is clear that the K_m of an enzyme becomes an important kinetic constraint to v when [S] is relatively low.

As enzyme-substrate interactions have evolved over time, catalytic constraints have emerged that limit the amount of divergence that is possible between the K_m and V_{max}. In theory, the evolution of greater affinity toward a substrate (i.e, reflecting a lesser K_m) in order to maximize catalysis at low [S], will require that the active site turn over substrate at a lesser rate (i.e., reflect a lesser V_{max}) when catalysis occurs at high [S]. The requisite relations between an active site and substrate, and the fact that efficient enzyme catalysis requires differential interactions of tight binding or fast release, depending on [S], can potentially cause mutual and antagonistic constraints on the evolution of an active site. For example, the K_m for CO_2 for the enzyme Rubisco is lower in C_3 plant chloroplasts, which must assimilate CO_2 from an atmosphere relatively depleted of CO_2, compared to the K_m for CO_2 for Rubisco from C_4 chloroplasts, which operates in a CO_2-rich atmosphere. Conversely, Rubisco from C_3 chloroplasts exhibits a lower k_{cat}, compared to Rubisco from C_4 chloroplasts.

The Michaelis–Menten assumption of rapid equilibrium among E, S, and ES, is not always valid. For the assumption to be valid k_3 must be considerably less than k_2. In some enzyme-catalyzed reactions k_3 is of similar magnitude to k_2, meaning that [ES] is controlled by the combined influence of k_3 and k_2 to produce a steady state, rather than on k_2 and k_1 to produce a rapid equilibrium. In that condition, a different kinetic model is required. The Briggs–Haldane model takes the same form as Eq. (3.19) except that K_m is defined as $K_m = (k_2 + k_3)/k_1$, rather than $K_m = k_2/k_1$. (Recall that this same relation between equilibrium and steady-state reaction coefficients was discussed in Section 3.1 within the context of chemical reactions in general.)

3.4.2 Modification of enzyme activity

Enzymes are subject to catalytic modification from a number of different environmental and biological factors; two of the most important are temperature and pH. Temperature affects enzyme-catalyzed reactions just as it does all chemical reactions, by increasing the frequency at which substrate and enzyme molecules interact with free energy exceeding E_a. Enzyme-catalyzed reactions exhibit an exponential dependence of reaction rate on temperature, as reflected in the Arrhenius model. At extremely high temperatures protein denaturation occurs, resulting in loss of catalytic function. Cellular pH affects the catalytic efficiency of enzymes by influencing electrical charge interactions within the structural framework of the protein and between the active site and substrates, and by amplifying the tendency for hydrophobic and hydrophilic domains of the protein to self-organize and distinguish themselves from one another. Depending on the prevailing pH, certain amino acid groups in a protein can be protonated or non-protonated, affecting their interaction with other polar, charged, or hydrophobic groups in the protein. These electrically induced interactions have the potential to affect the shape of the active site and enzyme-substrate binding affinity. Enzymes typically exhibit maximal activity within a narrow range of pH.

In addition to environmental influences, enzyme activity is often modified by the binding of alternative substrates to the active site (e.g., competitive inhibition), or the binding of effector ions or molecules to *allosteric sites*. Allosteric sites are alternative binding sites on the surface of a protein that, when engaged, cause the protein to change its shape or electrical properties and, in doing so, change the enzyme-substrate binding affinity. Allosteric modification can cause changes to the K_m and/or V_{max} of the active site with either a negative or positive influence on catalytic efficiency.

In a further example of complex regulation, enzymes can have multiple active sites that function in a cooperative manner. This is common for enzymes with multiple protein subunits. *Cooperativity* involves a change in the catalytic properties of one active site as the result of substrate binding to a different active site. Cooperativity is usually expressed through changes in the shape or charge properties of the protein. Finally, enzymes can be modified by covalent modification, including attachment of specific effectors, such as inorganic phosphate molecules or adenylate phosphate molecules (e.g., ADP or AMP), or cleavage of certain amino acids or amino acid side groups from the protein.

3.5 Stable isotopes and isotope effects

The reservoir of chemical compounds in the earth system is composed of both common and rare isotopic forms of elements. For example, the element carbon exists as three naturally occurring isotopes, including the most common, ^{12}C, and the rarer forms ^{13}C and ^{14}C. Isotopes of the same element possess the same numbers of electrons and protons, but different numbers of neutrons; thus, isotopes differ in atomic mass. In the case of the isotopes of carbon, ^{12}C has 6 protons and 6 neutrons giving it a total atomic mass of 12; the other two carbon isotopes, ^{13}C and ^{14}C, have 7 and 8 neutrons, respectively. The atmospheric CO_2 reservoir reflects approximately 98.9% $^{12}CO_2$, 1.1% $^{13}CO_2$, and a trace of $^{14}CO_2$. *Stable isotopes* refer to those forms that have not been observed to break down through the emission of high-energy particles and are thus stable over time. With regard to carbon, the ^{12}C and ^{13}C forms are stable, but the ^{14}C form is unstable (or "radioactive").

Whether an isotope is stable or radioactive depends on the relative numbers of protons and neutrons in the nucleus. Protons exert repulsive forces among themselves and this force increases as a decreasing function of separation distance. Neutrons that insert themselves among protons tend to increase the proton separation distance, and that by itself contributes to greater nuclear stability. Additionally, the presence of neutrons in close proximity to protons tends to enhance the *strong nuclear force* (or *strong inter-action*), a binding force between neutrons and protons that may be due to the presence of other subatomic particles such as gluons. Among the lighter elements there appears to be an optimal ratio of neutrons-to-protons of 1:1 in the nucleus. As the atomic mass of elements increases, more neutrons are needed per proton to achieve nuclear stability; closer to 1.5:1. The ratios of neutrons to protons present in stable isotopes, when plotted as a function of increasing atomic mass results in a well-characterized *band of stability* (Figure 3.8). For ratios above and to the left of the band of stability, elements would tend to emit nuclear mass as decaying neutrons (beta-emitters), shifting the ratio toward greater stability. Beta decay results in neutrons being converted into protons, electrons, or other types of subatomic particles. For ratios below and to the right of the line, elements would tend to emit mass as alpha particles (which originate from both protons and neutrons), a process that reduces the mass of the unstable atoms and shifts them toward greater stability. We will focus our discussions on the so-called stable isotopes – those within the band of stability.

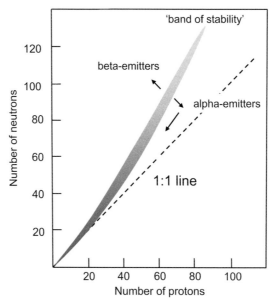

Figure 3.8 Graphical representation of the relationship between numbers of neutrons and protons that produce nuclear stability in elements as the overall mass of the elements increases. The lighter elements tend to be most stable with a neutron:proton ratio close to 1:1. As element mass increases, the ratio of greatest stability will also increase toward 1.5:1. Ratios outside the *band of stability* (indicated by gray shading) tend to be associated with unstable nuclei, which undergo radioactive decay either by loss of beta or alpha particles.

The abundance of an isotope is characterized by the *isotope abundance ratio* (R), which is defined as a ratio of molar concentrations:

$$R = \frac{[\text{rare isotope}]}{[\text{abundant isotope}]}. \tag{3.20}$$

A distinction must be made between the isotope abundance ratio and the isotope concentration; the latter is defined as the ratio of the rare isotope to *all* isotopic forms of the element, not only the most abundant isotopic form. It is common to express the isotope abundance ratio in terms of *delta notation*, in which the ratio of isotopes in an observed sample is related to the ratio of isotopes in a standard (reference) material:

$$\delta = \left(\frac{R_{\text{sample}}}{R_{\text{standard}}} - 1 \right) * 1000. \tag{3.21}$$

Delta notation reflects a "ratio of ratios." In the case of the stable isotopes of carbon, for example, R_{standard} is taken as the $^{13}C/^{12}C$ obtained in a specific belemnite limestone formation (Pee Dee) found in South Carolina, USA (Craig 1957). The units in δ notation are expressed as parts per thousand (‰), or "per mil," which, once again, does not reflect a true unit measure, but rather a ratio.

Isotopes of the same element participate in the same types of chemical reactions, but their differences in atomic mass affect their reaction rates, as well as physical properties such as diffusive mobility. Differences in reaction rate or mobility due to differences in atomic mass are called *isotope effects*. Isotope effects have the potential to alter the relative abundances of isotopes in a system. Thus, by measuring ratios of isotopes we can gain insight into the kinetics of biochemical and biophysical fluxes that may have influenced the state of a biogeochemical system. We can quantify an isotope effect (α) as:

$$\alpha = \frac{R_a}{R_b}, \tag{3.22}$$

where R_a and R_b refer respectively to the molar isotope abundance ratios of a source (a) and a sink (b). If there were no differential effect on isotopes during the progression of a process from State (a) to State (b), then $R_a = R_b$, and $\alpha = 1$. One way to describe an isotope effect is that it results in "discrimination against" (or alternatively "fractionation of") certain isotopes. *Discrimination* (Δ) is defined as:

$$\Delta = (\alpha - 1) * 1000 = \left(\frac{R_a}{R_b} - 1 \right) * 1000. \tag{3.23}$$

Note that discrimination is scaled similarly to that for δ notation, but that Δ is not dependent on the isotopic ratio of a measurement standard. Discrimination can be expressed in terms of δ notation as:

$$\Delta = \frac{\delta_a - \delta_b}{1 + \delta_b/1000}, \tag{3.24}$$

where (δ_a) reflects the source isotope ratio and (δ_b) reflects the sink isotope ratio, with both referenced to a measurement standard.

In order to illustrate the relation between Δ and δ, let's consider the case of CO_2 assimilation by plants that possess the C_3 photosynthetic pathway. The C_3 photosynthetic pathway uses CO_2 from the atmosphere as a substrate, which has a $^{13}C/^{12}C$ that is ~ 8‰ lower than that in the Pee Dee belemnite limestone standard. We will consider CO_2 in the atmosphere as our source, such that $\delta_a \approx -8‰$. The bulk biomass of C_3 plants has a $^{13}C/^{12}C$ ratio that is ~ 27‰ lower than the limestone standard. We will consider plant biomass as our sink, such that $\delta_b \approx -27‰$. When we put these values into Eq. (3.24), we calculate $\Delta \approx 19.5‰$. It is important to note that for work with ^{13}C and ^{12}C related to photosynthetic CO_2 assimilation, δ values for the $^{13}C/^{12}C$ ratio in atmospheric CO_2 or in the biomass of plant tissues are typically negative in sign, whereas Δ for the process of photosynthetic CO_2 assimilation is positive in sign.

Isotope effects that occur during diffusion, due to the differential mobility of isotopic forms, are directly proportional to differences in mass; heavier isotopes have relatively less mobility than lighter isotopes. Differences in the chemical reaction rates of isotopes are also a result of differences in atomic mass. Chemical reactions require the breakage and formation of bonds between atoms. In the case of covalent bonds, the bonded atoms are separated by a distance that is not fixed, but rather changes as the atoms vibrate back and forth along the axis of bonding. The vibrations are caused by a dynamic equilibrium between the attractive and repulsive forces among the nuclei and electron clouds participating in the bonds. The mass of

| Box 3.1 | Why do catalyzed chemical reactions result in discrimination against some isotopes compared to others? |

The kinetic characteristics of chemical reactions depend on the free energy transitions that occur when bonds are broken or formed in reactant molecules, compared to product molecules. Covalent bonds sustain thermodynamically stable intra-molecular interactions; breakage of a covalent bond requires that energy be added to the point that the bond is strained and broken. Enzyme catalysts lower the activation energy required to strain the bond to the critical transition state, and thus allow the bonded entities to dissociate with an accompanying loss of free energy. The stretching frequency of bonds is dependent on atomic mass; heavier isotopes exhibit lesser vibrational frequencies, which translates to shorter bond lengths, stronger bonds, and a greater activation energy required for bond cleavage or formation; the weaker bonding energies of atoms from lighter isotopes, along with lower activation energies, facilitate greater reaction rates. Bond breakage, and therefore primary isotope effects, occurs in quantized free energy changes. Differences in the vibrational frequencies of isotopes are responsible for primary isotope effects, defined as the ratio of reaction rates for molecules containing different isotopic constituents. The primary isotope effect is most easily measured in reactions that result in only bond breakage; i.e., not involving the formation of secondary bonding relations as the transition state moves to form new products. In the case of transitions to new bonding relationships, incomplete breakage of primary bonds in the transition state, prior to the initiation of novel bonds involved in product formation, can dampen primary isotope effects. In those reactions, where secondary kinetic effects are involved, determination of discrimination patterns can be complex.

the participating nuclei influence this dynamic equilibrium, creating discrimination against isotopes during chemical reactions (Box 3.1).

There are two types of isotope effects: thermodynamic and kinetic. *Thermodynamic isotope effects* occur when differential isotopic distributions exist between reactants and products within a reaction system that is at equilibrium. *Kinetic isotope effects* reflect differences in rate constants for a reaction when different isotopes take part in the reaction; kinetic effects occur when there is continuous production of a product (or diffusion to a sink) from reactants (or diffusion from a source), involving compounds of different isotopic form. Thermodynamic isotope effects are smaller than kinetic effects. An example of a thermodynamic isotope effect involves the dissolution of CO_2 from air ($CO_{2(a)}$) into water (to form $CO_{2(w)}$), and subsequent reaction with water to form carbonic acid and the various forms of carbonate:

$$CO_{2(a)} + H_2O \leftrightarrow CO_{2(w)} + H_2O \leftrightarrow H_2CO_3 \leftrightarrow H^+ + HCO_3{}^- \leftrightarrow 2H^+ + CO_3{}^{2-}.$$

$$(3.25)$$

In tracing the steps through this equilibrium, molecules of CO_2 will dissolve in water with a thermodynamic isotope effect that favors $^{12}CO_2$, $\alpha = 1.0011$ ($\Delta = 1.1‰$). The isotope effect that accompanies the dissolution of CO_2 in H_2O is actually the equilibrium sum of two kinetic effects – CO_2 going into the water solution and becoming hydrated with a shell of polar water molecules and CO_2 coming back out of solution after shedding the hydration shell. The net effect of the equilibrium between these fluxes favors more $^{12}CO_2$ in the water phase and more $^{13}CO_2$ in the air phase. Once dissolved in water, the actual reaction of CO_2 with H_2O to form $HCO_3{}^-$ exhibits an equilibrium isotope effect that favors ^{13}C in $HCO_3{}^-$ and ^{12}C in $CO_{2(w)}$. The equilibrium between $CO_{2(w)}$ and $HCO_3{}^-$ exhibits an isotope effect less than 1, $\alpha = 0.991$ ($\Delta = -9‰$). In this case, the covalent bonds that exist in $HCO_3{}^-$ are stronger with the ^{13}C isotope compared to the ^{12}C isotope. Thus, when the equilibrium is established between $CO_{2(w)} + H_2O$ and $HCO_3{}^- + H^+$, more of the ^{13}C remains in the $HCO_3{}^-$ compared to $CO_{2(w)}$. As a second example of a thermodynamic isotope effect consider the evaporation of H_2O from liquid to vapor into a closed volume. Net evaporation from the liquid will increase atmospheric humidity in the airspace of the volume until the air is saturated with water vapor, at which point phase equilibrium will exist. The $^{18}O/^{16}O$ stable isotope ratio of the water in the air of the volume will be different from that in the liquid phase because $H_2{}^{18}O$ partitions into the vapor phase with a different equilibrium constant that that for $H_2{}^{16}O$; this is also a case of thermodynamic fractionation.

In order to illustrate the kinetic isotope effect, we can turn once again to the process of CO_2 assimilation in plants with the C_3 photosynthetic pathway. The assimilation of CO_2 is catalyzed by the enzyme ribulose 1,5-bisphosphate carboxylase/oxygenase (Rubisco), which favors the assimilation of $^{12}CO_2$ relative to $^{13}CO_2$ with $\alpha = 1.030$ ($\Delta = 30‰$). The fractionation in this reaction is kinetic and is due to the differences in the potential for $^{13}CO_2$ versus $^{12}CO_2$ to form a covalent bond with a second substrate, ribulose 1,5 bisphosphate (RuBP). For kinetic isotope effects, we can rewrite Eq. (3.22) to reflect reaction rates, rather than isotope abundance ratios:

$$\alpha = \frac{k_{\text{abundant}}}{k_{\text{rare}}}, \qquad (3.26)$$

where $k_{abundant}$ is the reaction rate coefficient for a reaction using the abundant isotopic form ($^{12}CO_2$) as reactant and k_{rare} is the reaction rate coefficient for use of the rare isotopic form ($^{13}CO_2$) as reactant.

Kinetic isotope effects will only occur if unused reactant molecules – i.e., reactant molecules left behind by discrimination – are allowed to leave the reaction, either through physical leakage or utilization in a connected, but branched reaction. If a reactant is completely used to form product, then there is no potential for isotopic discrimination. Thus, we can write:

$$\Delta_e = \Delta_{max} \phi, \tag{3.27}$$

where Δ_e is the "expressed" discrimination, Δ_{max} is the maximum potential discrimination, and ϕ is the *branching factor* or *"leakage"* which facilitates the removal of unused reactant from the site of reaction.

Appendix 3.1 Formal derivations of the Arrhenius equation and the Q_{10} model

During the late nineteenth century, chemists were focused on gaining a better conceptual understanding of k and its dependence on temperature. These chemists, such as Arrhenius, began their work from the foundational relation derived by van't Hoff (van't Hoff 1884, see Laidler and King 1983):

$$\frac{d \ln K_e}{dT} = \frac{\Delta H}{RT^2}. \tag{3.28}$$

van't Hoff's model is based on thermodynamic considerations and was originally derived as a means of better defining K_e, the equilibrium coefficient. Recognizing that $K_e = k_1/k_2$, and that the enthalpy difference (ΔH) between two opposing reactions represents the difference in activation energies (i.e., $\Delta H = \Delta E_a = E_{a1} - E_{a2}$), we can write:

$$\frac{d \ln \left(\frac{k_1}{k_2}\right)}{dT} = \frac{\Delta E_a}{RT^2}. \tag{3.29}$$

Recognizing that $\ln (k_1/k_2) = \ln k_1 - \ln k_2$, and expanding Eq. (3.29), we can write:

$$\frac{d(\ln k_1)}{dT} - \frac{d(\ln k_2)}{dT} = \frac{E_{a1}}{RT^2} - \frac{E_{a2}}{RT^2}. \tag{3.30}$$

Resolving Eq. (3.30) as two separate parts, we can write:

$$\frac{d(\ln(k_1))}{dT} = \frac{E_{a1}}{RT^2} \pm I \quad \text{and} \quad \frac{d(\ln(k_2))}{dT} = \frac{E_{a2}}{RT^2} \pm I, \tag{3.31}$$

where I is a constant. Arrhenius noted that for most reactions, I can be set to zero and he used the assumption that E_a is independent of temperature to propose the general relation:

$$\frac{d(\ln k)}{d\,T} = \frac{E_a}{RT^2}. \tag{3.32}$$

Integrating both sides of Eq. (3.32) with respect to temperature, and recognizing the rules of integration that state $\int \frac{dx}{x} = \ln x$ and $\int \frac{1}{x^n} dx = -\frac{1}{x^{(n-1)}}$, we can write:

$$\ln k = -\frac{E_a}{RT} + C', \tag{3.33}$$

where C' is the constant of integration. Simplifying using the rules of logarithms, we can write:

$$k = e^{-E_a/RT+C'} = e^{-E_a/RT}e^{C'}. \tag{3.34}$$

Redefining $e^{C'}$ as a new constant C, we can write the more familiar form of the Arrhenius model as:

$$k = Ce^{-E_a/RT}. \tag{3.35}$$

It is important to note that the Arrhenius model is an empirically based model. There is no fundamental definition of E_a, it must be defined empirically within the context of specific reactions.

In order to achieve a more fundamental understanding of E_a, Henry Eyring, working at Princeton University in the 1930s, approached the definition of k from a different perspective. By focusing on molecular relations within the transition-state complex, rather than the collisional energies of reactants, Eyring (1935) proposed:

$$k = \kappa_E \frac{k_B T}{h} e^{-\Delta G^{\ddagger}/RT}, \tag{3.36}$$

where κ_E is called the transmission coefficient and fractionally accounts for the fact that some transition-state complexes do not go on to form product, k_B is the Boltzmann constant, h is Planck's constant, and the symbol $\ddagger$ indicates that the free energy difference (ΔG) is determined between reactants and an intermediate state (the transition-state complex). Both the Arrhenius and Eyring models have been used to describe the response of biological processes, such as respiration, to temperature; though both tend to underestimate flux densities at low temperature and overestimate flux densities at high temperature (Lloyd and Taylor 1994).

To develop an expression for the Q_{10} model, we begin with the ratio of reaction constants at two temperatures (k_{T1} and k_{T2}) defined according to the Arrhenius relation at each temperature:

$$\frac{k_{T2}}{k_{T1}} = \frac{C\exp\left(-\frac{E_a}{RT_2}\right)}{C\exp\left(-\frac{E_a}{RT_1}\right)} = \exp\left(\frac{E_a}{R}\left(\frac{1}{T_2} - \frac{1}{T_1}\right)\right) = \exp\left(\frac{E_a}{R}\left(\frac{T_2 - T_1}{T_2 T_1}\right)\right). \tag{3.37}$$

Thus, for $\Delta T = 10K$ we can write:

$$Q_{10} = \left(\frac{k_{T2}}{k_{T1}}\right)^{10/(T_2 - T_1),}$$
(3.38)

which can be rearranged to:

$$\ln \frac{k_{T1}}{k_{T2}} = \frac{T_2 - T_1}{10} \ln Q_{10},$$
(3.39)

and, finally, to:

$$\frac{k_{T2}}{k_{T1}} = \exp\left(\frac{\ln Q_{10}}{10}\right)(T_2 - T_1).$$
(3.40)

It is clear in comparing Eqs. (3.37) and (3.40) that the Q_{10} model is not equivalent in form to the Arrhenius model. We can equate forms of Eqs. (3.37) and (3.40) to state:

$$\exp\left(\frac{E_a}{R}\left(\frac{T_2 - T_1}{T_2 T_1}\right)\right) = \exp\left(\frac{\ln Q_{10}}{10}\right)(T_2 - T_1).$$
(3.41)

Recognizing that the exponential form cancels on each side of Eq. (3.41) and rearranging to solve for E_a/R, we can state:

$$\frac{E_a}{R} = \frac{\ln Q_{10}}{10}(T_2 T_1).$$
(3.42)

Further rearrangement leads to:

$$Q_{10} = e^{10 E_a / R(T_2 T_1)}.$$
(3.43)

Evaluating Eq. (3.43) at $\Delta T = 10\,K = (T_2 - T_1)$, we can compare the Arrhenius and Q_{10} models as:

$$k_{(T)} = C e^{-E_a/RT},$$
(3.44)

$$Q_{10} \approx e^{10 E_a / R T_2 T_1}.$$
(3.45)

Appendix 3.2 Derivation of the Michaelis–Menten model of enzyme kinetics

Consider a simple enzyme-catalyzed reaction in which a single substrate molecule is converted to a single product molecule:

$$E + S \underset{k_2}{\overset{k_1}{\rightleftharpoons}} ES \overset{k_3}{\longrightarrow} E + P, \tag{3.46}$$

where ES represents the enzyme-substrate complex which is formed according to rate constant k_1, and can either dissociate to form product and free enzyme (according to rate constant k_3) or decompose back into the original substrate and free enzyme (according to rate constant k_2). If we assume that E, S, and ES come to rapid equilibrium, compared to the rate of conversion of ES to E + P, then the velocity (v) of reaction (3.46) can be defined as:

$$v = k_3[ES]. \tag{3.47}$$

The total concentration of enzyme (E_t) can be defined as:

$$[E_t] = [E] + [ES]. \tag{3.48}$$

Using these relationships we can derive an expression for the rate of the reaction per unit of total enzyme, which should reflect the catalytic efficiency of the enzyme:

$$\frac{v}{[E_t]} = \frac{k_3[ES]}{[E] + [ES]}. \tag{3.49}$$

An equilibrium enzyme-substrate dissociation constant (K_d) can be defined in terms of the concentrations of E, S, and ES and the rate constants k_1 and k_2 as follows:

$$K_d = \frac{[ES]}{[ES]} = \frac{k_2}{k_1}. \tag{3.50}$$

Rearranging Eq. (3.50) and substituting for ES in Eq. (3.49), we obtain:

$$\frac{v}{[E_t]} = \frac{k_3[E]\dfrac{[S]}{K_d}}{[E] + [E]\dfrac{[S]}{K_d}}. \tag{3.51}$$

At rapid equilibrium the rate-limiting step in reaction (3.46) will be k_3. Here, we define a term, V_{max}, as the velocity of the reaction when $[S] \to \infty$ and $[E_t] \to [ES]$. Within this assumption, $V_{max} = k_3 [E_t]$. With algebraic rearrangement of Eq. (3.51) the rapid-equilibrium form of the Michaelis–Menten model is obtained as:

$$v = \frac{V_{max}[S]}{K_m + [S]}, \tag{3.52}$$

where K_d has been replaced by the more commonly used nomenclature of K_m, the Michaelis constant; $K_d = K_m = k_2/k_1$.

4   Control over metabolic fluxes

Few scientists acquainted with the chemistry of biological systems at the molecular level can avoid being inspired. Evolution has produced chemical compounds exquisitely organized to accomplish the most complicated and delicate of tasks.

Donald J. Cram, Nobel Prize Lecture, 1987

The inspiration referenced in Donald Cram's Nobel Prize lecture results from a sense of awe at the intricate control and complex design reflected in cellular genetics and metabolism. An understanding of metabolism is requisite to prediction of the exchange of many trace gases between organisms and the atmosphere. Metabolism has evolved as a complex web of intersecting biochemical pathways. Control over the flux of metabolites through these pathways requires that they be channeled in the proper direction, partitioned to alternative reactions at pathway intersections, and processed at the proper rate. Metabolic complexity includes network integration, adaptable control, and feedback, which in turn produce non-linear dependencies in the coupling of flux to the cellular environment. In recent decades a new framework, called *systems biology*, has emerged from the fields of metabolic biology and biochemistry to provide new approaches to describing the quantitative complexity of metabolic networks. Systems biology borrows heavily from concepts in the fields of engineering and mathematics, especially those involved with control theory and network integration. New fields of inquiry have emerged from systems biology, carrying names like genomics, transcriptomics, metabolomics, and fluxomics (Sweetlove and Fernie 2005, Steuer 2007).

   In this chapter we will not take up the full topic of systems biology. However, there are clear lessons emerging from this field that are relevant to the control over metabolic fluxes, and that have had an important influence on our ability to understand and predict plant-atmosphere exchanges. For example, it is becoming clear that important feedbacks exist between the rate of CO_2 exchange in the photosynthetic metabolism of chloroplasts and the CO_2 concentration of the atmosphere. These feedbacks are in turn regulated by metabolic interactions that determine the activity of a key enzyme in photosynthetic metabolism known as ribulose 1,5-bisphosphate carboxylase/oxygenase, or Rubisco. Most global-scale models of the carbon cycle now include explicit quantitative descriptions of metabolic control over Rubisco; extending our ability to work with quantitative precision on carbon cycle projections from the scale of chloroplasts to the globe. Our discussion of metabolism will emphasize these connections between metabolic control and the uptake or emission of important constituents of the atmosphere using the processes of photosynthesis and respiration as case studies. By the end of this chapter it should be clear to the reader that it is not possible to develop a useful understanding of earth system exchanges of CO_2 between the land surface and atmosphere without an understanding of the metabolic processes that occur in cellular chloroplasts and mitochondria.

4.1 The principle of shared metabolic control

We begin our discussion of metabolic control with the topic of enzymes, which was introduced in the last chapter. Most enzymes carry out their catalytic roles within the context of metabolic pathways. Metabolic pathways consist of chains of enzyme-catalyzed reactions beginning with an overall pathway substrate and leading to a final pathway product. The field of biochemistry contains a rich history of investigations on the kinetic and regulatory characteristics of isolated enzymes and, on the basis of these studies, many enzymes in the most important pathways of metabolism have been classified as to their contribution toward control over pathway flux; typically as rate-limiting or non-rate limiting enzymes. The conventional view of control over metabolic flux has focused on one to a few key rate-limiting enzymes, or "pacemaker enzymes." These pacemaker enzymes are identified as those that catalyze the slowest steps in a pathway, and they often catalyze the initial reaction in a pathway. Traditionally, they have been identified as those steps across which the substrate to product concentration ratio differs greatly from that defining the equilibrium constant for the reaction (Section 3.1). In other words, during normal metabolic flow through a pathway, the reactions catalyzed by pacemaker enzymes tend to be highly displaced from equilibrium. Pacemaker enzymes also tend to be subject to allosteric control, either through feedback from metabolites of the same pathway or through feedback from metabolites in other intersecting pathways.

By studying pacemaker steps in detail, biochemists have been able to construct metabolic maps and identify key control nodes. The characterization of pacemaker kinetic coefficients, such as those used in the Michaelis–Menten model of enzyme catalysis, has facilitated a "reconstructive" type of analysis by which metabolic flux through pathways has been modeled. More recently it has been recognized that this reconstructive approach, in which metabolic control is assigned to one or a few pacemaker enzymes, causes a priori biases in understanding the overall, synthetic behavior of a pathway. Over the past few decades, new perspectives have emerged in which the "global" or holistic view of a pathway is employed. These perspectives have largely emerged from computational approaches. Using these new approaches, both theoretical and empirical, a concept of *shared metabolic control* has emerged.

One of the more informative computational frameworks concerning metabolic control arose from theory developed by Kacser and Burns (1973) and Heinrich and Rapoport (1974), and is known as *metabolic control analysis* (MCA). As an introduction to this theory, consider a simple, hypothetical pathway in the steady state condition with regard to metabolite pool sizes (Figure 4.1). Metabolites, or substrates (here represented as S_1, S_2, S_3, S_4), are interspersed between coupled reactions catalyzed by a chain of enzymes (here represented as E_1, E_2, E_3) each catalyzing reactions that progress at a defined directional rate (here represented as v_1, v_2, v_3). Reverse reactions are possible for each catalyzed step

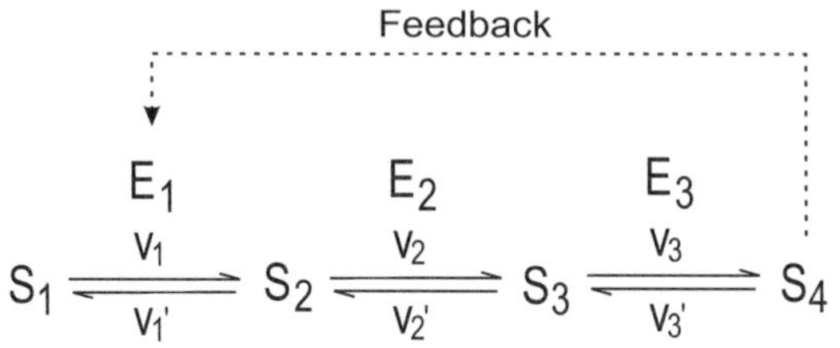

Figure 4.1 A hypothetical pathway of sequential metabolic steps demonstrating the connected relationship between successive reactions.

(here represented as v_1', v_2', v_3'). Positive or negative feedback is also possible and is often due to allosteric interactions. We assume that flux through the system is controlled entirely by biochemical properties of the enzymes, not by diffusion of metabolites to enzyme active sites. We also assume that all metabolites are freely mobile; that is, they are not bound to components of the cell in a way that would alter their availability to enzyme active sites. At steady state the net molar flux through each step of the pathway will equal the flux through the entire pathway (i.e., the rate at which S_4 is formed in units mol s^{-1}). Because each enzyme relies on an upstream reaction to provide substrate and a downstream reaction to remove product, at steady state, no single enzyme can operate faster than the entire pathway. The steps in such a pathway exhibit a high degree of *connectivity*, providing an integrated, systemic nature that reflects the combined contributions of all of its individual reactions.

Perturbation to the velocity of any single reaction in the pathway will be transmitted through the system until the original steady state is established (an outcome referred to as *dynamic stability*), or a new steady state is reached (an outcome referred to as *structural stability*). Consider a hypothetical perturbation whereby v_1 is increased. This would cause the concentration of S_2 to increase, thereby causing v_2 to increase. The increase in v_2, however, will eventually reach a limit. It will only be able to increase up to a rate that is equal to the new v_1. At that point, the flux through E_1 and E_2 will be at a new steady state and the concentration of S_2 will stabilize. This same effect will be transmitted through the chain of enzymes until all reaction rates in the pathway settle on the new steady state. One important assumption required for this scenario is that boundary conditions for the pathway must not limit its flux. In other words, the pathways upstream from S_1 and downstream from S_4 must be sufficient to sustain steady-state concentrations in S_1 and S_4.

Despite equality in reaction rate, every enzyme in a pathway exhibiting steady-state flux does not exert the same degree of control over the flux. The amount of control exerted by an enzyme can be quantified as the degree to which a perturbation to the concentration of the enzyme affects total flux through the pathway. Expressed formally, this relation defines the *flux control coefficient* (C_{Ex}):

$$C_{Ex} = \frac{\partial F/F}{\partial c_{Ex}/c_{Ex}},$$

(4.1)

where F is the molar flux through the pathway (mol s^{-1}) and c_{Ex} is the concentration (mol m^{-3}) of a specific enzyme in the pathway (e.g., E_1, E_2, E_3, etc.). It is assumed that a change in the concentration of the enzyme causes a proportional change in the catalytic activity of that enzyme. By convention, the nomenclature of C_{Ex} is such that the subscript refers to enzyme x (denoted as E_x), where x represents any sequential enzyme in the pathway, e.g., 1, 2, 3, etc. Because the relation described in Eq. (4.1) is defined in terms of a fractional change in pathway flux, control coefficients are dimensionless.

To be faithful to the original derivation of metabolic control analysis, we are using an infinitesimal change in enzyme concentration as the hypothetical perturbation required to quantify C_{Ex}. A change in enzyme activity, rather than concentration, could just as easily be used since enzyme concentration is proportional to enzyme activity under most conditions. In some studies of metabolic control analysis, genetically engineered plants with variable enzyme concentrations have been produced and used to define C_{Ex} in terms of a finite change in c_{Ex} (see Stitt 1994); while useful in its approach it formally violates the calculus of the theory.

The flux control coefficient is dependent on two components of flux. The *local* flux is defined by the individual response of an enzyme to the milieu of metabolites that surround it, and the *systemic* flux is defined by the connected influences of one enzyme on all other enzymes in the pathway (Fell 1992). Local control is ultimately expressed through characteristics of the active and regulatory sites of an enzyme (e.g., the substrate binding affinity, allosteric interactions, and substrate turnover rate). These properties underlie the *elasticity coefficient* ($\varepsilon_j^{v_x}$):

$$\varepsilon_j^{v_x} = \frac{\partial v_x / v_x}{\partial c_j / c_j}, \qquad (4.2)$$

where c_j represents the mole fraction of some metabolite (j) with which the enzyme associates. The nomenclature for the elasticity coefficient, $\varepsilon_j^{v_x}$, follows convention whereby the superscript refers to the rate of the specific reaction being characterized and the subscript refers to the metabolite being considered. The elasticity coefficient is an expression of the relative change in reaction rate given a change in metabolite concentration, holding all other metabolite concentrations constant. Like C_{Ex}, $\varepsilon_j^{v_x}$ is dimensionless and will vary from 0 to 1.

Systemic control over flux is expressed through the combined influences of the flux control coefficients of all enzymes in a pathway. Given the condition of shared control it might be expected that all flux control coefficients would sum to a limit. In fact, it can be shown that for any given pathway the flux control coefficients for the entire pathway must sum to a limit with $\partial F / F = 1$. This is expressed formally as the *summation theorem*:

$$\sum_{x=1}^{n} C_{Ex} = 1, \qquad (4.3)$$

where C_{Ex} represents the control coefficient for each successive enzyme in the pathway, and n is the total number of enzymes in the pathway.

From Eqs. (4.2) and (4.3) it is clear that a required interdependency exists between the systemic properties of pathway flux, as represented by the control coefficient, and the local properties of enzyme function, as represented by the elasticity coefficient. Using this concept, one can define a formal relation between C_{Ex} and $\varepsilon_j^{v_x}$ for sequential reactions that share common metabolites:

$$\frac{C_{E1}}{C_{E2}} = \frac{|\varepsilon_j^2|}{|\varepsilon_j^{1'}|}, \qquad (4.4)$$

where C_{E1} and C_{E2} are the control coefficients for two successive enzymes in a pathway (e.g., E_1 and E_2), and $\varepsilon_j^{1'}$ and ε_j^2 refer to the elasticity coefficients for the reverse reaction of E_1 (indicated as v_1' in Eq. (4.5)) and forward reaction of E_2 (indicated as v_2 in Eq. (4.5)), respectively, with respect to a single shared metabolite S_2.

$$\begin{array}{cc} E_1 & E_2 \\ & v_2 \\ S_1 \underset{v_{1'}}{\rightleftarrows} S_2 & \rightleftarrows S_3 \end{array} \qquad (4.5)$$

Equation (4.4) is known as the *connectivity theorem*. The connectivity theorem reveals that there is an inverse, proportional relationship between the control coefficient for an enzyme and its elasticity coefficient with respect to substrate. Greater control over pathway flux comes with a greater tendency to resist changes in substrate concentration. Taking this interpretation a bit further, enzymes with greater control coefficients tend to be less sensitive to mass action influences; that is, they tend to be further displaced from equilibrium than enzymes with lesser control coefficients.

The quantitative framework of MCA provides the basis for shared control among all enzymes in a pathway. In every case to date where MCA has been applied to a specific pathway calculated control coefficients are less than 1 for each step, indicating shared control. The theoretical underpinnings of MCA do not, however, require *equal* sharing. In most cases where MCA has been applied, the calculated control coefficients are high for relatively few enzymes, and in some cases the control coefficient approaches 1 for a single enzyme. Thus, control may indeed be shared, but it is unequally shared, and to a large extent focused on traditionally identified pacemaker steps. Thus, our journey through the theory of metabolic control analysis has brought us "full circle" back to the conclusions of traditional, reductionist approaches to understanding metabolism – "metabolic control is often concentrated on one or a few key pacemaker enzymes." However, now we are able to base this conclusion on the foundation of formal theory with no a priori assumptions about focused pathway control.

4.2 Control over photosynthetic metabolism

Few pathways are as crucial to understanding the metabolic regulation of plant-atmosphere interactions as that for photosynthetic CO_2 assimilation. Photosynthesis is controlled at a

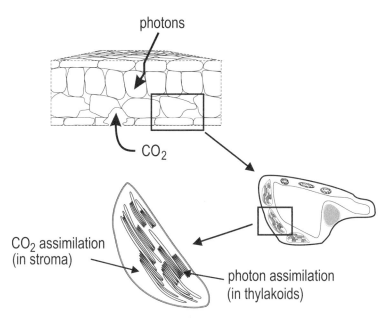

photons

CO_2

CO_2 assimilation
(in stroma)

photon assimilation
(in thylakoids)

Figure 4.2   The chloroplast of terrestrial plants shown within the context of a C_3 leaf mesophyll cell.

number of scales, including those within canopies, leaves, cells, and chloroplasts. In this chapter, we will be most concerned with interactions at the chloroplast scale (Figure 4.2).

4.2.1 The biochemistry of photosynthetic CO_2 assimilation

Photosynthetic CO_2 assimilation depends on the interaction between photon-assimilating and CO_2-assimilating reactions which occur in different parts of the chloroplast (Figure 4.3). Through this interaction the quantized energy carried in solar photons is used to synthesize adenosine triphosphate (ATP) molecules and chemically reduce the compound nicotinamide-adenine dinucleotide phosphate ($NADP^+$) to form NADPH. Ultimately, the energy contained in ATP and NADPH is used to convert CO_2 molecules into sugar-phosphate molecules. The electrons that participate in the C-O bonds of CO_2 exist at low potential energy due to the high electronegativity of O. The enzymatic conversion of three CO_2 molecules to a triose sugar-phosphate requires an increase in the free energy of electrons by 12.5 kJ mol^{-1}.

The rates at which the photon-assimilating and CO_2-assimilating pathways operate are constrained by their mutual interdependencies. The CO_2-assimilating pathway can only operate as fast as the photon-assimilating pathway provides it with ATP and NADPH, the source of potential energy and reductant electrons. The photon assimilating pathway can only operate as fast as the CO_2-assimilating pathway returns $NADP^+$, ADP, and inorganic phosphate (P_i) to the photon-assimilating reactions. The photon-assimilating reactions are composed of (1) *light-harvesting components* (including the chlorophyll protein complexes), which absorb photons of electromagnetic radiation and use the energy to excite electrons, and (2) *electron-transport components*, which facilitate transfer of the excited

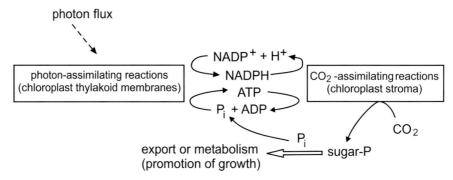

Figure 4.3 A general scheme showing interactions between the photon-assimilating reactions and CO_2-assimilating reactions in photosynthetic CO_2 assimilation.

electrons across thylakoid membranes, and energize the synthesis of ATP, and ultimately NADPH. In sum, the photon-assimilating reactions: (1) extract two electrons with relatively low potential energy from H_2O (producing half of an O_2 molecule); (2) energize the two electrons using photons of electromagnetic radiation; and (3) facilitate an oxidation-reduction reaction in which the energized electrons are transferred to a molecule of $NADP^+$, thus forming NADPH. The electrons carried in NADPH exist at a relatively high potential energy level. In the process of moving the electrons from H_2O to $NADP^+$ some of the free energy from the absorbed photons is extracted and used to synthesize ATP, a second potential energy source used in CO_2 assimilation.

The CO_2-assimilating reactions are known variably as the reductive pentose phosphate pathway, the C_3 pathway, or the Calvin–Benson cycle. In this book, these reactions will be referred to as the *reductive pentose phosphate pathway*, or RPP pathway (Figure 4.4A). The overall result of the RPP pathway is that one triose-phosphate sugar molecule is produced as a net gain for every three molecules of assimilated CO_2. The initial assimilation of each molecule of CO_2 occurs through a reaction catalyzed by the enzyme RuBP carboxylase/oxygenase (Rubisco) (represented by step 1 in Figure 4.4A) which resides in the stroma of the chloroplast. In this initial reaction, Rubisco facilitates the formation of a covalent bond between CO_2 and a five-carbon, sugar-phosphate molecule, ribulose 1,5-bisphosphate (RuBP). The initial product formed from the carboxylation of RuBP is an unstable, six-carbon "intermediate" compound that is spontaneously and immediately rearranged to form two molecules of the *three-carbon*, 3-phosphoglyceric acid (PGA), the first *stable* product of the Rubisco step; thus, the label "C_3-cycle" that is often used to describe the pathway. Following the initial assimilation of CO_2, other enzymes in the pathway utilize one ATP molecule and one NADPH molecule produced from the photon-assimilating reactions to chemically reduce each PGA molecule and form glyceraldehyde 3-phosphate (G3P), a triose-phosphate sugar (represented by steps 2 and 3). Some of the triose-phosphate sugar molecules are exported from the chloroplast to produce molecules of the disaccharide sugar sucrose (composed of a glucose and fructose molecule bonded together) in the cytoplasm and, in the process, release inorganic phosphate (P_i) that is recycled and transported back

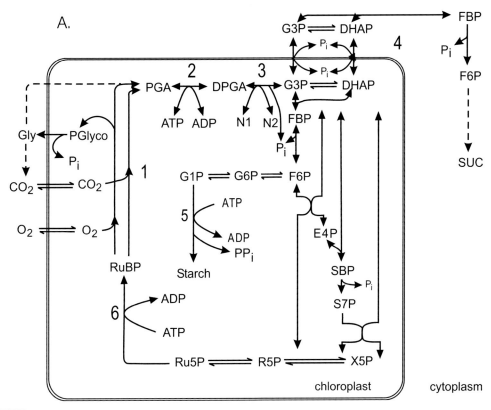

Figure 4.4   **A**. The reductive pentose phosphate pathway. The numbered steps (1 through 6) coincide with the descriptive sequence provided in the text. Abbreviations for intermediate metabolites are as follows: Gly = glycine, PGlyco = phosphoglycolate, PGA = 3-phosphoglyceric acid, DPGA = 1,3-diphosphoglyceric acid, G3P = glyceraldehyde 3-phosphate, DHAP = dihydroxyacetone phosphate, FBP = fructose 1,6-bisphosphate, F6P = fructose 6-phosphate, E4P = erythrose 4-phosphate, SBP = sedoheptulose 1,7-bisphosphate, S7P = sedoheptulase 7-phosphate, X5P = xyulose 5-phosphate, R5P = ribose 5-phosphate, Ru5P = ribulose 5-phosphate, RuBP = ribulose 1,5-bisphosphate, G1P = glucose 1-phosphate, G6P = glucose 6-phosphate, SUC = sucrose, N1 = NADP$^+$, N2 = NADPH, P$_i$ = inorganic phosphate, PP$_i$ = inorganic pyrophosphate. Redrawn from Woodrow and Mott (1993), with kind permission from Springer Science + Business Media. **B**. Simplified view of the RPP pathway showing its cyclic nature and the role of the photon-assimilating reactions in regenerating the original CO_2 acceptor molecule, ribulose 1,5-bisphosphate (RuBP). Balanced stoichiometry requires that one mole equivalent of product be extracted from the cycle (as sugar) for each mole of CO_2 assimilated.

into the chloroplast for further use (represented by step 4). The triose phosphate molecules (each containing three carbon atoms) that remain in the chloroplast and are not converted to sucrose are restructured in a series of enzyme-catalyzed reactions to form a variety of intermediate compounds, and eventually reform RuBP, the original CO_2 acceptor. An additional ATP molecule is used in the final step of the process (represented by step 6), bringing the total ATP cost to 3 and the total NADPH cost to 2 for each CO_2 molecule that is assimilated. Thus, the RPP pathway is truly "cyclic" in the sense that the original acceptor of

B.

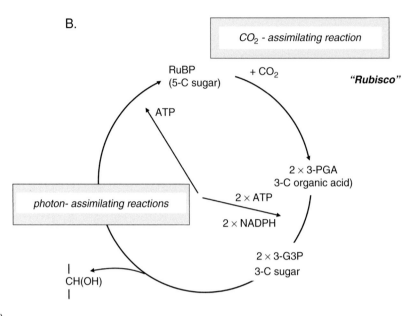

Figure 4.4 (cont.)

CO_2 (RuBP) is ultimately regenerated (Figure 4.4B) and each cycle resets the pathway in preparation for assimilating another CO_2 molecule. The maintenance of flux through the cycle requires the proper partitioning of sugar phosphate product between that exported to the cytoplasm (to form sucrose) and that remaining in the chloroplast (to regenerate RuBP) (see Woodrow and Berry 1988 for a review of the primary controls over flux through the RPP pathway).

4.2.2 The transfer of CO_2 between the gaseous (atmospheric) and aqueous (cellular) phases during photosynthesis

Photosynthetic fluxes require the transfer of CO_2 molecules from the gaseous phase in the atmosphere to the aqueous phase of leaf photosynthetic cells and ultimately to the active site of Rubisco in the chloroplast. Dissolved CO_2 is the primary form of inorganic carbon assimilated by Rubisco. Gases dissolve in the aqueous phase of cells according to *Henry's Law*. At equilibrium, Henry's Law predicts that the dissolved concentration of a constituent in the aqueous phase is proportional to its partial pressure or mole fraction in the adjacent gas phase. Thus, for CO_2:

$$\frac{RT c_{wc}}{P} = \frac{p_{gc}}{P k_H} = \frac{c_{gc}}{k_H}, \tag{4.6}$$

where c_{wc} is the molar concentration (mol liter^{-1}) of CO_2 dissolved in the liquid (water) phase, p_{gc} is the partial pressure of CO_2 in the gas phase (kPa), k_H is the Henry's Law coefficient for CO_2 (kPa liter mol^{-1}), P is total (atmospheric) pressure in the gas phase (kPa), and c_{gc} is the mole fraction of CO_2 in the gas phase (mol CO_2 mol^{-1} dry air). Compounds that have a higher k_H value partition into the aqueous phase at higher dissolved concentrations than those with a lower k_H value, when compared at equal concentrations in the adjacent gaseous phase. When written in the form of Eq. (4.6) and at 25 °C, k_H carries the value of 29.4×10^2 kPa liter mol^{-1}. In studies of photosynthesis, Eq. (4.6) is used to express the *molar solubility* of gaseous CO_2 into the aqueous phases of a cell. As the temperature of a leaf is increased, the k_H for CO_2 also increases and its molar solubility decreases.

Once in the cell a fraction of dissolved CO_2 ($CO_{2(w)}$) reacts with water, forming bicarbonate according to $CO_{2(w)} + H_2O \Leftrightarrow HCO_3^- + H^+ \Leftrightarrow CO_3^{2-} + 2H^+$. The dissociation reactions on the right-hand side of this equation occur nearly instantaneously. The hydration-dehydration reactions on the left-hand side occur more slowly. The final equilibrium with respect to $CO_{2(w)}$ and HCO_3^- depends on pH, favoring $CO_{2(w)}$ at acidic pH and HCO_3^- at basic pH (at pH 6 approximately 75% of the total inorganic carbon exists as $CO_{2(w)}$ and at pH 8 approximately 80% of the total inorganic carbon exists as HCO_3^-). The stroma of the chloroplast, where Rubisco occurs and $CO_{2(w)}$ is assimilated, typically exists at a pH near 8 during photosynthesis, meaning that most of the inorganic carbon exists as HCO_3^-. Given this situation, the conversion of HCO_3^- to $CO_{2(w)}$ plus H_2O, and subsequent diffusion of $CO_{2(w)}$ to the active site of Rubisco, is a rate-limiting photosynthetic process. In order to relieve this limitation, catalytic potential has evolved in the form of *carbonic anhydrase*, a zinc-containing enzyme capable of accelerating the $CO_{2(w)}/HCO_3^-$ equilibrium. Carbonic anhydrase occurs at relatively high activities in the chloroplasts of C_3 plants and ensures a constant supply of $CO_{2(w)}$ from HCO_3^- to the active sites of Rubisco.

4.2.3 The nature of Rubisco

The entrance of $CO_{2(w)}$ into the RPP pathway occurs by way of Rubisco. As enzymes go, Rubisco is large. It is composed of eight "large" subunits encoded by chloroplast genes and eight "small" subunits encoded by nuclear genes. Each large subunit contains an active site and allosteric regulatory sites. The small subunits are not catalytically active, but rather modify the activity of the large subunits. As enzymes go, Rubisco is also catalytically inefficient, at least with regard to its activity at current atmospheric CO_2 concentrations (Morell *et al.* 1992). The active site of Rubisco assimilates CO_2 with a Michaelis–Menten coefficient (analogous to the K_m, see Section 3.4.1, but often denoted as K_c) that is slightly less than the current atmospheric CO_2 concentration. The CO_2 concentration in the chloroplast is even less than that in the atmosphere. This means that in normal atmospheric conditions Rubisco is functioning at less than half its maximum catalytic velocity. Additionally, the turnover rate at each active site of Rubisco is slow (approximately 1–3 conversions per second depending on temperature and CO_2 partial pressure). The large size and inefficient catalytic function of Rubisco, along with its central role in assimilating CO_2, requires that large amounts of the enzyme exist in plants. In C_3 plants up to 35% of a leaf's N can be accounted for in Rubisco. In pure crystalline form Rubisco active sites exist at a concentration of approximately 10 mM.

In a typical C_3 plant chloroplast Rubisco active sites exist at a concentration of 1–4 mM. Although some storage proteins exist at concentrations in the mM range, it is rare for this to be the case for catalytic proteins; once again highlighting the importance of this large and inefficient protein to C_3 plant function. For a review of Rubisco, including its structural and regulatory properties, see Spreitzer and Salvucci (2002).

Rubisco is a highly regulated enzyme as might be expected given its role as the "gatekeeper" to photosynthetic CO_2 assimilation. Rubisco exists in active and inactive states. The inactive form is converted to the active form through the activity of an ATP-dependent regulatory enzyme, *Rubisco activase*, followed by the addition of a CO_2 molecule (carbamylation) to a critical amino acid residue in the active site, and finally the addition of a Mg^{2+} ion to the active site. When both CO_2 and Mg^{2+} are present, the enzyme is in the final active form. The active form has a finite lifetime and will spontaneously revert to the inactive form after several catalytic cycles when the active site loses its CO_2 activator molecule. Steady-state photosynthetic CO_2 assimilation can only be sustained through continual re-activation of Rubisco. (It is important to recognize that the CO_2 activator molecule is different than the CO_2 molecules used for substrate at the active site of Rubisco.) There is evidence that Rubisco activation state, as controlled by Rubisco activase enzyme, has a role in limiting the overall rate of CO_2 assimilation at high leaf temperatures (e.g., greater than 30 °C; Sage *et al.* 2008).

Once Rubisco is active, RuBP substrate is coordinated through electrical charge interactions to the Mg^{2+} ion residing in the enzyme's active site. Binding of RuBP, and the strain placed on its bonds as it aligns with the Mg^{2+}, causes it to deprotonate and form a reactive *enediolate intermediate structure*. This intermediate form of RuBP is receptive to forming a new covalent bond with a substrate CO_2 molecule, thus initiating CO_2 assimilation. Amino acid side chains in the active site do not actually bind the substrate CO_2 molecule and transfer it to the enediolate intermediate (which is the more typical "Michaelis–Menten type" of catalysis); rather, the substrate CO_2 adds spontaneously to the reactive enediolate intermediate structure. In essence, the enzyme creates a reactive form of RuBP, holds it in place, and allows it to react with dissolved CO_2. As a biochemical entity, a CO_2 molecule has few properties (e.g., polarity, asymmetric structure) which might permit formation of a properly bound enzyme-substrate complex. In the words of some researchers, the CO_2 molecule is biochemically "featureless" (Pierce *et al.* 1986, Tcherkez *et al.* 2006). This means that the only catalytic mechanism by which Rubisco can favor carboxylation of the enediolate intermediate is through the placement of amino acid side chains in the active site that favor a stable, carboxylated transition state. A thermodynamic analysis has revealed that stabilization of the carboxylated RuBP-CO_2 transition state requires such tight binding of the RuBP-enediolate intermediate to the active site that it slows cleavage to form the ultimate three-carbon products (two PGA molecules) (Tchkerz *et al.* 2006). Thus, evolutionary modification of the active site has been constrained by the tradeoff between "favoring" reaction with dissolved CO_2 substrate and turning over the carboxylated transition state to form product (see Section 3.4.1 for a discussion of tradeoffs and compromises in enzyme K_m and k_{cat}; also see Gutteridge and Pierce 2006).

In some cases the reactive RuBP-enediolate intermediate forms a bond with O_2, rather than CO_2. Oxygenation of the enediolate is due to the evolution of an imperfect catalytic mechanism at the active site; it is an unintended consequence of creating a reactive form of RuBP and then allowing it to react with atmospheric constituents. The potential for O_2 assimilation within the active site of Rubisco explains the word "oxygenase" in ribulose 1,5-bisphosphate carboxylase/oxygenase (i.e., the "o" in Rubisco; see Portis and Parry 2007). As substrates, CO_2 and O_2 have competitive access to the enediolate intermediate at Rubisco's active site. However, the catalytic affinity for CO_2 ($K_c \sim 300$ µmol mol^{-1} at 25 °C) is considerably greater than that for O_2 ($K_o \sim 420$ mmol mol^{-1} at 25 °C). The maximum rate of oxygenation catalyzed by Rubisco is approximately 25% the maximum rate of carboxylation. Thus, the rate of oxygenation can approach the rate of carboxylation only in circumstances when the CO_2 concentration is quite low compared to the O_2 concentration. For plants that utilize the RPP pathway to assimilate CO_2 in the normal atmosphere (~ 380 µmol mol^{-1} CO_2, 210 mmol mol^{-1} O_2), and at moderate temperatures (25–30 °C), carboxylation proceeds at 4–5 times the rate of oxygenation.

The degree to which the active site of Rubisco favors carboxylation over oxygenation can be described through a ratio of the relative kinetic parameters governing both reactions. The *relative specificity* of Rubisco activity for CO_2 versus O_2 (denoted as S_{rel}) is expressed as:

$$S_{rel} = \frac{V_{cmax}K_o}{V_{omax}K_c} \qquad (4.7)$$

where V_{omax} and V_{cmax} are the maximum rates of oxygenation and carboxylation by Rubisco. (As an exercise in the application of Michaelis–Menten kinetics, and to better understand the foundation for Eq. (4.7), the reader should attempt to derive S_{rel} as the ratio of first-order dependencies of carboxylation rate on CO_2 mole fraction and oxygenation rate on O_2 mole fraction.) During the past history of the earth, terrestrial plants have evolved changes in the active site of Rubisco that allow S_{rel} to track changes in atmospheric CO_2 and O_2 concentrations (Jordan and Ogren 1981). Current models suggest that the earth's atmosphere contained CO_2 concentrations that were many times greater than today's atmosphere and O_2 concentrations that were low until approximately 2–3 billion years ago. Rubisco from cyanobacteria, which retains many aspects of the primitive state of the enzyme, exhibits an S_{rel} value of 40–60. Rubisco from terrestrial C_3 plants, which reflects evolutionary change since atmospheric CO_2 concentrations have decreased and O_2 concentrations have increased, exhibits an S_{rel} value of 75–85. Thus, while the oxygenase activity has not been abolished through evolutionary modification, the active site of Rubisco has at least evolved in a way that increases its capacity to assimilate CO_2 relative to O_2 in response to historic atmospheric changes in the relative concentrations of these compounds. The evolutionary modifications required of the active site to alter CO_2 and O_2 specificity are likely to be small. One study has shown that substitution of a single amino acid within the large subunit of Rubisco, through site-directed mutagenesis, is sufficient to alter the value of S_{rel} by up to 4-fold (Whitney *et al.* 1999). Temperature is known to have a strong influence on the relative rates of carboxylation and oxygenation catalyzed by Rubisco. An increase in temperature causes a decrease in S_{rel}. The temperature effect is due to greater increase in K_c relative to K_o, and greater reduction in the solubility of CO_2 compared to O_2, as temperature is increased (Figure 4.5).

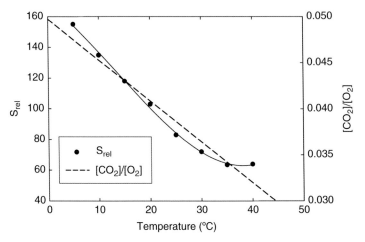

Figure 4.5   Relations among temperature, the Rubisco specificity factor (S_{rel}), and the relative concentrations of CO_2 and O_2 in solution. Compiled from data reported in Jordan and Ogren (1984) and Sage and Reid (1994).

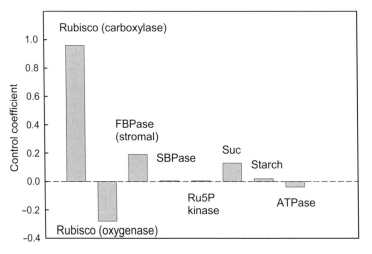

Figure 4.6   Calculated steady state, flux control coefficients for several steps of the RPP pathway in the presence of normal atmospheric CO_2 concentration ($\sim$ 345 µmol mol^{-1}). Suc and Starch refer to the composite processes of sucrose synthesis and starch synthesis, respectively, rather than to single enzyme steps. Redrawn from Woodrow and Mott (1993), with kind permission from Springer Science + Business Media.

4.2.4   Rubisco activity and RuBP regeneration as primary controls over the photosynthetic CO_2 flux

Metabolic control analysis has been applied to the RPP pathway in a handful of cases (Woodrow 1986, Stitt 1996, Poolman *et al.* 2000, Kreim and Giersch 2007). One key finding of these studies is that at the current atmospheric CO_2 concentration, control over flux is focused on one enzyme, Rubisco (Figure 4.6). In this condition, the control coefficient for carboxylation approaches unity. Smaller, positive control contributions are provided by stromal fructose 1,6-bisphosphatase and the cytoplasmic synthesis of sucrose. The oxygenase activity of Rubisco exerts negative control; meaning that the velocity of the net CO_2 assimilation rate decreases as the velocity of the oxygenase reaction increases.

The high degree of control by Rubisco is sensitive to the relative availabilities of CO_2 and RuBP. As CO_2 concentration is increased the velocity of the CO_2 assimilation rate will increase, the availability of RuBP will decrease, and carboxylation rate will be progressively limited by RuBP availability (Figure 4.7). As CO_2 concentration increases a shift occurs in the calculated control coefficients of the RPP pathway from a state of dominant control by

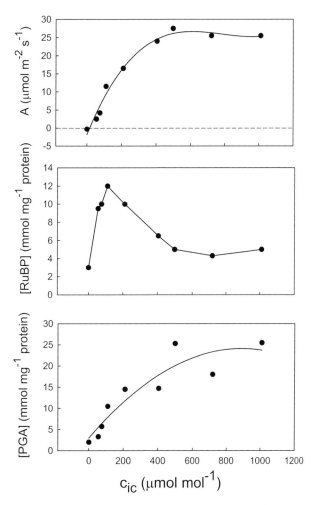

Figure 4.7 The relationship of leaf net CO_2 assimilation rate (A), extracted RuBP concentration, and extracted PGA concentration to the intercellular CO_2 mole fraction (c_{ic}) for leaves of *Phaseolus vulgaris* (common pole bean). As CO_2 mole fraction increases, the velocity of RuBP carboxylation increases (as reflected in greater values for A), the concentration of RuBP, the CO_2 acceptor molecule, decreases, and the concentration of PGA, the first stable product of RuBP carboxylation, increases. Redrawn from Badger *et al.* (1984), with kind permission from Springer Science + Business Media.

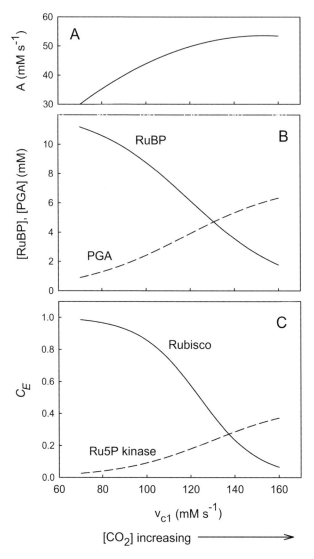

Figure 4.8 Modeled changes in the net CO_2 assimilation rate (A) by the RPP pathway (panel A), the stromal concentrations of RuBP and PGA (panel B), and control coefficient (C_E) for Rubisco and Ru5P kinase (panel C) as a function of the Rubisco carboxylation rate (v_{c1}). The model was driven by increasing CO_2 concentration, which forced the increase in v_{c1}. Redrawn from Woodrow (1986), with permission from Elsevier B.V. Publishers.

Rubisco to a state of dominant control by other enzymes, particularly those that control the rate of RuBP regeneration (Figure 4.8). For example, the enzyme ribulose 5-phosphate (Ru5P) kinase, which catalyzes the synthesis of RuBP, exhibits a greater control coefficient as CO_2 concentration increases and RuBP concentration decreases, and Rubisco exhibits a lower control coefficient. Transgenic experiments with tobacco plants that have produced an increased activity of the enzyme sedoheptulose 1,7-bisphosphatase (SBPase), which is also involved in the regeneration of RuBP, have shown an increase in overall plant production of 23% when compared to wild-type plants grown at elevated atmospheric CO_2 concentrations (585 µmol mol^{-1}; Rosenthal *et al.* 2011). (Interestingly, transgenic increases in SBPase activity also increased production (by 12%) at normal, ambient CO_2 concentrations (385 µmol mol^{-1}), suggesting that the CO_2 concentration of the current atmosphere has already increased to a level that causes greater control by RuBP regeneration than may have occurred in the past.)

One interesting control dynamic is evident in the large swing in the control coefficient of cytosolic sucrose synthesis, from negative to positive, as a result of changes in RuBP concentration (Figure 4.9). When RuBP concentration is low it is likely the result of high rates of CO_2 assimilation, which consumes RuBP and produces triose-phosphate sugars. In this condition an increase in sucrose synthesis rate will cause a decrease in flux through the RPP pathway. This is due to the fact that sucrose synthesis releases P_i as triose-phosphate sugars are condensed to form sucrose. The freed P_i is available to be transported back into the chloroplast in exchange for the continued export of triose-phosphate sugars and continued synthesis of sucrose (see step 4 in Figure 4.4A). The outward flux of triose-P sugars drains the chloroplast of carbon compounds that are needed for the RuBP regeneration phase of the RPP pathway, and thus provides the basis for the negative control coefficient. The state of high RuBP concentrations likely occurs when CO_2 assimilation rate is low, relative to the rate of generation of ATP and NADPH by the photon-harvesting reactions. In this condition the availability of carbon compounds required for RuBP regeneration is not as likely to limit the rate of RPP flux as is P_i (which is tied up in the RuBP pool). When P_i is tied up in photosynthetic metabolite pools the rate of ATP synthesis from ADP + P_i will be subject to strong feedback control, and eventually the steady-state flux of compounds

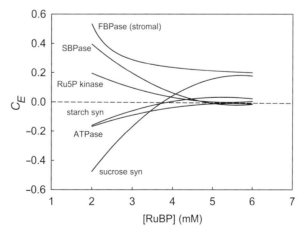

Figure 4.9   Changes in the control coefficients (C_E) of enzymes in the RPP pathway associated with processes other than Rubisco carboxylation and oxygenation as a function of stromal RuBP concentration. Note that the control coefficients of those enzymes that are involved in the regeneration of RuBP (FBPase, SBPase, and Ru5P kinase) exhibit high control over flux at low RuBP concentrations. Redrawn from Woodrow and Mott (1993), with kind permission from Springer Science + Business Media.

through the RPP pathway will be limited by the rate at which P_i is freed and made available for ATP biosynthesis. Thus, an increase in sucrose synthesis at high concentrations of RuBP will provide a supply of rate-limiting P_i to be transported back into the chloroplast and increase flux through the RPP pathway; providing the basis for the positive control coefficient. A similar control scenario can be provided for starch synthesis, although in that case the most relevant processes reside in the chloroplast.

4.3 Photorespiratory metabolism

The metabolic product of each oxygenation at the active site of Rubisco is one molecule of PGA (which is processed through the RPP pathway just as for the PGA produced by carboxylation) and one molecule of the two-carbon compound, phosphoglycolate (PGL). In terrestrial plants an independent metabolic pathway involving three different cellular organelles (chloroplasts, peroxisomes, and mitochondria) has evolved to process the PGL (Figure 4.10). This pathway is referred to as *photorespiration*. Through photorespiration

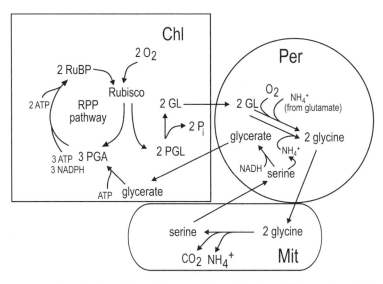

Figure 4.10 The photorespiratory pathway involving three cellular organelles – chloroplasts (Chl), peroxisomes (Per), and mitochondria (Mit). Photorespiration is initiated when the oxygenation of RuBP, which is catalyzed by Rubisco, produces phosphoglycolate (PGL) in the chloroplast. The phosphoglycolate is converted to glycolate and transported to the peroxisome where it is converted to the amino acid, glycine. Glycine is transported to the mitochondrion where two glycines (each with two carbons) are combined to form one serine (with three carbon atoms) and one CO_2 molecule. The serine is then transported back through the peroxisome where it is converted to glycerate, and then back into the chloroplast, where it is converted to phosphoglycerate (PGA), which then enters the RPP pathway. Thus, photorespiration provides an effective metabolic means of recycling some of the carbon atoms in PGL, which is produced through oxygenation of RuBP, to form a compound, PGA, capable of re-entering the RPP pathway.

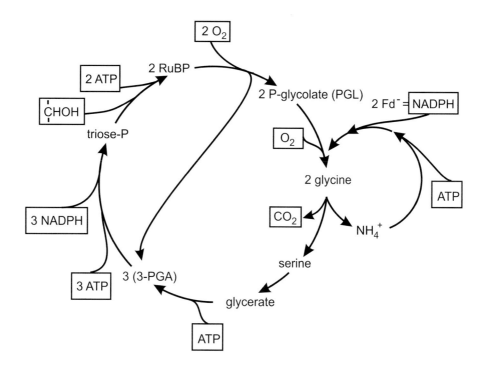

Summary: -3.5 ATP, -2 NADPH, -1.5 O_2, + 0.5 CO_2 per oxygenation

Figure 4.11 Stoichiometric relationships of the photorespiratory carbon oxidation (PCO) cycle. For summation purposes, the stoichiometry is determined on the basis of two oxygenations. Important stoichiometric steps are: (1) the consumption of 1 ATP and 1 NADPH to reassimilate NH_4^+; (2) the production of 1 CO_2 in the conversion of two glycines to one serine; (3) the consumption of 1 ATP to convert glycerate to P-glycerate (PGA); (4) the consumption of 3 ATP and 3 NADPH to reduce the 3 PGA molecules (two produced during the oxygenation and one produced from the oxidation of two PGL molecules to form one PGA molecule); (5) the consumption of 2 ATP in the final regeneration of RuBP; and (6) the required consumption of one unit of reduced carbon to maintain carbon balance in the cycle. This latter step reflects the fact that 25% of the carbon that enters the photorespiratory pathway is lost as CO_2; only 75% is recycled back to the RPP pathway. Thus, the stoichiometry must reflect the continual input of 1 unit of reduced carbon per two oxygenations to maintain steady-state balance. Redrawn from Berry and Farquhar (1978).

three-quarters of the carbon contained in PGL are converted to PGA, which then re-enters the RPP pathway (Figure 4.11). Thus, photorespiration can be viewed as a *carbon recovery process*. For every two oxygenations of RuBP the plant recovers one molecule of PGA through photorespiration. The photorespiratory cost of converting the two PGLs formed from the two oxygenations back to one PGA molecule is 2 ATP and 1 NADPH. If the two PGL molecules produced by the two oxygenations were discarded (an alternative used by some algae), and newly assimilated CO_2 (from the atmosphere) were used instead of two PGL molecules to construct a single PGA molecule, the cost would be 9 ATP and 6 NADPH (the cost of assimilating each of the three CO_2 molecules required to build a single PGA molecule is 3 ATP and 2 NADPH). Furthermore, some amount of H_2O would have to be lost from the leaf in order to facilitate the uptake of atmospheric CO_2. Thus, the photorespiratory recovery of PGL is advantageous to the carbon, water, and energy balance of terrestrial plants, at least compared to the alternative of discarding the PGL; but the recovery is not ideal in that one-quarter of the carbon contained in the PGL is lost as CO_2. If we take this cost analysis several steps further, and consider the costs of recycling three PGA molecules (produced through two oxygenations) all the way back to RuBP, thus completely correcting the "mistake" of oxygenation, the energetic tally sums to 7 ATP and 4 NADPH, plus one CO_2 (refer to Figure 4.11). In that case, one carbon equivalent must be added from outside the pathway for stoichiometric balance. (Keep in mind for the purpose of comparison that it takes only 5 ATP and 3 NADPH, plus one additional carbon equivalent, to send three molecules of PGA produced through carboxylation back into the RuBP pool.) It is clear that, in the overall balance, the "mistake" of oxygenation is costly to the plant, in terms of energy, reducing equivalents, and reduced carbon, but that photorespiration at least reduces that cost compared to the case of not correcting the "mistake." In this context, photorespiration can be viewed as a plant adaptation that has resulted from positive selection on productivity in the presence of an imperfect Rubisco enzyme.

Photorespiration rates increase with increased temperature and decrease with increased atmospheric CO_2 concentration (Figure 4.12), and these effects are explained by the CO_2- and temperature-dependent changes in S_{rel} that were discussed above. As the earth's atmosphere continues to change, with increases in both CO_2 concentration and temperature, it will be difficult to predict the ultimate effect on photorespiration and global primary productivity. However, any predictions that are made will have to rely on our understanding of the kinetic properties of the carboxylase and oxygenase reactions.

4.4 Tricarboxylic acid cycle respiration ("dark respiration") in plants

"Dark respiration" (R_d) refers to non-photorespiratory mitochondrial respiration which, despite the name, occurs in both the dark and light. During R_d, CO_2 is derived from the enzymatic oxidation of carbon substrate by pyruvate dehydrogenase, a large mitochondrial protein complex that links the glycolytic metabolic pathway to the tricarboxylic acid (TCA) pathway (also called the Krebs cycle), and two enzymatically catalyzed steps in the TCA pathway itself (Figure 4.13). Several possible substrates can be utilized by enzymes in the TCA cycle, including carbohydrates, proteins, and fatty acids; although carbohydrates are used most frequently.

Terrestrial plant R_d represents a large flux of CO_2, both in terms of absolute magnitude and as a fraction of photosynthesis. In a survey of past literature, Amthor (2000) concluded that R_d represents 30–50% of gross photosynthesis minus photorespiration in crop ecosystems and 50–75% in natural ecosystems. The amount of CO_2 lost through R_d is dependent on plant size, the ratio of photosynthetic to non-photosynthetic biomass, growth rate, and

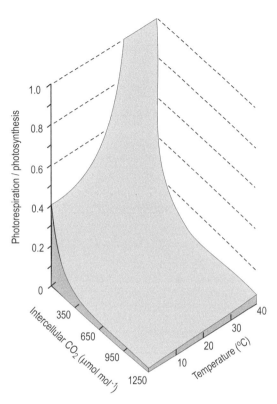

Figure 4.12 The relationships among the ratio of photorespiration to photosynthesis, leaf temperature, and leaf intercellular CO_2 mole fraction. The photorespiration to photosynthesis ratio reaches a limit of 1.0 because it is assumed that photorespiration rates cannot be sustained over time if the rate of CO_2 loss is greater than the rate of CO_2 uptake. Redrawn from Sage and Reid (1994).

recent environmental history. Large, actively growing plants, with a high proportion of non-photosynthetic, but metabolically active biomass (e.g., stems and roots), growing in warm climates will generally exhibit high whole-plant respiration rates. Plants exposed to environmental stress can exhibit increased or decreased R_d during the stress itself, depending on the stress, but they typically exhibit increased R_d following stress due to enhanced rates of cellular repair. The highest respiration-to-photosynthesis ratios are found in saline ecosystems where plants exhibit high rates of energy-driven transport as they sustain ionic balance within cells.

The CO_2 efflux from plant R_d is often divided into two components: growth respiration (R_g) and maintenance respiration (R_m), where $R_d = R_g + R_m$. *Growth respiration* represents CO_2 lost during the generation of energy used to support the addition of new biomass. *Maintenance respiration* represents CO_2 lost during the generation of energy required for the maintenance of tissues. In fully expanded leaves the CO_2 efflux due to growth respiration will be minimal; in such leaves, $R_d \approx R_m$. In cellular terms, R_m is the CO_2 cost of providing energy for protein synthesis, ion transport, and repair of cellular constituents. Enzymes have a finite lifetime,

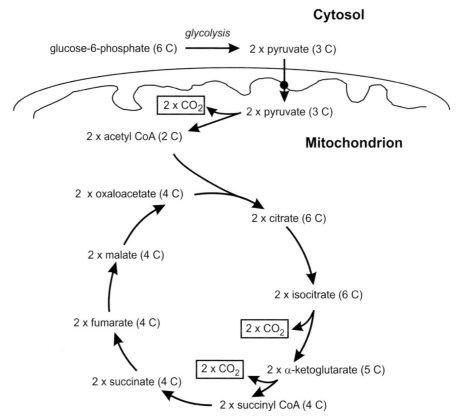

Figure 4.13 Carbon fluxes in the mitochondrial TCA pathway (Krebs cycle) of plant cells. Beginning with glucose 6-phosphate, the catalyzed reactions of glycolysis in the cytosol, and pyruvate dehydrogenase and the TCA cycle in the mitochondrion result in the oxidation of six carbon atoms to form CO_2. Energetic fluxes in the pathway are not explicitly shown, although it should be recognized that primary energetic products of the pathway include high-energy electron carriers (NADH and $FADH_2$) and ATP, both of which reflect oxidative free energy extraction from pyruvate.

which is partly influenced by genetically programmed cellular processes that break down old proteins and replace them with new proteins, and in part by damage that occurs to proteins due to the oxidative environment of cells. Enzyme replacement requires expenditure of energy, and in leaves protein turnover is the primary sink for the ATP energy generated by R_m.

The respiration rate (R_d) for whole plants has been shown to vary with biomass according to a simple scaling function across a broad range of plant species and functional types (Reich *et al.* 2006, Mori *et al.* 2010). This relation typically takes the form:

$$R_d = \frac{k_1 W_p^f}{W_p} \tag{4.8}$$

where R_d is in units of $\mu mol\ CO_2\ g^{-1}$ biomass s^{-1}, given that k_1 is a proportionality coefficient ($\mu mol\ CO_2\ s^{-1}\ g^{-f}$), f is referred to as the scaling exponent and is unitless, and W_p is total plant biomass (g). In small plants, particularly herbaceous plants, in which most of the biomass is metabolically active, $f \sim 1.0$, thus reflecting convergence of the unit-biomass metabolic rate of living tissues to k_1 across a broad range of plant types (Reich *et al.* 2006). In larger plants, such as trees and woody shrubs, in which the plant's biomass is distributed between both living and non-living tissues, f has been estimated to be ~ 0.75, reflecting the absence of metabolic activity and respiration from the non-living fraction of the biomass (Mori *et al.* 2010). The scaling of metabolic rates with biomass according to ¾ exponential relationships is consistent with general patterns of metabolic scaling within the biological realm, including those for animals. Near-universal scaling patterns, such as those discussed for respiration reflect natural allometric and biochemical constraints and have tremendous potential for simplifying the theoretical underpinnings of quantitative flux models.

Respiratory CO_2 losses from ecosystems occur during the microbial decomposition of soil organic matter, in addition to plant respiration. Given that the carbon in soil organic matter is ultimately derived from the gross primary productivity of plants, there are clear conceptual connections between respiration and photosynthesis. However, because the topic of microbial decomposition, and its associated carbon flux, is so fundamentally different from that of plant photosynthesis and respiration, we will address the topic of CO_2 fluxes that originate from soil decomposition, and linkages to climate, in a separate chapter (Chapter 16).

4.5 C_4 photosynthesis

To this point, we have considered plants that solely possess the C_3 photosynthetic pathway. Plants of this photosynthetic type account for most of the terrestrial plant biomass on the earth. However, plants with the C_4 photosynthetic pathway are also important in key ecosystems. Approximately 20% of the earth's gross primary productivity is channeled through C_4 photosynthesis (Still *et al.* 2003). The C_4 photosynthetic pathway has evolved in flowering plants in response to historical declines in the atmospheric CO_2 concentration and increase in the atmospheric O_2 concentration, both of which promote high photorespiration rates in low latitude ecosystems with frequent warm temperatures (Ehleringer and Monson 1993). The C_4 pathway has evolved over 60 times independently, a fact which by itself suggests that the selective pressures of enhanced photorespiration rates have had a significant negative influence on plant fitness in certain environments.

The fundamental advantage of C_4 photosynthesis is that Rubisco is localized within a special type of cell (the bundle-sheath cells), into which CO_2 is "pumped" using an energy-driven metabolic process. This metabolic "pumping" results in CO_2 concentrations in bundle-sheath cells that are several times greater than in the ambient atmosphere. Thus, in C_4 leaves, Rubisco exists in a cellular microenvironment that resembles the earth's primitive atmosphere of high CO_2 concentration. The elevated CO_2 concentration of bundle-sheath cells enhances the CO_2 assimilation rate by C_4 Rubisco, compared to C_3 Rubisco.

The C_4 pathway reflects coordinated metabolic function between two leaf cell types, mesophyll cells (MCs) and bundle-sheath cells (BSCs). Some of the key steps in C_4

photosynthesis include: (1) assimilation of inorganic carbon (as HCO_3^-) in MCs by the enzyme phosphoenolpyruvate (PEP) carboxylase; (2) transfer of the resulting 4-C organic acids to the BSCs; (3) decarboxylation of the 4-C acids in the BSCs to form an internal source of dissolved CO_2; (4) assimilation of the decarboxylated CO_2 by Rubisco, which is

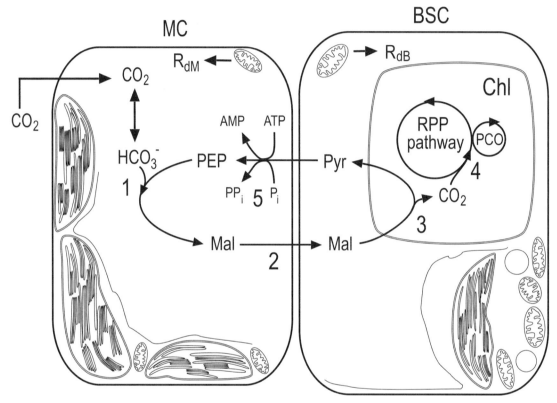

Figure 4.14   The C_4 pathway of CO_2 assimilation as represented by a C_4 plant exhibiting the NADP-malic enzyme type of decarboxylation. The steps are: (1) inorganic carbon is initially assimilated as HCO_3^- by the enzyme PEP carboxylase in the cytosol of the mesophyll cell (MC) (the equilibrium between CO_2 and HCO_3^- in the mesophyll cell is maintained by the enzyme carbonic anhydrase); (2) the 4-C dicarboxylic acid malate (Mal) is transported through plasmodesmatal connections to the bundle-sheath cell (BSC); (3) in this type of C_4 plant the malate is decarboxylated by the chloroplastic enzyme NADP-malic enzyme in the BSC; (4) the decarboxylated CO_2 is re-assimilated in the chloroplast (Chl) by Rubisco in the RPP pathway; (5) the 3-C product of decarboxylation, pyruvate (Pyr), is returned to the MC, where the energy and phosphate from one ATP molecule is used to produce phosphoenolpyruvate (PEP), the HCO_3^- acceptor molecule in the PEP carboxylase reaction. A second ATP molecule is used to convert the AMP formed during the production of PEP, to ADP, a form of the energy molecule capable of circulating back into the photophosphorylation reactions of photosynthesis. Thus, operation of the C_4 cycle requires two ATP molecules for steady-state operation. Overall, the C_4 cycle functions as a "CO_2 pump" whereby ATP energy is used to "pump" CO_2 into the BSC. R_{dM} = mesophyll cell respiration due to the tricarboxylic acid cycle; R_{dB} = bundle-sheath cell respiration due to the tricarboxylic acid cycle.

isolated to the BSCs; and (5) regeneration of a three-carbon CO_2 acceptor molecule by the expenditure of ATP energy in the MCs (Figure 4.14). In essence, ATP energy is used to transport CO_2 into the BSCs where it accumulates to relatively high concentrations (see Monson and Collatz 2012).

C_4 photosynthesis is most often found in the grasses of tropical and subtropical savannas, and in dicot species from hot, arid environments (Sage *et al.* 2011). In all cases, the CO_2-concentrating mechanism of the C_4 pathway facilitates high rates of CO_2 assimilation under conditions of high temperature and low CO_2 availability (due to drought and stomatal closure), both of which promote potentially high rates of photorespiration in C_3 plants. The C_4 CO_2-concentrating mechanism also provides C_4 plants with potentially greater rates of CO_2 assimilation per unit of N invested in Rubisco and per unit of water transpired (i.e., greater photosynthetic nitrogen- and water-use efficiencies, respectively).

Box 4.1   **Changes in the global atmosphere, climate, and the expansion of C_4 grasslands**

The C_4 photosynthetic pathway is known to be an evolutionary "embellishment" of existing metabolic processes in ancestral plants with the RPP (or C_3 pathway). This type of embellishment has evolved over 60 times in angiosperm plant lineages, independently (Sage *et al.* 2011). This fact alone informs us on two well-constrained inferences: (1) the environmental selection regimes that promote the evolution of C_4 photosynthesis are widespread, and highly selective as to those phenotypes that promote reproductive fitness; and (2) plants with the ancestral RPP pathway have few evolutionary options in terms of "re-wiring" metabolism as a response to those selection regimes. More than 60% of identified C_4 species are in the grass family, and the historic centers of C_4 grass evolution are in biomes that now support semi-arid, subtropical grasslands. Stable isotope analysis of numerous sites across the globe have established that 3–8 million years ago (Ma), high rates of evolutionary diversification and ecological expansion occurred in grassland ecosystems (Cerling *et al.* 1997, Edwards *et al.* 2010). Tectonic events involving uplift of the Himalayan Mountains and associated geochemical weathering likely led to global reductions in the atmospheric CO_2 concentration during the early Oligocene (30–40 Ma). Climate changes associated with these processes included greater overall cooling and aridity in subtropical woodlands, with the emergence of regional monsoon systems that delivered warm-season precipitation. These climate changes fostered wide-spread shifts in the dominance of certain ecosystems, with the overall result being a transition of forests to savanna woodlands dominated by C_3 grasses. Approximately 3–8 Ma, these C_3-dominated savanna ecosystems began to change in their species composition as C_4 grasses emerged and gained competitive advantages in a world characterized by a reduced atmospheric CO_2 concentration, more frequent disturbances due to wildfires and grazing, and greater aridity (Keeley and Rundel 2005). As grasses evolved in response to these new ecological situations, existing genetic and metabolic pathways in C_3 plants were modified to form new metabolic and physiological phenotypes – what is now recognized as broad convergence toward similar types of C_4 photosynthesis. The entire story of C_4 evolution and the expansion of C_4 grasslands over the past 3–8 million years provide a fascinating lesson in the synergies that have emerged among changes in climate, vegetation, and cellular biochemistry, with important ramifications for the effects and causes of variations in the global atmospheric CO_2 concentration.

The C_4 photosynthetic pathway provides one of the most impressive examples known of the power for evolution to change the form and function of organisms – of any type! The fact that this pathway requires numerous genetic modifications, and has evolved over 60 times independently within the span of a few million years, with the final anatomical and metabolic products being of similar form (with some notable variations), provides striking testimony to the power of biophysical and biogeochemical stresses to drive adaptive change in plants. Clearly, maintenance of a balanced carbon budget in the face of a changing earth and atmosphere is a critical factor shaping the evolution of plant traits (Box 4.1).

5 Modeling the metabolic CO$_2$ flux

> It is perhaps a matter of taste, but I find analytical solutions, as opposed to numerical ones, more enlightening. Unfortunately, the complexity of photosynthesis means that analytical descriptions can only be achieved at the expense of gross simplification ... [these models] can be useful aids to understanding, and for prediction, but are also potential hazards when the simplifications involved are forgotten.
>
> Graham Farquhar (1989)

Creating a computer model for processes as complex as photosynthesis, photorespiration, and dark respiration has proven to be an exceptional challenge. The various feedbacks that are engaged within the processes at specific metabolic steps, and that are triggered by specific sets of conditions, render the numerical modeling framework most applicable. Within the numerical framework, processes can be modeled as a series of sequential biochemical reactions, which can then be integrated across finite time intervals to produce pathway fluxes. While numerical solutions have their advantages, the path to the ultimate outcome tends to be less transparent than the alternative, an analytical solution. Numerical models are seldom amenable to complete mathematical closure, requiring that approximations be made to close the numerical iteration and reach a stable, steady-state, solution. An analytical solution is not constructed from sequential, time-dependent steps, but rather represents a closed set of equations (i.e., no variables left to approximation) that fully accounts for all components of a process or problem. For processes as complex as metabolic pathways, however, analytical solutions are difficult. Thus, even for this approach, researchers have sought approximations and simplifications that must be accommodated in order to achieve a stable, steady-state solution. The danger of relying on such approximations, as stated in the quote above by Graham Farquhar, a noted analyst in the field of photosynthetic modeling, is that these simplifications tend be forgotten and the analytical models migrate with time toward uncritical acceptance.

In this chapter we consider the most widely used analytical model of leaf CO$_2$ exchange, which was originally developed as a means to extract insight from leaf gas-exchange observations, but which has now become the quantitative scheme that defines biosphere-atmosphere CO$_2$ exchange in models ranging from the chloroplast up to global scales. The model was developed as a "simplification" of the processes controlling the reductive pentose phosphate pathway by Graham Farquhar, Susanne von Caemmerer, and Joe Berry. One can sense the regretful caution advocated by Farquhar with regard to how much confidence is now placed in the model, in the quote offered above. However, it has withstood the test of time and application across scales; it accurately represents many of the fundamental patterns and dependencies in biosphere-atmosphere CO$_2$ exchange, and so it has been rightfully embraced by a range of earth systems scientists. After considering the modeling of photosynthetic CO_2 fluxes, we will take up the topic of models of plant respiratory CO_2 exchange. Unfortunately, our ability to model respiration has not advanced as far as the case for photosynthesis. In the final sections of the chapter, we put the photosynthetic and respiratory processes together to derive a single expression of leaf net CO_2 exchange.

5.1 Modeling the gross rate of CO_2 assimilation and photorespiration

As knowledge emerged in the 1970s about the role of Rubisco as a primary control over CO_2 assimilation rate, modeling strategies switched from numerical descriptions of sequential reactions, to analytical descriptions that focused on two alternative rate-limiting conditions (see Farquhar *et al.* 1980b). As described in a published history of photosynthetic modeling (Farquhar *et al.* 2001), in the late 1970s, Farquhar, von Caemmerer, and Berry came to recognize the process of CO_2 assimilation as "see-sawing" back and forth between two potential limitations, the catalytic capacity of Rubisco and the availability of RuBP, dependent on the absorbed photon flux density and the chloroplast CO_2 concentration. In later studies, a third limiting case was added, in which the availability of inorganic phosphate (P_i), especially at low atmospheric O_2 concentrations and/or high atmospheric CO_2 concentrations, limited the CO_2 assimilation rate (Sharkey 1985). The most widely accepted models of the RPP pathway are still based on these three fundamental controls. Stated explicitly, the primary control over RPP flux is modeled as the minimum of three photosynthetic states: (1) the *Rubisco-limited state*, in which the availability of CO_2, the kinetic properties of the active site of Rubisco, and the amount of Rubisco limit the flux; (2) the *RuBP-limited state*, in which the availability of absorbed photons with which to power the production of ATP and NADPH, and thus ultimately the regeneration of RuBP, limits the flux; and (3) the *triose-phosphate utilization (TPU)-limited state*, in which the rate of P_i recycling from sucrose and/or starch synthesis limits the rate of ATP production, which is ultimately expressed as another form of the limitation on RuBP regeneration (Figure 5.1).

The interaction of these constraints can be illustrated through a graph of the rate of RuBP carboxylation plotted against the CO_2 mole fraction inside the chloroplast (c_{cc}) or the incident photosynthetic photon flux density (PPFD) (Figure 5.2). In assessing the response pattern in the upper panel of Figure 5.2, at low c_{cc} and high PPFD, the catalytic activity of Rubisco represents the principal limitation to the overall rate of CO_2 assimilation. As c_{cc} increases, RuBP availability, as controlled by its rate of regeneration in the RPP pathway, takes over as the principal limitation. At high c_{cc}, the rate of carboxylation will exceed the rate of triose-phosphate utilization for sucrose and starch synthesis and the third limitation, the rate at which P_i is recycled, becomes the principal limitation. In assessing the response pattern in the lower panel, the rate of carboxylation at low PPFD is limited by the rate of RuBP production. As PPFD increases, RuBP regeneration capacity becomes less of a limitation and the capacity of Rubisco to utilize its other substrate, CO_2, becomes more of a limitation.

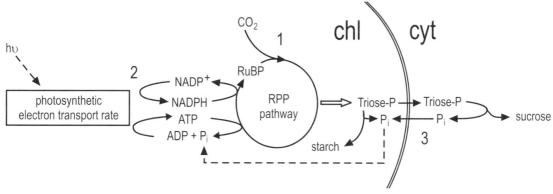

Figure 5.1 Schematic relationships within the RPP pathway demonstrating the three primary constraints to RuBP carboxylation rate upon which the most commonly used biochemical models of CO_2 assimilation rate are based. (1) The constraint reflected in CO_2 availability and the kinetic capacity of RuBP carboxylase to assimilate CO_2 (i.e., Rubisco-limited state); (2) the constraint reflected in the photosynthetic electron transport rate and its capacity to provide ATP and NADPH in order to regenerate RuBP (i.e., the RuBP-limited state); (3) the constraint reflected in the capacity for the assimilation of triose phosphate sugars and the freeing of inorganic phosphate (P_i) (i.e., the triose-phosphate utilization (TPU)-limited state).

5.1.1 Modeling the Rubisco-limited state

The Rubisco-limited state of photosynthesis is realized when RuBP is available at the active site of Rubisco in ample supply. In this case, it is the availability of the second substrate, CO_2, the inherent capacity of the active site of Rubisco to utilize this substrate, and the total amount of Rubisco enzyme protein that ultimately determines the rate of carboxylation. Following the general form of the Michaelis–Menten model the enzymatically catalyzed rate of RuBP carboxylation can be represented as:

$$v_{c1} = \frac{V_{cmax}c_{cc}}{c_{cc} + K_c(1 + c_{co}/K_o)}, \tag{5.1}$$

where v_{c1} represents the rate of carboxylation (μmol CO_2 m^{-2} leaf area s^{-1}) in the presence of adequate RuBP supply, V_{cmax} is the maximum velocity of carboxylation (μmol CO_2 m^{-2} leaf area s^{-1}), c_{cc} is the mole fraction of CO_2 in the chloroplast at the active site of Rubisco (mol CO_2 mol^{-1} dry air), K_c is the Michaelis–Menten coefficient describing the affinity of the active site for CO_2 (analogous to the K_m with respect to CO_2) (mol CO_2 mol^{-1} dry air), c_{co} is the mole fraction of O_2 in the chloroplast at the active site of Rubisco (mol O_2 mol^{-1} dry air), and K_o is the Michaelis–Menten coefficient describing the affinity of the Rubisco active site for O_2 (mol O_2 mol^{-1} dry air). The term $(1 + c_{co}/K_o)$ modifies K_c according to the presence of an alternative substrate that acts as a competitive inhibitor; in this case, O_2.

During the early phases of photosynthetic modeling, the Michaelis–Menten coefficients for CO_2 and O_2 that were available were determined from in vitro assays of extracted Rubisco which, by necessity, were conducted in test tubes in a relatively diluted state. In the

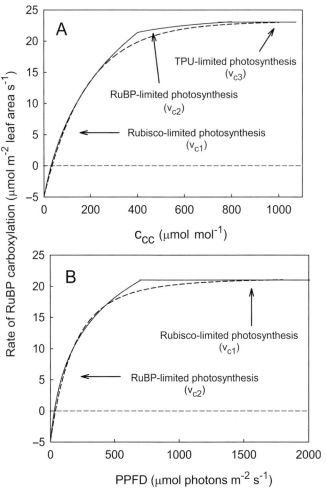

Figure 5.2 Diagrammatic representations of the RuBP carboxylation rate as it responds to CO_2 mole fraction in the chloroplast (c_{cc}) or photosynthetic photon flux density (PPFD). The terms v_{c1}, v_{c2}, and v_{c3} refer to the rate of RuBP carboxylation under conditions of Rubisco limitation, RuBP-regeneration limitation, or triose-phosphate utilization limitation, respectively (see text for details). The broken lines represent the type of response typically observed when observations are made with C_3 leaves. The solid lines represent the minimum of three possible limitations, which reflects the modeling approach of Farquhar *et al.* (1980b) in its simplest form.

chloroplast, however, Rubisco occurs in a more concentrated state (Section 4.2.3). The question as to the relevance of kinetic coefficients determined in vitro to the in vivo condition has remained open to debate. Using genetically engineered tobacco plants, von Caemmerer *et al.* (1994) were able to reduce the Rubisco content of leaves and approximate the diluted state that has been studied in vitro. Kinetic characterization of the transgenic plants confirmed that the Michaelis–Menten coefficients of Rubisco determined from in vitro studies are similar to those derived from in vivo measurements. This research has provided additional confidence in the application of Eq. (5.1).

5.1.2 Modeling the RuBP-limited state

In the presence of high c_{cc} or low PPFD the rate of net CO_2 assimilation is limited by the rate at which RuBP is regenerated in the RPP pathway, utilizing the ATP and NADPH produced from the photon-assimilating reactions. In general terms, the degree to which electron transport capacity limits the rate of carboxylation (v_{c2}) can be expressed as follows (derivation taken from Farquhar *et al.* 1980b):

$$v_{c2} = \frac{\text{molar electron transport rate}}{\text{moles of transported electrons required to regenerate RuBP per carboxylation}}.$$

(5.2)

The development of a quantitative model from this conceptual relation requires consideration of the amount of NADPH or ATP consumed in converting two moles of PGA (which are produced for each mole of CO_2 assimilated) to one mole of RuBP (which is consumed for each mole of CO_2 assimilated). For simplicity, we will derive the model on the basis of NADPH consumption. (A similar treatment has been developed for the case of ATP consumption; Long and Bernacchi 2003.)

In initiating the derivation we will delay consideration of the numerator in Eq. (5.2) and focus on the denominator. The rate of RuBP regeneration will depend on the rate of NADPH consumption, which is the reductant "power" required to recycle PGA and phosphoglycolate (the first products following carboxylation or oxygenation of RuBP) back to RuBP. In the presence of CO_2 and O_2, the total rate of NADPH consumption will equal the rate of PGA production (in both photosynthesis and photorespiration) plus the rate at which NH_4^+ is re-assimilated after release from glycine in the photorespiratory process (see Figures 4.10 and 4.11 in Chapter 4). Each mole of PGA requires one mole of NADPH for reduction to triose phosphate, and each mole of NH_4^+ released during photorespiration requires one mole of NADPH for re-assimilation into the amino acid pool of the cell. From the perspective of PGA production, we can write:

$$\text{rate of PGA production} = 2v_c + 1.5v_o,$$

(5.3)

where v_c is the rate of Rubisco-catalyzed carboxylation (mol CO_2 m^{-2} leaf area s^{-1}) and v_o is the rate of Rubisco-catalyzed oxygenation (mol O_2 m^{-2} leaf area s^{-1}). From the perspective of NADPH consumption, we can write:

$$\text{rate of NADPH consumption} = (2v_c + 1.5v_o) + 0.5v_o.$$

(5.4)

It is convenient to use an expression relating the rates of carboxylation and oxygenation as follows:

$$\phi = \frac{v_o}{v_c} = \frac{V_{omax}}{V_{cmax}} \frac{c_{co}}{c_{cc}} \frac{K_c}{K_o} = \frac{c_{co}}{S_{rel}c_{cc}}.$$

(5.5)

Equation (5.5) is derived by dividing the Michaelis–Menten model for oxygenation in the presence of CO_2 as a competitive inhibitor (the analog of Eq. (5.1) written for the case of

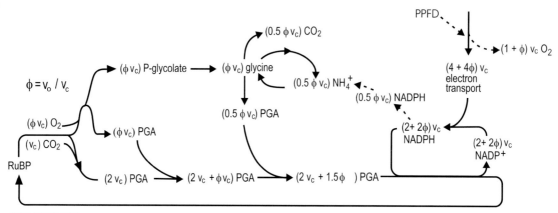

Figure 5.3 Stoichiometric relations in the RPP pathway expressed as functions of the rate of carboxylation (v_c) and the ratio (ϕ) of oxygenation to carboxylation rates. The stoichiometry is used to relate the rate of production of major products in the RPP and PCO pathways to v_c. As an example, the rate of PGA production directly from carboxylation by Rubisco, early in the pathway, is ($2\,v_c$). The rate of total PGA production, from both carboxylation and oxygenation is ($2v_c + 1.5\phi v_c$). Redrawn from Farquhar and von Caemmerer (1982), with kind permission from Springer Science + Business Media.

oxygenation) by the Michaelis–Menten model for carboxylation in the presence of O_2 as a competitive inhibitor (Eq. (5.1)). The stoichiometric relations underlying ϕ are illustrated in Figure 5.3. The relations shown in Figure 5.3 are often used in modeling the RPP pathway, as they show how the rates of formation or utilization of all major metabolites scale with the rates of Rubisco carboxylation (v_c) and oxygenation (v_o). Using Eq. (5.5) for substitution and applying algebraic rearrangement Eq. (5.4) can be simplified to:

$$\frac{\text{rate of NADPH consumption}}{v_c} = (2 + 2\phi). \tag{5.6}$$

When the rate of NADPH consumption is expressed in μmol NADPH m^{-2} (leaf area) s^{-1}, and v_c is in μmol CO_2 assimilated m^{-2} s^{-1}, Eq. (5.6) provides a steady-state expression for the moles of NADPH consumed during the regeneration of RuBP per unit CO_2 assimilated. In order to shift from consideration of NADPH consumption to electron transport rate, we note that the production of each mole of NADPH requires two moles of electrons to move through the photosynthetic electron transport chain. We can write an equation for the moles of electrons required for the regeneration of RuBP per mole of CO_2 assimilated by carboxylation as follows:

$$\frac{\text{moles of transported electrons required for RuBP regeneration}}{v_c} = (4 + 4\phi). \tag{5.7}$$

In essence, Eq. (5.7) defines the conditional denominator we stated for Eq. (5.2). Returning now to the fundamental relation presented in Eq. (5.2), the RuBP-limited rate of carboxylation (v_{c2}) (μmol CO_2 m^{-2} leaf area s^{-1}) can be quantitatively expressed as:

$$v_{c2} = \frac{F_J}{4 + 4\,\phi},\qquad(5.8)$$

where F_J is the rate of photosynthetic electron transport (μmol electrons m^{-2} leaf area s^{-1}).

5.1.3 Modeling the TPU-limited state

At high c_{cc} the rate of carboxylation can exceed the rate at which triose-phosphate is converted into sucrose and starch; the regeneration of RuBP is dependent on these latter processes to turnover organically bound P_i. In the absence of such turnover, P_i can remain sequestered in the triose-phosphate pool, potentially limiting the rate of photophosphorylation (ADP + $P_i \rightarrow$ ATP), and consequently limiting the overall rate of CO_2 assimilation. This limitation is most apparent in conditions such as low atmospheric O_2 concentrations or high atmospheric CO_2 concentrations, in which the rate of CO_2 assimilation is high relative to the combined rates of starch storage and sugar utilization for growth and respiration (which recycle P_i back into chloroplast metabolism). Limitations by P_i can also occur in some conditions when the utilization of triose-phosphates is constrained by plant growth; that is, when the rate at which triose phosphates are utilized for new plant growth is lower than the rate of triose phosphate production through whole-plant photosynthesis. This condition can also lead to P_i being bound up in sugar-phosphate compounds and cause a negative feedback on the RPP pathway. Theoretically, the rate of RuBP carboxylation will have to be three times the rate of triose phosphate utilization plus one-half the rate of RuBP oxygenation in order to maintain steady-state recycling of P_i and flux through the RPP pathway (Sharkey 1985):

$$v_{c3} = 3\text{TPU} + \frac{v_o}{2},\qquad(5.9)$$

where v_{c3} is the rate of carboxylation (μmol CO_2 m^{-2} leaf area s^{-1}) in the face of P_i limitation and TPU is the rate of triose phosphate utilization (μmol triose phosphate m^{-2} leaf area s^{-1}). The stoichiometry of Eq. (5.9) derives from the fact that each triose phosphate sugar utilized for growth or respiration by a plant will have to be replaced through three carboxylations plus the use of ATP and NADPH to produce one molecule of G3P. The inclusion of $v_o/2$ on the right-hand side of Eq. (5.9) can be understood if it is remembered that for every two moles of phosphoglycolate produced by oxygenation, one mole of PGA (containing three moles of carbon) is returned to the chloroplast by photorespiration and one mole of carbon is lost as CO_2 (see Figure 4.11 in Chapter 4). Thus, for every two oxygenations, one CO_2 is lost, which must be replaced by one carboxylation. From the perspective of recycling P_i, every two oxygenations recycles one P_i from RuBP back into PGA production, and frees up one P_i for use in subsequent ATP production. In this way, photorespiration actually has the potential to increase the rate of RuBP regeneration, and thus the rate of CO_2 uptake in the P_i-limited state (Harley and Sharkey 1991). It is relatively rare to observe v_{c3} as the primary limitation to the rate of carboxylation; though it can occur at relatively high chloroplast CO_2 concentrations, at low temperatures, and in the face of limited plant growth due to inadequate nutrient availability (Figure 5.2; Harley *et al.* 1992b).

5.2 Modeling dark respiration (R_d)

Considerable progress has been made in understanding biochemical controls over the flux of carbon through the steps of glycolysis and the TCA cycle (Plaxton 1996, Fernie *et al.* 2004). The derivation of an analytical model of R_d, however, has proven problematic. A primary obstacle involves coupling the loss of CO_2 and heat to the energetic and carbon demands of cellular maintenance and growth (e.g., the energy required for ion transport, protein turn-over, membrane repair, or cell wall construction). Such coupling requires knowledge about substrate-use efficiency, cellular enthalpy, and carbon channeling to specific pathways, which has not been resolved in a way that would permit analytical modeling. In the absence of knowledge about the mechanisms linking cellular energetics to CO_2 loss, R_d has been modeled as a scaled function of the rate of photosynthate utilization, the rate of photosynthesis, or the general effect of temperature on the rate of metabolism. The first approach has proven most useful in modeling respiration within the context of plant growth and longer-term plant carbon budgets. The second and third approaches have proven most useful in modeling the instantaneous rate of respiration and its response to short-term changes in the environment.

The *empirical* basis for scaling R_d to the rate of photosynthate utilization was originally provided by McCree (1970) and the *theoretical* basis was developed by Thornley (1970, 1971). In both sets of studies it was reasoned that at steady state all of the sugar phosphates produced from photosynthesis would (1) be converted to respired CO_2 as they are utilized in the production of energy for cellular maintenance, (2) be converted to respired CO_2 as they are used to produce energy for cellular growth, or (3) be converted to various carbon substrates that support new plant biomass. This logic is expressed in formal terms as:

$$\Delta S = \Delta S_m + \Delta S_{gr} + \Delta S_{gb}, \qquad (5.10)$$

where ΔS represents the total respiratory substrate (glucose) used during a specified time period (Δt), ΔS_m represents the substrate respired as CO_2 during maintenance respiration, ΔS_{gr} represents the substrate respired as CO_2 during growth respiration, and ΔS_{gb} is the substrate that is used to produce new biomass during growth; all expressed in units of moles. With regard to estimating the respiratory CO_2 yield, only S_m and S_{gr} are relevant. Using Eq. (5.10) as a foundation we can write:

$$R_d = \frac{6(\Delta S_m + \Delta S_{gr})}{W_p(\Delta t)}, \qquad (5.11)$$

where R_d represents the average dark respiration rate (mol CO_2 g^{-1} biomass s^{-1}), the factor 6 represents the molar equivalent of CO_2 produced from the carbohydrate substrate (in this case hexose equivalents) used in $\Delta S_m + \Delta S_{gr}$, W_p is plant biomass (g), and Δt is a finite time interval (s). The McCree–Thornley model was derived for whole plants (which accommodated the respiratory use of photosynthate that is exported to roots, stems, and growing apices, and for time intervals on the order of a complete day (which accommodated the use of stored photosynthate during the night). The case has been made for adding a third category of respiratory CO_2 loss; that due to "wastage" (see Amthor 2000). Wastage refers to CO_2 that is released during the generation of energy and/or substrates that do not directly contribute to growth or maintenance. This would include futile cycles of ATP production and hydrolysis and the production of heat through the alternative respiration oxidase (cyanide-resistant) pathway.

In order to make Eq. (5.11) tractable for estimating instantaneous respiration rates, flux coefficients have been formulated as:

$$R_d = R_g + R_m = \frac{g_R G + m_R}{W_p},$$

(5.12)

where R_{gr} and R_m are the growth and maintenance respiration rates (mol CO_2 g^{-1} biomass s^{-1}), respectively, G is growth due to new biomass (g s^{-1}), W_p is existing plant biomass (g), and g_R and m_R are scaling coefficients with units mol CO_2 g^{-1} and mol CO_2 s^{-1}, respectively. Experimental approaches to measure g_R and m_R have included (1) component partitioning in which the theoretical energetic costs of individual metabolic processes are summed to provide composite CO_2 fluxes, and (2) regression analyses in which R_d is regressed against $1/W_p$ in plants that are no longer growing (thus assuming that the term $(g_R G) \rightarrow 0$) to derive a value for m_R. These approaches have not been entirely satisfactory as gaps in our knowledge cause the first approach to suffer from undefined (and therefore missing) metabolic components and the second approach relies on assumptions about the constancy of energetic costs as a function of tissue age.

In some past models g_R has been related to "growth efficiency" using an analog equation originally developed for predictions of microbial growth and respiration. According to Thornley (1970) biomass-specific respiration can be written as:

$$R_d = \frac{(1 - Y_g) \, g_R G + m_R}{W_p},$$

(5.13)

where Y_g is the growth yield (the fraction of photosynthate that is converted to biomass and thus is not available for respiratory CO_2 loss). Equation (5.13) provides a context within which to explore tradeoffs between growth efficiency and respiration. For example, an increase in the efficiency of carbon-substrate utilization for increase in biomass must occur concomitantly with a decrease in respiratory CO_2 loss. Equation (5.13) does not solve the issue of intractability that was raised for Eq. (5.12), however, since we are now faced with a different unknown coefficient, Y_g.

More recently, some workers have argued that the theoretical foundation for coupling respiration to growth is best developed through a complete thermodynamic framework that takes account of not only the conservation of mass and chemical energy, but also heat. The *enthalpy-balance model* is written as:

$$-R_{SG} \, \Delta H_B \quad = \quad R_d \, \Delta H_{CO2} \quad + \quad q,$$
$$\text{Term I} \qquad\qquad \text{II} \qquad\qquad \text{III}$$

(5.14)

where R_{SG} is the biomass-specific rate of conversion of substrate into biomass (mol C g^{-1} biomass s^{-1}), ΔH_B is the change in enthalpy when a mole of substrate C is converted into biomass C (kJ mol^{-1} C), R_d is the rate of respiratory CO_2 production (mol CO_2 g^{-1} biomass s^{-1}), ΔH_{CO2} is the change in enthalpy for the conversion of substrate C to CO_2 (kJ mol^{-1} C), and q is the metabolic heat rate (kJ g^{-1} biomass s^{-1}) (Hansen *et al.* 1994). Equation (5.14) balances the total change in internal energy channeled into the conversion of substrate to biomass (i.e., growth) against the loss of enthalpy due to respiratory oxidation of substrate. The equation can be rearranged to solve for R_d. The terms in Eq. (5.14) are resolved through a combination of chemical composition analyses, calorimetric combustion, and measurement of respiratory energy-efficiency quotients. The advantage to using Eq. (5.14) is that it places respiratory flux into a thermodynamic context capable of accommodating different enthalpy conversion efficiencies among diverse processes (e.g., O_2-dependent respiration versus cyanide-resistant respiration) and is thus better able to couple R_d to the energetic demands of growth and maintenance.

None of the modeling approaches described to this point provides the direct link to photosynthetic carbon assimilation that would permit representation of R_d in net CO_2 exchange models. In order to accommodate such linkage researchers have often rejected formal models of R_d, and instead, they have represented R_d through linear scaling with V_{cmax}. Past studies have revealed that R_d measured on mature foliar tissues for a number of plant species is positively correlated with N concentration (Figure 5.4), and this relationship generally holds for roots and stems (Reich *et al.* 2008). This relationship is not unexpected given that up to 60% of maintenance respiration can be devoted to supporting protein turnover (Penning de Vries 1975), and that proteins are the primary sink for tissue N. Similar linear dependencies have been observed for V_{cmax} and F_{Jmax} (when regressed against leaf N concentration) (Figure 5.5), providing good biochemical justification for the linear scaling of R_d with V_{cmax} (and by way of caveat, the linear scaling of F_{Jmax} with V_{cmax}). There are cautionary issues that must be considered when applying these simple scaling relationships. For example,

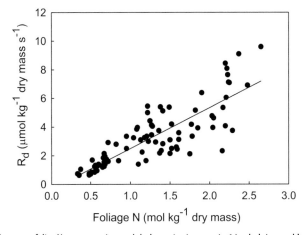

Figure 5.4 Linear relationship between foliar N concentration and dark respiration rate in 14 subalpine and boreal forest trees and shrubs. Redrawn from Ryan (1995).

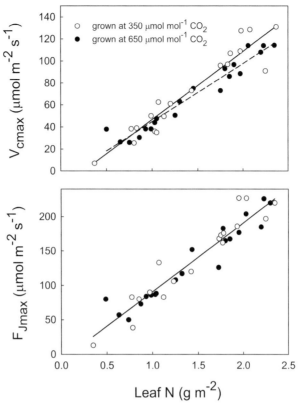

The relationship between the maximum rate of RuBP carboxylation (V$_{cmax}$) the maximum rate of electron transport (F$_{Jmax}$) and leaf N concentration in leaves of *Gossypium hirsutum* (cotton). The plants used in this analysis were grown at two different atmospheric CO$_2$ concentrations. Separate regressions are presented for the plants from the different CO$_2$ treatments in the case of V$_{cmax}$. In the case of F$_{Jmax}$, all values were analyzed with same regression. Regression lines were: V$_{cmax\,(350)}$ = 60.0 N − 9.6 (r^2 = 0.90); V$_{cmax\,(650)}$ = 52.8 N − 6.6 (r^2 = 0.94); F$_{Jmax}$ = 98.1 N − 4.6 (r^2 = 0.91). Redrawn from Harley *et al.* (1992b).

in those species (e.g., conifers) in which a significant fraction of foliar N occurs in storage or structural compounds, rather than enzymes, the positive correlation between R$_d$ and tissue N concentration may still be present, but difficult to detect (Kruse and Adams 2008). Nonetheless, in past studies that combine measurements of photosynthesis rate and R$_d$, it has been shown that these two variables are linearly correlated across a broad range of species (see Gifford 1994). Photosynthetic models parameterized with the linear constant R$_d$/V$_{cmax}$ ≅ 0.015 are generally well-validated with observations of leaf gas exchange.

5.2.1 Modeling the temperature dependence of R$_d$

Maintenance respiration rate is highly sensitive to temperature. On theoretical grounds the temperature dependence of R$_m$ should follow the same exponential form that holds for other chemical reactions, e.g., the temperature-dependent rate constant is defined according to the

Arrhenius or Eyring relations (see Section 3.2). By convention, however, biologists have preferred the Q_{10} model. In applying the Q_{10} relation to R_m we can write:

$$Q_{10} = \left(\frac{R_m}{R_0}\right)^{10/(T_m - T_0)}, \tag{5.15}$$

where R_m is the respiration rate at a measured temperature, R_0 is the respiration rate at a standard temperature, T_m is the temperature (K) at which R_m is measured, and T_0 is the standard temperature (K). As an example of application of the Q_{10} model, if R_m doubles in response to a 10 °C increase in temperature, then $10/(T_m - T_0) = 1$ and the Q_{10} is 2.

In leaves, the Q_{10} for respiration rate ranges between 1.5 and 3.0; a value of 2.0 is often used as "typical" for plant tissues. However, the Q_{10} varies with temperature and values near 2.0 are only found over a narrow temperature range (20–30 °C). Over a broader range the Q_{10} decreases with increasing temperature, from values near 3.0 at temperatures less than 10 °C to values near 1.5 at temperatures greater than 35 °C (Atkin and Tjoelker 2003). The decrease in Q_{10} with increasing temperature is at least in part due to shifts in the Boltzmann distribution of reactant kinetic energy as temperature increases (see Section 3.2). As temperature increases, the fraction of molecules with kinetic energy exceeding the energy of activation (E_a) increases; but it increases more in reactions at lower temperatures compared to reactions at higher temperatures. Other factors may also be involved in the temperature dependence of Q_{10}, including shifts in the fractional limitations imposed by protein versus substrate concentration and changes in the fluidity of membranes and associated efficiencies of electron transport processes. Fortunately for modeling efforts the temperature dependence of the Q_{10} is linear across the biologically relevant temperature range from 10–40 °C, and highly conserved across a broad range of plant species (Atkin and Tjoelker 2003). In this case, the temperature dependence of the Q_{10} refers to short-term (e.g., hours) changes in temperature.

In addition to dependence on short-term dynamics in temperature (seconds-to-minutes), R_d is also dependent on longer-term temperature history (hours-to-days). Thus, a process known as *temperature acclimation* occurs in the respiratory patterns of plants. The temperature dependence of R_d can exhibit a higher or lower thermal optimum when plants have experienced higher or lower temperatures in their recent past, respectively (Atkin *et al.* 2005). The potential for respiratory acclimation varies among species, and general trends on which predictions can be based have been difficult to derive. The causes of acclimation are not fully understood, although it is known that expanding leaves have greater potential to adjust to a new temperature regime than fully expanded leaves and that acclimation in both R_d and A facilitates a nearly constant ratio of the two CO_2 fluxes within the temperature ranges normally encountered by plants (see Ow *et al.* 2008).

5.3 Net versus gross CO_2 assimilation rate

With the understanding that the leaf-atmosphere flux of CO_2 reflects the balance of several processes, we can define net CO_2 assimilation rate as the difference between the gross rates of CO_2 uptake and loss. In this book we will use the symbol A in general terms for CO_2

assimilation rate without assignment as to gross or net fluxes, whereas A_n and A_g will be used when we need to explicitly represent net CO_2 assimilation rate and gross CO_2 assimilation rate, respectively. Furthermore, we will use the term *gross CO_2 assimilation rate* within the context of biochemical discussions to mean the rate of RuBP carboxylation, whereas *net CO_2 assimilation rate* will be used to mean the difference between CO_2 uptake due to RuBP carboxylation and CO_2 loss due to photorespiration plus dark respiration.

5.3.1 Modeling the net CO_2 assimilation rate

The equations described to this point are founded on biochemical processes of the chloroplast. There are numerous assumptions that must be accepted in order to make the transition from chloroplast biochemistry to leaf physiology in order to describe the net photosynthesis rate. Gradients in the availability of photons, CO_2, and O_2 within the leaf and dynamics in the availability of metabolite co-factors within the cell represent uncertainties that make this transition difficult. Nonetheless, there are several "checks" and "balances" that can be evaluated in making the transition and, to a large degree, the scientific community has accepted the validity of chloroplast-to-leaf scaling. Some of the concepts presented in this section, such as the CO_2 concentration in the intercellular air spaces of a leaf (c_{ic}) are more appropriately considered within the context of diffusion across leaf surfaces (e.g., in Chapter 7). However, we present them here in abbreviated form because the topic of biochemical modeling of photosynthesis has traditionally been integrated with observations of leaf CO_2 exchange patterns (e.g., von Caemmerer and Farquhar 1981), and thus some of the simplifying parameters used in the biochemical models are derived from leaf-scale measurements.

One of the more important synthetic parameters used in modeling the net CO_2 assimilation rate is the *CO_2 compensation point*. In general terms the CO_2 compensation point represents the CO_2 mole fraction in the atmosphere outside a leaf (c_{ac}) that allows the rate of photosynthetic CO_2 uptake to be balanced by the rate of respiratory CO_2 loss. Given that the net flux of CO_2 across the leaf surface is zero at the CO_2 compensation point, and in the interest of simplicity, we assume for the moment that the CO_2 concentrations of the air outside the leaf and inside the leaf are equal (i.e., $c_{ac} = c_{ic}$). (This assumption is dependent on binary gas composition; in this case CO_2 mixed with dry air (assuming that N_2 and O_2 represent a single diffusional entity that we call "dry air"). The ternary situation, which is actually more representative of leaf diffusion, in which CO_2 exchange is considered within the context of ternary gas mixtures (e.g., CO_2, H_2O, and dry air) is considered in Chapter 6. In the ternary case, $c_{ac} \neq c_{ic}$ at the compensation point.) There are two types of CO_2 compensation points. The value denoted as Γ is the CO_2 compensation point in the presence of CO_2 efflux from both dark respiration and photorespiration. The value denoted as Γ_* is the CO_2 compensation point in the presence of CO_2 efflux from only photorespiration; this is often called the *photocompensation point*. It is important to recognize how the differences between Γ and Γ_* are reflected in their responses to changes in the atmospheric CO_2 concentration and PPFD, the two principal drivers of A_n.

Let's consider the responses of Γ and Γ_* to changes in PPFD within the condition of v_{c2} limitation (i.e., RuBP-regeneration limitation). For the case of Γ we must consider the values of v_{c2}, v_o, and R_d as PPFD is increased, whereas for the case of Γ_* we must consider only the values of v_{c2} and v_o. We begin with the condition that PPFD is relatively low such that $A_n <$ 0, and therefore $c_{ic} > c_{ac}$ and $c_{ic} > \Gamma$. It is important to note that Γ is not a conserved value with regard to PPFD; any change in PPFD will cause both v_{c2} and v_o to increase (because both are limited by RuBP availability), but dark respiration rate (R_d) will not increase. This means that as PPFD changes, the c_{ic} that exists when $A_n = 0$ will also change; i.e., Γ will change. Now, let's assume that PPFD is increased above that of the initial condition. Because v_{c2} is greater than v_o, the net result of an increase in PPFD will be an increase in A_n. In order to cause c_{ic} to converge to Γ, and thus be at a state where the CO_2 fluxes into and out of the leaf are perfectly balanced, c_{ic} would have to decrease as PPFD is increased. This is seen as a decrease in c_{ic} at $A_n = 0$ in Figure 5.6; in other words, Γ has decreased as PPFD has increased. For the case of Γ_*, an increase in PPFD will cause proportional increases in v_{c2} and v_o at any given c_{ic}. Because R_d is not a component of Γ_*, the c_{ic} that defines Γ_* will remain constant in the face of changing PPFD; it follows that Γ_* can be quantified using the relation between v_{c2} and c_{ac} (assuming constant diffusion resistances) observed at different PPFD levels (Figure 5.6). At Γ_*, the assimilation of one molecule of CO_2 through carboxylation is balanced by the loss of one molecule of CO_2 through two oxygenations (i.e., $\phi = 2$). Thus, Γ_* is the value of c_{cc} when $\phi = 2$, and we can state:

$$\Gamma_* = \phi \frac{c_{cc}}{2} = \frac{0.5 V_{omax} K_c c_{co}}{V_{cmax} K_o} = \frac{0.5 \, c_{co}}{S_{rel}}, \qquad (5.16)$$

where Γ_* carries units of mole fraction ($\mu mol \, CO_2 \, mol^{-1}$ air).

Using Γ_*, and now assuming the condition of v_{c1} limitation (Rubisco limitation), we can derive an expression containing V_{cmax} according to:

$$\frac{dA_n}{dc_{cc}} = \frac{V_{cmax}}{\Gamma_* + K_c(1 + c_{co}/K_o)}. \qquad (5.17)$$

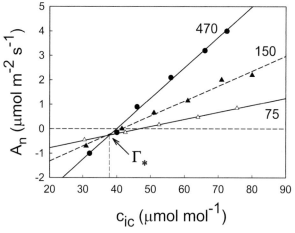

Figure 5.6 Rate of net CO_2 assimilation (A_n) in spinach leaves as a function of intercellular CO_2 mole fraction (c_{ic}) at three different PPFD values. The light-independent intersection of the three lines is taken as the CO_2 compensation point in the absence of TCA-cycle respiration (Γ_*). Redrawn from Brooks and Farquhar (1985), with kind permission from Springer Science + Business Media.

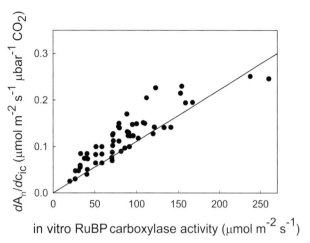

Figure 5.7 The relation between the initial slope of the A_n:c_{ic} response measured on intact leaves of *Phaseolus vulgaris* and the maximum RuBP carboxylase activity extracted from the same leaves. The line represents the relation modeled from Eq. (5.17) assuming that the extracted activity measured in vitro equals V_{cmax}. The difference between the modeled and measured values for dA_n/dc_{ic} is probably due to incomplete extraction of the enzyme activity from the leaves. Redrawn from von Caemmerer and Farquhar (1981), with kind permission from Springer Science + Business Media.

The relationship dA_n/dc_{cc} is often approximated as the initial slope of the A_n:c_{ic} relation (accepting the assumption that $c_{ic} \approx c_{cc}$) in an intact leaf, and is sometimes called the *carboxylation efficiency*. (Equation (5.17) is obtained after differentiation of Eq. (5.1) with respect to c_{cc}. At $c_{cc} \approx \Gamma_*$, Eq. (5.17) is obtained.) Equation (5.17) can be rearranged to solve for V_{cmax}, thus providing an expression for the V_{cmax} of Rubisco within the context of Γ_* and for an intact leaf. Equation (5.17) has been validated through the observation that dA_n/dc_{ci} is highly correlated with the extracted activity of RuBP carboxylase (Figure 5.7), thus justifying the foundation of this relation within the definition for v_{c1}.

Using the derived value for V_{cmax}, and Eq. (4.8) (from Chapter 4), the rate of CO_2 assimilation for an intact leaf when constrained by Rubisco kinetics (v_{c1}) can be determined as:

$$v_{c1} = \frac{V_{cmax}c_{ic}}{c_{ic} + K_c(1 + c_{io}/K_o)}. \tag{5.18}$$

Note that in using this relation the relevant CO_2 and O_2 mole fractions are those in the intercellular air spaces of the leaf (c_{ic} and c_{io}, respectively); these are the values most easily determined during observations of CO_2 exchange in intact leaves. The most relevant values with regard to Rubisco carboxylation, however, are those in the chloroplast (c_{cc} and c_{co}, respectively). In the case of O_2, the difference between c_{io} and c_{co} is so small compared to their absolute values, there is no significant error in using c_{io}. In the case of CO_2, however, the use of c_{ic} is a simplification that carries significant error. The assumption that $c_{ic} = c_{cc}$ ignores resistance to CO_2 diffusion between the intercellular air spaces of the leaf and the chloroplast stroma. The fact that c_{cc} is normally lower than c_{ic} means that estimates of v_{c1}

using Eq. (5.18) will be erroneously high when using c_{ic}. Studies using a broad range of species have shown that the presence of diffusive resistances within a leaf cause c_{cc} to be as much as 30% lower than c_{ic} (Evans and von Caemmerer, 1996).

Γ_* can also be used to derive an expression for the RuBP-limited rate of carboxylation (v_{c2}) within the context of an intact leaf. Once again, within the caveat of the assumption $c_{cc} = c_{ic}$, and using Eq. (5.16) to derive a combined expression for Γ_* and ϕ, we obtain:

$$\phi = 2\Gamma_*/c_{ic}. \tag{5.19}$$

Using Eq. (5.7) as a starting point, and substituting from Eq. (5.18), v_{c2} can now be expressed in terms of Γ_*:

$$v_{c2} = \frac{F_J \, c_{ic}}{4c_{ci} + 8\Gamma_*}. \tag{5.20}$$

The whole-chain electron transport rate (F_J) is one of the more poorly understood components in the current generation of photosynthesis models. In the original version of the Farquhar–von Caemmerer–Berry model (Farquhar $et\ al.$ 1980b), F_J was entered as an empirical function derived from in vitro studies on isolated chloroplasts. The situation has not improved much during subsequent decades, and we are still left without an electron-transport rate equation derived from first principles. In light of these deficiencies, F_J continues to be modeled as an empirically derived, quadratic function, which takes a form similar to:

$$F_J = \frac{I_2 + F_{Jmax} - \sqrt{(I_2 + F_{Jmax})^2 - 4\Theta I_2 F_{Jmax}}}{2\Theta}, \tag{5.21}$$

where I_2 is the maximum (potential) photon flux density (mol photons m^{-2} leaf area s^{-1}) that could be used to drive whole-chain electron transport, F_{Jmax} is the maximum possible electron transport rate (mol electrons m^{-2} leaf area s^{-1}), and Θ is a unitless curvature factor intended to "tune" the function with a smooth transition between the PPFD-limited and PPFD-saturated parts of the curve. I_2 is defined as:

$$I_2 = I \cdot a_L \cdot \phi_{PSII,max} \cdot \beta, \tag{5.22}$$

where I is the PPFD incident on the upper surface of the leaf (mol photons m^{-2} leaf area s^{-1}), a_L is the fractional absorptance of the leaf, $\phi_{PSII,max}$ is the maximum fractional quantum yield for whole-chain electron transport (i.e., the molar ratio of electron transport rate to absorbed photon flux density), and β is the fraction of absorbed light that reaches PSII light-harvesting complexes (assumed to be 0.5; Ögren and Evans 1993). The values for I, a_L, and $\phi_{PSII,max}$ can be measured. The values for F_{Jmax} and Θ are typically determined from empirical fitting of the A_n:PPFD response curve (see Bernacchi $et\ al.$ 2003).

It should be noted that Eq. (5.20) has been derived to reflect the requirement for four electrons to be transported through the electron transport chain to reduce ferredoxin and produce two reduced NADPH molecules which then drive the assimilation of one CO_2

molecule. Alternative forms of the model have been published that assume ATP as the primary limitation to RuBP regeneration (Bernacchi *et al.* 2003), or a combination of ATP and NADPH (Yin *et al.* 2004), with the principal difference being slight adjustments to the denominator of Eq. (5.20).

The influence of temperature on A_n has been modeled through various approaches, all of which derive from the fundamental exponential relations between reaction rate and temperature as required by the Maxwell–Boltzmann relation (see Section 3.2). As noted above, a modified version of the original Arrhenius relation, referred to as the Q_{10} model, is commonly used to predict maintenance respiration rates in leaves as a function of temperature. The temperature dependencies of variables such as ϕ, Γ_*, and α_c are ultimately determined by the temperature dependencies of K_c, K_o, V_{cmax}, V_{omax}. Because these parameters are associated with enzyme function, which declines at extremely high temperatures, the temperature response has been modeled according to a modified version of the Eyring equation (see Section 3.2.1). Following treatments provided in Johnson *et al.* (1942) and Sharpe and deMichele (1977), a model that is commonly used to model temperature dependence in photosynthesis is:

$$k = \frac{e^{c - \frac{\Delta H_a}{RT}}}{1 + e^{\frac{\Delta S}{R} - \frac{\Delta H_d}{RT}}}, \quad (5.23)$$

where k (analogous to a reaction rate constant) is taken as the temperature scaling coefficient, c is a scaling constant, H_a is the enthalpy of activation, H_d is the enthalpy of deactivation (due to denaturation), and S is an entropy term (see Harley *et al.* 1992b and Bernacchi *et al.* 2001 for more discussion on how to apply this temperature dependence model). An analysis by Medlyn *et al.* (2002) provides values for H_a and H_d derived from observed temperature responses for V_{cmax} and F_{Jmax} for different plant growth forms; values are approximately 85 kJ mol^{-1} and 200 kJ mol^{-1} for H_a and H_d, respectively, in the temperature response of V_{cmax} and 70 kJ mol^{-1} and 200 kJ mol^{-1} for H_a and H_d, respectively, in the temperature response of F_{Jmax}. The temperature dependence of F_J, in relation to F_{Jmax}, has been modeled as a simple empirically determined exponential function as:

$$F_J = F_{Jmax} e^{-\left(\frac{T_L - T_{opt}}{\Omega}\right)^2}, \quad (5.24)$$

where T_L is the leaf temperature at which F_J is determined, T_{opt} is the leaf temperature at which F_{Jmax} is determined, and Ω is the range of temperature in which F_J decreases to e^{-1} relative to F_{Jmax} (June *et al.* 2004).

Under any set of defined environmental conditions, A_n is assumed to equal the minimum of v_{c1}, v_{c2}, or v_{c3} (gross flux of CO_2 uptake) minus the rates of photorespiration and R_d (gross fluxes of CO_2 loss):

$$A_n = \min\{v_{c1}, v_{c2}, v_{c3}\} - 0.5 v_o - R_d, \quad (5.25)$$

where all fluxes carry units of μmol CO_2 m^{-2} leaf area s^{-1} and min{ } represents the "minimum of."

Recognizing that $v_o = v_c \phi$, and that ϕ can be related to Γ_* according to Eq. (5.16), we can rewrite Eq. (5.25) as:

$$A_n = (1 - \Gamma_*/c_{ic}) \min\{v_{c1}, v_{c2}, v_{c3}\} - R_d. \qquad (5.26)$$

When modeled according to these relations, one obtains response curves similar to those shown in Figure 5.2, with a prominent feature being a sharp, angular transition between the regions limited by v_{c1}, v_{c2}, and v_{c3}. The sharp transition is an artifact created by the structure of the model as the minimum of three processes. In a real leaf individual chloroplasts will differ in the degree to which v_{c1}, v_{c2}, or v_{c3} limit A_n at any instant in time, and they will face intra-leaf gradients in PPFD and CO_2 mole fraction, all of which will blend, or smooth, the response of the leaf-averaged A_n to both forcing variables. The transitions are often "softened" in the model through quadratic "tuning" using the following relationship:

$$\theta A_n{}^2 - A_n(v_{c1} + v_{c2}) + (v_{c1}v_{c2}) = 0, \qquad (5.27)$$

where θ is a convexity factor. As $\theta \to 0$, the modeled response of A_n to PPFD or c_{ic} will exhibit a non-rectangular hyperbola. As $\theta \to 1$, the response will exhibit a sharp transition. The presence of θ in Eq. (5.27) can be thought of as calibrating the degree to which v_{c1} and v_{c2} combine to influence the shape of the transition. Similar "tuning" can be applied to transitions involving v_{c3}.

Many of the parameter values required to run the model of A_n in C_3 plants are presented in Table 5.1. For variables such as the K_c and K_o for Rubisco, which exhibit less interspecific variation in C_3 plants, a single value can be used in model parameterization. Variables such as c_{ic} and Γ_* are also somewhat conserved among different C_3 species. The variable V_{cmax} is related to the allocation pattern of N, a growth-limiting resource, and is thus strongly dependent on plant growth form, seasonal variation in environment, and nutrient characteristics of a plant's habitat. The mechanisms that control the allocation of nitrogen to V_{cmax} in different plants are not well understood. As a result, V_{cmax} is typically prescribed based on its tendency to fall within certain bounds for various plant growth forms, or it is determined empirically from the carboxylation efficiency using Eq. (5.17) (Rogers 2013). Among plant growth forms, values for F_{Jmax} tend to scale positively with values for V_{cmax}. A general analysis showed that F_{Jmax} is, on average, 1.67 times higher than V_{cmax}, and this relationship holds across a range of plant growth forms (Medlyn *et al.* 2002). Within a growth form, some variation in the parameterization of V_{cmax} and F_{Jmax} can be expected when the model is applied to different species. Interspecific variation for these variables within the "crop plant" growth form has been shown to be in the range 15–35% (Archontoulis *et al.* 2012).

Also problematic is our limited ability to predict dynamics in V_{cmax} and R_d in the face of ontogenetic change and longer-term environmental stresses such as occur during drought or periods of extreme temperature. We know that mechanisms exist in plants to acclimate photosynthetic processes to changes in environment, but we don't know yet how to represent those acclimation processes. One of the complicating aspects of these responses is that they depend on stochastic features of the environment, such as how quickly stress occurs. Dynamics in model parameters due to these influences are currently represented by empirically derived relations. Recently, breakthroughs have been made in our understanding

Table 5.1 Approximate parameter values for the C_3 model of CO_2 assimilation rate*

Parameter	Value range	Description
K_c (from in vitro assay)*	250–300 μmol mol^{-1}	Michaelis–Menten coefficient for CO_2 for C_3 Rubisco
K_c (from in vivo assay)**	405 μmol mol^{-1}	Michaelis–Menten coefficient for CO_2 for C_3 Rubisco
K_o (from in vitro assay)*	400–450 mmol mol^{-1}	Michaelis–Menten coefficient for O_2 for C_3 Rubisco
K_o (from in vivo assay)**	278 mmol mol^{-1}	Michaelis–Menten coefficient for O_2 for C_3 Rubisco
V_{cmax}***		Maximum carboxylation rate
agricultural crops	90–180 μmol CO_2 m^{-2} s^{-1}	
deciduous trees	50–60 μmol CO_2 m^{-2} s^{-1}	
coniferous trees	25–63 μmol CO_2 m^{-2} s^{-1}	
tropical trees	29–50 μmol CO_2 m^{-2} s^{-1}	
understory herbs	~ 70 μmol CO_2 m^{-2} s^{-1}	
desert annuals	~ 150 μmol CO_2 m^{-2} s^{-1}	
sclerophyllous shrubs	~ 50 μmol CO_2 m^{-2} s^{-1}	
V_{omax}	0.25 V_{cmax}	Maximum oxygenation rate
F_{Jmax}****		Maximum electron transport rate
agricultural crops	132–218 μmol m^{-2} s^{-1}	
deciduous trees	45–147 μmol m^{-2} s^{-1}	
coniferous trees	71–175 μmol m^{-2} s^{-1}	
TPU	20–60 μmol CO_2 m^{-2} s^{-1}	Triose-phosphate utilization rate
α	0.050 mol CO_2 mol^{-1} photons absorbed	quantum yield for CO_2 assimilation in normal air
α_c	0.075 mol CO_2 mol^{-1} photons absorbed	CO_2-saturated quantum yield for CO_2 assimilation
Γ_*	35 μmol CO_2 mol^{-1}	CO_2 photocompensation point
c_{co}	210 mmol mol^{-1}	O_2 concentration in chloroplast
c_{ic}	235–250 μmol CO_2 mol^{-1}	Intercellular CO_2 concentration
c_{cc}	210–225 μmol CO_2 mol^{-1}	CO_2 concentration in chloroplast
R_d	0.015 V_{cmax}	Whole-leaf respiration rate

* All values are estimated for 25°C

** Determined from analysis of a Rubisco-antisense line of tobacco with reduced in vivo Rubisco concentrations (from Bernacchi *et al.* (2001)).

*** Values for V_{cmax} for C_3 plants were taken from Wullschleger (1993), Medlyn *et al.* (2002), and Kattge *et al.* (2009). See Rogers (2013) for an updated analysis.

**** Values for F_{Jmax} are estimated for 25 °C and are from Medlyn *et al.* (2002).

of the longer-term effects of elevated atmospheric CO_2 mole fraction on photosynthetic biochemistry and control over CO_2 assimilation rate (Box 5.1). This has provided the promise that we will soon be able to derive fundamental, biochemically based models of the response of plants to a changing atmosphere.

Box 5.1   **Response of photosynthesis in C_3 plants to elevated atmospheric CO_2 concentration**

Over the past 130 years the earth system has experienced a significant increase in the atmospheric CO_2 mole fraction (c_{ac}) due to human activities. Since the late nineteenth century, the production of CO_2 as a pollutant from fossil fuel combustion and biomass burning has caused c_{ac} to increase by approximately 40%, and it continues to increase at the rate of 2–3 μmol mol^{-1} per year (about 4.1 Pg of CO_2 per year) (Canadell *et al.* 2007). Given that the K_c of Rubisco is in the same range as c_{ac}, past increases in c_{ac} have undoubtedly stimulated C_3 plant photosynthesis. Using the biochemical model of photosynthesis described in this chapter we can estimate that in the past 130 years C_3 photosynthesis rates have increased by 30% and 9% when limited by Rubisco (v_{c1}) or RuBP regeneration (v_{c2}), respectively. Because the competition between carboxylation and oxygenation shifts to favor oxygenation at higher temperatures, the effect of an increase in c_{ac} on atmospheric CO_2 assimilation has been enhanced in warmer climates (Figure B5.1).

In addition to direct effects on Rubisco activity, the growth of plants at an elevated c_{ac} has longer-term influences on photosynthesis. For example, growth at a c_{ac} equal to twice the current level causes leaf carbohydrate concentrations to increase by 50% in some species (Long and Drake 1992). Growth at elevated c_{ac} also causes a reduction in the genetic transcription of genes for photosynthetic enzymes, including Rubisco activase (van Oosten *et al.* 1994), carbonic anhydrase (Majeau and Coleman 1996), and Rubisco (Sage *et al.* 1989, Moore *et al.* 1998). In fact, it is the enzymes that support the RPP pathway that are most reduced in activity when plants are grown under elevated c_{ac} (Nie *et al.* 1995a, 1995b). A model has been proposed in

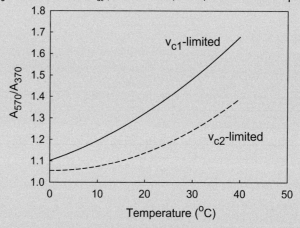

Figure B5.1   The potential increase in net CO_2 assimilation rate (A) in C_3 plants with an assumed atmospheric CO_2 concentration (c_{ac}) of 370 μmol mol^{-1} versus 570 μmol mol^{-1} across a range of temperatures. V_{cmax} is taken as 120 μmol m^{-2} s^{-1} at 25 °C and the ratio of intercellular CO_2 mole fraction to ambient CO_2 mole fraction (c_{ic}/c_{ac}) is taken as 0.6. The increase in v_{c1} at higher CO_2 mole fraction is due to the effect of a higher substrate concentration on the velocity of Rubisco and a decrease in the competitive interaction between CO_2 and O_2 at the active site of Rubisco. The increase in v_{c2} at higher CO_2 concentration is due to the reduction in ATP and NADPH required for photorespiration at the higher CO_2 concentration, and therefore an increase in the potential to regenerate RuBP. Redrawn from Long *et al.* (2004).

which an imbalance between the rate of photosynthetic CO_2 assimilation and the rate at which sugars are utilized for growth triggers a negative feedback that controls the expression of photosynthetic genes (Moore *et al.* 1999, Figure B5.2). Despite a downregulation in the expression of photosynthetic enzymes when plants are grown at elevated c_{ac}, the overall growth rate of the plants is increased. When plants are grown at elevated CO_2 concentrations in Free Air CO_2 Exchange (FACE) experiments, and therefore under relatively natural environmental conditions, the reduction in Rubisco protein is more than compensated by the greater availability of CO_2, the primary substrate for Rubisco (Long *et al.* 2004, Nowak *et al.* 2004). Thus, on theoretical grounds, global increases in c_{ac} should cause increases in primary production. There are reasons why these theoretical predictions may not match observations in real ecosystems. For example, nutrient and climate limitations to plant growth may accompany elevated c_{ac}, causing unanticipated stress or nutrient imbalances.

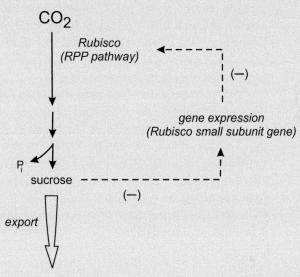

Figure B5.2 Conceptual model showing negative feedback between sucrose concentration in a photosynthetic cell and expression of the gene for the small protein subunit of Rubisco. If the photosynthetic assimilation of CO_2 exceeds the capacity of the plant to accommodate the photosynthate (sucrose) that is produced, then the negative feedback loop will function to reduce the production of more Rubisco protein. As natural turnover of Rubisco occurs, the result will be a gradual reduction in Rubisco protein levels that permits better balance between the rates of photosynthetic CO_2 assimilation and photosynthate utilization.

5.4 The scaled connections among photosynthetic processes

The relations reflected in Eq. (5.25) have become the cornerstones of how we understand photosynthetic carbon cycling at scales ranging from leaves to the globe. Virtually every model of the global carbon cycle will intersect with the relations of Eq. (5.25). In essence, the global carbon cycle is modeled as a "see-saw" tilting toward first-order control by CO_2, and those biophysical processes that control the flow of CO_2 to the chloroplast, or first-order control by RuBP, and those biophysical processes that control the capacity of the chloroplast to regenerate RuBP. In some models, the additional constraint imposed by limited recycling of P_i in the chloroplast (and thus v_{c3}) has been added and this has been critical to understanding the upper limits that exist on adjusting to increases in atmospheric CO_2 concentration.

The challenge, since recognizing these fundamental relations, has been to develop them in forms that are appropriate for higher scales of organization. For example, representation of v_{c2} as a photosynthetic limitation at the scale of a whole canopy is difficult. Most treatments will regard the canopy as one enormous leaf and apply Eqs. (5.25) or (5.26) just as would be done for a single leaf, or they independently consider different vertical layers in the canopy with respect to these equations, and then sum the time-integrated CO_2 flux for all layers. Both approaches require assumptions and simplifications that erode our ability to apply the theory at higher scales of organization. One of the very useful conveniences for purposes of scaling that have emerged in recent years is recognition of the linear proportionalities that have evolved in plants – e.g., linear relations among V_{cmax}, F_{Jmax}, R_d, and leaf N concentration. These proportionalities exist because they reflect efficient allocation of limiting resources among those plant functions critical to growth and reproduction. They have provided mathematical "short-cuts" to the scaling of processes across complex mosaics of vegetation and climatic stresses, though they possess internally and externally imposed limitations, which have not always been appreciated in scaled modeling efforts (see Lloyd *et al.* 2010). We will come back to the challenges of simplifications and their role(s) in scaled modeling as we move to future chapters. For now, we note that the quest for more fundamental, analytical means of expressing the relations of Eq. (5.25) as a function of scale is an extremely important one.

6 Diffusion and continuity

The observation that a mixture of gas molecules of different masses does not behave as expected when subjected to gravity was systematically described by Thomas Graham early in the nineteenth century. Rather than sorting according to their masses, Graham observed that a true mixture was achieved. It was clear that a force must exist to oppose gravity and facilitate the intermingling of gases, but what could be the nature of that force? Even earlier, in 1827, Robert Brown had used a microscope to observe that pollen grains suspended in water exhibit random patterns of motion. Brown reasoned that these motions must be caused by randomly arranged forces in the molecular realm, but once again the nature of such forces was not apparent. It was not until several decades later that Albert Einstein, working on issues concerned with thermal and kinetic energy, provided a theoretical explanation. Einstein reasoned that the thermal energy contained within microscopic bodies is transformed into kinetic energy providing a means for velocity. In the case of Graham's gases we can use Einstein's theory to explain how suspended molecules collide with one another in "random walks," providing the potential for an upward force vector that opposes gravity and sustains a random mixture. In the case of Brown's pollen grains we can use the theory to explain the perpetual motion of water molecules with random collisions occurring between molecules and grains, forcing the grains to vibrate and move through the water. We now understand that Einstein's theories on energy and motion explain one of the fundamental processes by which mass is transported at the microscopic scale – *molecular diffusion*.

We begin this chapter with a discussion of diffusion as a process; defining the length and time scales at which it operates and the framework of interacting forces that defines diffusive flux density. We will start with the simple case of binary diffusion; two constituents transported through opposing density gradients in a common mixture. The case of binary diffusion provides an opportunity to develop a simple phenomenological model, Fick's First Law, which illustrates the dominant driving and resistive forces that determine diffusive flux density. We will then consider the case of diffusion in multi-constituent mixtures; a case where Fick's model is violated. Multi-component diffusion is typically described by an alternative model, the Stefan–Maxwell model, which emerges from fundamental molecular kinetic theory; we will consider it briefly. We will proceed from the multi-component case to that for diffusion through porous media, which brings into play novel interactions between diffusing constituents and the walls of pores, and creates emergent properties that modify the relevant dimensions of diffusion. This will set the stage for treatments in later chapters where we will consider the cases of molecules diffusing through the porous air spaces of soil and leaves. Following the discussion of diffusion through pores we will take note of a fundamental mathematical relation that emerges from diffusion theory; the linkage between divergence in flux density across spans of space and time-dependent changes in scalar concentration above that space. This linkage is required by the conservation of mass and establishes one of the most fundamental relations in flux theory; the concept of continuity. Continuity is the basis from which we analyze the mass and momentum balances of defined control volumes in micrometeorology and the connection between fluxes and gradients in the biogeochemical sciences. We will establish the theory of continuity in this chapter and return to it frequently in future chapters as we develop the concepts of turbulent and advective transport.

6.1 Molecular diffusion

Molecular diffusion is the movement of molecules along a defined flux vector determined by the balance between *driving forces* that cause acceleration and opposing, *resistive forces* that cause deceleration. Motion is propelled by the thermal energy content of molecules. As thermal energy is converted to kinetic energy, molecules take on a free-flight velocity; *the kinetic driving force*. Collision among "unlike" molecules provides a resistive force, or drag, that reduces the potential for molecules of any single constituent to move along sustained, flux vectors. At the local scale (clusters of molecules), this combination of sequential collision and acceleration/deceleration along random velocity vectors creates a type of molecular chaos. At the broader scale of the entire diffusion system, gradients in constituent density, however, produce systematic, organized fluxes that can be described reliably by statistical and physical characterization. It is in the locations and directions of these gradients that we find diffusion linked to biogeochemical sources and sinks. It is at the scale of molecular gradients that we will begin our exploration of diffusion.

6.1.1 A physical description of molecular diffusion

The purpose of this section is to provide the reader with a general impression of diffusive processes. This treatment has been derived in concept from Cunningham and Williams (1980, Chapter 1), and the reader is referred to that book for a mathematically nuanced discussion. We start by defining a simple system in which an ideal gas consisting of a single type of molecule occupies unbounded space. We assume that when averaged across time, molecular density (and thus pressure) is uniform across the space. Molecules will move randomly through the system, sustained by their thermal energy, colliding with other molecules. We can define the *thermal speed* for any single molecule, and a *reference velocity* for the entire population of molecules. We assume that the reference velocity will be zero because of the large number of molecules and the random nature of their motions; any molecule exhibiting velocity in any single direction is on average going to be balanced by a different molecule exhibiting equal velocity in an opposing direction. We can define the spatial domain of the system according to a set of reference coordinates that we will refer to as the *reference frame*. Given the existence of zero reference velocity, the reference frame is motionless. We would say that there is *no diffusive flux* within this system, despite the fact that molecules are moving and colliding. The only means by which to establish a flux would be to move the reference frame with an accompanying non-zero velocity.

Now, imagine a cluster of molecules of a different type that are introduced into the system. Furthermore, imagine that these molecules are released such that they are initially clustered in one localized part of the system, and then allowed to disperse. The new molecules will have their own reference frame, and they will exist at a non-zero reference velocity; there will be a systematic, directional pattern to their dispersion. We would say that these molecules exhibit a *diffusive flux* and a *diffusive velocity* as they disperse from their center of mass to more remote domains of the system. The diffusive velocity is defined by *diffusive vectors* that characterize the flux as having speed and direction. The center of mass of the diffusing constituent will change with time, but the center of mass of the entire system will remain uniform. We assume that as molecules of the diffusing constituent disperse outward, the space that they previously occupied will be backfilled through binary exchange with unlike molecules from the surroundings. Thus, the diffusive flux of the newly introduced constituent causes an equal and opposite flux in its binary complement. The diffusive vectors for both constituents will, on average, sum to zero.

The diffusive process that we just described is *segregative* in the sense that the flux of each constituent can be characterized by its own individual velocity vector. It is possible to distinguish and segregate constituents according to their diffusive velocities. Now, let us consider the case whereby the two constituents differ significantly in their masses, and therefore in their molecular speeds. Let us assume that the newly introduced constituent is considerably lighter than the constituent composing the original reference frame. As molecules of the new constituent are released they will move through the system at higher velocities, on average, than the original constituent, creating momentary lags in the rate at which their positions can be backfilled. This condition will establish a local pressure gradient with lower pressure near the cluster of introduced molecules and higher pressure in the expanding domains of dispersal. As this local pressure gradient increases, it will accelerate molecules in the vicinity of the dispersing cloud toward the center of mass, thus opposing the diffusive velocity vectors of the lighter constituent, and dissipating the pressure gradient. This pressure-driven acceleration is *non-segregative* in that it carries both constituents with the same velocity toward the region of backfill, and it is nearly instantaneous, being tightly coupled in time with expansion of the dispersal cloud of the lighter constituent. This non-segregative flux is referred to as the *non-equimolar flux*, because it results from a difference that forms in the molar density gradients of opposing constituents; that is, it is due to the overall *non-equimolar* state of the system. (The non-equimolar flux is sometimes called the *diffusive slip flux*.) The local pressure gradients that form and drive non-equimolar fluxes are internal; that is, they are embedded within the diffusion system. These pressure gradients are different than those that might develop external to the system, such as those that are sustained by point-to-point differences in air temperature or atmospheric pressure, and thus drive advective flows. The segregative and non-segregative components of diffusion are inseparable – one component creates the other. Thus, we tend to refer to both as diffusive fluxes. We will refer to the combined segregative and non-segregative components as *ordinary, molecular diffusion*.

6.1.2 Diffusion time and space scales and the concept of diffusivity

Ordinary, molecular diffusion is defined by characteristic length and time scales. It is important to understand these scales in order to appreciate the limits of diffusion as a transport mechanism. The *mean free path* is defined as the average distance covered by a diffusing particle before it makes contact with another particle (either of the same or different type), and *mean free time* is the average time between contacts. In an ideal gas, the mean free path and mean free time will depend on constituent mass and density, as well as temperature and pressure. For a binary system we can describe the diffusion potential for individual molecules of each constituent according their *diffusivities*. Diffusivity is defined with units $m^2 s^{-1}$. From these units it is clear that a constituent's diffusivity is proportional to the product between velocity ($m s^{-1}$) and distance (m) between contacts. In other words, diffusivity reflects the product between free flight velocity and mean free path. Diffusivity will be uniquely dependent on the nature of constituents. The diffusivity of one constituent, for example constituent i, in a binary mixture with a second constituent, for example constituent j, will not be equal to the diffusivity of constituent i in a binary mixture with a different, third type of constituent, for example constituent q.

   Diffusivity of a constituent is influenced by the overall molecular density of the surrounding medium. The diffusivity of CO_2 molecules in air, for example, is 10 000 times greater than that for CO_2 in water (16 $mm^2 s^{-1}$ in air compared to 0.0016 $mm^2 s^{-1}$ in water). At a constant temperature the average kinetic energies of CO_2 molecules in air versus water are constant; meaning that free flight velocity should also be constant. Thus, by default, the lower diffusivity of CO_2 in water versus air is due to a lesser mean free path in water. That is, the average frequency by which collisions occur between CO_2 molecules and water molecules is greater in liquid water, than between CO_2 molecules and N_2 or O_2 molecules in air.

   Diffusivity is also influenced by temperature and pressure. In gases the change in diffusivity of a constituent is directly proportional to a change in temperature (T) and inversely proportional to a change in pressure (P). The relation between diffusivity and T is explained by the fact that an increase in the average thermal energy of a system will cause an increase in the kinetic free-flight velocities of system constituents. The dependence of diffusivity on T for constituents in a gaseous mixture scales approximately as T^2 (Brown and Escombe 1900, Fuller *et al.* 1966).

Diffusivity of a constituent is inversely proportional to the pressure (P) of the medium through which it is diffusing. This is explained by an increase in the mean free path as P is decreased. Recall from the ideal gas law (Section 2.5) that the pressure and molar density (ρ_m; mol m^{-3}) of a gas are proportional to one another according to P = ρ_m RT. Assuming constant temperature, a decrease in pressure must be accompanied by a decrease in molar density, which in turn must be accompanied by a decrease in collision frequency and an increase in diffusivity. The overall influence of T (in K) and P (in kPa) on diffusivity (K_d) for constituents in relatively diffuse gas mixtures can be expressed as:

$$K_{dij} \approx K_{d0ij} \left(\frac{T}{273} \right)^2 \left(\frac{101.3}{P} \right), \tag{6.1}$$

where K_{dij} is the binary diffusivity for constituent i with respect to j, K_{d0ij} is the standard, or reference, diffusivity for i with respect to j determined at 273 K and 101.3 kPa ambient pressure. The dependencies of diffusivity on temperature and pressure have potentially important ramifications for studies of plant-atmosphere gas exchange across altitudinal gradients (Box 6.1). While the partial pressure gradients for gases between leaves and the atmosphere decreases as altitude increases, diffusivity for those gases will increase in a compensatory manner. *In fact, the inverse proportionality between K_d and P, and the direct proportionality between ρ_m and P, mean that while K_d*

Box 6.1	Photosynthetic diffusive relations at high elevation

The partial pressure of an atmospheric constituent will decrease as elevation increases and, concomitantly, atmospheric pressure decreases. By Henry's Law the mole fraction of CO_2 in the aqueous phase of the chloroplast (c_{cc}) depends on the partial pressure of CO_2 in the intercellular air spaces (p_{ic}), meaning that c_{cc} should also decrease with increasing elevation. Past studies have reported that many plants at high elevation have likely adjusted to the lower c_{cc} by increasing the concentration of leaf Rubisco, thus increasing the net CO_2 uptake rate per unit of leaf area (e.g., Körner and Diemer 1987, Friend *et al.* 1989). In an early theoretical study, however, it was noted that the decrease in p_{ic} due to an increase in elevation should be accompanied by an increase in the diffusion coefficient for CO_2 in air (K_{dc}) (Gale 1973). In other words, although CO_2 becomes "scarcer" at high elevations, it also has greater diffusive mobility. The photosynthetic relations of high-elevation plants are made even more complex by the fact that most alpine plants occupy microhabitats near the ground where the air temperature can be several degrees higher than the air further above the ground. It is known that the Michaelis–Menten affinity coefficients of Rubisco for CO_2 (K_c) and O_2 (K_o) increase differentially with temperature (Jordan and Ogren 1984), and the Henry's Law coefficients for the solubility of CO_2 and O_2 in water decrease differentially with temperature. Given the complexity of these interactions, Terashima *et al.* (1995) used models of gas diffusion and photosynthetic biochemistry to estimate responses of photosynthesis rate to increases in elevation. The model predictions showed that reductions in p_{ic} as elevation increased caused the gross photosynthesis rate (A_g) to decrease, even with consideration of the concomitant increase in diffusion coefficient. The greatest decreases in A_g occurred at the highest leaf temperatures (Figure B6.1).

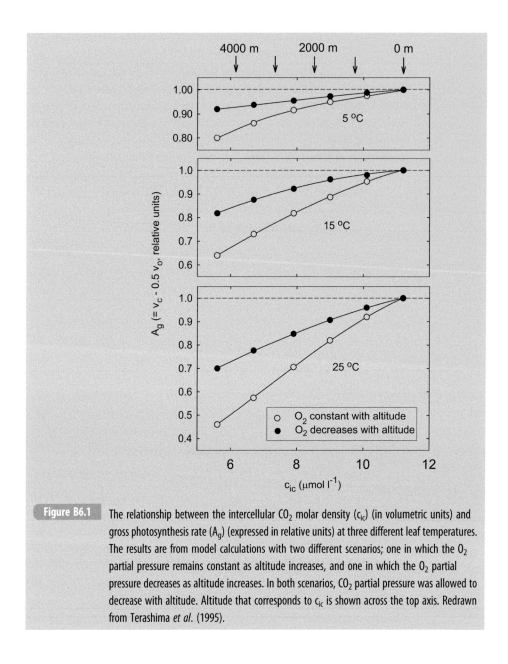

Figure B6.1 The relationship between the intercellular CO_2 molar density (c_{ic}) (in volumetric units) and gross photosynthesis rate (A_g) (expressed in relative units) at three different leaf temperatures. The results are from model calculations with two different scenarios; one in which the O_2 partial pressure remains constant as altitude increases, and one in which the O_2 partial pressure decreases as altitude increases. In both scenarios, CO_2 partial pressure was allowed to decrease with altitude. Altitude that corresponds to c_{ic} is shown across the top axis. Redrawn from Terashima *et al.* (1995).

is dependent on P, the diffusive flux density for a constituent within the system, is generally not.

Finally, we note that diffusivity is dependent on the mass of diffusing constituents. Thus, heavier molecules will diffuse more slowly than lighter molecules with respect to the same binary counterpart. The diffusion coefficients for two molecules of different masses (*i* and *j*) diffusing with regard to the same binary complement (e.g., constituent *q*) can be related to one another according to:

$$K_{diq} = K_{djq}\sqrt{\frac{\mathrm{m}_j}{\mathrm{m}_i}}, \qquad (6.2)$$

where m_j and m_i refer to the molecular masses of constituents j and i, respectively. Equation (6.2) is derived from Graham's Law, and it provides the kinetic basis for stable isotope fractionation by diffusion, which will be discussed in more detail in Chapter 18.

The influences of T and P on the diffusivity of dense gases (i.e., non-ideal gases), liquids, or dissolved gases are more difficult to predict. In dense gases and liquids the close packing of molecules introduces inter-molecular forces, including polar and non-polar forces that influence viscosity of the diffusion medium and reduce the activities and potentials for elastic collisions among diffusing constituents. In general, as T increases diffusivity in liquid systems also increases, both due to an increase in the average kinetic energy of particles and a decrease in viscosity of the medium. Pressure has less influence on diffusivity in liquids, compared to gases, due to the already high densities of particles in liquids and therefore lesser margin for a change in mean free path as a function of changing P. For diffusion in water the temperature dependence of diffusivity can be approximated by a form of the Einstein–Stokes equation, and written as:

$$K_{diw} \approx K_{d0iw}\left(\frac{\mathrm{T}}{273}\right)\left(\frac{1.79 \times 10^{-3}}{\mu}\right), \qquad (6.3)$$

where K_{diw} is the binary diffusion coefficient for constituent i in water, T is in K, and μ is the dynamic viscosity of water (at temperature T) in units of kg m^{-1} s^{-1}.

6.1.3 Diffusion length and the effective scale for diffusive fluxes

Now, having developed the concept of diffusivity according to its underlying forces and the influences of state parameters, let us address the question: across what scale of distance is molecular diffusion most effective as a transport process? *Diffusion length* describes the distance that a unit of density for a constituent is propagated in a given amount of time (t). Mathematically, the diffusion length is defined as $2\sqrt{K_d t}$. In order to provide an example we return to our case of CO_2 diffusing in air versus water. The binary diffusion coefficient for CO_2 in air (K_{dca}) is 16 mm^2 s^{-1}. Using a time scale of 1 second (a typical time scale for leaf-atmosphere exchange), the diffusion length is calculated as 8 mm. The diffusion coefficient for CO_2 in water (K_{dcw}) is 0.0016 mm^2 s^{-1} and the diffusion length is calculated as 0.08 mm. Thus, as a transport process, diffusion is most effective at distances that range from a few nanometers to a few millimeters; distances that characterize flux through a cell or across a leaf epidermis, but not through a forest canopy.

The word "diffusion" does indeed occur in descriptions of transport at larger scales. For example, the term "eddy diffusion" is often applied to turbulent transport within and above canopies. It is important to recognize that this application of the word is solely for purposes of convenience. It is intended to build an analogy between two processes (diffusion and turbulent transport) in which atmospheric constituents are moved from regions of greater to lesser density. Turbulent fluxes, however, are unique. They are better described as viscous flows because they are driven by the bulk flow of a fluid (in this case air), not random collisions of molecules with velocity derived from thermal energy. As a fundamental concept, the word *diffusion* is best reserved for interactions at the molecular scale.

6.1.4 Fick's diffusion laws

[A]ccording to this law [Graham's Law], the transfer of salt and water occurring in a unit of time between two elements of space filled with two different solutions of the same salt, must be, ceteris paribus, directly proportional to the difference of concentrations, and inversely proportional to the distance of the elements from one another. (Adolph Fick (1855))

Molecular diffusion within a mixture of two constituents can be described by a statistical framework, based on linear probability, which is capable of predicting both the direction and rate of flux. In the middle of the nineteenth century, Adolf Fick, a German physiologist working from the foundations of Thomas Graham's observations a few years earlier, conducted studies on diffusion that led to the quantitative model that bears his name. Fick's model relates the diffusive flux density (F_{dj}) of a constituent (represented here as j) to its difference in concentration (Δc_j) per unit change in distance (Δx). Given a two-component mixture (j and i), the proportionality coefficient that relates the flux density of j to its concentration gradient is defined by K_{dji}. When stated within the limit as $\Delta x \to 0$ the concentration gradient is expressed as $\partial c_j/\partial x$ and Fick's model takes the form:

$$F_{dj} = -K_{dji} \; \rho_m \left(\frac{\partial c_j}{\partial x} \right), \qquad (6.4)$$

where F_{dj} is expressed in units mol m^{-2} s^{-1}, when K_{dji} is in units m^2 s^{-1}, ρ_m is molar density of the total mixture in units mol m^{-3}, c_j is mole fraction of constituent j, and x is in m. The negative sign on the right-hand side of the equation indicates that the direction of net flux is opposite the direction in which the gradient increases. One condition that is implicit in Eq. (6.4) is that P, and therefore ρ_m, is uniform in space and time. Equation (6.4) is commonly referred to as *Fick's First Law of Diffusion*. We emphasize that Fick's First Law is *only relevant to diffusion in binary mixtures*.

When restricted to the binary case, the mole fraction gradient of one constituent (e.g., j) can be expressed in relation to that of its complement (e.g., $c_j = 1 - c_i$). Furthermore, the condition $F_{dj} = -F_{di}$ must be satisfied in order to preserve the center of mass in the diffusion system. Thus, binary diffusion can be viewed as a process of mutual displacement. On average, the flux of constituent j to a new position along coordinate vector (x) must be compensated by an equal and opposite flux of constituent i along coordinate vector ($-x$). To a first approximation we can also state that $K_{dij} \approx K_{dji}$. Within the context of the discussion in Section 6.1.1, the Fickian diffusion coefficient reflects the combined influences of the segregative and non-segregative components of molecular diffusion. When used in Fick's First Law, K_d is best described as a phenomenological coefficient. However, it does have a thermodynamic basis. A thermodynamic derivation of Fick's First Law, including consideration of K_d is presented in Appendix 6.1. Within the framework of

Fick's First Law, knowledge of the concentration gradient and flux density of only one of the two constituents is required in order to solve the diffusive state of the entire system. This means that Fick's First Law is an analytical (complete) solution to diffusion in a binary system.

Recognition that Fickian diffusion describes the time-dependent process of "dispersion" allows us to combine Fick's First Law with the law of conservation of mass and derive a second relation, commonly referred to as *Fick's Second Law*:

$$\rho_m \frac{\partial c_j}{\partial t} = K_{dji} \ \rho_m \frac{\partial^2 c_j}{\partial x^2}. \tag{6.5}$$

Fick's Second Law states that the one-dimensional (*spatial*) change in the mole fraction gradient (the second derivative of the mole fraction gradient with respect to distance) scales with the time-dependent (*temporal*) change in mole fraction along that gradient. Fick's Second Law forms the foundation of the concept of continuity, which is discussed in more detail below.

In practice, the Fickian concentration gradient for a single constituent ($\partial c_j/\partial x$) is expressed with a variety of units. We have used *mole fraction* in this book because it is a unit of measure that is independent of pressure and temperature. However, one of the most commonly used units has c_j in terms of *mass density* (mass per unit volume). In those cases where mass density is used to define the concentration gradient, Fick's First Law must be expressed as:

$$F_{dj} = -K_{dji} \ \frac{\partial c_j}{\partial x}, \tag{6.6}$$

where F_{dj} resolves to units of $g \ m^{-2} \ s^{-1}$ when K_{dji} is in $m^2 \ s^{-1}$, c_j is in $g \ m^{-3}$, and x is in m.

6.1.5 Fick's First Law as an "electrical analog"

As a conceptual convenience, many studies of the diffusive exchanges of CO_2 and H_2O, especially in the older literature of plant physiology, have noted analogies in the form of Fick's First Law and *Ohm's Law*, a fundamental relation in electrical physics. In Ohm's Law, electrical current and resistance are mathematically related to the difference in electrical potential according to $I = \Delta V/r$, where I is electrical current, ΔV is the voltage difference (potential) between two points, and r is electrical resistance. In the Fickian analogy, the diffusive flux density represents the electron flux of an electric current, the difference in concentration between two points (Δc) represents the difference in electrical potential (ΔV) and diffusive resistance (defined as $\Delta x/K_d$) represents the electrical resistance (r). In some cases, *diffusive conductance* (represented by the lower-case letter g) is used rather than resistance. Conductance is the reciprocal of resistance (g = 1/r). Conductance has become especially popular in describing the theory underlying plant-atmosphere fluxes because it scales positively with diffusive flux density (i.e., $F_d = (\Delta c) \ g$), which simplifies the form of some mathematical derivations.

6.1.6 The Stefan–Maxwell model for molecular diffusion

Fick's First Law is limited to the case of binary mixtures. When considering the case for a multi-component mixture, an alternative model is often used, the *Stefan–Maxwell Diffusion Model*. The principal tenet of the Stefan–Maxwell model is that the one-dimensional driving force for steady-state diffusion of a constituent must be balanced by loss of momentum due to collisions with other (unlike) components of the mixture. The Stefan–Maxwell model is derived from molecular kinetic theory, and is therefore more fundamental in the physics that underlie its form. Diffusivity in the Stefan–Maxwell model is based on differences in the diffusive velocity vectors of interacting constituents, and the need to maintain conservation of momentum in the system by balancing those differences against molecular drag due to particle-to-particle collisions. In unbounded diffusion, the magnitude of the diffusion resistance assigned to the flux of species j will depend on the net velocities and gradient densities of all other constituents in the mixture, collectively referred to as i. The Stefan–Maxwell relation can then be written as:

$$-\rho_m \frac{\partial c_j}{\partial x} = \sum_{\substack{i=1 \\ j \neq i}}^{N} \frac{1}{K_{dji}} \left(c_i F_{dj} - c_j F_{di} \right), \tag{6.7}$$

where c_j and c_i are respectively the mole fractions of j and the multi-species mixture represented by i, F_{di} and F_{dj} are respectively the diffusive flux densities of species i and j, and K_{dji} is the diffusivity of species j with respect to the multi-species mixture represented by i. Summation is used to account for all components in the mixture from $i = 1$ to N. In the binary condition, wherein i represents a single constituent, the Stefan–Maxwell equation simplifies mathematically to Fick's First Law. An examination of Eq. (6.7) reveals that the quantity resolved on the left of the relation is the divergence in molar density along the x-coordinate. This divergence is balanced on the right-hand side of the equation by the difference in flux density of constituent j (scaled to the concentration of all other constituents) and flux density of all the other constituents (scaled to the mole fraction of constituent j). In order to better understand the kinetic relations on the right-hand side of the relation, we note that the mean diffusion velocity ($\bar{v}$) of a diffusing species can be expressed as $F_d/(\rho_m c)$, with resulting units of m s^{-1}. With this relation in place and following Leuning (1983), we can modify Eq. (6.7) to:

$$-\rho_m \frac{\partial c_j}{\partial x} = \sum_{\substack{i=1 \\ j \neq i}}^{N} \frac{1}{K_{dji}} \rho_m \, c_i c_j \left(\bar{v}_j - \bar{v}_i \right). \tag{6.8}$$

In this form, it is clear that the divergence in molar density on the left-hand side of the relation is dependent on the differences in mean velocity of the mixture constituents and the relative mole fractions of the constituents. The Stefan–Maxwell model is the foundation on which quantitative relations are derived for interactions among the ternary constituents (CO_2, H_2O, and dry air) involved in leaf-atmosphere gas exchange (Jarman 1974, Leuning 1983).

6.2 Diffusion through pores and in multi-constituent gas mixtures

As we move from the general topic of diffusion to its specific application in describing the transport of gases across leaf surfaces and within soils, we need to broaden the scope of systems and cases that are considered. In particular, we need to take up the cases of diffusion through pores and diffusion in multi-component gas mixtures. The presence of pores imposes bounds on the diffusive system and creates the condition for novel, emergent influences on the diffusive flux vector. Depending on the diameter of the pores, diffusing constituents will interact to greater or lesser extent with the pore walls, thus altering diffusive resistances. Pores also alter the shapes and spatial distributions of constituent concentration gradients, affecting the overall rate of diffusion, and sustaining non-equimolar diffusion in multi-component gas mixtures. Finally, pores provide the opportunity for pressure gradients to develop across a surface, thus permitting bulk fluid flows, which complement or oppose diffusive fluxes, and directly influence the free energy gradients that drive ordinary, molecular diffusion. In order to illustrate the ways that pores and multi-component mixtures affect the diffusive process, and the novel processes and states that emerge, we will begin with an analysis of two extreme and opposite conditions – one in which mean pore diameter ($\overline{d}_p$) is large (i.e., with $\overline{d}_p \gg \lambda_a$, where λ_a is the mean free path of a constituent diffusing in moist air), and one in which the pore is relatively narrow (i.e., $\overline{d}_p \approx \lambda_a$). We will refer to the first case as "molecular diffusion through pores" and the second case as "Knudsen diffusion through pores." After considering the roles of pores and multi-component mixtures in these two extreme cases, we will take up the issue of viscous flows. Viscous flows, also called advective flows, are driven by pressure gradients and they form the basis for non-segregative, non-diffusive fluxes across porous surfaces. At the end of this section we will put all of these processes together in the form of a diffusion model, derived from the Stefan–Maxwell model, which can be applied to cases beyond the binary context of Fick's Laws and to the specific case of diffusion through porous media.

6.2.1 Molecular diffusion of a multi-component gas through pores

In the case with $\overline{d}_p \gg \lambda_a$, and continuing for the moment with our assumption of a binary diffusion system with constituents i and j, collisions among molecules will be frequent compared to collisions with the walls of the pore. Diffusive fluxes for both constituents will be driven by their respective mole fraction gradients and momentum will be extracted from each respective flux through collisions between molecules of unlike constituents. (Collisions among "like" molecules do not change the overall momentum balance of a constituent and therefore do not represent a resistive force to the diffusive flux.) In a two-constituent system, with uniform pressure, the mole fraction gradients of each constituent must be equal and opposite. Thus, $F_{dj} = - F_{di}$. We have already considered this case in some detail, and we have established that Fick's First Law is adequate to describe the segregative and non-segregative components of ordinary, molecular diffusion in this type of system.

Now, let's broaden the scope of our discussion and consider the case of molecular diffusion in a three-component (ternary) system. In a ternary system, inequalities can exist in the magnitude of opposing mole fraction gradients; in other words, the non-equimolar state can be sustained. Such is the case for the diffusion of CO_2, H_2O, and dry air ($N_2 + O_2$) across the porous surfaces of leaves (Jarman 1974, Leuning 1983). In this book, we will not derive a model for ternary diffusion from first principles (see Cunningham and Williams 1980 for that treatment). However, we will consider the ramifications of ternary interactions to leaf-atmosphere exchanges of CO_2 and H_2O, and we will derive ternary flux equations in simplified form in the next chapter when we discuss the specific case of diffusion through stomatal pores. In order to lay a foundation for our future discussions, here we will consider the processes underlying the CO_2 compensation point for a leaf. This example will also serve to illustrate why binary models, such as Fick's First Law, are inadequate to describe a ternary diffusive flux.

In analyzing leaf-gas exchange, we typically focus on the opposing fluxes of CO_2 and H_2O. At one limit of the net CO_2 assimilation rate we can define the CO_2 compensation point (Section 5.3.1). As the atmospheric CO_2 concentration is progressively reduced in the boundary layer of a leaf, a point will be reached at which the net diffusive CO_2 flux density (F_{dc}) will equal zero; that is, the inward flux of CO_2 due to photosynthetic assimilation will be exactly balanced by the outward flux of CO_2 due to respiration (both photorespiration and mitochondrial, dark respiration). Using Fick's First Law, and falsely (for the moment) assuming that we are dealing with two independent, but parallel, binary flux systems (CO_2 diffusion in dry air and H_2O diffusion in dry air), we would predict that if $F_{dc} = 0$, the CO_2 mole fraction gradient across the leaf ($\Delta c_c/\Delta z$) must also equal zero. We know from observations and theory, however, that in the presence of transpiration it is possible for $\Delta c_c/\Delta z > 0$, despite no net CO_2 flux (Jarman 1974, von Caemmerer and Farquhar 1981, Leuning 1983). What is the basis for this apparent paradox? The answer lies in the fact that the parallel diffusion systems involving CO_2, H_2O, and dry air (where N_2 and O_2 are grouped together as one "dry air" constituent) are actually one interacting ternary system. All three constituents collide and interact simultaneously during diffusion. The CO_2 mole fraction gradient across a leaf, and its associated flux density, is approximately three orders of magnitude less than that for H_2O. This state emerges from the relative mole fractions of CO_2 (in the $\mu mol\ mol^{-1}$ range) and H_2O (in the $mmol\ mol^{-1}$ range) in the ambient atmosphere. The ternary nature of the system permits the opposing mole fraction gradients for CO_2 and H_2O to be of such different magnitudes because dry air is present to provide the compensatory balance and sustain uniform pressure across the system. Molecules of CO_2 diffusing *into the leaf* will encounter a higher frequency of collisions with "unlike" molecules that possess opposing flux vectors, compared to CO_2 molecules diffusing *out of the leaf*. This condition creates differences in the diffusive resistances encountered by the gross fluxes of CO_2 across the leaf surface. Thus, with $F_{dc} = 0$, and in the presence of transpiration, the CO_2 mole fraction gradient required to balance the inward and outward fluxes of CO_2 must be greater than zero, and we must invoke the existence of a density (local pressure) gradient in dry air that opposes that of H_2O. Dry air will diffuse into the leaf following this gradient, and thus establish a non-segregative, non-equimolar air flux (or diffusive slip flux).

6.2.2 Knudsen diffusion through pores

Now, we will consider the case of molecular transport across a surface with mean pore diameter of approximately the same length as the mean free path of constituents diffusing in air. (The value of λ_a for constituents in air is $\sim 0.07\ \mu m$ at 20 °C and 101 kPa of atmospheric pressure.) A dimensionless value that is used to define the ratio of length scales involved in diffusion, and is especially useful to the study of diffusion through pores is the *Knudsen number* (*Kn*):

$$Kn = \frac{\lambda_a}{L},\qquad(6.9)$$

where L is a representative length scale (m), in this case the diameter of the pore (i.e., $L = d_p$). Diffusion that occurs when $Kn \approx 1$ is known as *Knudsen diffusion*.

In Knudsen diffusion the types of collisions that resist the flux are different than those that occur during ordinary, molecular diffusion; collisions among molecules are infrequent, and collisions with the walls of the pore are frequent (Figure 6.1). Knudsen diffusion can be visualized as a narrow stream of molecules moving through a pore with frequent molecule-to-wall collisions. As with molecular diffusion, at steady state the driving force provided by the mole fraction gradient is balanced by the resistive force of molecular collisions between unlike constituents, but in this case the unlike constituent is not a complementary component of the gas mixture, it is the pore walls. Unlike the case for ordinary, molecular diffusion, in Knudsen diffusion the potential exists for significant pressure gradients to form across a porous surface; such gradients can be sustained in the steady state by the high diffusive resistance imposed by the narrow pores. Knudsen diffusion is composed of both segregative and non-segregative (non-equimolar) components, like ordinary, molecular diffusion.

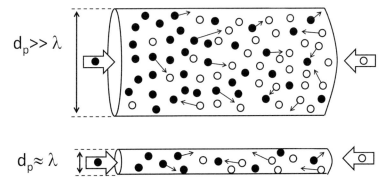

Figure 6.1 Representation of two contrasting patterns of diffusion through pores. In the upper figure, pore diameter (d_p) is greater than the mean free path of molecular motion (λ). Most collisions occur between molecules and diffusion is dominated by ordinary, molecular diffusion. In the lower figure, pore diameter is approximately equal to the mean free path of molecular motion. Most collisions occur between molecules and the walls, and diffusion is dominated by Knudsen diffusion.

6.2.3 The pressure gradient and its role in driving viscous fluxes across porous surfaces

At $Kn \approx 1$, diffusive resistance is great enough for pressure gradients to be sustained across a porous surface. In leaves, such pressure gradients can form theoretically when pore sizes are extremely narrow and solar heating causes an increase in the temperature of internal leaf gases. This type of process has been proposed as a force driving advective flows through the connected culms of rhizomatous wetland plants, allowing for ventilation and the transport of oxygen to submerged tissues in otherwise hypoxic zones of the soil (discussed further in Chapter 16). In the leaves of terrestrial plants pores are rarely of such a narrow diameter that large pressure gradients can be sustained. Thus, many past treatments have ignored pressure-driven flows as a component of leaf-atmosphere gas exchange. Nonetheless, the stomatal pores of leaves do pass through transition phases from nearly closed (at night) to fully open (at midday), and there are times during those transitions when the pressure gradient across a leaf can increase and potentially affect rates of leaf-atmosphere gas exchange. Thus, there are situations that justify consideration of pressure-driven fluxes in the processes of leaf gas transport. Even more importantly, however, we note that if our aim is to develop a complete theoretical treatment of gas transport across leaves, we cannot arbitrarily ignore processes, such as pressure-driven flows, because of a priori biases against their potential contributions. We must include them in the treatment, and then demonstrate their potential for contribution, or lack thereof, in a comprehensive assessment along with all other contributing factors. Thus, we will introduce the topic of pressure-driven viscous fluxes here, even while knowing that they will be resolved as minor components to the overall rate of gas transport across leaf surfaces. There is another, indirect advantage to laying the theoretical foundation for describing viscous flows at this point in the book. Viscous flows will be more important to our future discussions of liquid transport in whole-plant and soil water relations. Thus, we will lay the foundation here, and then pick up the topic once again in Chapter 9.

At pore diameters in which $Kn \approx 1$, diffusive resistances are so great that viscous flows in response to a pressure gradient must be minimal. This self-evident inference reflects the fundamental paradox of viscous flows through pores – the pores must be narrow enough to sustain a pressure gradient across the pore, but wide enough to permit gas to flow through the pore in response to the gradient. The molar viscous flow of dry air through a capillary (F_v; mol m^{-2} s^{-1}), assuming laminar flow, can be written as:

$$F_v = -\frac{\rho_m}{A}\left(\frac{\pi r^4}{8\mu}\frac{\partial p}{\partial z}\right), \tag{6.10}$$

where ρ_m is the molar density of dry air (mol m^{-3}), A is the cross-sectional area of the capillary (m^2), r is the radius of the capillary (m), μ is dynamic viscosity (kg m^{-1} s^{-1}), p is air pressure (Pa), and z (m) defines the primary flow axis as parallel to the vertical coordinate. The parenthetical component of Eq. (6.10) is the one-dimensional form of the *Hagen–Poiseuille Law*. The finite form of this relation is frequently used in the field of fluid

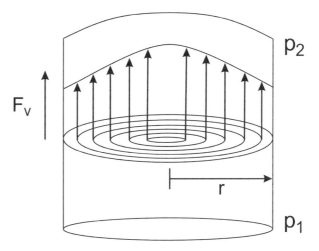

Figure 6.2 Conceptual representation of the Hagen–Poiseuille relation whereby laminar flow of a fluid through a cylindrical pore is presented as thin layers of fluid moving through viscous flow in response to a pressure gradient (i.e., $p_1 > p_2$).

mechanics to predict flows through pipes given a known pressure gradient ($\Delta p/\Delta z$). The relation can be visualized as concentric layers of a fluid moving in parallel in response to ($\Delta p/\Delta z$), with the flow velocity dependent on the radius of the pipe (Figure 6.2). The term ($\pi r^4/8\mu$) can be interpreted as the conductive element in the relation, which facilitates flow given a driving force ($\partial p/\partial z$).

For the case of a porous medium, an analogy can be made of numerous capillaries, but with the inclusion of obstructing mass occupying the space between pores. In that case, the term ($\pi r^4/8$) is replaced with a phenomenological "permeability" coefficient (B_k; unitless), which represents the fraction of total cross-sectional surface area (A; m^2) available for flow. Pressure-driven flow through porous media is modeled according to *Darcy's Law*, which is expressed in its one-dimensional form as:

$$F_v = -\rho_m \, A \, \frac{B_k}{\mu} \left(\frac{\partial p}{\partial z} \right). \tag{6.11}$$

Darcy's Law is often used to describe the flow of water through porous media, such as soil, but it has also been used as the basis for describing viscous fluxes across leaf surfaces.

The potential for viscous flows to contribute to gas transport across a porous surface will be determined within a narrow margin of opportunity as pore diameter increases to the point that the resistive forces of Knudsen flow give way to the inertial forces induced by the pressure gradient. As pore diameter continues to increase the pressure gradient will be gradually dissipated along with the inertial forces that propel the viscous flow. When considered as an entire transition from extremely narrow pores to extremely wide pores, the components of gas transport will progress from purely Knudsen diffusion, through a phase with combined Knudsen diffusion, ordinary, molecular diffusion, and viscous flux, to a phase of pure ordinary, molecular diffusion.

6.2.4 Modeling gas transport through pores

We now have three foundational concepts in place from which we can develop a formal mathematical description of constituent transport across porous surfaces: (1) a phenomenological understanding of binary diffusion and its limits; (2) an understanding of Knudsen diffusion and its importance within the context of pore wall effects; and (3) an understanding of how pressure gradients form across porous boundaries and how those pressure gradients affect ordinary, molecular diffusion and drive viscous flows. Consider the case of a single gaseous compound (j), which is one constituent of a multi-compound mixture (collectively referred to as i) diffusing across a porous surface, with pores of an intermediate diameter, allowing both ordinary, molecular diffusion and viscous flux. By treating all constituents other than j as a single entity we will simplify our initial treatment to that of binary transport. The total flux density of constituent j (F_j) will equal the sum of its diffusive (F_{dj}) and viscous (F_{vj}) fluxes:

$$F_j = F_{dj} + F_{vj}. \tag{6.12}$$

The diffusive flux of Eq. (6.12) can be written with reference to the vertical (z) coordinate as:

$$F_{dj} = \underbrace{-K_{d1}\rho_m \left(\frac{\partial c_j}{\partial z}\right)}_{\text{Term} \quad \text{I}} - \underbrace{K_{d2} \frac{c_j}{RT}\left(\frac{\partial p}{\partial z}\right)}_{\text{II}}, \tag{6.13}$$

where Term I accounts for ordinary, molecular diffusion in the absence of a pressure gradient, and Term II accounts for the effect of any pressure gradients across the pores on the free energy available for molecular diffusion. K_{d1} and K_{d2} are composite diffusion coefficients that include the potential for both Knudsen and ordinary, molecular diffusion according to:

$$K_{d1} = \frac{K_{dji}\,{}^kK_{dj}}{K_{dji} + \left(c_j\,{}^kK_{di} + c_i\,{}^kK_{dj}\right)} \quad \text{and} \quad K_{d2} = \frac{{}^kK_{dj}\left(K_{dji} + {}^kK_{di}\right)}{K_{dji} + \left(c_j\,{}^kK_{di} + c_i\,{}^kK_{dj}\right)}. \tag{6.14}$$

The use of modified diffusion coefficients is justified by the influence of pore wall effects; in other words, the presence of some Knudsen flow properties in this intermediate state between bounded Knudsen diffusion and unbounded ordinary, molecular diffusion, must be accommodated. Modification of the coefficients is accomplished with the composite relations shown in Eq. (6.14), where ${}^kK_{dj}$ and ${}^kK_{di}$ represent Knudsen diffusion coefficients for i and j, respectively, and defined according to:

$$^kK_d = \frac{\overline{d}_p}{3}\left(\frac{8RT}{\pi m}\right)^{1/2} = \frac{\overline{d}_p}{3}\overline{v}. \tag{6.15}$$

From Eq. (6.15) it is clear that as $\overline{d}_p$ decreases, so do ${}^kK_d$ and $\overline{v}$, and $\overline{v}$ in Knudsen flow is equal to $\sqrt{8RT/\pi m}$. The derivations of K_{d1} and K_{d2} are relatively complex, but they have been described in detail in Leuning (1983). We can now use a form of Darcy's Law to represent the viscous flux and obtain a complete description of the transport of j across a porous surface:

$$F_j = -K_{d1}\rho_m \left(\frac{\partial c_j}{\partial z}\right) - K_{d2} \frac{c_j}{RT} \left(\frac{\partial p}{\partial z}\right) - \rho_m c_j A \frac{B_k}{\mu} \left(\frac{\partial p}{\partial z}\right).$$ (6.16)

$$\phantom{F_j = -K_{d1}\rho_m \left(\frac{\partial c_j}{\partial z}\right)}\mathrm{I}\phantom{-K_{d2}}\mathrm{II}\phantom{\frac{c_j}{RT}}\mathrm{III}$$

It is clear from this analysis that relations among pore diameter, the pressure force, and the relative contributions of Knudsen diffusion, ordinary, molecular diffusion, and viscous flow are complex and interrelated. As pore diameter decreases, the pressure differential across a porous surface increases toward an asymptotic limit as $\overline{d}_p \to 0$ (Figure 6.3A). As the mean

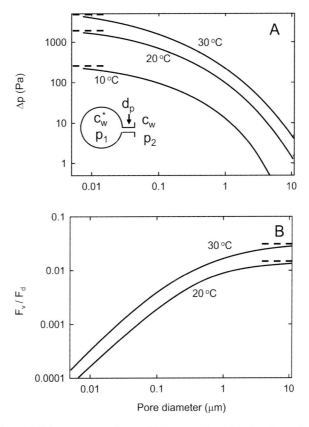

Figure 6.3 **A**. The pressure differential (Δp) across a pore that would be expected as a function of pore diameter at three temperatures, assuming the system is at standard atmospheric pressure. **B**. The ratio of viscous flow (F_v) to diffusive flow (F_d) through a pore with viscous flows driven by the pressure gradients shown in panel A as a function of pore diameter at two temperatures. The dashed lines indicate asymptotic limits for the trends. For reference, leaf stomatal pores tend to exhibit diameters in the range 20–50 μm when fully open. Redrawn from analyses presented in Leuning (1983).

pore diameter approaches 0.1 μm, however, transport through the pores approaches the limit defined by Knudsen diffusion, and even though the pressure force is large, viscous flow represents a small fraction of the total flux density (Figure 6.3B). As pore diameter increases, the fraction of viscous flow to diffusive flow also increases. However, because the pressure differential across the pore progressively decreases, the fractional contribution of viscous flow to the total flux density remains small ($F_v/F_d < 0.05$), even for relatively large pores.

Stomatal pores in leaves typically have diameters of 20–50 μm when fully open (although it should be noted that these pores are seldom circular in shape, but rather elliptical) (Meidner and Mansfield 1968). When closed, the residual opening of the stomatal pore is < 1–2 μm. *Thus, when leaves are transpiring at high rates, the pores exhibit diameters that are beyond those that facilitate Knudsen diffusion, and while some viscous flow is likely to occur, 95% or more of the total flux density through the stomata is due to ordinary, molecular diffusion.*

To some degree, we could have predicted this outcome using the knowledge that stable isotope fractionation during photosynthesis and transpiration is significant. The transport of constituents composed of stable isotope variants should not be fractionated in true viscous flow – all molecular masses should be transported with nearly equal efficacy in an advected fluid. The differences in mass among stable isotope variants are simply too small to matter in fluids advected by all but the very lowest pressure gradients. The fact that we find significant fractionation of $^{13}CO_2$ and $H_2{}^{18}O$, compared to $^{12}CO_2$ and $H_2{}^{16}O$, during leaf photosynthesis and transpiration, respectively, indicates that most transport occurs by diffusion, not viscous flow.

6.2.5 Paradoxical diffusive flux densities across a porous surface

Diffusion across a surface with pores is capable of occurring at considerably higher flux densities compared to diffusion from an open surface. For example, the maximum diffusive flux density of water vapor across a porous leaf surface is several times higher than that from an open pan of water, despite the fact that the pore area of a leaf surface is a small fraction of the equivalent surface area of a pan of water. This apparent "paradox of pores" can be explained purely in terms of physics, but it is through biology that it provides advantages to the regulation of trace gas exchange between organisms and the atmosphere. Natural selection is apparently sensitive to those advantages, because we find porous surfaces to be one of the most common types of diffusive architectures used in both the animal (e.g., the porous shells of bird eggs) and plant (e.g., the porous surfaces of leaves) kingdoms.

To explain the paradox of pores, we begin with recognition that diffusion from a bounded, circular pore scales with the area of that pore much differently than for an open diffusion source. Recalling Fick's First Law, we note that the flux density of a constituent in a binary diffusion system scales proportionally and positively according to the surface area of the diffusion front; this is seen in the units of K_d as m^2 s^{-1}. In the case of diffusion from a bounded pore, however, flux density scales with the *diameter* of the diffusion front, *not its cross-sectional area*. This can be demonstrated using dishes of water with different surface diameters (Figure 6.4). The rate of evaporation per unit area decreases exponentially as the diameter of the dish increases. In other words, the total amount of water that is evaporated

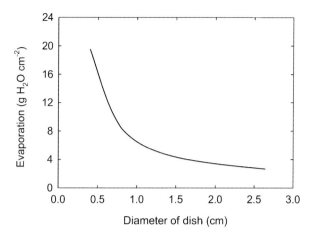

Figure 6.4 Daily evaporation into still air from dishes of open water of different diameter. From data provided in Sayre (1926).

increases as the surface area of the evaporation surface increases, but evaporation efficiency, defined per unit area decreases. This is because the highest evaporation efficiency occurs at the perimeter of the dishes, which is defined by circumference. The cross-sectional area of dishes will increase as a function of $\pi(d/2)^2$, but the circumference, which defines the perimeter length of the dish, will only increase as a function of πd, where d is diameter. For each incremental increase in cross-sectional area as diameter is increased, the circumference will increase less, and the diffusive flux per unit of cross-sectional area will decrease. The difference between diffusion from a pore, as opposed to an open surface, derives directly from this so-called *perimeter effect*.

What is the explanation for the perimeter effect in terms of physics? In still air, as molecules diffuse through a pore and disperse into space, a shell of contours forms with concentration gradients extending outward from the pore (Figure 6.5). The steepest part of the gradients will occur along vectors fanning out from the pore's perimeter. In the case of small, open pores the ratio of pore circumference to surface area will influence the diffusion rate in a way not possible for an unbounded surface. Thus, the existence of small, bounded pores in a surface provides an effective means by which to control diffusive flux. The advantage of a high diffusive efficiency is amplified when the diameter of pores is adjustable and subject to control, such as in the case of plant stomata. The paradox of pores provides a wonderful example of one of our favorite "tongue-in-cheek" phrases – "*if physics is truth, biology is how it's done.*"

6.2.6 End effects on diffusion through pores

It is clear that the rate of diffusion of a constituent gas through pores is difficult to predict because of the shape and depth of the concentration contours (isopleths) in the shells of constituent that accumulate at the outlet of the pores. To develop this concept further let us

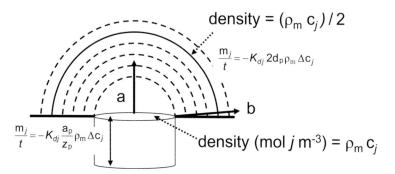

Figure 6.5 Scheme of diffusion for constituent j through a pore and into the atmosphere. The flux expressed as moles of j (m_j) per unit time (t) is dependent on Fick's First Law within the pore cylinder, and is linearly, directly dependent on the cross-sectional area of the pore (a_p) and the mole fraction gradient (Δc_j), and linearly, inversely dependent on the length of the pore (z_p). The molar flux (m_j/t) through the atmosphere above the pore is linearly, directly dependent on the diameter of the pore (d_p), not the cross-sectional area. The broken lines indicate contours of equal density and the solid curve indicates the cross-sectional spherical surface at which the molar density is equal to half that at the pore surface. The arrows labeled a and b are of equal length, showing that the shortest path through the density gradient is from the perimeter, outward at the lowest possible angle. Thus, the greatest flux occurs at the perimeter of the pore, defined by its circumference.

consider our pores to be composed of two separate components, short narrow tubes with diffusive resistance across the porous surface indicated as r_{tube}, and circular openings present at both ends of the tube with diffusive resistance indicated as r_{pore}. For the moment, we will consider this system as the stomatal pores in the surface of a leaf. Further, we will assume that the average diameter of the tubes is considerably greater than the mean free path of air, and therefore most of the diffusive flux through the system is by ordinary, molecular diffusion. Considering diffusion in one dimension, parallel to the cross-sectional axis of the tubes (defined as z), resistance to the flux will be greater than predicted solely on the basis of the average tube length (defined in Figure 6.5 as z_p). This is because the concentration of the diffusing constituent at the ends of the tubes will differ from that in the boundary layer above and below the leaf surface. These end effects create an "effective tube length" that is dependent on the depth of the concentration contours that fan out from the openings at the end of the tubes. Correct reconciliation of diffusive resistance must take account of an *end correction*. For simplicity, we will derive the end correction for the case of binary diffusion (according to Meidner and Mansfield 1968), rather than ternary diffusion. We begin with Fick's First Law and recognition that the average diffusive resistance due to the stomatal tubes ($\bar{r}_{tube}$), and referenced to the entire area of the leaf, is defined as $\rho_m (\Delta c_j/F_{dj})$, where Δc_j is expressed as mole fraction, F_{dj} is the molar diffusive flux density of a constituent expressed per unit leaf area, and $\bar{r}_{tube}$ is reconciled to units of s m^{-1}:

$$F_{dj} = -K_{dj}\frac{\overline{A_p}}{A_L \bar{z}_p}\rho_m \Delta c_j; \text{ and therefore } \bar{r}_{tube} = \left|\rho_m \frac{\Delta c_j}{F_{dj}}\right| = \left|\frac{A_L \bar{z}_p}{\overline{A}_p K_{dj}}\right|, \qquad (6.17)$$

where $\overline{A}_p$ is the average area of the stomatal pore openings in the surface (m^2), A_L is the total leaf area contributing to the flux (m^2), $\overline{z}_p$ is the average depth of the stomatal tubes, ρ_m is molar density relative to dry air (mol air m^{-3}), Δc_j is mole fraction of the diffusing constituent relative to dry air (mol mol^{-1} dry air), and F_{dj} is expressed in units mol m^{-2} s^{-1}. Recognizing that diffusive flux through the pores at the ends of the tubes is a function of average pore diameter, we can define the diffusive resistance due solely to the pores ($\overline{r}_{pore}$), and referenced to the entire leaf surface area, as:

$$F_{dj} = -K_{dj}\frac{2\overline{d}_p}{A_L}\rho_m \Delta c_j; \text{ and therefore } r_{pore} = \left|\rho_m \frac{\Delta c_j}{F_{dj}}\right| \approx \left|\frac{A_L}{K_{dj}2\overline{d}_p}\right|, \qquad (6.18)$$

where $\overline{d}_p$ is the average diameter of the pores (assuming circular pores). Recognizing that $\overline{r}_{tube}$ and $\overline{r}_{pore}$ represent resistances in series with respect to one another (and can therefore be added as a simple sum), taking $\overline{A}_p = \pi\overline{d}_p^2/4$ to account for the cross-sectional area of the tube, and applying algebra, we can write:

$$r_{tube} + r_{pore} = \frac{A_L 4\overline{z}_p}{K_{dj}\pi\overline{d}_p^2} + \frac{A_L}{2K_{dj}\overline{d}_p} = \frac{A_L 4}{K_{dj}\pi\overline{d}_p^2}\left(\overline{z}_p + \frac{\pi\overline{d}_p}{8}\right) = \frac{A_L}{K_{dj}\overline{A}_p}\left(\overline{z}_p + \frac{\pi\overline{d}_p}{8}\right). \quad (6.19)$$

The end correction, taken as $\pi\overline{d}_p/8$, accounts for the effect of concentration contours at the end of the tube on the diffusive resistance; the term $\overline{z}_p + \pi\overline{d}_p/8$ is sometimes called the *effective length* of the diffusive path.

6.3 Flux divergence, continuity, and mass balance

One of the more useful relations for defining the transport of a conserved quantity through a finite volume emerges from Fick's Second Law and its inherent recognition that the time-dependent change in concentration of the quantity can only occur as the result of imbalance between fluxes of the quantity into and out of a defined volume. To illustrate this concept consider a hypothetical volume into which dry air is flowing at molar flow rate u_{in} (mol air s^{-1}) and which contains scalar j at mole fraction $c_{j\ in}$ (Figure 6.6). Thus the overall molar flux (mol s^{-1}) of j into the volume is $F_{j\ in} = u_{in}c_{j\ in}$. Furthermore, let's assume that a leaf with area A_L (m^2) exists within the volume such that some fraction of F_j is assimilated by the leaf as air flows over its surface. This means that the flux of j will progressively decrease as air passes through the volume and the flux of j that eventually exits the volume ($F_{j\ out}$) can be expressed as:

$$F_{j\,out} = F_{j\,in} + \int_{x_1}^{x_1 + \Delta x}\frac{\partial F_j}{\partial x}dx = u_{out}c_{j\,out}, \qquad (6.20)$$

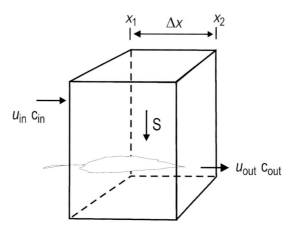

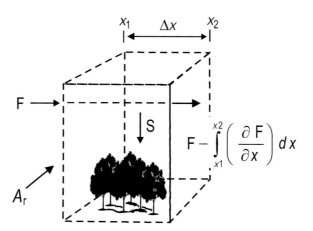

Figure 6.6   Volumes in which the principals of mass balance and continuity are illustrated. In the upper figure mass balance is applied to a leaf chamber in which the flux of a scalar into and out of the chamber is measured as the product between the molar flow rate (u) and mole fraction (c); S is the sink strength of the leaf, expressed as an area dependent flux. In the lower figure mass balance is applied to a hypothetical volume representing the flux footprint for a stand of trees. In this case, the flux of the scalar occurs as the wind advects air through the volume. In the presence of a sink (S), flux divergence $\partial F/\partial x$ is produced.

where u_{out} is the flow rate out of the volume and $c_{j\,out}$ is the mole fraction of j in air exiting the volume. In this case, with the leaf representing a sink for j, $\partial F_j/\partial x$ carries a negative sign. The sink per unit leaf area for j (S_j) can be expressed as:

$$S_j = \frac{F_{j\,in} - F_{j\,out}}{A_L} = \frac{u_{in}\,c_{j\,in} - u_{out}\,c_{j\,out}}{A_L}, \qquad (6.21)$$

where the sink represents a molar flux density (mol m^{-2} s^{-1}). We can use Fick's Second Law to state the *mass balance* of any point within the volume (with the condition $\Delta x \to 0$). We start with a re-statement of Fick's Second Law as:

$$\rho_m \frac{\partial c_j}{\partial t} = K_{dj}\rho_m \left(\frac{\partial^2 c_j}{\partial x^2}\right) = K_{dj}\rho_m \left(\frac{\partial}{\partial x}\left(\frac{\partial c_j}{\partial x}\right)\right) = -\frac{\partial F_j}{\partial x}, \tag{6.22}$$

with the recognition that $F_j = -K_{dj}\rho_m (\partial c_j/\partial x)$. Multiplying both sides by dx, and returning to the context of fluxes through a control volume, we can write:

$$F_{j\,in} - \left(F_{j\,in} - \frac{\partial F_j}{\partial x}dx\right) = \rho_m \frac{\partial c_j}{\partial t}dx. \tag{6.23}$$

Simplifying Eqs. (6.22) and (6.23), we obtain a succinct expression that relates time- and space-dependent interactions in the flux of constituent j through a control volume:

$$-\frac{\partial F_j}{\partial x} = \rho_m \frac{\partial c_j}{\partial t}. \tag{6.24}$$

Equation (6.24) is known as the *continuity equation*. The left-hand side of Eq. (6.24) is the space-dependent *flux divergence* and the right-hand side is the time-dependent *concentration divergence*. In essence, the continuity equation states that for a conserved quantity any spatial divergence in flux must be balanced by a temporal divergence in concentration. For control volumes of finite dimensions and lacking spatially explicit definition of the flux and concentration divergences, the continuity equation is expressed in terms of average fluxes and constituent concentrations within the volume. The continuity equation provides the foundation for mass balance in many biogeochemical models. The continuity equation also provides the basis for calculating leaf-scale, or soil-plot scale, flux rates, which are often measured using leaves or a small plot of soil enclosed by a chamber. In those cases Eq. (6.21) can be used to determine the source or sink term from the air flow rates and scalar concentrations going into and out of the chamber.

Appendix 6.1 A thermodynamic derivation of Fick's First Law

To this point, we have emphasized the statistical nature of Fick's First Law: it is a linear expression of proportional dependency between flux density and the mole fraction gradient. The Fickian diffusion coefficient represents a first-order scaling parameter, but the form of the law provides no insight into the physical determinants of K_d. Now, we will explore a more fundamental thermodynamic treatment within which to interpret K_d. We begin by recognizing diffusion as a response to the non-equilibrium nature of a system defined by a gradient in chemical potential (μ). For a population of molecules of chemical constituent j diffusing with respect to its binary complement i, we can write:

$$F_{dj} = -\frac{d\mu_j}{dx}, \tag{6.25}$$

where F_{dj} is used to designate the molar "diffusive force" and carries units of kg m s^{-2} mol^{-1} (or N mol^{-1}). The molar diffusive force is the "driving force" that reflects the free energy available to move chemical constituent j by diffusion down its gradient in chemical potential in the direction of coordinate x. As was the case for Fick's First Law, the negative sign is introduced to the right-hand side of Eq. (6.25) to indicate that the diffusive force occurs in the direction of decreasing free energy (i.e., diffusion is forced from a point of higher free energy to a point of lower free energy). In an ideal gas, and assuming constant pressure, chemical potential can be defined in terms of the mole fraction of j as:

$$\mu_j = \mu^* + RT \ln c_j, \tag{6.26}$$

where μ^* is the standard molar chemical potential in J mol^{-1} (Section 2.3). Differentiating Eq. (6.26) with respect to distance (x), provides us with:

$$F_{dj} = -\frac{\partial \mu_j}{\partial x} = -RT \frac{\partial \ln c_j}{\partial x} = -\frac{RT}{c_j} \frac{\partial c_j}{\partial x}. \tag{6.27}$$

(Inherent in the derivation of Eq. (6.27) is the recognition that μ^* is a constant, and therefore $d\mu^*/dx = 0$. We used the chain rule for differentiating natural log functions to derive: $(\partial \ln c_j)/\partial x = (1/c_j)(\partial c_j/\partial x)$.) Equation (6.27) links the gradient in chemical potential to that in mole fraction for constituent j.

As a driving "force," $\partial c/\partial x$ is opposed by a resistive "force," which in large part represents collisional interaction. Molecular collisions will tend to redistribute momentum to vectors other than those associated with the net flux. In essence, collisions increase entropy in the diffusion system. Using the thermodynamic concepts we have developed to this point, we can frame our discussion within the context of "mobility," where *diffusive mobility* refers to the capacity for a unit of molecules to move in the face of the resistive collisional forces they encounter. We will define the mean mobility of the molecules of constituent j ($\bar{u}_j$) in mathematical terms according to the Einstein–Smoluchowski relation (Einstein 1905, Smoluchowski 1906):

$$\bar{u}_j = \frac{\bar{v}_j}{F_{dj}}, \tag{6.28}$$

where $\bar{v}_j$ is the mean free-flight velocity of constituent j (m s^{-1}), referenced to coordinate x, and $\bar{u}_j$ carries units of s kg^{-1}. Mobility, as represented in the vector quantity $\bar{u}_j$, reflects the ratio of a population of particles' mean drift velocity (in m s^{-1}) in coordinate direction x, to a mean applied diffusive force (in N). After rearranging we can state:

$$\bar{F}_{dj} = \frac{\bar{v}_j}{\bar{u}_j} = -\frac{RT}{c_j} \frac{\partial c_j}{\partial x}. \tag{6.29}$$

Multiplying all terms by $\bar{u}_j$ and the molar density of constituent j ($\rho_m c_j$) we derive the diffusive flux density (F_{dj} in mol m^{-2} s^{-1}) as:

$$F_{dj} = \rho_m c_j \bar{v}_j = -\rho_m \bar{u}_j RT \frac{\partial c_j}{\partial x}. \tag{6.30}$$

(While potentially confusing, please note that we have used F_d, in italicized form, to indicate diffusive force, and F_d, in non-italicized form, to indicate diffusive flux.) Relying further on Einstein's analysis we can equate $(\bar{u}_j RT)$ with the binary diffusion coefficient originally derived by Fick (K_{dji}), with units $m^2\ s^{-1}$. Thus, the nature of K_{dji} as a "mobility" term is clearly evident:

$$F_{dj} = \rho_m c_j \bar{v}_j = -K_{dji}\ \rho_m \frac{\partial c_j}{\partial x}. \qquad (6.31)$$

It is important to note that the relations presented in Eq. (6.31) are predicated on the assumption of uniform temperature and pressure within the diffusion system. The presence of a temperature and pressure gradients within the system will influence the chemical potential gradient that defines the molar "diffusive force," and thus diffusive mobility. It is also important to note that the mean values used for F_{dj}, u_j, and v_j represent ensemble means determined for all molecules that make up c_j at a single point x.

7

Boundary layer and stomatal control over leaf fluxes

> The resistances encountered by molecules of carbon dioxide in moving into the leaf from the source in the ambient air to the sink at the sites of reaction in the chloroplasts may be used to describe quantitatively specific anatomical and physiological responses to environment ... Similarly, the resistances to the transfer of water through the leaf from the source, which can be considered to be at the termination of the xylem, to the sink in the ambient air, first as a liquid and then as a vapour, describe adaptations and responses of the leaf to control water loss ...
>
> Paul Jarvis (1971)

The quote by Paul Jarvis, which he offered in a synopsis paper concerning leaf diffusive resistances, contains implicit reference to the fact that controls over plant-atmosphere fluxes reflect not only the processes that drive the exchanges of H_2O and CO_2, but also past evolutionary modification of the leaf form and function. Thus, an understanding of fluxes at the plant and leaf scales requires perspectives on *adaptation*, in addition to *biophysical processes*. In fact, recognition that leaf and plant function can be best explained when both of these principles are integrated into a common framework has served as the intellectual cornerstone for the discipline of plant physiological ecology for over four decades. In this chapter we develop this integration with regard to the specific case of leaf processes and their underlying diffusive fluxes. In the next chapter, we will consider explicitly the process of adaptation with regard to leaf function, and the concept of adaptation as an organizing principle from which we can predict patterns of covariance between environmental change and traits that control leaf-atmosphere gas exchanges. Although we will focus on the leaf scale in both chapters, we will also begin to introduce concepts associated with atmospheric pressure gradients, turbulent transport, and eddy diffusivity, all of which will be valuable as we move into future chapters.

Leaf surfaces represent fundamental exchange boundaries between plants and the atmosphere. We can detect the effect of exchange processes at the leaf scale in observations of ecosystem-to-regional scale fluxes made from towers or aircraft; leaf processes leave their "fingerprints" on processes that we observe at larger scales. For example, one of the more commonly observed patterns of surface-atmosphere H_2O exchange is a midday decrease in latent heat loss from landscapes, concomitant with an increase in the incident net radiation flux (e.g., Davis *et al*. 1997). Pure physical principles, operating at the landscape scale, would lead us to predict an increase, not decrease, in midday latent heat loss; higher net radiation loads at the surface should force higher rates of latent energy dissipation. Stomata in the leaf surfaces of many plants, however, close in response to midday increases in atmospheric water vapor deficit. As the day progresses and turbulence intensity increases, drier air from aloft is entrained into more humid air near the canopy surface, and stomata respond by closing to some degree. Stomatal closing in response to an increase in the leaf-to-air water vapor deficit is an adaptive trait that has been introduced into the physiological constitution of plants through the process of natural selection. Thus, the reduction in ecosystem latent heat loss during the middle of the day, coincident with increases in the surface radiative heat load, can be traced to processes that have evolved at the scale of individual leaves. This continuity of processes across spatiotemporal scales allows us to construct mechanistic models that can be applied to broad aspects of hydrology, biogeochemistry, and ecology.

In this chapter, we put the "living plant" back into the mixture of physics and chemistry that has been the focus of the past few chapters. We begin, however, with a continuation of physics as we develop the concept of leaf boundary layers. Shortly thereafter, we will introduce the perspective of leaf form, its effect on boundary layer thickness, and the potential for boundary layer resistances to control leaf temperature. Following our consideration of leaf boundary layers, much of our discussion will be devoted to stomatal function and the quantitative models that have been developed to predict stomatal resistance, or more commonly its inverse, stomatal conductance. As we develop the biological components that control leaf-atmosphere fluxes we will be setting the stage for the introduction of biological evolution and its potential to influence tradeoffs in the exchanges of key plant resources, such as CO_2 and H_2O. This evolutionary perspective can be effectively integrated with biophysical perspectives to provide a framework within which to make predictions and design models around principles that converge toward common patterns of biological design and constraint. Such models of convergent function provide an even broader set of tools from which to generate universal predictions of biosphere-atmosphere exchange across global environmental gradients, and provide a segue to the next chapter on the coupling between leaf form and function.

7.1 Diffusive driving forces and resistances in leaves

During periods of high diffusive flux between leaves and the atmosphere the principal driving force determining the flux density of a scalar constituent is the relevant mole fraction gradient between the well-mixed air outside the leaf and the site of the sink or source within the leaf. In the case of transpiration, the diffusive flux will be mitigated by resistances due to the stomatal pores and the boundary layers at the surfaces of the leaf. In the case of photosynthesis, the diffusive flux will be mitigated by resistances due to the boundary layers, the stomatal pores, and the tissues of the leaf. Thus, we can write an equation for leaf flux density (F_j, mol m^{-2} s^{-1}), considering any given atmospheric constituent j, as:

$$F_j = \rho_m \frac{\left(c_{aj} - c_{ij}\right)}{r_{bj} + r_{sj}\left(+r_{ij}\right)}, \tag{7.1}$$

where ρ_m is molar density referenced to dry air (mol dry air m^{-3}), c_{aj} and c_{ij} are mole fractions (mol j mol^{-1} dry air) for the ambient atmosphere and intercellular air spaces, respectively, and r_{bj}, r_{sj}, and r_{ij} are the leaf area-averaged diffusive resistances (s m^{-1}) of the surface boundary

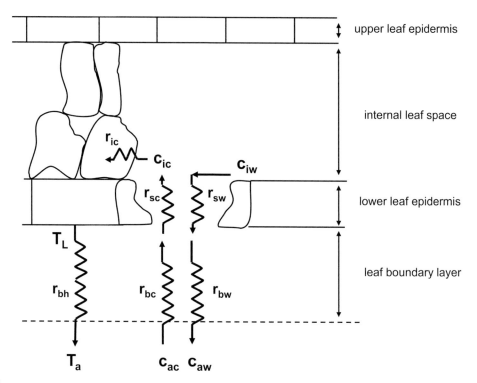

Figure 7.1 Schematic showing the major diffusive paths and resistances (as Ohm's Law analogs), and concentration of sources and sinks for CO_2, water, and heat for a hypothetical leaf. r_{ic} = internal resistance to CO_2; r_{sc} and r_{sw} = stomatal resistances to CO_2 and H_2O, respectively; r_{bc}, r_{bw}, and r_{bh} = boundary layer resistances to CO_2, H_2O, and heat, respectively; c_{ic} and c_{iw} = internal mole fractions of CO_2 and H_2O, respectively; c_{ac} and c_{aw} = ambient mole fractions of CO_2 and H_2O, respectively; T_L and T_a = temperatures of the leaf surface and ambient air, respectively.

layer, stomata, and internal liquid phase of the leaf, respectively. The internal resistance is placed in parentheses to indicate that it is a relevant component of only those flux pathways (e.g., photosynthesis) that include cellular structure. Energy exchange between leaves and the atmosphere is affected by some of the same resistances that affect the exchange of mass. The leaf boundary layer resistance for heat (r_{bh}) influences the exchange of sensible heat with the atmosphere, and the boundary layer and stomatal resistances for water vapor diffusion (r_{bw} and r_{sw}, respectively) influence the latent heat flux. The relevant diffusive pathways and resistances for leaf-atmosphere exchange are shown in Figure 7.1.

7.2 Fluid-surface interactions and boundary layer resistance

As mass and energy are exchanged across the surfaces of a leaf, to and from the atmosphere, a transport resistance is encountered due to the presence of the boundary layers next to leaf surfaces. A boundary layer results from mechanical stress that develops within a flowing fluid in the vicinity of a surface. The mechanical stress is in turn due to friction and pressure forces near the surface that oppose the inertia of the flowing fluid. These concepts, involving mechanical stress and the resultant instabilities that lead to the formation of boundary layers are best discussed within the context of fluid mechanics as applied to compressible fluids.

Flows in compressible fluids are either *laminar* or *turbulent*. We can think of laminar flow in an idealized sense as stacked layers of moving fluid parcels with velocity vectors of the different "layers" oriented parallel to one another and equal in magnitude. In this ideal condition we refer to the flow as *inviscid*. Real fluid flows, however, are not inviscid. Adjacent layers of real fluids interact with one another, as parcels of one layer move up or down into a different layer; colliding with parcels of that layer and transferring momentum. This interaction among fluid layers causes *shear stress*. Shear stress is observed as a change in the local velocities of the stacked, fluid layers, and it is highest at the boundary between the fluid and the surface. In fact, we assume that the fluid has zero velocity exactly at the surface; a condition called the "*no-slip*" *assumption*. Shear stress is proportional to the magnitude of deformation in the fluid, and the mathematical constant of proportionality between shear stress and rate of deformation is known as *viscosity*. Viscosity can be viewed as a metric of "fluid friction"; a force that opposes the inertia of a fluid and together with other forces, such as the pressure force created by molecular collision, contributes to *aerodynamic drag*, a force on the surface in the direction of the flow (derived from the fact that the surface would be "dragged" or carried with the fluid flow if the surface were moveable). (Friction is not a fundamental "force" in the Newtonian sense of the term. However, friction does resist the inertia of a moving fluid, and so it produces a result similar to an opposing Newtonian force. We will use the term "friction" in the same sense as other forces that oppose fluid flow.) Drag on a surface can be demonstrated through sliding a single card from the top of an entire deck of cards by applying pressure and pushing in a horizontal direction; the card beneath the top card may also be dragged in the same direction by shear forces due to downward momentum transfer.

Shear stress causes turbulence in an otherwise laminar flow. Turbulence is manifest in fluid flows as *eddies*. Eddies form when shear stress in the fluid disrupts laminar streamlines and redirects momentum into new velocity vectors. In an apparent enigma, turbulent eddies simultaneously exhibit chaotic and coherent tendencies. Turbulent eddies exhibit coherence as fluid parcels track similar vectors in the vortices that form from the eddy motion; at the same time, eddies move with stochastic tendencies as they break chaotically from the laminar streamlines. Turbulent eddies form from small, shear-induced perturbations in the flow (or *Kelvin–Helmholtz instabilities*). Kelvin–Helmholtz (KH) instabilities organize into KH waves. Kelvin–Helmholtz instabilities are common features of fluids undergoing deformation. When strain within a flowing fluid reaches a critical threshold, layers will separate from the streamlines, rise up and roll forward into a tumbling vortex. It is the KH waves that initiate the property of "coherence," which is propagated as parcels in the fluid follow rotating vectors while moving forward.

The formation of turbulent eddies is clearly visible in Figure 7.2, which shows the formation of a boundary layer in an orderly array of bubble trails as water is forced across a stationary surface. From this photograph we can discern some consistent patterns in the relation between laminar and turbulent flows. First, despite the fact that drag begins working

Direction of fluid flow →

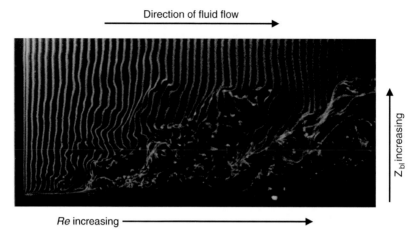

Z_{bl} increasing ↑

Re increasing ——————————→

Figure 7.2   Visualization of a turbulent boundary layer forming above a flat plate containing flowing water with vertical reference lines marked by hydrogen bubbles produced from a pulsed wire along the horizontal axis. As water moves across the surface it is progressively more influenced by the drag force. The orderly lines of hydrogen bubbles are disturbed by the chaotic motions of the turbulent flow. The Reynolds number (*Re*) increases from left-to-right, accompanying the transition from laminar flow to turbulent flow. The thickness of the boundary layer (z_{bl}) is observed to increase as *Re* increases. Image copyright 1980 by Y. Iritani, N. Kasagi, and M. Hirata. Reprinted with permission by courtesy of Y. Iritani, N. Kasagi and M. Hirata.

on the surface when the fluid first contacts it, a finite distance is required before instabilities develop. The emerging instabilities are progressive and the layer of unstable air thickens as a function of distance. Thickening of the turbulent flow gives rise to the boundary layer (Figure 7.3).

The physical extent of a boundary layer is defined by two limits; one at the surface where the no-slip assumption sets the condition of zero velocity, and one at the outer edge where mean fluid velocity is equal to that of the well-mixed atmosphere. Physical description and demarcation of these boundaries is difficult, so we often rely on phenomenological consideration of relations involving heat transfer rate, mean fluid velocity, and surface dimensions to derive boundary layer thickness. We begin a phenomenological derivation of boundary layer thickness with Fourier's Law (Section 2.4). In the case for heat transfer in the vertical (z) coordinate, and integrating the vertical temperature gradient across the finite boundary layer, we can state:

$$H_{se(z)} = -K_{dh}c_p\rho_m \int_0^z \frac{\partial T}{\partial z} dz = -K_{dh}c_p\rho_m \frac{\Delta T}{z_{bl}}, \tag{7.2}$$

where K_{dh} is the diffusion analog coefficient for sensible heat ($m^2\ s^{-1}$), c_p is the specific heat of dry air ($J\ mol^{-1}\ K^{-1}$), ρ_m is the molar density of dry air ($mol\ m^{-3}$), T is absolute temperature (K), and z_{bl} is the depth of the boundary layer (m), such that the relation reconciles to a steady-state, sensible heat flux (H_{se}, $J\ m^{-2}\ s^{-1}$). Now, we can define the heat transfer coefficient (h_c) for the boundary layer (again along the z-coordinate) according to:

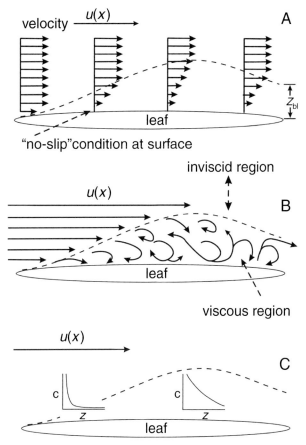

Figure 7.3 Diagrammatic representations of a leaf boundary layer and its influence on scalar concentration gradients. **A.** Velocity vectors for horizontal wind demonstrating the progressive reduction in velocity at the leaf surface due to drag. z_{bl} = boundary layer thickness. **B.** Illustration of how laminar flow breaks into turbulent eddies within the boundary layer. The "inviscid region" is an approximation in which viscous forces in the fluid are assumed to have a negligible influence on flow vectors. **C.** Hypothetical profiles of concentration (c) across the vertical distance of the boundary layer (z), at the leading edge of the leaf and the middle of the leaf. $u(x)$ = wind velocity in the direction of the x Cartesian coordinate.

$$h_c = \left| \frac{H_{se}}{\Delta T} \right| = \left| \frac{K_{dh}c_p\rho_m}{z_{bl}} \right|, \tag{7.3}$$

where h_c ($J\,m^{-2}\,s^{-1}\,K^{-1}$) represents the inherent capacity for the fluid (in this case dry air) to transport sensible heat across distance z_{bl}.

In order to continue, we identify two useful terms: the *Nusselt number* (Nu) and the *Reynolds number* (Re). The Nusselt number is dimensionless and is defined as:

$$Nu = \frac{h_c d}{K_{dh}c_p\rho_m}, \tag{7.4}$$

where d is the length of the surface in the direction of the fluid flow (m), and the denominator term $(K_{dh}c_p\rho_m)$ is *thermal conductivity* ($\mathrm{J\,m^{-1}\,s^{-1}\,K^{-1}}$). Combining Eqs. (7.3) and (7.4), we see that in essence, Nu is a ratio of the horizontal and vertical distances involved in the heat flux; i.e., $Nu = d/z_{bl}$. The Reynolds number is also dimensionless and can be used to define the point along a surface where laminar flow gives way to turbulent flow:

$$Re = u\,d/\nu, \tag{7.5}$$

where u is wind speed ($\mathrm{m\,s^{-1}}$) in the coordinate defined by d, and ν is the kinematic viscosity ($\mathrm{m^2\,s^{-1}}$), in this case for dry air. From empirical studies of the forced convection of air across flat plate surfaces, the following general relationship between Nu and Re has been developed, which is valid for leaves and for mean wind speeds ($\bar{u}$) in the range 0.5 to 5 $\mathrm{m\,s^{-1}}$ (Pearman *et al.* 1972):

$$Nu = 0.97\,Re^{0.5}. \tag{7.6}$$

Taking Nu as d/z_{bl}, expressing ν as a constant ($1.53 \times 10^{-5}\ \mathrm{m^2\,s^{-1}}$, the value in dry air at 20 °C), and using some algebra, the following empirically derived relation can be used to estimate the average boundary layer thickness of a leaf:

$$\bar{z}_{bl} = 4 \times 10^{-3}\sqrt{d/\bar{u}}, \tag{7.7}$$

where the averaging operator, from which $\bar{z}_{bl}$ is determined, is time (Gates 1980, Nobel 1999).

It is useful to take some time to fully appreciate the relations described in Eq. (7.7). As mean wind speed increases, the mean depth of the boundary layer will decrease. In a turbulent boundary layer, increased wind speed leads to greater momentum transfer from the fluid to the surface in the streamwise direction; this "enables" the flowing fluid to more easily overcome viscous forces at the surface, and thus increases the distance from the leading edge at which the boundary layer begins to thicken. Also according to Eq. (7.7), at any given mean wind speed, increases in d, the distance across which inertial forces are permitted to interact with viscous forces, the mean depth of the boundary layer will increase. Keep in mind that boundary layer thickness (z_{bl}) scales in inverse manner with h_c (see Eq. (7.3)); thus, as z_{bl} increases, the capacity for sensible heat transfer from the leaf surface into the well-mixed air above the boundary layer will decrease. In general, greater boundary layer depth is more resistive to diffusive fluxes away from the surface.

The boundary layer is thinnest at the leading (windward) edge of a surface. Moving across the surface from the leading edge the wind will induce greater drag and boundary layer thickness will increase. Spatial variation in boundary layer thickness accounts in part for steady-state gradients in leaf temperature (Figure 7.4). This can be appreciated by rearranging components of Eqs. (7.2) and (7.3):

$$\Delta T = -\frac{H_{se}z_{bl}}{K_{dh}c_p\rho_m}. \tag{7.8}$$

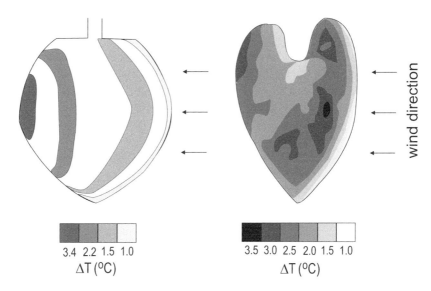

Figure 7.4 Topographical distribution of the difference between leaf and air temperature (ΔT) for (left) an artificial leaf and (right) a real leaf of *Phaseolus vulgaris* (pole bean). Most of the differences in ΔT across the leaves reflect variability in boundary layer thickness. Redrawn from Wigley and Clark (1974), with kind permission from Springer Science + Business Media.

In the leaves depicted in Figure 7.4 the coolest regions occur at the leading edges where the boundary layer is thinnest. The difference between leaf and air temperature is small at the leading edge, and sensible heat exchange will occur with minimal resistance. Leaf temperature tends to be warmer near the middle and far edges of the leaf where the boundary layer is thickest. The complex variation in ΔT that is observed in the real leaf, compared to the artificial leaf in Figure 7.4, is likely due to spatial variation in stomatal conductance and its control over latent heat loss.

Using the phenomenological context of Fick's First Law, we can define boundary layer conductance for binary diffusion as:

$$g_{b_j} = \frac{1}{\rho_m} \frac{F_j}{\Delta c_j}, \tag{7.9}$$

where g_{bj} is boundary layer conductance with respect to constituent j (m s^{-1}), F_j is the molar flux density (mol m^{-2} s^{-1}) across the boundary layer, ρ_m is the molar density of air (mol m^{-3}), and Δc_j is the mole fraction difference across the boundary layer. In this form, g_{bj} is analogous to ($K_{dj}/\Delta x$). It is common practice for researchers to express g_j as the simple ratio $F_j/\Delta c_j$, in which case g_j will carry units of mol (air) m^{-2} s^{-1}, the same dimensional units as F_j. This form of expression is convenient because it reflects conductance as a simple scaled value of the flux density. However, it must be made clear that when expressed with these units, the conductance is actually a mathematical composite of ($g_j\rho_m$), and so the effects of pressure and temperature on ρ_m must be factored into the calculation.

We will be mostly concerned with the case of CO_2 and H_2O diffusing across leaf boundary layers. Due to its higher mass, CO_2 molecules diffuse more slowly than H_2O

molecules in dry air. The ratio of conductances for the diffusion of CO_2 and H_2O can be determined as the ratio of their respective molecular diffusivity coefficients: $K_{dw}/K_{dc} = 1.58$ (Massman 1998). In the leaf boundary layer, the nature of CO_2 and H_2O transport varies from pure diffusion at the surface to mostly turbulent transport at the outer edge; thus, the ratio g_{bw}/g_{bc}, when integrated across the entire depth of the boundary layer will be less than 1.58; a good estimate that has been used in past studies is $g_{bw}/g_{bc} \approx 1.37$ (Schlicting and Gersten 2004).

7.3 Stomatal resistance and conductance

> Gas exchange with the atmosphere is made possible through the evolution in the epidermis of mechanical devices of microscopic dimensions, the stomata. They are able to regulate the exchange of gases between the interior of leaves and the atmosphere. To them the plant has delegated the task of providing food while preventing thirst. (K. Raschke (1976))

The overlying waxy cuticle and tightly connected cells of a leaf epidermis force most of the diffusive flux between the leaf and atmosphere to the stomatal pores. Each stomatal pore is formed by the interaction of bounding guard cells, producing a dynamic aperture. In response to various environmental cues the guard cells undergo changes in hydrostatic pressure, which are translated into changes in cellular volume, and consequently cause them to expand and work against the surrounding cells of the epidermis, thus forming stomatal pores (Figure 7.5). The total area of open stomatal apertures on a leaf surface, relative to the total leaf surface area, should scale proportionately with transpiration rate per unit leaf area. As a surrogate for aperture diameter, and recognizing that diffusive fluxes through stomata are the processes of most interest to the study of leaf-atmosphere gas exchange, we often use stomatal resistance, or its reciprocal, stomatal conductance, rather than stomatal diameter, to define stomatal control over leaf fluxes.

7.3.1 Mathematical representations of stomatal conductance

The concept of diffusion through pores was discussed in detail in Section 6.2 and in Section 6.2.6 we developed an expression of stomatal resistance as a function of pore area, depth, and density. On theoretical grounds, the equations developed in those sections provide clear insight into the dimensional determinants of r_s and, by reciprocal, g_s. When expressed within the context of observed fluxes and gradients, rather than pore dimensions, g_s is calculated as:

$$g_{s_j} = \frac{1}{\rho_m} \frac{F_j}{\Delta c_j}, \tag{7.10}$$

where, in this case, F_j and Δc_j represent the flux density and mole fraction difference from the intercellular air spaces to the near-surface edge of the boundary layer, respectively. As in

A

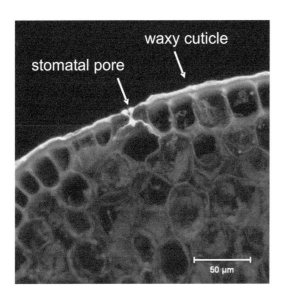

B

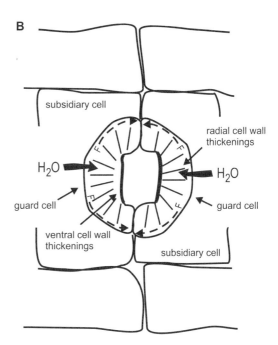

 Figure 7.5 **A**. Confocal microscope image of a stomatal pore in a leaf of *Wachendorfia thyrsiflora* (red root plant). Mesophyll cells beneath the epidermis contain red-fluorescent chloroplasts. **B**. Top view of a pair of guard cells working against one another to form a stomatal opening. The guard cells possess radial thickenings of the cell wall. As water enters the cells and turgor pressure increases, the force of the turgor (F) is directed to right angles to the radial thickenings, causing the guard cell pair to press against each other at the ends of the cells. This radially directed force, coupled to longitudinal thickening along the inner (ventral) walls of the cells, causes the guard cells to work against the turgor pressures of neighboring subsidiary cells, displacing a finite volume and forming a characteristic pore. Micrograph courtesy of Professor Joerg-Peter Schnitzler (Helmholtz Zentrum München, Munich, Germany). See color plates section.

the case for Eq. (7.9), which defines the boundary layer conductance, g_{sj} in Eq. (7.10) will carry units of m s^{-1}. Also as discussed above in the case for boundary layer conductance, it is common for researchers to use the convention $g_{sj} = F_j/\Delta c_j$, with g_{sj} defined in units of mol (air) m^{-2} s^{-1}, recognizing that g_{sj} actually represents the composite ($g_{sj}\,\rho_m$). It is difficult to measure g_{sj} for most atmospheric constituents. In practice, we use observations to estimate g_s for water vapor (g_{sw}), and then scale g_{sw} to that for other constituents according to ratios of diffusivities, e.g., $g_{sj} = (K_{dj}/K_{dw})\,g_{sw}$. As an example of the application of Eq. (7.10), let us consider the definition of the stomatal conductances to H_2O vapor (g_{sw}) and CO_2 (g_{sc}) with units mol m^{-2} s^{-1}:

$$g_{sw} = \frac{E}{\Delta c_w},$$ (7.11)

$$g_{sc} = \frac{A_n}{\Delta c_c},$$ (7.12)

where, by convention, E is the flux density of water through the stomatal pores (transpiration rate), Δc_w is the difference in the mole fraction of water vapor between the intercellular air spaces (c_{iw}) and the outer leaf surface (c_{sw}) (i.e., $\Delta c_w = c_{iw} - c_{sw}$), A_n is the rate of net CO_2 uptake by the leaf, and Δc_c is the difference in atmospheric CO_2 mole fraction between the outer leaf surface (c_{sc}) and the intercellular air spaces (c_{ic}) (i.e., $\Delta c_c = c_{sc} - c_{ic}$). Leaf transpiration rate (E) is typically calculated from observations using a mass balance approach after measuring the flow rates of moist air in and out of a well-stirred leaf chamber and the mole fraction of water vapor in each flow (Section 6.3). The value of Δc_w is calculated from the difference in water vapor mole fraction between air inside the leaf (assumed to be saturated at leaf temperature) and the air of the chamber, following the assumption that rapid mixing in the chamber reduces the boundary layer of the leaf to a negligible thickness. (Alternatively, g_{bw} can be estimated for each chamber condition using a moist leaf replica constructed from blotter paper, thus providing a condition in which the only conductance affecting the observed E is g_{bw}.) Once g_{sw} has been calculated, g_{sc} can be determined as $g_{sc} = g_{sw}/1.58$. Although it is most common to use the phenomenological approaches described here to define diffusive conductances, it is possible to justify the form of these relations on thermodynamic grounds (Appendix 7.1).

In leaves, stomatal conductance occurs *in series* with the boundary layer conductance. That is, the rate by which molecules move in or out of a leaf is serially affected first by one, then by the other, of these two conductances. Relying on electrical analog models (Section 6.1.5), serial resistances are added as a simple sum, whereas serial conductances are added as the reciprocal sum. The summed influence of g_b and g_s is often termed *total leaf conductance*, or simply *leaf conductance*. Total leaf conductance (g_t) can be represented as:

$$\frac{1}{g_t} = \frac{1}{g_b} + \frac{1}{g_s}. \tag{7.13}$$

Using algebra to express a common denominator, Eq. (7.13) can be rewritten as:

$$g_t = \frac{g_b g_s}{g_b + g_s}. \tag{7.14}$$

The forms of Eqs. (7.11) and (7.12) can be traced to Fick's First Law, and they reflect diffusive interactions driven by binary mole fraction gradients (CO_2 in dry air or H_2O in dry air). As discussed in Chapter 6, however, binary relations alone cannot adequately describe diffusive processes in leaves. In leaves, interactions among all three constituents (CO_2, H_2O, and dry air) occur simultaneously. For example, CO_2 molecules will interact with H_2O molecules as they collide in the stomatal channel and in the leaf boundary layer. At a maximum, transpiration rates can be 2–3 orders of magnitude greater than photosynthesis rates. Thus, the diffusion of CO_2 molecules into a leaf has little resistive effect on the diffusion of H_2O molecules out of a leaf. Of more importance is the tendency for outward diffusing H_2O molecules to impede inward diffusing CO_2 molecules. The interaction between CO_2 and H_2O will cause the actual diffusivity of CO_2 in the moist air within and just above a leaf to diverge from the binary diffusion coefficient of CO_2 in dry air. In calculating the stomatal conductance based on *ternary diffusive interactions* (involving CO_2, H_2O, and "air") the effect of E on A, at the least, has to be taken into account. Ternary diffusive interactions are typically considered with the Stefan–Maxwell diffusion model as a starting point (Section 6.1.6). A derivation of ternary diffusive interactions for the case of leaves is presented in Appendix 7.2.

7.3.2 Stomatal guard cell function

The openings and closings of stomatal pores are driven by swelling and shrinking, respectively, of the guard cells that bound the pore. Stomatal opening is triggered when environmental cues activate proton pumps in the guard cell membrane resulting in the efflux of H^+ to the extracellular space (Figure 7.6). The pumping of protons from the guard cell requires ATP, which is obtained from photosynthetic electron transport in guard cell chloroplasts. (Unlike other epidermal cells, guard cells possess the capability for photosynthetic CO_2 assimilation, including an active photosynthetic electron transport system.) As H^+ ions leave the guard cell the cytoplasm becomes more negatively charged. As the negative charge increases, K^+ ions flow from the subsidiary cells through protein channels, and into the guard cell. The inward flux of K^+, however, is dependent on an electrochemical gradient across the cell membrane, and so the outward displacement of H^+ must be well established before a significant K^+ flux develops. To compensate for electrical imbalances caused by the exchange of H^+ for K^+, some Cl^- ions will also flow into the guard cells through dedicated anion channels. The inward flow of Cl^-, which goes against the electrochemical gradient (i.e., from the extracellular space with lesser negative charge to the cytoplasm with greater negative charge), is coupled to the inward flow of H^+ through a dual-function protein carrier; protons are then pumped back out, leaving Cl^- inside the cell. The K^+ that flows into the guard cells is stored in the central vacuole, and in many plants malate (an organic acid that carries a net negative charge in its dissociated state) is synthesized in the cytoplasm and transported into the vacuole to balance the charge of the K^+. The malate is synthesized using carbon substrates mobilized from starch stored in guard cell chloroplasts. The inward flow of Cl^- and synthesis of negatively charged malate maintains the electrochemical difference across the guard cell membrane, and thus sustains the uptake of K^+. By sustaining the electrochemical gradient through the influx of K^+ and Cl^-, as well as the synthesis of malate, the ionic concentration of the guard cells can reach high values relative to that of the neighboring subsidiary cells. The high ionic concentration in the guard cells creates relatively high (and negative) osmotic potential gradients, thus forcing an osmotic flux of water into the guard cells and increasing their turgor pressure (see Section 2.3.1 for a discussion of water potential gradients and water fluxes).

The cellulosic walls of guard cells are unevenly thickened, with microfibrils of cellulose extending radially from the pore surface outward (Figure 7.5). This unique feature produces a mechanical consequence as guard cell turgor pressure increases. An increase in the internal pressure force will be channeled longitudinally toward the ends of the guard cells, rather than laterally; the microfibril thickenings will resist radial stretching. As the force is directed to the joined ends of the guard cell pair, a pore will open between the pair. Thus, the opening

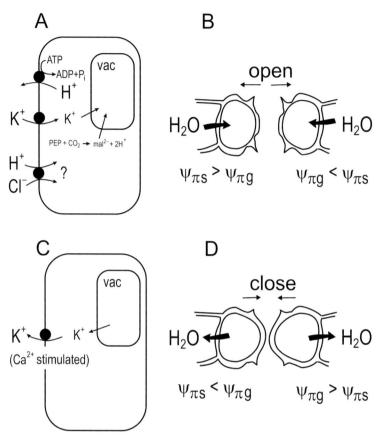

Figure 7.6 **A**. Primary features of ion exchange across the plasma membrane of a guard cell during stomatal opening. Key features of the exchange are described in the text. The question mark reflects a lack of knowledge about the proposed co-transport of Cl^- as H^+ moves back into the guard cell. vac = vacuole; mal = malate; PEP = phosphoenolpyruvate. **B**. Water flux from neighboring subsidiary cells into guard cells during stomatal opening in response to a difference in osmotic potential (i.e., a difference in free energy due to the accumulation of osmotically active solutes in the guard cell). $\psi_{\pi s}$ = osmotic potential of subsidiary cell; $\psi_{\pi g}$ = osmotic potential of guard cell. **C**. K^+ efflux from the guard cell during stomatal closure. **D**. Water flux from guard cells into neighboring subsidiary cells in response to reversal in the gradient in osmotic potential as K^+ leaves the guard cells.

of stomata is ultimately driven by hydromechanics – the creation of an osmotic gradient interacting with mechanical features of guard cell walls.

Stomatal closure involves depolarization (loss of charge) across the guard cell membrane through the activation of K^+ efflux channels and inactivation of K^+ influx channels. An unequal distribution of Ca^{2+} ions across the guard cell membranes appears to be involved in the depolarization, but the exact nature of the interaction is unknown. As the electrochemical gradient across the guard cell membrane is dissipated, water will be redistributed between the guard cells and subsidiary cells, turgor pressure within the guard cells will decrease, and the pore will close.

7.3.3 The short-term (near-instantaneous) response of stomatal conductance to isolated environmental variables

Despite a century's worth of studies, stomata and their responses to environmental variation remain one of the more mysterious topics of plant-atmosphere interactions. Variation in g_s as a function of PPFD, atmospheric humidity, leaf temperature, and atmospheric CO_2 concentration can be observed by placing a leaf in a gas-exchange chamber and changing one factor at a time (Figure 7.7). Responses to changes in each of these factors can occur on relatively short time scales (seconds to minutes) and when considered together they determine most of the natural diurnal variation that occurs in g_s. The shapes and magnitudes of such responses have been described for a broad range of plant species, although underlying cellular and leaf mechanisms are only slowly being discovered. Guard cells integrate multiple, often conflicting, signals from the environments both inside and outside the leaf in order to establish the appropriate stomatal aperture. Input from the leaf's environment is used to engage a variety of signal transduction pathways that modulate the transport of ions and water in and out of guard cells, and thus elicit the stomatal response.

Stomatal conductance is controlled in part by the PPFD absorbed by a leaf (Figure 7.7A). PPFD-dependent control over g_s is facilitated by two photoreceptors. One receptor responds to visible (400–700 nm) light, though this receptor has traditionally been called the "red-light" receptor because the original observations of its activity involved studies using high-intensity red light. The second receptor responds to weak-intensity blue light, and has an action spectrum that is unique, compared to that for the red-light receptor. The action spectrum of the *red-light receptor* is consistent with that of chlorophyll and most studies to date have concluded that chloroplasts in either guard cells or mesophyll cells function as this receptor (Mott 2009). The *blue-light receptor* is most likely two photosensitive proteins, called photropins, located in guard cells, and working in tandem (Shimazaki *et al.* 2007). In response to an increase in PPFD, both the red- and blue-light receptors induce an increase in the rate at which H^+ ions are pumped from the guard cells. The exact mechanisms of the light modulation of H^+ pumping are unclear but they likely involve phosphorylation and dephosphorylation of key transport proteins in the guard cell plasma membrane. The blue- and red-light sensors work in complementary fashion. The blue-light sensor is considerably more sensitive to changes in PPFD, and may be the primary control over g_s during periods of low PPFD, such as just after dawn or during cloudy days with variable and highly diffuse light. The red-light sensor controls responses at higher PPFD levels. Beyond these simple characterizations the significance of utilizing two independent photoreceptors remains unclear. In a broad adaptive context, however, it is clear that the positive response of g_s to increases in PPFD allows the leaf to increase its diffusive potential for CO_2 uptake at the same time that the photon flux is favorable for photosynthetic CO_2 assimilation.

A second principal control over g_s regulates the diffusive potential for transpiration as a function of atmospheric humidity and serves as a physiological adaptation by maintaining a favorable H_2O balance in the leaf. As the humidity of the atmosphere decreases, g_s also decreases; thus, reducing potential losses of H_2O as the evaporative demand of the atmosphere increases. This response is exemplified in the observations reported in Figure 7.7B in

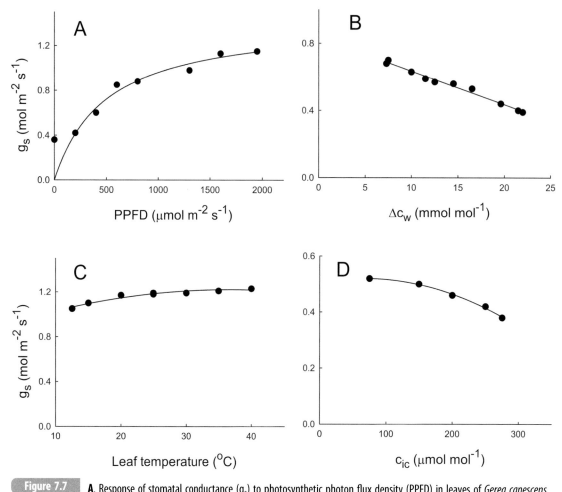

Figure 7.7 **A**. Response of stomatal conductance (g_s) to photosynthetic photon flux density (PPFD) in leaves of *Gerea canescens* (a desert annual plant). Redrawn from Ball and Berry (1982). **B**. Response of g_s to leaf-to-air water vapor mole fraction gradient (Δc_w) in leaves of *Glycine max* (soybean). Redrawn from Ball *et al*. (1987). **C**. Response of g_s to leaf temperature at constant Δc_w in leaves of *Agropyron smithii* (a prairie grass). Redrawn from Monson *et al*. (1982). **D**. Response of g_s to intercellular CO_2 mole fraction (c_{ic}) in leaves of *Eucalyptus pauciflora* (a rapidly growing tree). Redrawn from Wong *et al*. (1978). In each case, all environmental variables except that being studied were maintained constant.

which an increase in the difference in leaf-to-air water vapor mole fraction (Δc_w) caused g_s to decrease; the increase in Δc_w was achieved by reducing ambient humidity while adjusting air temperature to keep leaf temperature relatively constant in the face of changes in the rate of latent heat loss. It is difficult to separate the effects of atmospheric humidity and atmospheric temperature on g_s, because the former is physically dependent on the latter; as the atmosphere warms, it is capable of reaching a higher maximum humidity. Further complicating this interaction is the fact that changes in leaf temperature (T_L) also affect Δc_w. Increases in T_L, at constant atmospheric humidity will cause an increase in Δc_w because the water vapor mole fraction of the intercellular air spaces of the leaf will increase. Studies, in which atmospheric humidity was adjusted upward to maintain a constant Δc_w in the face of increased T_L, have generally shown a range from insignificant to modest increases in g_s (see Mott and Peak 2010) (Figure 7.7C). These increases have been interpreted as a direct response of g_s to T_L. Within the context of overall changes in the state of the atmosphere, an increase in g_s in response to an increase in temperature, would favor increased leaf latent heat loss during the middle of the day, at the same time that Δc_w is likely to increase and favor stomatal closure. These responses underlie competing constraints placed on stomatal function – the requirement for maintenance of a favorable leaf energy balance *and* the requirement for conservation of H_2O. The interactions among leaf temperature, Δc_w, g_{sw}, and E are complex and non-linear due to multiple mutual influences and feedbacks (Figure 7.8). Generally, the direct response of g_s to diurnal changes in T_L is less than the direct response of g_s to Δc_w.

The negative response of g_s to increases in Δc_w is controlled by a direct response of g_s to E or some internal leaf factor coupled to E, not to the ambient atmospheric humidity. This conclusion was derived from experiments in which the composition of the atmosphere outside the leaf was changed from air (N_2 and O_2) to helox (He and O_2), but the humidity of the atmosphere and temperature of the leaf were kept constant (Mott and Parkhurst 1990).

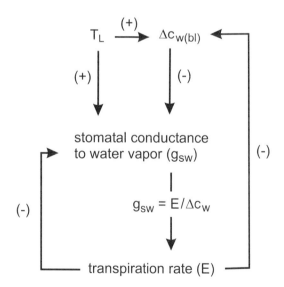

Figure 7.8 Interacting controls and feedbacks between leaf temperature (T_L) and the difference in water vapor mole fraction between the intercellular air spaces of the leaf and the air in the boundary layer of the leaf, adjacent to the leaf surface ($\Delta c_{w(bl)}$), and stomatal conductance to water vapor (g_{sw}). The value of g_{sw} directly affects leaf transpiration rate (E) through the diffusive flux relationship. The value of E directly affects g_{sw} through a hydrologic feedback, possibly related to increased evaporation from the guard cells themselves, which would reduce guard cell turgor pressure. The value of E also increases the humidity of the boundary layer, causing a reduction in $\Delta c_{w(bl)}$. T_L directly affects g_{sw} through a positive control, and indirectly affects g_{sw} through a negative control imposed by $\Delta c_{w(bl)}$. $\Delta c_{w(bl)}$ exerts a negative direct influence on g_{sw}.

Due to the lesser mass of He, compared to N_2, the diffusivity of H_2O in helox is 2–3 times higher than in air. When the atmospheric composition was switched to helox, E initially increased, but then g_s decreased, forcing E back to the same value that had originally been observed in air. These results can only be explained if g_s responded directly to E or to a change in the humidity of the internal air of the leaf, which changed in response to the experimental treatment, not to ambient atmospheric humidity, which remained constant.

Stomatal conductance is negatively correlated with increases in atmospheric CO_2 concentration (Figure 7.7D). The CO_2 pool that is sensed by the epidermis appears to be in the intercellular air spaces of the leaf (i.e., c_{ic}) (Mott 1988). This conclusion was derived through experiments with amphistomatous leaves (i.e., leaves with stomata in both the upper and lower epidermi), in which intercellular air spaces provided continuity between the chamber atmosphere on either side of the leaf. Thus, the CO_2 concentration of the air flowing through the chamber on the upper surface of the leaf could be altered, causing the intercellular CO_2 concentration to also be altered, despite keeping constant the CO_2 concentration in air flowing through the chamber on the lower side of the leaf. When this treatment was imposed, g_s of the lower epidermis decreased in response to an increase in c_{ic}, despite constant c_{ac} at the lower leaf surface. Alternatively, when c_{ic} was maintained constant, but c_{ac} at the lower surface was varied, g_s remained constant. The conclusion derived from this experiment was that stomata respond to c_{ic} alone.

Whatever the exact mechanism of CO_2 sensing by the guard cells, the net effect is to control c_{ic} in a way that improves the leaf's capacity to assimilate CO_2, while at the same time reducing the leaf's tendency to lose H_2O. For example, when the sun emerges from behind a cloud, and the photon flux to the leaf increases, photosynthetic depletion of CO_2 in the intercellular air spaces will occur and g_s will increase, causing a concomitant increase in the potential for CO_2 to diffuse into the leaf. This series of events will provide the diffusive support needed to facilitate the PPFD-driven increase in photosynthetic potential. Conversely, when a cloud blocks the photon flux from the sun, the leaf's photosynthesis rate will decrease more quickly than stomatal conductance, causing c_{ic} to increase. The increase in c_{ic}, along with the direct stomatal response to a reduced photon flux, will work in tandem to reduce g_s, bringing c_{ic} into balance with the reduced photosynthesis rate, and reducing the rate of transpiration at the same time.

In all of the short-term responses of g_s to environment, our knowledge of underlying mechanisms remains uncertain. The history of this field shows that each time we have begun to coalesce around a theory that accurately explains stomatal dynamics, new experimental evidence forces us to partially or entirely abandon the theory. Even past theories that have served as cornerstones of our understanding of stomatal function for several years have been challenged in recent studies (Mott 2009, Mott and Peak 2010, Pieruschka *et al.* 2010, Peak and Mott 2011). Despite our lack of knowledge about underlying mechanisms, it has still been possible to describe dynamics in g_s through a combination of response functions that are dependent on atmospheric state variables. All of the responses reported in Figure 7.7 represent observations of leaves isolated in gas-exchange chambers and characterized with regard to changes in one or another environmental variable. It is these empirical responses that have formed the foundation of the models that have been developed to describe stomatal dynamics. We can only look forward in anticipation of future discoveries that reveal the mechanisms underlying the models.

7.3.4 Integration of the stomatal response to multiple, and simultaneous, environmental controls

In the interest of simplifying experimental observations on stomatal function, researchers often place leaves in a climate-controlled chamber and change one environmental variable while keeping all others constant. This is not, however, representative of meteorological dynamics in the natural environment. In that case, stomata respond to simultaneous variation in numerous forcing variables. Synergistic effects occur as guard cells integrate the influence of multiple signals. In this section, two responses to co-varying factors are considered for the purpose of exemplifying the interactive nature of stomatal control. We will consider stomatal responses to simultaneous changes in c_{ic} and PPFD, and stomatal responses to simultaneous changes in Δc_w and leaf temperature. A number of other interactions could have been chosen, though few have attracted as much experimentation and interpretation as those described here.

In response to increases in c_{ic} with constant PPFD, stomata close, and in response to increases in PPFD with constant c_{ic}, stomata open (Figure 7.9). These responses are inherently coupled through parallel feedbacks connected by the influences of PPFD and c_{ic} on A (Figure 7.10). To illustrate the nature of the feedbacks consider that PPFD determines the "demand potential" and c_{ic} determines the "supply potential" of photosynthesis rate. In other words, PPFD determines the potential for chloroplasts to *utilize* CO_2, whereas c_{ic} determines the potential for chloroplasts to *receive* CO_2. Imagine an arbitrary steady-state flux rate with its associated balance between the supply and demand potentials. If the magnitude of either potential is perturbed g_{sc} will respond in a fashion that at least partially, if not wholly, diminishes the resulting imbalance. Thus, if PPFD is increased, forcing the demand potential higher and increasing the limitation imposed by c_{ic}, stomata will open, increasing the supply potential and re-establishing a balance between the limitations. Alternatively, if c_{ic} is increased the supply potential will be out of balance with the demand potential, and stomata will close, reducing the supply potential and once again re-establishing a balance between the two control processes.

As a second example we consider the case of stomatal responses to simultaneous changes in humidity and leaf temperature. As leaf temperature increases, stomata become less sensitive to changes in Δc_w (Figure 7.11). Originally, this response was used as evidence that guard cells sense temperature and humidity through a common mechanism, perhaps as *relative humidity* (Ball *et al.* 1987). Alternative mechanisms for explaining temperature-Δc_w interactions involve temperature-dependent changes in the hydraulic transfer of liquid water through the epidermis of the leaf (Matzner and Comstock 2001), or differential hydration states of the intercellular air spaces of the leaf (Pieruschka *et al.* 2010, Peak and Mott 2011). In the study by Peak and Mott (2011), it was proposed that guard cells come to equilibrium with the thermodynamic potential of water vapor in the air near the bottom (intra-leaf) of the stomatal chamber, causing the guard cells to expand or contract,

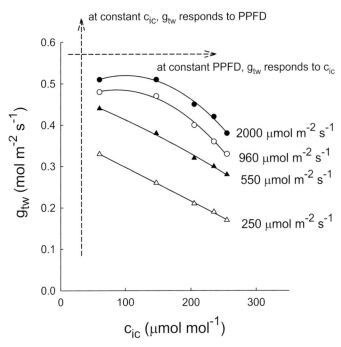

Figure 7.9 The relationship between total leaf conductance to water vapor (g_{tw}) and intercellular CO_2 mole fraction (c_{ic}) at different photosynthetic photon flux densities (PPFD) in leaves of *Eucalyptus pauciflora*. At constant PPFD, g_{tw} decreases as c_{ic} increases; at constant c_{ic}, g_{tw} increases as PPFD increases. Redrawn from Wong *et al.* (1978).

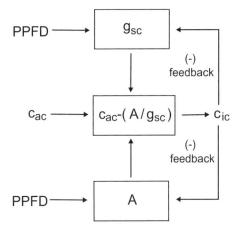

Figure 7.10 Schematic representation of interactions between PPFD and ambient CO_2 mole fraction (c_{ac}) as they influence stomatal conductance to CO_2 (g_{sc}). The influences of these variables are integrated through $g_{sc} = A/\Delta c_c$; where $\Delta c_c = c_{ac} - c_{ic}$. Thus, any positive influence of PPFD on the net CO_2 assimilation rate (A) will tend to decrease c_{ic} and cause a negative feedback (i.e., a feedback that opposes the increase in A). Similarly, a positive influence of PPFD on g_{sc} will tend to increase c_{ic} and cause a negative feedback on g_{sc}. This illustrates the interwoven dependencies of PPFD and c_{ic} as they influence g_{sc}. Redrawn from Wong *et al.* (1978).

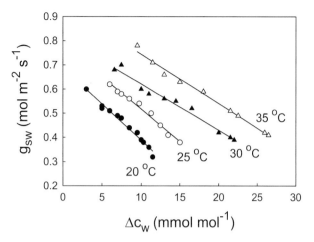

The relationship between stomatal conductance to water vapor (g_{sw}) and the leaf-to-air difference in water vapor mole fraction (Δc_w) at different leaf temperatures for a leaf of *Glycine max* (soybean). Redrawn from Ball *et al.* (1987).

hydromechanically, in a way that is initially, at least, uncoupled from liquid water transport in the epidermis (Peak and Mott 2011). The theory proposed in the latter study takes us back to a re-evaluation of the original hypothesis offered in the late 1980s that stomata respond to humidity and temperature together, not separately. Whatever, the exact mechanism, the research into these phenomena has made it clear that synergistic sensing of the environment by guard cells, rather than the sensing of individual meteorological factors, is likely to reflect the most accurate mode of function.

One perspective that underlies virtually all interpretations of stomatal response to the environment involves the recognition that stomata have evolved in a way that balances the fluxes of CO_2 and H_2O to maximize photosynthesis, growth, and ultimately fitness. Stomatal responses have been subjected to the same processes of natural selection as all other physiological and morphological traits, and therefore they should reveal evidence of selection toward optimal function. Optimality is often a condition that demands compromises among various competing traits. Many of the compromises evident in plant function exhibit broad universal patterns, especially with regard to those traits governing CO_2 and H_2O exchange (Reich *et al.* 1997, 2003). In this book we have, with some exceptions, focused on the proximate, not ultimate, causes of plant-atmosphere exchange, and so we have chosen not to devote too much space to a discussion of evolutionary trends in stomatal function. However, there are important organizing principles that can be drawn from studies of plant evolution and applied to the issue of proximate biosphere-atmosphere interactions. Some of these principles are discussed in Box 7.1, which takes up the topic of optimality in stomatal function and the tradeoffs between plant H_2O and CO_2 fluxes.

Box 7.1   **The evolution of stomatal compromises in the control over H$_2$O and CO$_2$ fluxes**

It is proper to enquire, first, what stomata do; secondly how they do it; and only then, if the question is allowed at all, why they do it. (Cowan and Farquhar (1977))

Efforts to address the "why" questions in biology can provide insight into the evolutionary costs and benefits that have shaped the response of plants to the environment. One of the paradigms in the field of evolutionary ecology is that plants have evolved in a way that spreads the limitations to resource acquisition across all potential resources. This is founded on the argument that natural selection works to minimize costs while maximizing gains, thus producing adaptations that maximize fitness. The paradigm has analogies to anthropocentric concepts such as the theory of economic optimization or maximization of industrial efficiency.

When applied to the evolution of stomatal function the paradigm predicts that stomata will respond to the environment in a way that minimizes water loss and maximizes carbon gain. By necessity, these are contrasting outcomes; complete mitigation of leaf water loss by stomatal closure would also require complete mitigation of leaf carbon gain. Thus, stomatal function has evolved in a way that balances these opposing outcomes. Another way of stating the problem is that stomata are required to function in a way that maximizes photosynthetic water-use efficiency (i.e., minimizes the molar ratio of transpiration to photosynthesis) while at the same time maximizing photosynthetic nitrogen-use efficiency (i.e., maximizing the photosynthetic CO$_2$ assimilation rate per unit of nitrogen invested in photosynthetic enzymes). This requires that stomata function to maintain the intercellular CO$_2$ concentration (c_{ic}) at a level that simultaneously maximizes the marginal gains for photosynthesis and transpiration. Stated, in those terms, the problem is clearly seen as one of economic optimization.

Cowan and Farquhar (1977) argued that with regard to stomatal function and the requisite tradeoff between water loss and carbon gain, the criterion of optimization will be met when the integral of g_s across a day is such that daily CO$_2$ assimilation rate is maximized; in other words, no alternative distributions of g_s across the day will further increase the CO$_2$ assimilation rate. Mathematically, this theory is represented as:

$$\partial E / \partial A = \lambda, \tag{B7.1.1}$$

where λ is the *constant of optimal proportionality*. Equation (B7.1.1) formalizes the assumption that if stomata function to minimize transpiration for a fixed amount of CO$_2$ assimilation, then the proportioning between dynamics in E and A should be constant. For the situation where $g_b \rightarrow \infty$, and following Lloyd (1991), it can be shown that:

$$\lambda = \frac{\Delta c_w (1.58 r_{sc} + k)^2}{1.58 (c_{ac} - \Gamma) r_{sc}^2} \tag{B7.1.2}$$

where r_{sc} is the stomatal resistance to diffusion of CO$_2$, k is a scaling coefficient and is taken as the initial (linear) slope of the A:c_{ic} relationship (often called the "carboxylation efficiency," see Section 5.3.1), and 1.58 is the ratio of the diffusion coefficients for H$_2$O and CO$_2$ (K_{dw}/K_{dc}). This relationship expresses λ in terms of the fundamental drivers of E and A, namely the gradients in H$_2$O and CO$_2$ concentration across the leaf and the sensitivity of A and E to those gradients. Using Eq. (B7.1.2) and again following Lloyd (1991) we can write:

$$g_{sw} = 1.58k\left(\sqrt{\frac{\lambda c_{ac} - \Gamma}{1.56\Delta c_w}}\right). \hspace{2cm} \text{(B7.1.3)}$$

Equation (B7.1.3) provides an expression for g_{sw} given the a priori assumption that stomata function in an optimal manner when g_s changes as a function of Δc_w. It further reveals that within this assumption, $g_{sw} \propto 1/\sqrt{\Delta c_w}$.

The Cowan–Farquhar theory addresses the issue of why stomata respond the way that they do to the complexity of environmental variation that occurs across a day. Use of λ in models of leaf CO_2 and H_2O exchange provides an a priori criterion (the criterion of optimal function) within which specific patterns of stomatal dynamics are permitted. The theory is ideal for understanding stomatal function within the context of compromises between water loss and carbon gain. However, it has shortcomings in that it was developed in *relative* terms to describe the tradeoffs that are required between two limiting resources. *The theory predicts the existence of constant λ, but it doesn't require that λ take a specific value.* In fact, the theory allows for an infinite range of "optimal" states. Additional conditions can be imposed on λ in order to limit the range of possible values. For example, Cowan (1986) coupled λ to the balance between precipitation and evaporation (i.e., soil moisture deficit) and a probability function linking plant survival to soil moisture deficit. Givnish (1986) defined biomass allocation constraints to leaves versus roots within the context of finite water resources. Friend (1991) used a similar approach to that developed by Givnish. The models developed by Friend, however, contain deeper mechanistic details, particularly with respect to nitrogen partitioning within the chloroplast and the relative constraints of c_{ic} and PPFD on A.

Optimality is one perspective by which to view stomatal function. As a foundation for constructing stomatal models, the optimality criterion has the advantage of integration; i.e., with one overarching theory stomatal control over both E and A can be simultaneously solved. This is a simpler approach to that provided by "bottom-up," mechanistic theories in which functions are developed from principles of guard cell processes and dynamics. Stomatal function *in situ* shows many of the expected trends in optimality theory (Williams 1983, Comstock and Ehleringer 1992, Hari *et al.* 1999). However, tests of optimality have not yet been subjected to the full range of environmental interactions that exist in natural habitats.

7.3.5 The long-term (seasonal) response of stomatal conductance to environmental variation

On the time scale of days to weeks, as the soil dries and the maximum midday leaf water potential (ψ_w) decreases, midday g_s will also decrease (Figure 7.12). This longer-term response is mediated by stomatal interactions with mesophyll processes, as well as processes in the roots (Jones 1998). The longer-term response reflects the tight coupling of g_s to soil water status; a coupling that functions to reduce the water usage of the plant at the same time that soil water supplies are slowly diminished through the growing season and between precipitation events.

The signal that induces the longer-term reduction in midday g_s is an increase in production of the plant hormone, abscisic acid, or ABA. ABA is produced in large quantities in the chloroplasts of mesophyll cells in water-stressed leaves, and in the roots of water-stressed

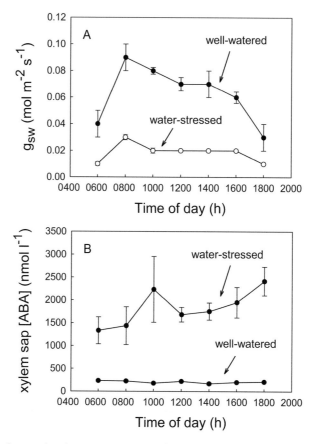

Figure 7.12 **A**. Diurnal pattern of stomatal conductance to water vapor (g_{sw}) in *Ceanothus thyrsiflorus* (a chaparral shrub) in well-watered plants or plants exposed to several days without watering. **B**. The concentration of abscisic acid (ABA) in the xylem sap of *C. thyrsiflorus* plants that were well-watered or exposed to several days without water. Note the decrease in g_{sw} that is correlated with the accumulation of ABA. Redrawn from Tenhunen *et al*. (1994), with kind permission from Springer Science + Business Media.

plants. Once produced, ABA moves through the plant to the guard cells where it inactivates the inward flux of K^+. Concomitant with the loss of K^+ is a loss of hydrostatic pressure and closure of stomata. In plants that are native to environments with frequent drought, chronic exposure to high tissue ABA concentrations probably causes developmental and anatomical adjustments in the guard cells and surrounding epidermal issues that reduces overall water loss rates (Franks and Farquhar 2001).

From diffusion theory, g_s has the potential to influence c_{ic}, and thus A. There is also evidence, however, that control can be exerted in the opposite direction, allowing changes in A to influence g_s. Intuitively, it would make sense for such a relationship to exist; i.e., as the photosynthetic capacity of the mesophyll tissue is increased the potential for CO_2 diffusion into the leaf should also increase. This is a required response since an increase in A cannot be sustained without an increase in g_s. (Note that the inverse is not necessarily true – a decrease in A can be met with no decrease in g_s, though it will be costly in terms of H_2O loss.)

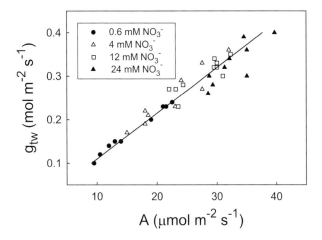

Figure 7.13 The relationship between total leaf conductance to H_2O (g_{tw}) and CO_2 assimilation rate (A) in leaves of *Gossypium hirsutum* (cotton) from plants grown with different nitrogen availability. Because boundary layer conductance is relatively constant for the experimental leaves, most of the variation in g_{tw} can be attributed to variation in stomatal conductance. Redrawn from Wong *et al.* (1985a).

Experimentally, a correlation between A and g_s has been well-documented (Figure 7.13). Experiments such as those depicted in Figure 7.13 rely on a manipulated factor of the growth environment (e.g., nitrogen availability) to perturb A, and concomitantly g_s. It is tempting to conclude that the results of such experiments reflect coordinated developmental events. That is, as leaves develop in different N regimes, hormonal cues cause the coordinated production of chloroplasts in mesophyll cells and stomata on the leaf surface – thus maintaining proportional scaling in the expression of photosynthetic capacity and g_s. However, as discussed above, experiments using genetically engineered tobacco plants with reduced amounts of Rubisco showed no effects on g_s, suggesting that the linear scaling between these variables may be the result of independent developmental processes (von Caemmerer *et al.* 2004).

7.3.6 Models of stomatal conductance

The most widely used models of g_s and its response to the environment are derived from empirical data, not underlying theory; we will refer to these types of models as *phenomenological*. The most widely used phenomenological model of g_s, the so-called Ball–Berry model, is based on dimensional analysis; it is grounded in the observation that most dynamics in g_s can be explained by the integrated response to three fundamental variables, CO_2 assimilation rate (A), relative humidity at the external surface of the leaf (h_s), and CO_2 mole fraction at the external surface of the leaf (c_{sc}) (Ball *et al.* 1987). When expressed in the mathematical form of the Ball–Berry model the response of g_s to these three variables collapses to a single linear function (Figure 7.14). In its original form, the Ball–Berry model (which is more correctly referred to as the Ball–Woodrow–Berry model and is abbreviated hereafter as the BWB model) is stated as:

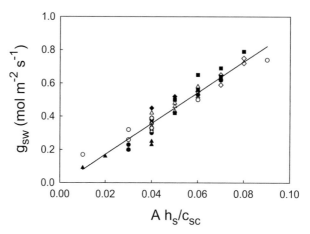

Figure 7.14 The relationship between stomatal conductance to water vapor (g_{sw}) and the mathematical index calculated as $A\,h_s/c_{sc}$ (the Ball–Woodrow–Berry index). The data were taken from several different experiments in which PPFD, c_a, or Δc_w were individually varied. Redrawn from Ball *et al.* (1987).

$$g_s = g_{s0} + \frac{k\,A\,h_s}{c_{sc}}, \tag{7.15}$$

where g_{s0} is a residual stomatal conductance (the *y*-intercept when g_s is plotted against Ah_s/c_{sc}), and *k* is a scaling coefficient (the slope of the relation between g_s and Ah_s/c_s). The BWB model has several advantageous attributes compared to alternative models that preceded it. In the BWB model, the variable A acts as an integrator for some of the environmental interactions that influence g_s. For example, the influences of leaf temperature and PPFD on g_s are empirically subsumed into the model as integrated influences on A. In a similar manner, use of the variable h_s at the leaf surface provides an integrated influence of leaf temperature and atmospheric water vapor concentration on dynamics in g_s. In these ways, the BWB model achieved a level of multi-variable integration that had been missing from the models that preceded it. In most of the earlier models, g_s was determined as the product of fractional, and independent adjustments in a base value (analogous to g_{s0}) to individual environmental factors. One additional advantage of the BWB model is that g_s is linked, as a dependent variable, to A. Though there is still no known process that links changes in A to changes in g_s, *mathematical* linkage of the two is of tremendous use in modeling responses of g_s to the environment. Finally, Eq. (7.15) emphasizes the fact that it is the environment at the surface of the leaf that is most important to stomatal dynamics, not the well-mixed atmosphere above the leaf. This perspective provides allowance for the role of the leaf boundary layer in stomatal modeling and makes clear the need to consider feedbacks between boundary layer and stomatal conductances (Collatz *et al.* 1991). As an example of feedbacks that are possible when the leaf boundary layer is included in the model consider the case where g_{sw} and E decrease, yet g_{bw} remains constant. The reduced flux of H_2O into the boundary layer will cause a reduction in h_s, which in turn will feed back to further reduce g_{sw}. This is a case of positive feedback that is not possible when the boundary layer and stomatal conductances are considered separately in models.

The relationships depicted in Eq. (7.15) are empirically robust. They have been validated in numerous species, including representatives from diverse growth habits and C_3 and C_4 species, as well as for plants growing under a diversity of environmental conditions (e.g., Collatz *et al.* 1992, Katul *et al.* 2000, Gutschick and Simonneau 2002). The scaling parameter, k, varies significantly in the various observations that have been made, reflecting the influence of factors such as soil moisture and other longer-term characteristics of the growth regime.

The original derivation of the BWB model failed to address one key observation concerning stomatal response to the environment; that g_s may respond most directly to E or c_{iw}, not h_s (Mott and Parkhurst 1991, Aphalo and Jarvis 1991, Peak and Mott 2011). Leuning (1995) concluded that one gets a better fit of the model to empirical data if the response of g_s to humidity is expressed on the basis of Δc_w, which scales directly with E. Taking the analysis even further, Leuning (1995) reasoned that the response of g_s to CO_2 would be better represented by the factor $(c_{sc} - \Gamma)$ in the denominator of the BWB model; where Γ is the CO_2 compensation point (i.e., the intercellular CO_2 mole fraction where A → 0). This modification was required to improve performance of the model at low CO_2 concentrations. Without consideration of Γ, as in the original BWB model, g_s approaches 0 at Γ, which we know from observations, is not a true response. These changes resulted in a different form of the original BWB model:

$$g_s = g_{s0} + \frac{k\mathrm{A}}{(c_{sc} - \Gamma)(1 + \Delta c_w/b)}. \tag{7.16}$$

Equation (7.16) is often referred to as the "Leuning form of the Ball–Woodrow–Berry model" (hereafter abbreviated as BWBL). Further modification of the BWB model was made by Monteith (1995) in an effort to accommodate emerging observations that g_s responds directly to E, rather than Δc_w; we will refer to this version as the BWBM model. Derivations of the BWBL and BWBM versions of the Ball–Woodrow–Berry model are provided in Appendix 7.3.

The BWB model and its modified versions were originally derived from specific, case-dependent, empirical relations. However, these models are able to accurately predict g_s in a broad variety of plants in response to a broad range of environmental conditions. (It should be emphasized that all of the models that we have discussed for calculating g_s are restricted to the steady-state condition.) Accuracy in predicting the steady-state condition must reflect some degree of adherence to fundamental stomatal mechanisms. In an analysis of the various phenomenological models used to predict g_s, Dewar (1995) presented a compelling derivation of mechanistic principles that was based on guard cell water relations and the synthesis of photosynthetic metabolites by the guard cells or nearby mesophyll cells. In a separate analysis, Buckley *et al.* (2003) derived quantitative descriptions of the hydromechanical feedbacks that control stomatal dynamics. Peak and Mott (2011) evaluated the models with regard to humidity responses and provided new model components to explain the tendency for leaf temperature to modify the response of g_s to Δc_w. These mechanistic derivations converge in several places with the phenomenological models presented above. It is likely that Eqs. (7.15) and (7.16) reflect more about the first principles of stomatal function than might be thought on initial inspection, and their phenomenological form has provided "intellectual space" for the development of new hypotheses concerning underlying mechanisms. Discussion of mechanistic developments in the interpretation of Eqs. (7.15) and (7.16) is provided in Box 7.2.

Box 7.2   **Toward a mechanistic understanding of stomatal conductance models**

The stomatal conductance models discussed to this point are based on observations and despite their accurate portrayal of the responses of g_s to A, E, Δc_w, and c_{ic}, they reflect no obvious mechanistic components in their mathematical formulas. Using a different approach in which a priori assumptions are made with regard to biochemical and physiological mechanisms, several alternative models have been developed (Dewar 1995, 2002, Haefner *et al.* 1997, Buckley *et al.* 2003, Peak and Mott 2011). While being more informative on mechanistic grounds, these models require extensive parameterization, particularly with regard to biochemical parameters, and may be less useful in applications at larger spatiotemporal scales.

To illustrate the mechanistic approach we consider certain components of the model originally developed by Dewar (2002) and further derived by Buckley *et al.* (2003). Stomatal aperture and thus g_s reflect a linear combination of the turgor pressures within guard cells and adjoining epidermal subsidiary cells; outwardly oriented turgor forces produced by the guard cells work against inwardly oriented turgor forces produced by the subsidiary cells, which together define the resultant stomatal aperture. Empirical studies have shown that the force provided by the subsidiary cells exerts greater mechanical influence on the ultimate stomatal aperture (Wu *et al.* 1985), resulting in what is often termed *mechanical advantage*. Experiments using a microscopic pressure probe have revealed that turgor pressures do not vary significantly between subsidiary cells and non-subsidiary epidermal cells (Franks *et al.* 1995, 1998, Mott and Franks 2001), meaning that the mechanical advantage can in effect be extended to the entire epidermis. With these mechanistic principles in place we can state:

$$g_s = \chi(\psi_{pg} - \hat{m}\psi_{pe}), \tag{B7.2.1}$$

where χ is a mechanical coefficient (with units mmol H_2O m^{-2} s^{-1} MPa^{-1}), ψ_{pg} and ψ_{pe} are the spatially averaged pressure potentials (MPa) of the guard cells and epidermal cells, respectively, and $\hat{m}$ represents the mechanical advantage of the epidermis (which by definition is unitless and $\hat{m} > 1$). From empirical data, $\hat{m}$ has been estimated as 1.98. Relying on the fact that the pressure potential reflects the sum of the total water potential (ψ_w, taken as negative value) and the osmotic potential (ψ_π, taken as a positive value) we can rewrite Eq. (B7.2.1) as:

$$g_s = \chi[(\psi_{wg} + \psi_{\pi g}) - (\hat{m}\psi_{we} + \psi_{\pi e})]. \tag{B7.2.2}$$

The pressure potential difference between guard cells and epidermal cells reflects a free-energy gradient created by the active, coupled outward transport of H^+ and inward transport of K^+. As the K^+ ions are transported into the guard cells an osmotic gradient is created, water flows and guard cell turgor pressure increases. Buckley *et al.* (2003) have used the ATP requirement for guard cell turgor pressure as the basis for a biochemical model capable of simulating g_s in response to c_{ic}. An assumption of the model is that the ATP status of the guard cells is linked to the ATP status of mesophyll cells (Farquhar and Wong 1984). This assumption is consistent with several observations concerning the response of g_s to various environmental variables (Buckley *et al.* 2003), but to date the linkage has not been identified. Alternatively, the ATP status of the guard cells may be determined by their own photosynthetic processes. Whichever the case, the ATP status of the photosynthetic cells is assumed to be negatively influenced by increases in the rate of Rubisco carboxylation. Thus, when A responds positively to an increase in c_{ic} at high PPFD the amount of ATP available for actively transporting K^+ ions into the guard cells is assumed to decrease. We can state:

$$\rho_{ATP} = \rho_{At} - \rho_P \frac{v_{c1}}{v_{c2}}, \qquad (B7.2.3)$$

where ρ_{ATP} is the areal molar density (mol m^{-2} leaf area) of ATP in guard cell chloroplasts available for ion pumping, ρ_{At} is the areal molar density of total active adenylates (ATP + ADP) in the chloroplast, ρ_P is the areal molar density of photophosphorylation sites in the chloroplast, v_{c1} is the RuBP-saturated rate of carboxylation, and v_{c2} is the RuBP-limited rate of carboxylation. Equation (B7.2.3) is only valid when $v_{c1} < v_{c2}$. In cases where $v_{c1} > v_{c2}$, an alternative model can be derived as:

$$\rho_{ATP} = \frac{K v_{c2}}{\left(\dfrac{v_{c1}\rho_{RuBP}}{\rho_{Et}}\right) - v_{c2}}, \qquad (B7.2.4)$$

where ρ_{RuBP} is the areal molar density of RuBP in the chloroplasts of the guard cells, K is a combined term defined as $K = (\rho_{At} - \rho_P)(\rho_{RuBP} - \rho_{Et})/\rho_{Et}$, and ρ_{Et} is the areal molar density of Rubisco active sites (i.e., $\rho_{Et} k_{cat} = V_{cmax}$, where k_{cat} is the catalytic turnover number for Rubisco (mol CO_2 (mole catalytic sites)$^{-1}$ s^{-1}). Using Eqs. (B7.2.3) and (B7.2.4), an expression can be derived relating the steady-state osmotic gradient across the guard cell membrane ($\Delta\psi_\pi$) to guard cell ATP concentration and the sensitivity of $\Delta\psi_\pi$ to the turgor potential of epidermal cells:

$$\Delta\psi_\pi = \beta A_L \rho_{ATP} \psi_{pe}, \qquad (B7.2.5)$$

where β is a dimensionless sensitivity coefficient and A_L is leaf area (m^2). The full context for Eq. (B7.2.5) and its role in the fully expressed hydromechanical and biochemical model for g_s can be found in Buckley *et al.* (2003).

7.3.7 Conductance as a control over CO_2 assimilation rate

In well-watered plants at moderate temperatures, high humidities, and high PPFD, the controls over A exerted by g_{bc} and g_{sc} tend to be small relative to biochemical processes (Figure 7.15). Nonetheless, they do exert some degree of control and at low atmospheric humidity, low PPFD, or during episodes of drought or high leaf temperature, stomatal limitations increase. At the limit, as $g_{sc} \to 0$, the *stomatal control coefficient* will approach 1. Stomatal and boundary layer conductances to CO_2 affect A by affecting c_{ic} relative to c_{ac}. The ratio c_{ic}/c_{ac} is typically in the range 0.6–0.8 for C_3 leaves and 0.3–0.4 for C_4 leaves.

To better understand the roles of g_{sc} and g_{bc} as modifiers of A we return to an analysis of the A:c_{ic} curve which was so critical to understanding the biochemical limitations to photosynthesis discussed in Chapter 5. Let us assume, hypothetically, that g_{sc} and g_{bc}

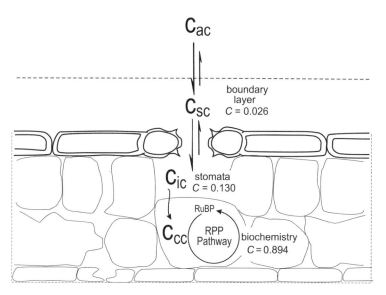

Figure 7.15 The series of four CO_2 pools leading to assimilation in the reductive pentose phosphate (RPP) pathway of the chloroplast. The ambient (c_{ac}) and leaf surface (c_{sc}) CO_2 pools are connected across the leaf boundary layer. The leaf surface and intercellular CO_2 (c_{ic}) pools are connected across the stomata. The intercellular CO_2 pool and the chloroplastic CO_2 pool (c_{cc}) are connected across the aqueous-phase of mesophyll cells. Representative control coefficients (C) at photosynthetic light saturation, 25 °C, and ambient CO_2 mole fraction (in this case 345 µmol mol^{-1}) are provided for the four primary determinants of flux through these CO_2 pools in *Glycine max* (soybean) leaves. The values for C were determined from independent evaluation of elasticity coefficients, and do not sum to 1 because of additional control factors that were not accounted for in this simple model. Redrawn from Woodrow *et al.* (1990).

impose no diffusive limitation on A. In that case, the A that is observed in the presence of a normal, atmospheric CO_2 concentration (e.g., ~ 390 µmol mol^{-1}) can be represented at point A_1 in Figure 7.16. Without stomatal and boundary layer limitations, c_{ic} would equal c_{ac}, and the observed A would correspond to the value extrapolated to c_{ac}. In that case, A_1 is determined solely by the leaf internal conductance to CO_2 diffusion and the biochemical capacity for photosynthesis at that given c_{ac}. Now, let us assume that g_{sc} and g_{bc} impose a limitation on A. Following Fick's First and Second Laws, any limitation would have to be expressed through a proportional reduction in c_{ic}, relative to c_{ac}. This is a result of continuity, any decrease in the flux of CO_2 into a leaf, relative to the rate of metabolic CO_2 assimilation, will have to be balanced by a reduction in c_{ic} relative to c_{ac}. The conductance-limited A (indicated as point A_2 in Figure 7.16) would be less than the non-conductance limited A. The degree of conductance limitation can be represented as a line from the c_{ic} that would exist at no limitation (i.e., at $c_{ic} = c_{ac}$) to point A_2. The slope of this line equals $-\Delta A/\Delta c_{ic}$. Invoking the linear relationships that exist among diffusive conductance, A, and c_{ic}, and recognizing that $\Delta c_{ic} = \Delta A/g_{tc}$, the slope of the line can be simplified to $-g_{tc}$ (once again, ignoring the internal, liquid phase conductance for the moment and defining g_{tc} only in terms of g_{bc} and g_{sc}). As the slope defined by $-\Delta A/\Delta c_{ic}$ decreases, the conductance

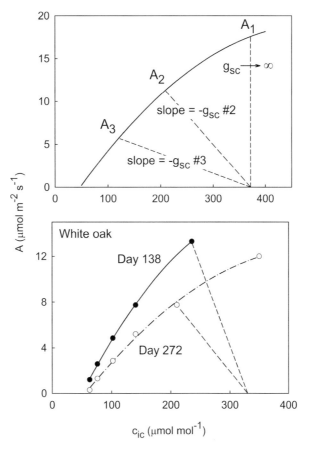

Figure 7.16 **Upper panel**. The A:c_{ic} relationship illustrating the dependence of A on g_{sc}. (For the simplicity of illustrating stomatal interactions with CO_2 assimilation rate, the potential limitation of g_{bc} to diffusive flux is ignored.) **Lower panel**. Data from a white oak leaf in Tennessee. The A:c_{ic} curve observed on Julian day 272 exhibits both lower biochemical capacity for photosynthesis (as evidenced by lower A at any given c_{ic}), and a greater stomatal limitation to photosynthesis (as evidenced by a lower slope to the line connecting $c_{ic} = c_{ac}$ with the observed A), compared to the A:c_{ic} curve observed on Julian day 138. The lower panel is redrawn from Wilson *et al.* (2000b), with permission from Oxford University Press.

limitation to A increases. Thus, the line intersecting point A_3 depicts a greater degree of conductance limitation than the line intersecting point A_2 (Farquhar and Sharkey 1982).

In modeling the interaction between conductance and CO_2 assimilation rate, models such as BWB and BWBL must be combined with biochemical models of A; producing an iterative framework that can be used to predict steady-state CO_2 fluxes (Figure 7.17). Initial values for the fundamental determinants of g_{sc}, g_{bc}, and A must be entered into each component of the combined models, and then the model outputs for g_{sc}, g_{bc}, and A must be used as inputs in iterative fashion to mutually resolve all outputs to a constant solution. It is worth noting that the modeling approach described in Figure 7.17 suffers from a fundamental limitation – lack of mathematical closure. In order to solve the mathematical relationships among A, g_{sc}, g_{bc},

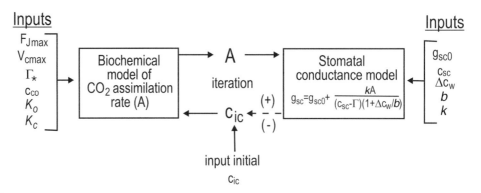

Figure 7.17   Schematic of how the models for determining A and g_{sc} fit together. An initial input for c_{ic} is entered into the model. The calculated value for A is then used to determine g_{sc}, which in turn modifies c_{ic} through positive (+) or negative (−) feedback. Through an iterative process, a stable balance between A and c_{ic} is determined, and in the process a final solution for g_{sc} is provided.

c_{ic}, and c_{ac} we can write $A = g_{tc}(c_{ac} - c_{ic})$, where $1/g_{tc} = 1/g_{bc} + 1/g_{sc}$. These equations can only be resolved if we have independent equations to define three of the four unknowns. We can define c_{ac} without a model because the ambient concentration of CO_2 is highly conserved. We can define g_{tc} independently from the other unknowns as $g_{tc} = E/1.58(c_{iw} - c_{aw})$. These assumptions would move us closer to a solution for A as long as E and $(c_{iw} - c_{aw})$ could be defined without reliance on further unknowns. Even if we could define g_{tc} with closure, however, we are still left with two unknowns (A and c_{ic}), and only one equation. This illustrates one of the most common challenges in modeling biosphere-atmosphere fluxes; deriving equations to define all but one variable in a way that does not create new unknowns in need of further equations. In the case of photosynthetic modeling, c_{ic} is known to be highly conserved among species and is generally constant in the face of moderate environmental change (Yoshie 1986). Using this knowledge, c_{ic} is often prescribed as a fixed value in the range 220–270 µmol mol^{-1}, thus facilitating closure. Alternatively, several models are available for independently estimating c_{ic} (Katul *et al.* 2000). Each of these models, however, rests on further assumptions required for the estimation of c_{ic}.

7.4 The leaf internal resistance and conductance to CO_2 flux

To this point, we have focused on the control over leaf CO_2 fluxes by boundary layer and stomatal conductances. Once CO_2 passes through the stomata into the intercellular air spaces of the leaf, however, it must enter liquid phases in order to continue diffusing toward the chloroplasts; the physical phase of the internal tissues of a leaf offers a *leaf internal resistance* (r_{ic}), or reciprocally, *leaf internal conductance* (g_{ic}), to the diffusive transport of CO_2 (see Figure 7.1). Similar to the cases for g_{sc} and g_{bc}, the leaf internal conductance places diffusive constraints on the CO_2 assimilation rate by reducing the CO_2 concentration within the chloroplast (c_{cc}), relative to c_{ic} and c_{ac}. Under the steady state assumption, and working within the context of the entire leaf, the net flux of CO_2 from the ambient atmosphere to the active site of Rubisco that is associated with the net photosynthetic CO_2 assimilation rate can be partitioned into three equal serial fluxes:

$$A_n = g_{bc}(c_{ac} - c_{sc}) = g_{sc}(c_{sc} - c_{ic}) = g_{ic}(c_{ic} - c_{cc}). \qquad (7.17)$$

In sum, the factors that contribute to g_{ic} include any conductive aspects of mesophyll cell walls (which are often hydrated with water), the plasma membrane, the cell cytosol, chloroplast envelope membranes and the chloroplast stroma. The magnitude of g_{ic} varies depending on species and growth conditions (Flexas *et al.* 2012). Those species with high photosynthesis rates also exhibit high g_{ic} compared to species with lower photosynthesis rates. Studies that have attempted to isolate the most limiting component of g_{ic} have generally concluded that it is the cytosolic and chloroplastic resistances associated with liquid-phase diffusion (e.g., Gorton *et al.* 2003).

Direct measurement of g_{ic} is difficult. One approach uses online measurement of $^{13}CO_2$ discrimination (Δ) along with the overall CO_2 assimilation rate (Evans *et al.* 1986). The magnitude of g_{ic} can be determined as the difference between theoretical values of Δ, assuming that $c_{ic} = c_{cc}$, and observed values whereby $c_{ic} \neq c_{cc}$. The difference will reflect the effect of g_{ic} on isotope discrimination, with the difference between c_{ic} and c_{cc} being proportional to g_{ic}. The estimation of g_{ic} by this approach, however, requires several assumptions about the discrimination against $^{13}CO_2$ by Rubisco and associated processes (e.g., compensating discrimination by the small amounts of PEP carboxylase in C_3 leaves and effects of photorespiration on c_{cc}), most of which have not been validated.

An alternate approach requires the use of chlorophyll fluorescence and gas-exchange techniques to estimate the chloroplast electron transport rate (F_J) (Di Marco *et al.* 1990, Harley *et al.* 1992a), and then using F_J, along with a model of photosynthesis and photorespiration to estimate c_{cc}:

$$F_J = v_{c2} \frac{4c_{cc} + 8\Gamma_*}{c_{cc} - \Gamma_*}. \qquad (7.18)$$

Equation (7.18) is a rearrangement of Eq. (5.20) in Chapter 5, but using ($c_{cc} - \Gamma_*$) as the relevant CO_2 concentration at the active site of Rubisco. Equation (7.18) is valid under those conditions when the overall rate of CO_2 assimilation is limited by the RuBP regeneration rate. When using gas-exchange measurements, v_{c2} represents ($A - R_d$); we have left the equation with v_{c2} as the relevant CO_2 assimilation rate in order to emphasize connections between Eq. (7.18) and Eq. (5.20). After substitution of c_{cc} with ($c_{ic} - v_{c2}/g_{ic}$), and rearrangement, we can write an equation relating g_{ic} to F_J (Warren 2006):

$$g_{ic} = \frac{A}{c_{ic} - \frac{\Gamma_*(F_J + 8v_{c2})}{F_J - 4v_{c2}}}. \qquad (7.19)$$

Some caution must be used in applying Eq. (7.19) because the determination of F_J using chlorophyll fluorescence includes measurement uncertainties that are carried through to the ultimate determination of g_{ic}.

To this point, we have focused on the diffusive constraints associated with the net flux of CO_2 into a leaf. However, a biochemical constraint also exists. The kinetic constraints associated with Rubisco and other reactions of the RPP pathway impose a non-diffusive "conductance" to CO_2 assimilation. We can develop a "biochemical conductance" that can be added in series to the diffusive conductances. Recall that the rate constant for a first-order chemical reactions (k, in s^{-1}) along with reactant molar concentration density (c, in mol m^{-3}) defines the ultimate rate of a reaction (v, in mol $m^{-3}\ s^{-1}$) (Section 3.1). Within this framework, the approximation can be made that in systems with combined diffusive and biochemical conductances, the biochemical component can be approximated as the biochemical rate constant (k) multiplied by a characteristic length scale (l, in m) (i.e., $g \approx kl$). Thus, for the chain of serial conductances for CO_2 diffusion involved with A, we can write:

$$\frac{1}{g_t} = \frac{1}{g_b \rho_m} + \frac{1}{g_s \rho_m} + \frac{1}{g_i \rho_{mw}} + \frac{1}{k_c l \rho_{mw}}. \tag{7.20}$$

In this case, conductances are in $m\ s^{-1}$, ρ_m is the molar density (mol m^{-3}) of dry air in the gas phase, ρ_{mw} is the molar density of water, k_c represents a composite rate constant for the net assimilation of CO_2, and l represents a length scale defining the span of CO_2 diffusion within the chloroplast. The value of k_c will include the kinetic definition of carboxylation by Rubisco (e.g., v_{c1}, v_{c2}, or v_{c3} as defined in Chapter 5) and the activity of carbonic anhydrase, which will influence the diffusive "mobility" of CO_2 by converting some fraction to bicarbonate, followed by conversion back to CO_2 at the site of carboxylation.

7.5 Evolutionary constraint on leaf diffusive potential

We started this chapter with a quote from Paul Jarvis in which biophysical limitations to the transport of CO_2 and H_2O across leaf surfaces were described as "adaptations" that control the rates of water loss and carbon uptake. Adaptations to a plant's environment emerge from the evolutionary process, in which plant function, growth, and ultimately reproductive fitness are maximized. In this chapter we devoted a considerable amount of text to establishing the principle that plant water loss can only be reduced at a cost to plant carbon uptake; they are controlled by tradeoffs linked through the same diffusive pathways. Similarly, the internal leaf conductance and stomatal conductances to CO_2 are linked through serial diffusive pathways. A proportionately lower internal conductance, relative to the stomatal conductance, will resist CO_2 diffusion to the chloroplast, at the cost of water loss through the stomata.

Consideration of the evolutionary compromises that must be made in balancing the need for effective physiological function takes us beyond the thermodynamic and biophysical principles that have been the focus of earlier chapters. In this chapter we have introduced new emergent scales of organization – those associated with plant tissues, their associated functions and the adaptations they represent. Leaf processes reflect the combined constraints of physics and evolution. As we leave this chapter on leaf function and move into the next chapter, we will develop a more direct consideration of the evolutionary process and its role in leaf design. It turns out that evolutionary outcomes are not entirely opaque to prediction. In fact, once we understand convergent patterns in evolution that are driven by tradeoffs in the face of common resource constraint, we can build such understanding into the prediction of plant-atmosphere exchange.

Appendix 7.1 A thermodynamic derivation of diffusive conductances

To this point we have, for convenience, defined conductance in terms of diffusive flux per unit of concentration gradient. Implicit in this definition is the concept of conductance as facilitating the *mobility* of mass under the influence of a thermodynamic driving force. In this appendix we will try to gain a better understanding of the term "conductance" within the context of the term "mobility." Much of the content of this section is founded on the thermodynamic treatment presented for Fick's Law in Appendix 6.1. The diffusive velocity of a scalar (v) is derived in thermodynamic terms as:

$$v = -u R T \frac{dc}{dx}, \qquad (7.21)$$

where u is "mobility" as defined by the Einstein–Smoluchowski relation (see Appendix 6.1). We define u as the terminal velocity reached by a unit mass given a diffusive driving force in the face of frictional resistance due to molecular collision. Equation (7.21) can be converted from velocity to flux density (units of mol m^{-2} s^{-1}) by multiplying both sides by molar density (ρ_m). Taking the case for diffusion across a finite distance (Δx), and recognizing the diffusive flux per unit of concentration gradient as conductance (g), we can write:

$$g = \frac{u R T}{\Delta x} = -\frac{v}{\Delta c}, \qquad (7.22)$$

with g reflecting a fundamental mobility coefficient scaled linearly to distance in the direction of the flux (Δx) and opposite the direction of the concentration gradient (Δc).

Appendix 7.2 Derivation of the ternary stomatal conductance to CO_2, H_2O, and dry air

In the steady state condition there is no *net flux* of dry air into a leaf, as deduced from the fact that there are no sources or sinks for dry air. However, there is a *gradient* for dry air ($\rho_m \Delta c_a / \Delta z$, where c_a is the mole fraction for dry air and ρ_m is the molar density of dry air), due to the fact that the humidity of air inside the leaf is higher than that outside the leaf. The gradient in c_a will force a *gross diffusive flux* of dry air into the leaf. It is assumed that a small pressure gradient exists across the leaf surface due to the difference in water vapor partial pressure. The pressure gradient is assumed to drive an *outward viscous flow* of H_2O (along with air) out of the leaf that balances the *inward diffusive flow* of dry air (Leuning 1983). The interaction between the outward flux of moist air and the inward flux of dry air can be represented as:

$$g_{tw} = \bar{c}_a \frac{E}{\Delta c_w},\qquad(7.23)$$

where $\bar{c}_a$ is the average mole fraction of dry air molecules across the distance from the intercellular air spaces to the outside limit of the leaf boundary layer. In this derivation, $\bar{c}_a$ is estimated as $c_{ia} + c_{aa}/2$, where c_{ia} and c_{aa} are the mole fractions of air in the intercellular air spaces and outside atmosphere, respectively. Because the primary molecular components of air are N_2, O_2, and H_2O, $\bar{c}_a$ can be represented as $(1 - \bar{c}_w)$, where $\bar{c}_w$ is the average mole fraction of water vapor (i.e., $c_{iw} + c_{aw}/2$). Using these relations for substitution, Eq. (7.23) can be expressed as:

$$g_{tw} = (1 - \bar{c}_w) \frac{E}{\Delta c_w}.\qquad(7.24)$$

The term g_{tw} in Eq. (7.24) differs from a binary conductance (i.e., $g_{tw} = E/\Delta c_w$) in that the factor $(1 - \bar{c}_w)$ accounts for the gradient in dry air that opposes the gradient in H_2O. Lack of consideration for this effect can result in a small (2–3%) but potentially significant overestimation of g_{tw}.

In deriving a conductance for CO_2, the effect of collisions between CO_2 and H_2O cannot be ignored. Thus, the total conductance to CO_2 (g_{tc}) is composed of two component conductances, $g_{t(ca)}$ and $g_{t(wc)}$, where $g_{t(ca)}$ is the component affected by CO_2 molecules colliding with air molecules and $g_{t(wc)}$ is the component affected by CO_2 molecules colliding with H_2O molecules. By coincidence, the binary diffusion coefficient (K_d) for CO_2 diffusing through air containing water vapor is approximately equal to that for CO_2 diffusing through dry air. Thus:

$$g_{t(ca)} \approx g_{t(wc)}.\qquad(7.25)$$

A ternary equation for g_{tc} can therefore be developed entirely in terms of $g_{t(ca)}$ (von Caemmerer and Farquhar 1981). To derive such an equation, the following relation can be stated:

$$c_{ic} - c_{ac} \approx -\left(\frac{\bar{c}_a}{g_{t(ca)}} + \frac{\bar{c}_w}{g_{t(wc)}}\right) A + \frac{\bar{c}_c E}{g_{t(wc)}},\qquad(7.26)$$

$$\text{Term}\qquad\quad \text{I}\qquad\quad \text{II}\qquad\quad \text{III}$$

where $\bar{c}_c$ represents the average CO_2 mole fraction between the intercellular air spaces of the leaf and the ambient atmosphere (i.e., $c_{ic} + c_{ac}/2$). In essence, Eq. (7.26) states that the CO_2 concentration gradient that exists across a leaf is a function of the CO_2 assimilation rate and diffusive interactions of CO_2 and H_2O with air molecules, as well as with each other. The first two terms represent the effects of air and water vapor molecules on the inward diffusive velocity of CO_2. Term III represents the effect of the outward diffusion of water molecules on the CO_2 concentration gradient. In sum, these effects represent the conductance to CO_2.

Recalling that $\bar{a} + \bar{w} \approx 1$ and using Eq. (7.25) for substitution, Terms I and II can be simplified to $1/g_{t(ca)}$, and the entire equation can be rearranged to:

$$g_{tc} \approx \frac{A}{\Delta c_c - [(c_{ac} + c_{ic})E/2]}. \tag{7.27}$$

Appendix 7.3 Derivation of the Leuning and Monteith forms of the Ball–Woodrow–Berry model

The stomatal conductance model developed by Ball *et al.* (1987; here called the Ball–Woodrow–Berry (BWB) model) and presented in Eq. (7.15) fails to address one key observation concerning stomatal response to the environment: that g_s may respond most directly to E, not h_s. Leuning (1995) concluded that one gets a better fit of the model to data if the response of g_s to humidity is expressed on the basis of Δc_w, which scales directly with E. The value for c_{iw}, the water vapor mole fraction inside the leaf, approaches saturation. When expressed in terms of the saturation mole fraction (c_w^*), c_{iw}/c_w^* approaches unity (i.e., the relative humidity approaches 100%) and Δc_w can be defined as $1 - (c_{sw}/c_w^*)$, where c_{sw} is the water vapor mole fraction at the outer leaf surface. In reconciling Eqs. (7.15) and (7.16), Δc_w can be related to h_s by $h_s = 1 - \Delta c_w/c_w^*$. If we assume a positive-linear relation between g_s and h_s, as implied in the original BWB model, Eq. (7.15) can be rewritten as:

$$g_s = g_{s0} + \frac{kA(1 - \Delta c_w/c_w^*)}{c_{sc}}. \tag{7.28}$$

Leuning (1995) determined that using a negative-hyperbolic relationship between g_s and Δc_w, rather than a positive linear relationship between g_s and $1 - \Delta c_w/c_w^*$, provided a better fit to the existing data. Thus, he inserted a different form for the humidity response, namely $(1 + \Delta c_w/b)^{-1}$. (The latter equation was obtained from Lohammer *et al.* (1980).) The parameter b is an empirically determined coefficient. This final change provides the full BWBL form of the model expressed in Eq. (7.16).

In an effort to modify the BWBL model further, Monteith (1995) recast the theory to accommodate the past observation that g_s responds directly to E, rather than Δc_w. Monteith began with the following relation:

$$g_{sw} = \frac{E}{\Delta c_w} = a(1 - mE), \tag{7.29}$$

where a and m are constants that can be found theoretically by regressing g_{sw} on E in the face of varying Δc_w, or by regressing $1/E$ against $1/\Delta c_w$. The relation depicted on the right-hand side of Eq. (7.29) represents a negative-linear response of g_{sw} to increasing E, which is consistent with empirical observations. The constant a of Eq. (7.29) is taken as the extrapolated y-intercept, when g_{sw} is regressed on E and E $\rightarrow 0$ (i.e., when $\Delta c_w \rightarrow 0$) (Figure 7.18).

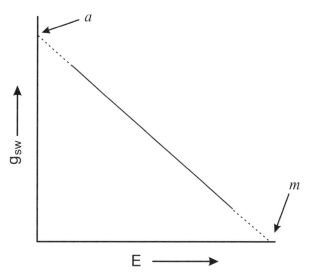

Figure 7.18 Theoretical relationship between stomatal conductance to water vapor (g_{sw}) and leaf transpiration rate (E) due to varying the leaf-to-air water vapor concentration difference (Δc_w). The extrapolated y-intercept (a) is taken as the maximum possible g_{sw} which occurs as E $\rightarrow$ 0 and $\Delta c_w \rightarrow$ 0. The x-intercept (m) is taken as the maximum possible E which occurs as $g_{sw} \rightarrow$ 0 and $\Delta c_w \rightarrow \infty$.

The constant a is thus equal to the maximum possible value for g_{sw}, and can be represented as g_{swmax}. In a similar way, the constant m is taken as the extrapolated x-intercept, as $g_{sw} \rightarrow$ 0 (i.e., when $\Delta c_w \rightarrow \infty$), and E approaches a maximum which can be represented as E_{max}. Rearrangement and analysis of the right-hand side of Eq. (7.29) will reveal that $E_{max} = 1/m$ as $g_{sw} \rightarrow$ 0. Substituting g_{swmax} for a and $1/E_{max}$ for m, Eq. (7.29) can be rewritten as:

$$g_{sw} = g_{swmax}(1 - E/E_{max}). \tag{7.30}$$

With Eq. (7.30) in place it is possible to modify the BWBL model and replace the dependence of g_s on Δc_w with a dependence on E. Working with $I_t \gg 0$, such that $g_s \gg g_{s0}$, and with $\Delta c_w \rightarrow$ 0, such that $g_s \rightarrow g_{smax}$, Eq. (7.16) reveals that:

$$g_{smax} = \frac{kA}{c_{sc} - \Gamma}. \tag{7.31}$$

Combining Eqs. (7.30) and (7.31):

$$g_s = \frac{kA(1 - E/E_{max})}{c_{sc} - \Gamma}. \tag{7.32}$$

Equation (7.32) is the BWBL model rewritten to accommodate the response of g_s to E, rather than Δc_w (and with the assumption that $g_s \gg g_{s0}$). Equation (7.32) is hereafter referred to as the Monteith form of the Ball Woodrow–Berry model, or BWBM.

8

Leaf structure and function

Form Follows Function.

Louis Henri Sullivan, architect

Form follows function . . . has been misunderstood. Form and function should be one, joined in a spiritual union.

Frank Lloyd Wright, protégé of Louis Henri Sullivan and architect

Leaves provide the infrastructure within which solar photons and CO_2 are channeled to photosynthetic mesophyll cells to support gross primary productivity, and from which absorbed energy is partitioned into latent and sensible heat loss. Both the internal environment and surface features of a leaf are the products of modification through natural selection to produce a form and function that provides advantages toward carbon and energy assimilation, and ultimately growth. For example, the structural arrangement of leaf cells affects the organized dispersion of the solar photon flux and the density and function of stomata affect the diffusive uptake of atmospheric CO_2 and its relationship to H_2O loss. As indicated in the quotes from two famous architects that we cite above, the concept that "form is married to function" is perpetuated across generations of multiple disciplines; it is not limited to the biological sciences. It is a fundamental philosophical relation. In fact, some of the earliest intellectual pursuits of this relation were conducted by very broad thinkers; practitioners of literature, music, and the visual arts. Johann Wolfgang von Goethe, best known for his literary works, spent many hours in discussions with Alexander von Humboldt, the "founder" of the discipline of phytogeography, discussing the concept that plant form could be used to discern plant-climate relations; discussions that no doubt influenced Humboldt's theories on the determinants of vegetation distribution across the globe. There is no requisite sequence by which "form mandates function" or "function mandates form"; evolutionary modification occurs to both anatomy and physiology, maintaining both within a coordinated set of resource use parameters. Leaf form and function must be viewed as integrative, and the coupled nature of form and function must be viewed, in and of itself, as a target of natural selection.

The integration of form and function is facilitated by cost-benefit constraints imposed by selection. Leaf size, for example, is determined to a large extent by tradeoffs between the benefit of capturing the solar photon flux in light-limited habitats and the cost of transpiring soil moisture in water-limited habitats. Large-leaved species are relatively rare in open, perennially hot, dry habitats. In those habitats, the benefit of large leaves is reduced because of the high solar photon flux and the cost of large leaves is increased because of the low availability of soil moisture. In that type of habitat, large leaves would cause internal temperatures to exceed those favorable for photosynthesis, and expose the leaf tissues to a greater risk of desiccation and death; natural selection would favor smaller leaves with tight coupling of stomatal conductance to soil moisture availability. Leaf thickness is also determined by cost-benefit tradeoffs reconciled through natural selection. In a resource-rich environment thick leaves will have a benefit in being able to conduct higher photosynthesis rates per unit of absorbed photon flux density, compared to thin leaves. However, at a certain limit, thick leaves will reduce the potential for penetration of the photon flux, and diffusive constraints will reduce the effective distance for CO_2 transport within the leaf, so that those cells farthest from stomatal pores and those near the bottom of the leaf may incur unfavorable carbon and energy budgets. Natural selection will favor leaf thicknesses that maximize the margin between the gain in carbon and energy assimilation versus the loss in carbon and energy during tissue construction and maintenance.

In this chapter we place these issues of form and function, cost and benefit, and the role of natural selection in directing plant design toward optimal function, within the context of our past discussions on photosynthetic metabolism and leaf diffusive processes. We will start the chapter with further elaboration of the topic of natural selection and leaf form. A primary aim of this section is to achieve a better understanding of why traits tend to converge toward common patterns among divergent taxonomic lineages of plants, and thus why we can organize plants into predictable functional types. We will then move to a discussion of how leaf form affects the movement of photons, CO_2 molecules, and H_2O molecules through, and within, leaves. We will focus on the pathways of flux within the context of leaf anatomy and the consequences of anatomical design to tradeoffs between carbon gain and water loss. Near the end of the chapter we will introduce one aspect of leaf form and function that influences mathematical modeling. That is, the tendency for leaf structure to force non-linearities in the relation between CO_2 and H_2O fluxes and the environmental drivers that sustain them. We often resolve this complexity by relying on evolutionary tendencies toward optimization. We will discuss how ecologists and earth system scientists have used those tendencies to develop mathematical "conveniences" with which to linearize otherwise non-linear patterns. Even within the realm of modeling, the most commonly used lines of logic take advantage of the tight evolutionary connection between form and function; recognizing the requisite and mutual proportionalities that emerge from coupled constraints.

8.1 Leaf structure

Leaves come in different shapes and sizes ranging from relatively flat, broad structures specialized for capturing photons to needle-like structures specialized for low surface-to-volume ratios. In most cases, leaf morphology has evolved in response to selection for maximum photosynthesis rate or minimum transpiration rate, depending on plant growth habit, climatic factors, and the availability of environmental resources. In some cases, such as the evolution of thorns, trichomes, or fibrous tissues, defense against herbivory has also been an important selective factor. Selection for increased defense against herbivory may have been especially important in determining the form of long-lived leaves in resource-poor environments. In these situations, leaves are exposed to the threat of herbivory for a longer period of time and the resources required to replace lost leaves are not easily obtained; thus, justifying the increased allocation of resources to the construction of morphological defenses.

An example of the unique relation between leaf size and shape and its ramifications for CO_2 and H_2O exchange is seen in the desert flora of the southwestern United States. In a study of the energy budget of desert plants, Gates and co-workers (Gates *et al.* 1968) noted that leaves tend to be small and leaf temperatures are closely coupled to air temperatures. They speculated that these two observations were related, and that the small leaf size facilitated thinner surface boundary layers and efficient exchange of sensible heat with the atmosphere. Ten years later this theory was modified in a paper by Smith (1978), who observed that some desert species have large, flat leaves, and these tend to be the species with greatest access to water; either because they occupy ecological niches in washes or basins with near-surface water tables and/or they possess tap roots that reach deep moisture reserves. Thus, it appeared that some species, with greater access to water in an otherwise arid ecosystem, had evolved a leaf morphology that favored latent heat loss at the expense of efficient sensible heat loss. Plants with large leaves and high transpiration rates could maintain leaf temperature close to air temperature (or even below air temperature), but in this case by increasing the potential for transpiration and latent heat loss. The cooler leaf temperatures that are achieved by high rates of transpiration in these species are close to the biochemical optimum for photosynthesis and thus they permit high rates of CO_2 assimilation.

It would be possible to dedicate an entire chapter to the topic of diversity in leaf structure. We will focus on only one type of leaf – the *bifacial leaf*. Bifacial leaves exhibit heterogeneous anatomies in cross section, with tightly packed cells just beneath the top epidermis and loosely packed cells just above the bottom epidermis (Figures 8.1A and 8.1B). The bulk of the cells between the two epidermal layers compose a tissue known as *mesophyll*. Leaf mesophyll cells contain chloroplasts and are specialized for conducting photosynthesis. In many bifacial leaves the mesophyll tissue is organized into two distinct zones: the tightly packed upper zone known as *palisade mesophyll*, and the loosely packed lower zone known as *spongy mesophyll*. In conifer needles, the mesophyll tissue is not differentiated into palisade and spongy layers, but rather occurs as tightly packed cells surrounding a vascular bundle (Figure 8.1C); a form that results in relatively low surface-to-volume ratios.

The vascular system of the leaf enters through the leaf *petiole*; the "stalk" of the leaf. Vascular bundles in the petiole and veins of the leaf contain *xylem*, the tissue of water transport, and *phloem*, the tissue of carbon transport (sugars and amino acids). (Organic molecules can also be found dissolved in xylem water, but the principal path of carbon transport is in the form of sugars transported through the phloem.) Xylem water transport is from roots to leaves, although water can also be moved among leaves if driven by appropriate water potential gradients. Phloem transport is from leaves to growth or storage sinks in the stem, roots, or meristematic tissues; either upward, for the case of transporting sugars to the growing apex of the shoot, or downward, for the case of transporting sugars to growing roots. Stomata can be located either in the upper or lower epidermis of leaves, or both, depending on the plant species.

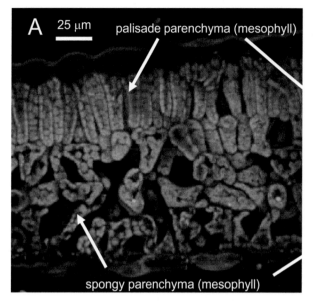

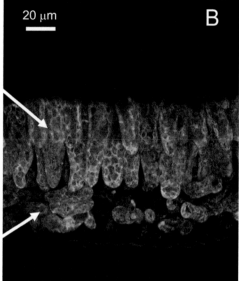

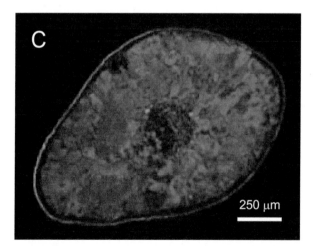

Figure 8.1   **A**. Confocal microscope picture of an oak leaf (*Quercus ilex*) showing yellow-fluorescent chloroplasts within a single layer of tightly packed palisade mesophyll cells just beneath the upper epidermis and loosely packed spongy mesophyll cells, with large, interspersed air channels, just above the lower epidermis. **B**. Two palisade mesophyll layers above spongy mesophyll tissue in a poplar leaf (*Populus trichocarpa*). **C**. Tightly packed mesophyll cells surrounding the vascular bundle in a Norway spruce needle (*Picea abies*). Micrographs courtesy of Professor Joerg-Peter Schnitzler (Helmholtz Zentrum München, Munich, Germany). See color plates section.

8.2 Convergent evolution as a source of common patterns in leaf structure and function

> As is well known, nearly all chemical energy and organic carbon and every bit of food that enter into any ecosystem on earth is provided by photosynthesis. As a result, all biological activity from the simplest virus to man is ultimately limited by how well photosynthesis is able to operate in all of the very diverse environments that exist on this earth. (Olle Björkman (1975))

We can recognize consistent relations between leaf form and function across broad taxonomic groups of terrestrial plants. We often refer to *plant functional types* (PFTs) as those groups that exhibit similar structural and functional attributes. The formulation of PFTs is founded on the recognition that natural selection works independently in different taxonomic lineages to produce convergent and predictable patterns in form and function (see Reich *et al.* 2003). The attributes by which PFTs are defined can refer to general growth habit, such as evergreen versus deciduous or woody versus herbaceous, or they can refer to physiological activity, such as maximum photosynthesis rate or leaf life span.

Natural selection works to modify species attributes in a manner that minimizes cost and maximizes benefit with regard to reproductive fitness. Because many attributes are linked in their effect on fitness, evolutionary modification of one attribute is likely to cause a change in the fitness value of a second attribute; and these coupled influences are likely to vary in different environments. Thus, selection often leads to compromises in the relative values of attributes, a phenomenon we refer to as *adaptive tradeoffs*. As an example, consider the case of plants native to dry ecosystems in which low stomatal densities impose a high leaf diffusive resistance, thus limiting H_2O loss; a clear benefit. At the same time, however, the high diffusive resistance will limit CO_2 uptake; a clear cost. Selection for maximum fitness in this environment will require a compromise between water loss and CO_2 uptake; those traits that ultimately emerge and are sustained in a lineage will reflect the product of adaptive tradeoffs. We can extend the concept of adaptive tradeoffs to other linked attributes. For example, the limited potential for photosynthesis in plants from arid ecosystems will also limit their capacity to sustain high rates of growth and concomitant leaf turnover; this should favor selection for high levels of defense and longer leaf life spans. One leaf attribute that has been used as an integrator of form and function is *specific leaf area* (leaf surface area per unit of leaf mass). This attribute is easily measured on dried leaf material, has been reported for many species in different ecosystems, and appears to be correlated with other attributes, such as photosynthesis, transpiration, and respiration rates (Wright *et al.* 2001, 2003, Reich *et al.* 2006, 2008). Thus, SLA can be used as a proxy in modeling studies to predict rates of leaf-atmosphere exchange. We have provided a more detailed discussion of SLA and its relevance to leaf CO_2 fluxes in Box 8.1. Through these several examples we find threads of logic that allow us to infer the forces of evolutionary optimization and predict common patterns of covariance in traits associated with form and function.

Inferred relations between form and function are intuitive within the economic framework of cost-benefit optimization, but why should we think that our intuition is consistent with

| Box 8.1 | Specific leaf area: an integrative attribute linking leaf form and function |

The thickness of a leaf reflects developmental processes that have been influenced in the long term by natural selection and in the short term by acclimation to the prevailing environment. Specific leaf area ($cm^2 g^{-1}$) is a measure of leaf thickness and is correlated with rates of leaf metabolism when assessed across broad taxonomic boundaries. This correlation can be explained within the context of cost-benefit tradeoffs. With low resource availability, specific leaf area tends to decreases (Reich *et al.* 2003); this is likely due to evolutionary constraints that link specific leaf area and life span. At low resource availability plant growth and metabolism may be constrained to the point where short-lived leaves are not sustainable. Longer-lived leaves are more likely to evolve protection against mechanical damage and herbivore loss through leaf fibers, secondary cell wall thickenings and chemical constituents; basically, producing tougher, thicker leaves with concomitantly lower specific leaf area. Thus, in plants native to resource-limited sites we tend to find a combination of low metabolic capacity, long leaf life span and low specific leaf area. The mutual constraints imposed on each of these attributes are expressed as positive, quantitative correlations (Figure B8.1A). These correlations are useful in modeling studies as they allow us to move along multiple axes of leaf form and function and synthesize dimensional parameters that combine aspects of form and function.

While a positive correlation between specific leaf area and metabolic rate has been observed across broad geographic resource gradients, we often see the opposite relation when observed across gradients of light (Figure B8.1B). Shade leaves are constructed with lower A_{mes}/A_L, or conversely higher specific leaf area, thus enhancing the surface area available for light absorption. Shade leaves also exhibit lower metabolic capacities. Thus, the correlation between metabolic rate and specific leaf area is typically negative when assessed across a collection of sun and shade leaves within the same plant or canopy. When assessed across broad taxonomic and geographic gradients, however, some studies have found a positive relation between adaptation to shade and specific leaf area (Veneklaas and Poorter 1998, Reich *et al.* 2003). As the number of species and habitats considered in an analysis increases, the relation between specific leaf area and net photosynthesis rate may converge for light-, water-, and nutrient-limited ecosystems. More research will be needed to resolve this issue.

One final issue to consider in the use of correlations such as those shown in Figure B8.1 is that past researchers have expressed net photosynthesis rate in units based on leaf area *or* leaf mass depending on the analysis. Expression on a leaf area basis is more informative as to the photosynthetic efficiency of solar photon use (what is often referred to as the quantum yield or light-use efficiency). Expression on a leaf mass basis is more informative as to the effect of cost-benefit constraints. The cost of constructing a leaf can be best evaluated against the benefit of photosynthetic return if both are expressed on the basis of mass; carbon is assimilated as mass, not area. In contrast, the metric most relevant to studies of plant-atmosphere exchange is flux density; recall that flux density is defined as flux per unit of cross-sectional area in the direction of the flux (see Section 1.2). Thus, surface area is the most appropriate means of expressing CO_2 assimilation rate in studies of surface-atmosphere exchange dynamics.

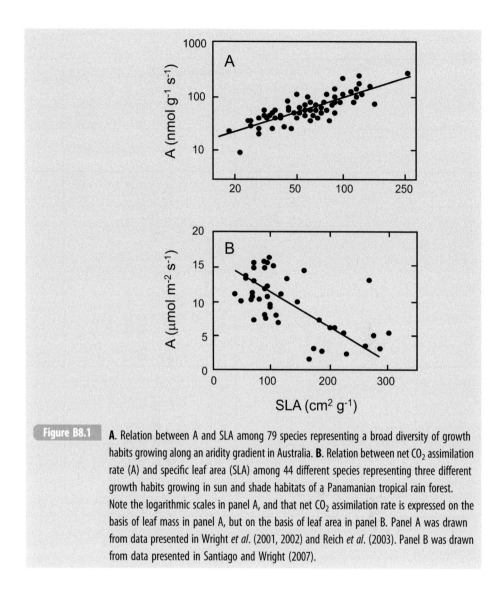

Figure B8.1 **A.** Relation between A and SLA among 79 species representing a broad diversity of growth habits growing along an aridity gradient in Australia. **B.** Relation between net CO_2 assimilation rate (A) and specific leaf area (SLA) among 44 different species representing three different growth habits growing in sun and shade habitats of a Panamanian tropical rain forest. Note the logarithmic scales in panel A, and that net CO_2 assimilation rate is expressed on the basis of leaf mass in panel A, but on the basis of leaf area in panel B. Panel A was drawn from data presented in Wright *et al.* (2001, 2002) and Reich *et al.* (2003). Panel B was drawn from data presented in Santiago and Wright (2007).

evolutionary outcomes? We must seek supporting observations. We can ask the question: do these generalizations hold up when examined across multiple habitats and ecosystems? Several studies have been conducted using a broad range of species and growth habits and it has been shown that even across ecosystem boundaries the structural and functional attributes of leaves are correlated, that they sort out according to PFTs, and they co-vary in patterns that follow the intuitive logic of cost-benefit arguments (Reich *et al.* 1997, Wright *et al.* 2004, Reich *et al.* 2006; although see Lusk *et al.* 2008). Species that exhibit the evergreen growth habit tend to have longer-lived leaves with lower metabolic rates, compared to species that exhibit the deciduous growth habit (Figure 8.2). Evergreen species with long leaf life spans also tend to be native to ecosystems with significant resource limitations

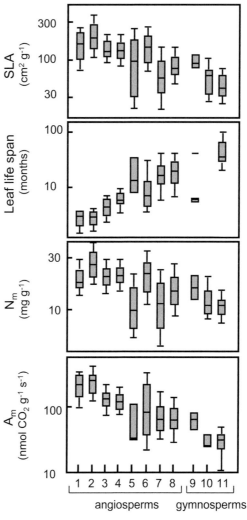

Figure 8.2 Leaf structural and functional attributes for 11 plant functional types (PFTs). Plant functional types are: (1) deciduous grasses, (2) deciduous herbs, (3) deciduous shrubs, (4) deciduous trees, (5) evergreen grasses, (6) evergreen herbs, (7) evergreen shrubs, (8) evergreen trees, (9) deciduous trees, (10) evergreen shrubs, and (11) evergreen trees. There are clear trends toward less specific leaf area (SLA) (i.e., thicker leaves), longer leaf life spans, lower leaf nitrogen concentrations per unit mass (N_m), and lower photosynthetic capacities per unit leaf mass (A_m) in gymnosperms compared to angiosperms, trees compared to shrubs, grasses, and herbs, and in evergreen plants compared to deciduous plants. Redrawn from Reich *et al.* (2007).

(dry, infertile, or shaded). The same types of correlations are found in sun and shade leaves. Shade leaves tend to have longer life spans, lower N concentrations, and lower rates of metabolism, compared to sun leaves (Ackerly and Bazzaz 1995, Montgomery 2004).

Taking account of the various observations that have been reported, we can conclude that there are indeed natural patterns of convergent organization within global vegetation. These natural patterns can be used as conveniences to produce vegetation assemblages of common ecosystem-atmosphere flux attributes. For example, we might group together boreal coniferous forests and subalpine coniferous forests as ecosystems with low photosynthetic capacities compared to mid-latitude broadleaf forests. Similarly, we might divide a single broadleaf forest into two vertical zones – one with sun leaves with higher flux densities, and one with shade leaves with lower flux densities. Recognition of these natural patterns of covariance provides us with a powerful tool in the modeling of ecosystem-atmosphere fluxes. Plant functional types are often organized to represent the vegetation in specific grid cells in a mapped geographical domain. This process of PFT organization can occur in way that directly maps PFTs onto grid cells, or that organizes PFTs into representative biomes, which are then mapped onto grid cells (Figure 8.3). In Figure 8.3, the results of six dynamic global vegetation models (DGVMs) are presented showing the global distribution of different biome types, which in turn reflect the differential distribution of vegetation-atmosphere exchanges of mass and energy. In Figure 8.4, climatic controls over net primary productivity are shown for different biome types using a combination of satellite remote sensing, to provide vegetation leaf area and canopy structure, climate variation, and modeled photon use efficiency. This map is used to show an example as to how the organization of PFTs into biomes, and differentiation of biomes on the basis of functional traits, is used to infer global-scale distributions of the interactions between functional traits and climate; in this case the difference between plant photosynthesis and respiration, which we designate as net primary production (NPP).

8.3 Photon transport in leaves

To better understand the photosynthetic assimilation of CO_2 within leaves, we begin with the analogy of a leaf as a photon-gathering "scaffold" within which the biochemical apparatus of photosynthesis is embedded. Solar photons are absorbed by, transmitted through, or reflected from leaves, depending on surface characteristics and the pattern by which the photosynthetic apparatus is distributed within the leaf. On average, leaves absorb 40–60% of the total incident solar photon flux density. This value can vary depending on the angle at which solar rays strike the leaf, being closer to 40% for horizontal leaves receiving photons at low solar angles and 60% for horizontal leaves receiving photons at high solar angles. Absorption of photons in the visible part of the electromagnetic spectrum (i.e., 0.4–0.7 μm), which is also the part of the spectrum utilized in photosynthesis, is higher than that for the total solar photon flux; generally being greater than 80% (Figure 8.5). In the near-infrared part of the solar spectrum (0.7–1.35 μm), absorption is less than 10%, with most of the energy transmitted through the leaf or reflected from its surface.

The reflection of solar photons that strike leaves includes *diffuse* and *specular* components. Diffuse reflection can occur from both internal and external surfaces; its primary source, however, is from photons that penetrate the leaf and are reflected among the various internal surfaces of mesophyll tissue. Internal reflections cause photon scattering, some of which are reflected back across the surface and exit the leaf. Diffuse reflection reaches a maximum when the solar angle is high and solar photons have the greatest opportunity to penetrate the leaf (Figure 8.6). Specular reflectance occurs through the interaction

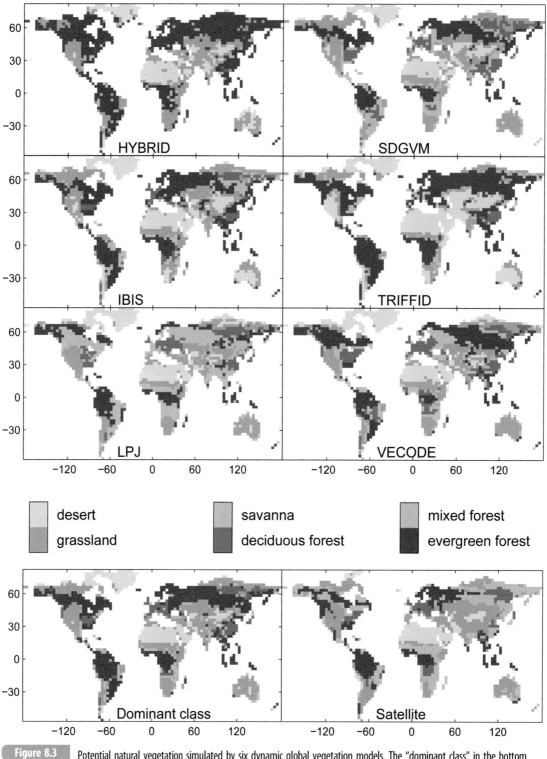

Figure 8.3 Potential natural vegetation simulated by six dynamic global vegetation models. The "dominant class" in the bottom left panel is the consensus vegetation class that was predicted by the largest number of models in each grid cell. The bottom right panel is a simplified vegetation map inferred from satellite remote sensing (see Cramer *et al*. 2001). Ecophysiological properties of global vegetation are inferred from the types of correlations displayed in Figure 8.2 for different biome types and can be used to predict functional traits, such as gross primary productivity (GPP), which reflects photosynthetic potential, and net ecosystem productivity (NEP), which reflects the sum of photosynthesis and ecosystem respiration (distinguished by opposing mathematical sign). From Cramer *et al*. (2001). See color plates section.

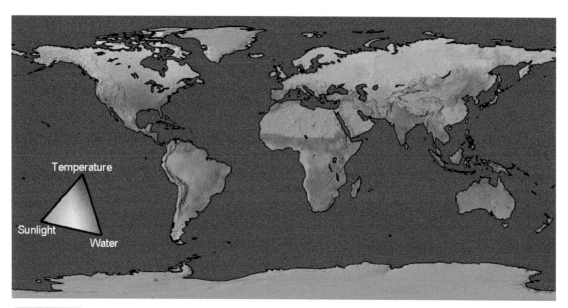

Figure 8.4 Geographic variation in climatic controls over net primary productivity (NPP) considering water, temperature, and sunlight availability. The patterns of variation in NPP were derived from satellite data on leaf area index, combined with climate data and consideration of plant functional type (PFT) differences in different biomes. Several PFTs can be found in a biome and the biomes were distinguished on the basis of their photon use efficiencies (NPP per unit of absorbed photon flux density). The biome types considered in this analysis were evergreen needleleaf forest, deciduous broadleaf forest, shrubland, savanna, grassland, and cropland. From Running *et al.* (2004). See color plates section.

of photons with epidermal surfaces ("specular" refers to the mirror-like quality of leaf surfaces) and involves reflection along angles that equal the angle of incidence. Specular reflectance is greatest at low solar angles.

The internal reflection of photons that enter a leaf is caused by the refractive discontinuities that occur between intercellular air spaces (refractive index of 1.0) and the surrounding mesophyll cell walls (refractive index of 1.47 for hydrated cells) (Walter-Shea *et al.* 1989). Generally, a leaf with a greater fraction of leaf intercellular air space has greater potential to scatter incident photons. This is especially obvious when comparing the reflectance patterns for cells near the top versus bottom of a bifacial leaf. When viewed with the human eye and held perpendicular to a light source, the top surface of a bifacial leaf appears darker, whereas the bottom surface appears lighter. This difference is due to greater scattering of photons by the air-cell interfaces of loosely packed cells in the spongy mesophyll tissue near the bottom surface. When the air spaces of a bifacial leaf are infiltrated with oil, which approximates the refractive properties of hydrated cell walls, total reflectance decreases by 15–20%.

Specular reflection is caused by the interaction of photons with outer leaf surfaces. Ignoring the effects of hairs, leaf surfaces are generally "optically smooth." However,

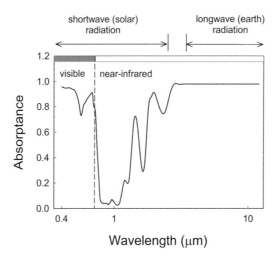

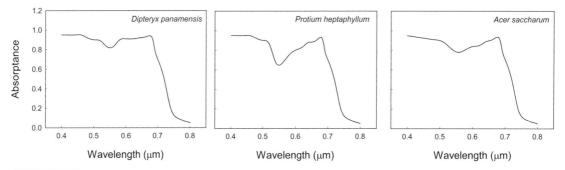

Figure 8.5 **Upper figure**. The spectral distribution by which a cottonwood leaf (*Populus deltoides*) absorbs solar and earth radiation. Absorptance is the fraction of incident radiation absorbed at each wavelength. Redrawn from Gates (1980). **Lower figure**. Absorption spectra of three representative leaves at those wavelengths within and just beyond the visible part of the spectrum. *D. panamensis* and *P. heptaphyllum* are tropical forest tree species. *A. saccharum* is sugar maple, a temperate forest tree species. Redrawn from Poorter *et al.* (1995), Roberts *et al.* (1998), and Carter *et al.* (2000), respectively, with permission from Elsevier B.V. Publishers.

even the smoothest facets of a leaf possess small microscopic irregularities that diffuse intercepted radiation. The specular and diffuse fractions reflected from the outer surfaces of a leaf can be separated on the basis of polarization. Radiation that is reflected in specular fashion is *polarized* (i.e., oscillations in the radiation field occur in the same direction), whereas diffuse radiation is non-polarized.

The presence of hairs on the leaf surface has a negative effect on specular reflection; leaf hairs increase photon scattering. The capacity of leaf hairs to reflect solar radiation is to some extent dependent on the nature of the hairs themselves. In the desert shrub *Encelia farinosa*, leaves produced during the cool winter months are covered with a sparse layer of living, fluid-filled hairs, whereas leaves produced during the hot

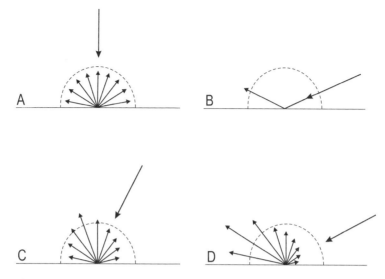

Figure 8.6   **A**. The hemispherical pattern of diffuse (or Lambertian) reflectance. **B**. Specular (or non-Lambertian) reflectance. **C**. Combined diffuse and specular reflectance when the angle of solar incidence is high. **D**. Combined diffuse and specular reflectance when the angle of solar incidence is low.

late-spring are covered with a dense layer of dead, air-filled hairs (Ehleringer and Björkman 1978). Given the prominent infrared absorption bands for water, fluid-filled hairs facilitate the *absorption* of infrared radiation during the seasonal period when leaf temperatures are coolest. Given the tendency for photon scattering at air-cell interfaces, air-filled hairs facilitate the *reflection* of infrared radiation during the seasonal period when leaf temperatures are warmest. The dense pubescence layer of *E. farinosa* leaves during the late-spring skews leaf reflectance to favor wavelengths of infrared radiation (0.7–3.0 μm) relative to photosynthetically active (0.4–0.7 μm) radiation (Figure 8.7). The differential reflectance potential of leaf hairs provides advantages that enhance carbon gain and facilitate a balanced energy budget in plants of *E. farinosa* leaves in their native hot, arid environments.

A quantitative understanding of photon scattering is required to assess the photon budget, quantify the amount of radiant energy absorbed, and couple photon flux to the exchange of mass and energy between leaves and the atmosphere. Additionally, the interpretation of radiometric data acquired with aircraft- or satellite-mounted sensors, which has become so fundamental to our understanding of vegetation structure and leaf chemical composition, requires knowledge about photon reflectance patterns. Theoretically, the angular distribution of the scattered photon flux from a leaf is described by the *leaf scattering phase function* (Ross 1981):

$$f(\Omega' \to \Omega; \Omega_L)d\Omega = \frac{dE_L}{dE'_L}, \tag{8.1}$$

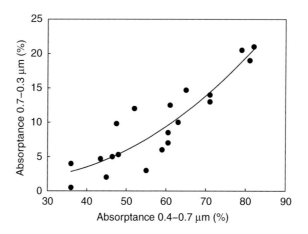

Figure 8.7   Relationship between the absorptance of photosynthetically active radiation (0.4–0.7 µm) and infrared radiation (0.7–3.0 µm) in leaves of *Encelia farinosa* (desert brittlebush). The points represent measurements taken at different times of the year. As the seasonal climate becomes hotter and drier absorptance of infrared radiation decreases more than absorptance of photosynthetically active radiation. Redrawn from Ehleringer and Björkman (1978), with kind permission from Springer Science + Business Media.

where $f(\Omega' \rightarrow \Omega; \Omega_L)$ describes the scattering of photons from the solid angle about direction Ω' into direction Ω given a leaf surface oriented at direction Ω_L. The right-hand side of Eq. (8.1) represents the fraction of total incident photon flux from direction Ω' (indicated as E'_L) that is reflected into direction Ω (indicated as E_L). The form of Eq. (8.1), written as a differential equation, accounts for changes in reflectance within the entire range of possible leaf viewing angles (Ω); integration of Eq. (8.1) with respect to $d\Omega$ provides the total leaf reflectance. In its integral form, Eq. (8.1) is a complex statement of a simple concept: *after accounting for absorption, the incident photon flux density is conserved when scattering is viewed from all possible angles*. Ideally, we would like to understand the underlying principles that go into defining $f(\Omega' \rightarrow \Omega; \Omega_L)$; this is key to understanding the controls over photon reflection. Unfortunately, limited insight exists into the physical processes that determine the vector distribution of scattered photons. We know that $f(\Omega' \rightarrow \Omega; \Omega_L)$ is determined by the nature of the leaf surface and the internal structure of the leaf, leaf age, pigmentation, and leaf hydration status (Gausman 1985); however, quantitative insight into the relations among these parameters and their cumulative influence on photon scattering is lacking. Equation (8.1) is often simplified by assuming that the reflected photon flux is entirely diffuse, being isotropically scattered into a solid hemispheric angle normal to the leaf (the *Lambertian assumption*) (Ross and Nilson 1968, Gutschick and Weigel 1984). This simplification does not move us much closer to an understanding of the mechanisms of leaf reflectance, as it perpetuates integral quantities and ignores anisotropic components of the reflected photon stream, which we know exist (Myneni *et al.* 1989). It does, however, provide a practical means of using the leaf scattering phase function, especially to retrieve leaf optical properties from remotely sensed reflectance (Pinty and Verstraete 1998). Now, as we proceed with our consideration of how photons are utilized by leaves to energize biochemical processes, we will leave the topic of reflectance and move to the fate of absorbed photons.

Photons that are trapped within the leaf contribute to photosynthesis and other photochemical processes, as well as adding thermal energy. Leaves have evolved various mechanisms to maximize photon trapping within the photosynthetically active waveband, and efficiently deliver them to the hundreds of thousands of chloroplasts that exist beneath each square centimeter of a leaf's surface. As photons penetrate the waxy cuticle of the leaf the first tissue they encounter is the epidermis. Except for guard cells, the epidermis does not contain chloroplasts and its cells exhibit a convex shape. The lens-like shape of epidermal cells has the potential to focus a *collimated* photon beam (i.e., a photon flux with parallel incident rays) on underlying mesophyll cells (Figure 8.8). (Direct solar rays are not truly collimated although when considered within the scale of leaf thickness they can be approximated as such since they have incidence angles within 1° of each other.) Epidermal focusing can cause photon flux densities in mesophyll cells that are 10–20 times greater than that incident on the leaf surface. Photon focusing creates "bright" spots in some areas of the mesophyll and "dark" spots in neighboring areas. Chloroplasts respond to the bright spots by migrating to them or away from them depending on the intensity of the photon flux and their inherent capacity for energy utilization (Brugnoli and Björkman 1992, Gorton *et al.* 1999). Chloroplast migration is facilitated by contractile proteins attached to the cellular cytoskeleton.

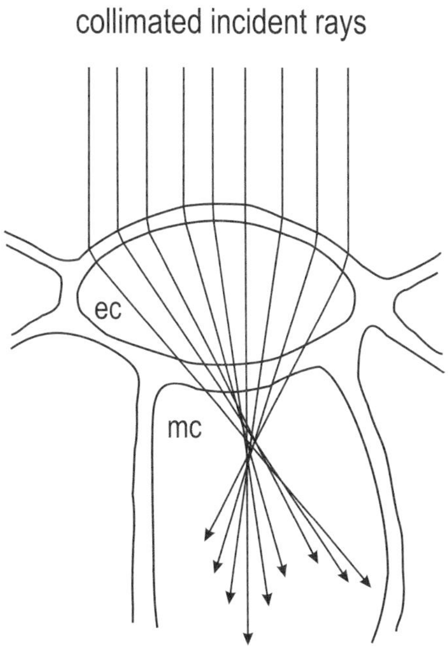

Figure 8.8   The focusing of collimated solar rays by an epidermal cell (ec) of the adaxial surface of a leaf to a focal region in the underlying palisade mesophyll cell (mc).

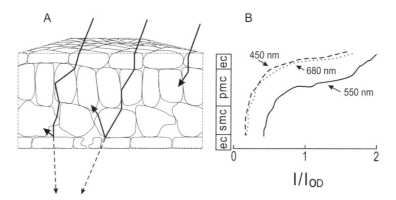

Figure 8.9 **A**. Schematic of a leaf cross-section illustrating light trapping through multiple reflections from cell surfaces. In this case, the paths for three different, hypothetical photons are presented. The dashed arrows show possible escape routes through the lower epidermis. **B**. Photosynthetic photon flux density (PPFD) profiles for three different wavelengths expressed as the ratio of measured direct beam PPFD to incident direct beam PPFD (I/I_{OD}) for a leaf of *Medicago sativa*. PPFD at different points along the vertical profile was measured with the use of a microscopic photon flux sensor that was progressively pushed through the leaf. Note that the PPFD within the leaf can be higher than that incident on the surface due to photon traping. Also note the strong extinction of photosynthetically active photons in the blue (0.45 μm) and red (0.68 μm) wavelengths in the region of the palisade mesophyll. ec = epidermal cells, pmc = palisade mesophyll cells, smc = spongy mesophyll cells. Adapted from Vogelmann *et al.* (1989).

The columnar shape and tight packing of palisade parenchyma cells, beneath the upper epidermis, facilitates the vertical penetration of photons through the leaf by producing a type of "light pipe" that channels collimated photon beams downward. Once in the lower layers of the leaf, photons are scattered by irregularly oriented cell walls, increasing their intercellular path lengths and decreasing the probability that they will be transmitted back out of the leaf before being absorbed by a chloroplast. Given this effective design for delivering and capturing photons leaves can be viewed as *light traps* (Figure 8.9A). One of the more important components of leaf light traps is the inside surface of the lower epidermis; cells of this surface often exhibit a lighter shade of color because of their lower amounts of pigmentation and capacity to back-scatter incident photons. Light trapping has been shown to increase the PPFD within a leaf by a factor of 2–5 in a number of plant species (Vogelmann and Martin 1993).

Light trapping increases the PPFD within a leaf according to the same principles associated with continuity and diffusive flux divergence that were discussed in Section 6.3. Light trapping increases the residence time of photons within a leaf, thus facilitating their retention and increasing the internal scattered PPFD. Imagine a bifacial leaf at an initial steady state wherein the flux of photons into the upper leaf surface equals the scattered flux of photons out of relatively transparent upper and lower leaf surfaces. If we now perturb the steady state by imposing the condition of enhanced back-scattering at the lower leaf surface, an initial increase will occur in the internal density of photons. A new steady state will be achieved whereby the scattered photon flux exiting the leaf plus any increase in the absorbed photon flux will come back into balance with that entering the

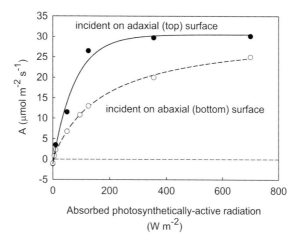

Figure 8.10 The response of net CO_2 assimilation rate (A) to absorbed photosynthetically active radiation in a leaf of *Syringa vulgaris* when light is provided to the adaxial (top) surface or the abaxial (bottom) surface. Note that the sharpest transition from light-limited to light-saturation occurs when the light energy enters from the adaxial surface. In this case, all chloroplasts became light saturated at approximately the same incident light intensity because the intra-leaf light gradient parallels the pattern of palisade and spongy mesophyll acclimation. Redrawn from Oya and Laisk (1976) and Terashima and Hikosaka (1995).

leaf, but this balance will occur at a greater density of photons within the leaf, compared to what was present in the initial state.

The photon reflections and scattering that occur within a leaf would appear to work against stratification of the PPFD; however, steep vertical gradients have been observed (Figure 8.9B). This is because the photosynthetic potential of the leaf decreases from top to bottom causing more photons to be absorbed near the upper surface compared to the lower surface. The photosynthetic systems of leaves are distributed such that cells near the upper surface are provisioned with higher densities of chloroplasts, and with each chloroplast containing higher concentrations of photosynthetic enzymes, chlorophyll, and light-harvesting proteins. The stratification of photosynthetic capacity within a horizontally positioned bifacial leaf can be viewed as adaptive given the downward delivery of solar photons. Evidence of this adaptation is seen in the fact that illumination of a sun-adapted leaf from the lower surface results in lower photosynthesis rates compared to illumination from the upper surface (Figure 8.10).

8.4 CO_2 transport in leaves

The path for CO_2 transport in leaves begins at the point where molecules diffuse through the stomatal pores (Figure 8.11A). Once inside the leaf, CO_2 molecules diffuse down a CO_2 concentration gradient, crossing a series of internal resistances which end at the active sites of

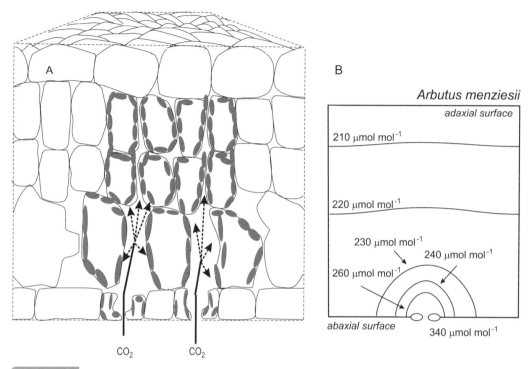

Figure 8.11   **A.** Representative diffusion paths for CO_2 crossing the stomatal pores of a hypostomatous leaf. The arrangement of chloroplasts at the periphery of the cytosol minimizes the diffusion path length within the cell. **B.** Isopleths of modeled intercellular CO_2 mole fraction using a three-dimensional anatomical model of a leaf of *Arbutus menziesii*, a thick-leaved species in the Ericaceae, showing progressive depletion of CO_2 as the diffusion pathlength from the stomatal pore increases. Redrawn from Parkhurst (1994).

Rubisco (Section 7.4). While most models of CO_2 transport in leaves consider this transport process in one-dimension – as a diffusive front that moves along the z-coordinate from the upper and lower epidermi inward – some models have developed the more appropriate three-dimensional framework – considering CO_2 transport as molecules disperse into the entire volume of the leaf (Parkhurst 1994). With regard to the internal components of the leaf the network of intercellular air spaces imposes a small resistance compared to that for the liquid phase (e.g., membranes, cytosol, and stroma). The most relevant resistive component of the liquid phase is the water-infused channels of the cell walls. The thickness of cell walls is greater than that of membranes, and aquaporin proteins in the membranes of cells may facilitate CO_2 diffusion, further reducing diffusive resistance. Resistance to CO_2 diffusion through the cytosol of the cell is also small, compared to that through the cell walls. In most leaves, protein fibers anchor chloroplasts to the cytosolic side of the plasma membrane, making the cytosolic diffusive path length relatively short. (In some cases, however, the chloroplasts can migrate to positions further from the cell wall, increasing the cytosolic path length and thereby increasing cytosolic resistance; Tholen *et al.* 2008.) Overall, one of the principal determinants of the internal resistance to CO_2 transport in leaves is the ratio of mesophyll cell surface area to overall leaf area (A_{mes}/A_L). The importance of this relation has been long highlighted in the plant physiological literature, extending back to discoveries by Park Nobel and his students in the 1970s (e.g., Nobel *et al.* 1975).

The intercellular CO_2 concentration within a leaf decreases as a function of distance from stomatal pores (Figure 8.11B). The intercellular CO_2 concentration at specific locations within a leaf will be determined by the balance between "supply" processes (the potential for diffusion to move CO_2 to a specific location within the leaf) and "demand" processes (the potential for photosynthesis to remove CO_2 from a specific location within the leaf). Quantitatively, supply processes are defined by the CO_2 concentration gradient multiplied by diffusive conductance, and demand-side processes are defined by the local biochemical capacity for photosynthesis. Using this type of supply and demand logic, isopleths of CO_2 concentration within a leaf can be drawn as shown in Figure 8.11B.

Consistent with a common theme we have emphasized in this chapter, the internal structure of leaves has been subjected to evolutionary modification through natural selection; most likely through selection to maximize the demand of CO_2 assimilation rate given a certain supply capacity for diffusion within the leaf. If the availability of water and nitrogen to a plant facilitates high CO_2 assimilation rates, those rates will be realized only if the internal structure of the leaf facilitates high diffusive rates to the site of carboxylation. This is best accomplished through high A_{mes}/A_L. Once again, structure and function must be integrated to facilitate high rates of CO_2 assimilation.

8.5 Water transport in leaves

Water moves through leaves according to thermodynamic gradients in water potential (Section 2.3.1). Leaf thermodynamic gradients take the form of diffusion gradients as water moves from vascular veins to the vicinity of the guard cells, where it evaporates to the atmosphere. The water potential of the atmosphere is large when its vapor pressure exists at a deficit, compared to the water potential of the leaf's intercellular air spaces, which exist near the state of water vapor saturation. Due to the large difference in water potential between an unsaturated atmosphere and the near-saturated air spaces of the leaf, we can cast the process of transpiration as one of water being "pulled" out of the leaf. The hydraulic paths from veins into the intercellular air spaces of the mesophyll are not well understood. Most observations support the conclusion that veins and the epidermis are hydraulically coupled through the liquid phase, not the vapor phase (Sack and Holbrook 2006). Thus, it is likely that much of the transpired water flows to the vicinity of the stomatal pore, and then evaporates into the air spaces of the substomatal cavity and from there, into the boundary layer outside the leaf's surfaces. Three possible paths exist for this type of water transport: (1) water can flow through the porous cell walls that surround mesophyll or epidermal cells, moving from cell-to-cell (an apoplastic route); (2) water can flow through the internal cytoplasm of mesophyll or epidermal cells, from cell-to-cell, crossing membranes through aquaporin proteins (a symplastic route); or (3) water can flow through plasmodesmatal channels (cytoplasmic connections) from cell-to-cell (also a symplastic route). Evidence exists to support both apoplastic and symplastic paths (Westgate and Steudle 1985, Ye *et al.* 2008), though details of the partitioning of water flows among the paths are not yet known. Whatever the exact partitioning it is clear that effective hydraulic coupling exists between the veins and epidermis, ensuring that changes in the water potential of the entire plant are efficiently communicated to stomatal guard cells. This allows quick response of guard cell turgor to changes in the xylem tension of the plant, and thus provides the plant with a means of balancing its transpiration rate against hydraulic stresses that occur within the plant. In those angiosperms that have been examined, complete rehydration of the epidermis in leaves released from moderate water stress can occur within one to two minutes (Zwieniecki *et al.* 2007).

The transport of water from the minor veins of the leaf to the epidermis requires passage through a layer of cells that surrounds the veins, called the *bundle sheath*. Bundle sheath cells often span the entire distance from the center of the leaf to the epidermis, forming *bundle-sheath extensions* (Figure 8.12). Many bundle sheath cells have a layer of waxy material in those cell walls perpendicular to the veins, which forces water to leave the veins by crossing a membrane, thus entering the symplasm. This causes hydraulic conductivity in the leaf to be sensitive to temperature; membrane viscosity is sensitive to temperature such that at lower temperatures, hydraulic conductivity decreases.

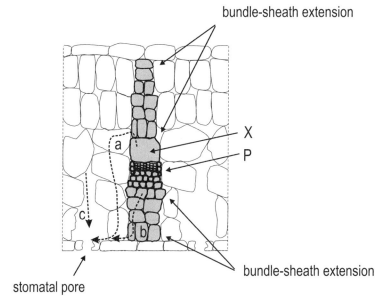

Figure 8.12 Possible paths for H_2O as it moves through the leaf in either the liquid or vapor phases and arrives at the stomatal pore where it exits the leaf. In paths a and b, the water moves symplastically (through cells) from the xylem (X), through bundle-sheath extensions, to the stomatal pore in the epidermis. This path requires liquid water to cross membranes or travel through protein pores in the membrane known as aquaporins. In path c, water evaporates from the outer cell wall of mesophyll cells and diffuses in vapor phase to the epidermis. P = phloem.

Bundle-sheath extensions can facilitate the movement of water from minor veins directly to the epidermis, by-passing the mesophyll tissue (Buckley *et al.* 2011). This direct path across the leaf may permit local uncoupling of the hydration states of the mesophyll and epidermis, focusing control over transpiration on coupling between the veins and epidermis. Contrary evidence also exists, however, which supports the existence of a direct evaporation path from mesophyll tissue to the atmosphere (Nonami and Schulze 1989), or from the mesophyll tissue, through the intercellular air spaces to the epidermis (Pieruschka *et al.* 2010). There is no consensus at the present time on the exact paths that water takes as it moves through leaves, nor is there reason to suspect that such paths are anything but variable among different species or among individual leaves of the same species growing in different microenvironments of the canopy (Sack and Holbrook 2006).

8.6 The error caused by averaging non-linearities in the flux relations of leaves

From the discussion to this point, it should be clear that leaf structure determines intercellular gradients in CO_2, H_2O, and PPFD, which in turn affect leaf-atmosphere fluxes. Leaf fluxes are often related to their environmental drivers by non-linear responses, and these non-linearities can be related to intra-leaf gradients in CO_2, H_2O, and PPFD. As an example, we will consider the case of CO_2 fluxes and their relation to PPFD and the intercellular CO_2 mole fraction (c_{ic}). If we characterize absorbed PPFD and c_{ic} with leaf-averaged values, and use those values to determine an average CO_2 assimilation rate (A), we have made the implicit assumption that the averaged gradients in PPFD and c_{ic} are related to A in the same form as that determined when A is related to individual, local values of PPFD and c_{ic} at specific points along the gradients, and then averaged. In other words, given a function (*f*) that relates an independent variable (PPFD or c_{ic}) to a dependent variable (A), we can state $\overline{f(x)} = f(\overline{x})$. This assumption is valid as long as the relation (*f*) between the flux and each variable is linear (Figure 8.13A). For non-linear functions, the assumption is not valid; a mathematical property that is referred to as *Jensen's inequality* (see Ruel and Ayers 1999).

We will explore the implications of Jensen's inequality in more detail using the A:c_{ic} relation. As an input to most biochemically based models of A (estimated as a dependent variable), the spatially averaged value for c_{ic} is prescribed (as the driving independent variable). The estimated value for A is assumed to represent the spatial average of all photosynthetic cells located beneath a unit area of leaf surface. However, in the hypothetical case where half the photosynthetic cells are functioning at a c_{ic} analogous to $x(1)$ and half are functioning at a c_{ic} analogous to $x(2)$, the value of A predicted from the average of c_{ic} ($\overline{y}_{app}$ in Figure 8.13B) will be less than the average of A predicted from independent calculation using each value of c_{ic} ($\overline{y}_{true}$ in Figure 8.13B). This is because the non-linear convexity of the curve allows c_{ic} to increase without a proportional increase in A at high values of c_{ic}. Averaging across non-linear functions creates errors when the average for a dependent

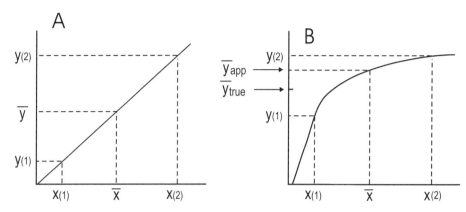

Figure 8.13   **A.** The concept of linear proportionality is illustrated with respect to determining the mean value for a dependent variable from the mean value of an independent variable. **B.** Convex non-linear relationships cause the apparent mean ($\overline{y}_{app}$) to be an overestimate of the true mean ($\overline{y}_{true}$).

variable is derived from values that are distributed heterogeneously within the non-linear domain of the response function. Heterogeneity in the values of c_{ic} from different portions of a leaf can be caused by vertical gradients in the supply demand relation of diffusion and photosynthetic capacity, or by the presence of anatomical barriers that isolate portions of the leaf and prevent thorough mixing of air in the intercellular spaces. Thick leaves with well-developed palisade and spongy mesophyll, and with stomates only on the lower surface (*hypostomatous*), will develop the largest vertical CO_2 gradients (Figure 8.14). Hypostomatous leaves are at a diffusive disadvantage due to the fact that palisade mesophyll cells, which receive the highest PPFD, are the greatest distance from the CO_2 source; leaves with stomates on both surfaces (*amphistomatous*) can more effectively supply CO_2 to palisade cells. The steep CO_2 concentration gradient across hypostomatous leaves increases their susceptibility to averaging errors.

In addition to heterogeneity along the vertical axis, horizontal heterogeneity in c_{ic} is characteristic of many broad-leaved species. In those cases in which leaves have bundle-sheath extensions that stretch from vascular bundles to the epidermis, a diffusive barrier exists that isolates patches of the leaf into autonomous or semi-autonomous gas-exchange units, each with their own characteristic stomatal conductance, photosynthesis rate, and c_{ic}. Horizontal mixing of air in the intercellular spaces is not possible. Leaves that are partitioned into lateral units by bundle-sheath extensions are referred to as *heterobaric*. Heterobaric leaves are particularly common in deciduous forest trees from temperate environments. Those leaves that lack bundle-sheath extensions and exhibit lateral continuity in their intercellular air spaces are called *homobaric* (Figure 8.15). Homobaric leaves are common in evergreen shrubs and trees from warm climates. The fundamental unit of gas exchange in heterobaric leaves is the isolated patch, or *alveolus*. Stomata of the same alveolus tend to function in a coordinated manner, whereas separate alveoli have the potential to respond

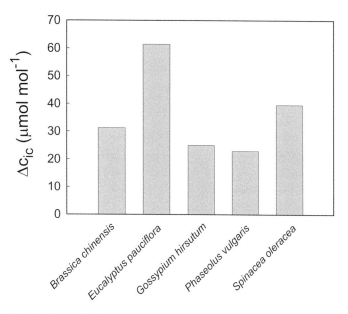

Figure 8.14 The difference in CO_2 mole fraction (Δc_{ic}) across the cross-sectional distance of amphistomatous leaves from different species measured when normal ambient CO_2 concentration was provided to the lower leaf surface with one leaf cuvette, and the concentration at the adaxial leaf surface was measured with a different cuvette. Thus, normally amphistomatous leaves were forced to behave like hypostomatous leaves. (It is not possible to conduct this type of measurement with hypostomatous leaves.) In this experiment, the lower side of the leaf (with normal ambient CO_2) is allowed to exchange CO_2 and H_2O with the air of the cuvette, and a value of c_{ic} can be calculated for that side of the leaf. On the other side of the leaf, the CO_2 mole fraction in the air of the cuvette will come to equilibrium with c_{ic} at that side of the leaf, and can be directly measured using infrared gas analysis. The difference in c_{ic} between the two sides of the leaf is taken as Δc_{ic}. Note the relatively large Δc_{ic} for the thick-leaved *Eucalyptus* species. Adapted from Parkhurst *et al.* (1988).

independently to environmental signals. In response to low atmospheric humidity or leaf water stress the stomata of some alveoli can be observed to close completely, whereas others remain open (Terashima *et al.* 1988). This pattern of heterogeneous distribution in stomatal opening has been referred to as *"patchy" stomatal behavior*. The occurrence of heterogeneous stomatal patches and the non-linear response of A to c_{ic} require that spatially explicit averaging be applied to the analysis of the A:c_{ic} relationship.

The non-linear averaging problem is also relevant to modeling the A:PPFD relationship. As demonstrated in Figure 8.16 there is an exponential extinction of PPFD across the vertical span of a leaf. The A:PPFD curve is also non-linear; thus, when an average PPFD is assumed for a leaf some of the cells near the upper surface may be at a higher PPFD than those near the lower surface. If the cells near the upper surface are assimilating CO_2 near light saturation, and those near the lower surface are below light saturation, then averaging errors will occur.

In many past models of CO_2 assimilation, averaging errors are acknowledged, but ultimately ignored. In the case of c_{ic}, amphistomatous leaves tend to exhibit across-the-

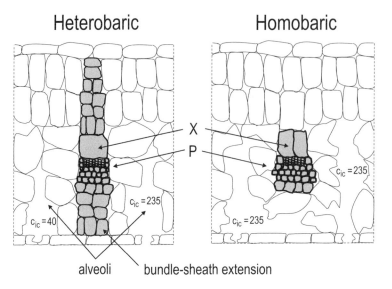

Figure 8.15   Cross-sectional depiction of heterobaric and homobaric leaves (see text for details of these terms). Bundle-sheath extensions extend from the vascular bundle to the upper and lower epidermi of the heterobaric leaf. The extensions divide the leaf into separate alveolar units. In the leaf on the left, the stomate that serves the left alveolus is closed and the c_{ic} exists near the CO_2 compensation point; the open stomate that serves the alveolus on the right maintains the c_{ic} at 235 μmol mol^{-1}). X = xylem, P = phloem.

leaf differences in CO_2 mole fraction that are less than 15 μmol mol^{-1} (approximately 6–7% the average c_{ic} value for a C_3 leaf). Uncertainties in averaging the A:c_{ic} relationship would be small in these leaves. However, thick, hypostomatous leaves can have CO_2 gradients that measure 60–70 μmol mol^{-1} (approximately 25–30% the average c_{ic} value for a C_3 leaf). Gradients of this magnitude could cause significant error in a linear averaging of the A:c_{ic} relationship. Across-the-leaf gradients in PPFD can also be significant (e.g., 1500 μmol photons m^{-2} s^{-1}, or 75% the magnitude of the maximum PPFD received by a leaf).

In the case of some non-linear flux responses, we can take advantage of the tendency for evolution and cost-benefit optimization to force otherwise non-linear relations into linear form (see Box 8.1). For example, returning to the A:PPFD relation, averaging errors are avoided if, in fact, linearity exists between the CO_2 assimilation rate at light saturation (A_{max}) and absorbed PPFD when assessed across the vertical span of the leaf (Figure 8.16). Thus, as mean PPFD decreases from the top to bottom surfaces, we assume that the photosynthetic capacity of the leaf decreases in linear proportion. The assumption of linearity is justified on the basis of evolutionary cost-benefit optimization; leaves are expected to allocate just enough resource (e.g., nitrogen) to the photosynthetic process as is allowed by available PPFD. As the mean PPFD decreases across a leaf, photosynthetic resources will be allocated in proportionately lower amounts. This assumption is generally valid for broad-leaved species and for growth in relatively sunny habitats. In such conditions, chloroplasts acclimate to gradients in PPFD, with those near the upper epidermis exhibiting greater rates of photosynthesis at high PPFD, and those near the lower epidermis

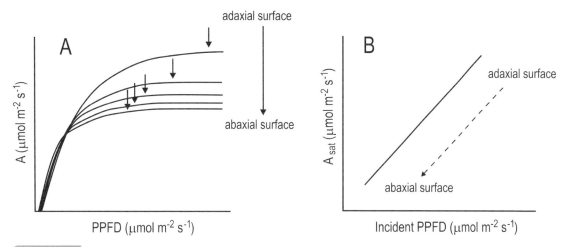

Figure 8.16   **A**. The A:PPFD relationship for hypothetical mesophyll cells at different layers within a leaf. The arrows represent the point of incipient PPFD saturation for A, as well as the PPFD that is incident at each respective layer. Thus, as the incident PPFD decreases through the leaf, A remains PPFD-saturated. **B**. The relationship between PPFD-saturated A (A_{sat}) at different hypothetical layers within a leaf and the PPFD that is incident at each respective layer. Note that as a result of the progressive acclimation of cells to lower incident PPFD (as illustrated in panel A), the A:PPFD relationship is approximately linear across the leaf.

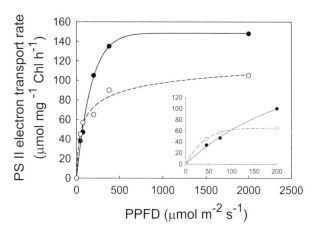

Figure 8.17   The response of photosystem II electron transport rate to PPFD in isolated cells from palisade mesophyll tissue or spongy mesophyll tissue of *Camilla japonica* leaves. **Inset**: The same response but for a limited range of low PPFD. Redrawn from Terashima and Inoue (1984), with permission from Oxford University Press.

exhibiting greater rates of photosynthesis at low PPFD (Figure 8.17). This acclimation increases the whole-leaf efficiency by which photons are absorbed and utilized. Chloroplast acclimation is caused, to some extent, by adjustments in Rubisco activity and chlorophyll content (Figure 8.18). With the assumption of linearity between A_{max} and PPFD, A can be

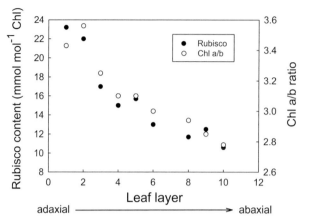

Figure 8.18 The decline in Rubisco content and chlorophyll a/b ratio measured in the cells from different vertical layers within a leaf of *Spinacia oleracea* (spinach). Chlorophyll a/b ratio can be used as an index to evaluate photosynthetic adjustment to growth at progressively lower PPFD, with lower Chl a/b occurring in cells acclimated to lower PPFD. Redrawn from Terashima and Inoue (1985), with permission from Oxford University Press.

modeled as though it occurs in one big chloroplast with a V_{cmax} and F_{Jmax} that reflect the arithmetic average of the entire leaf.

8.7 Models with explicit descriptions of leaf gradients

As an alternative to the "big chloroplast" simplification, some researchers have developed vertically explicit models to accommodate variation in PPFD and its effect on A (e.g., Terashima and Saeki 1983). These types of models involve partitioning the leaf into successive layers, which are homogeneous in the horizontal plane, but heterogeneous in the vertical plane. Using a "photon accounting" approach, and the assumption of conservation of photons, the cumulative photon absorption for layers between the leaf's surface and depth z can be represented as:

$$a[z]\,I \; = \; I\,(1 - \alpha - t[z] + \; r_u[1 - \beta - \tau]), \qquad (8.2)$$

where $a[z]$ is the fractional absorptance to PPFD of a leaf layer with depth z, I is the total incident PPFD, α is albedo from the upper surface of the layer, $t[z]$ is the transmittance of PPFD through the layer, r_u is the fractional reflectance from all layers below depth z that sends photons upward into the layer, β is the fraction of r_u that is reflected from the lower surface of the layer, and τ is the fraction of r_u that is transmitted out the upper surface of the layer (Figure 8.19). Higher-order reflections, beyond r_u, are ignored. After parameterization for the various reflectance and transmittance values, A for each vertical layer, and its response to absorbed PPFD, can be modeled according to the equations described in Chapter 5. The primary inputs for the photosynthesis model of

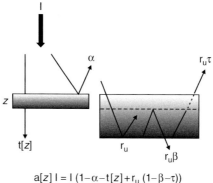

$$a[z]\, I = I\, (1-\alpha-t[z]+r_u\,(1-\beta-\tau))$$

Figure 8.19   A diagrammatic definition of the variables used in the model of PPFD gradients within leaves developed by Terashima and Saeki (1985). See the text for details of the variables.

each layer would be absorbed PPFD, F_{Jmax}, and V_{cmax}, with each adjusted for acclimation to the intra-leaf gradient in PPFD. The value of A for the entire leaf is then calculated as the cumulative sum for all layers.

8.7.1   A one-dimensional analytical model of leaf structure and its relation to the intercellular CO_2 concentration

Modeling within-leaf gradients in CO_2 availability, and thus the CO_2 constraint on photosynthetic capacity, has been developed through recognition of the tradeoffs between internal diffusive resistance and Rubisco activity as leaf (mesophyll) thickness changes. Using an analytical model to define the relation between A and mesophyll thickness, Terashima *et al.* (2001) showed that the capacity for diffusion can potentially limit photosynthesis to the point that diminishing carbon gain is an important determinant of cost-benefit analysis and the evolution of leaf form. Considering the simplest case of a hypostomatous leaf in which CO_2 diffuses from one side, these researchers showed that the cost of adding mesophyll layers beyond an asymptotic limit exceeds the benefit of enhanced CO_2 assimilation; thus, reducing the marginal adaptive gain. Changes in the cost-benefit ratio occur because as thickness increases, diffusive resistance also increases. The marginal gain that is possible for any given leaf thickness is increased as whole-leaf Rubisco concentration is increased. In fact, using the model it was shown that in bifacial leaves, the thickness of the leaf *must* increase as a function of Rubisco content, if a fitness advantage is to be sustained (Figure 8.20). That is, for each increment of additional resource that becomes available in the form of increased Rubisco concentration, greater photosynthetic benefit will accrue for an exponential increase in mesophyll thickness. This type of analysis begins to put the integrated constraints and benefits of leaf structure and photosynthetic potential into a common adaptive context. A derivation of the model by Terashima *et al.* (2001) is provided in Appendix 8.1.

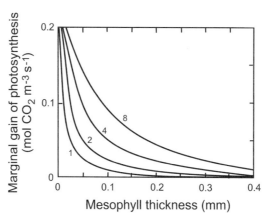

The relation between the marginal gain for photosynthetic CO_2 assimilation (taken as dA/dl) and mesophyll thickness (l). Each curve represents the relation for a different mesophyll Rubisco concentration (mol Rubisco m^{-3}). Redrawn from Terashima *et al.* (2001), with kind permission from Springer Science + Business Media.

Appendix 8.1 Derivation of the Terashima *et al.* (2001) model describing leaf structure and its relation to net CO_2 assimilation rate

Here, we consider the model by Terashima *et al.* (2001) in more detail. We start with recognition that this is a one-dimensional model and recognizing Fick's First Law, A can be defined by dc_{ic}/dx according to:

$$A = -(p/\tau)\ \rho_m\ K_{dc}\ \frac{dc_{ic}}{dx}, \tag{8.3}$$

where p is mesophyll porosity (the volumetric fraction of the leaf that is occupied by air space), τ is the relative tortuosity (unitless) of the internal air spaces, ρ_m is the molar density of dry air (mol air m^{-3}), K_{dc} is the molecular diffusivity of CO_2 in air ($m^2\ s^{-1}$), and c_{ic} is the mole fraction of CO_2 in the leaf referenced to dry air. The porosity and tortuosity terms begin to define components of the leaf's diffusive conductance due to anatomical features of the mesophyll; porosity scales directly with conductance and tortuosity scales inversely. The ratio of porosity to tortuosity is commonly used in models to define flow conductances in porous media, such as leaves and soils. Assuming continuity, we can state that the difference in A at x and $(x + \Delta x)$ must equal the divergence in c_{ic} across distance Δx:

$$A[x] - A[x + \Delta x] = (p/\tau)\rho_m\ K_{dc}\left(\frac{dc_{ic}[x]}{dx} - \frac{dc_{ic}[x + \Delta x]}{dx}\right)\ A_v \cdot (\Delta x), \tag{8.4}$$

where A_v is the net CO_2 assimilation rate per unit of mesophyll tissue volume (mol CO_2 m^{-3} s^{-1}). If we now use calculus to evaluate Eq. (8.4) in the limit as $\Delta x \to 0$, we can write:

$$A_v = \lim_{\Delta x \to 0} \frac{-(p/\tau)\,\rho_m K_{dc}\left(\dfrac{dc_{ic}[x]}{dx} - \dfrac{dc_{ic}[x+\Delta x]}{dx}\right)}{\Delta x}$$

$$= -(p/\tau)\rho_m K_{dc}\,\frac{d\left(\dfrac{dc_{ic}[x]}{dx}\right)}{dx} = -(p/\tau)\rho_m\, K_{dc}\,\frac{d^2 c_{ic}[x]}{dx^2}. \tag{8.5}$$

Equation (8.5) provides the theoretical linkage of A to dc_{ic} with explicit consideration of leaf anatomy. In other words, the theory develops a physical basis for defining the internal gas phase conductance to CO_2 assimilation rate.

Equation (8.5) can be developed further, recognizing several additional factors that determine A_v including the fraction of mesophyll cell wall surface area (A_{mes}) per unit of mesophyll cell volume (V_{mes}), which we will abbreviate as S (i.e., $A_{mes}/V_{mes} = S$; $m^2\ m^{-3}$), the total Rubisco enzyme content per unit mesophyll volume (E_{tv}; $mol\ m^{-3}$), the liquid phase conductance for CO_2 diffusion (g_{ic}, $mol\ H_2O\ m^{-2}\ s^{-1}$), the molar density of water (ρ_{mw}; $mol\ H_2O\ m^{-3}$), and the mitochondrial (dark) respiration rate expressed per mole of Rubisco (R_d; $mol\ CO_2\ mol^{-1}\ Rubisco\ s^{-1}$). Stated formally:

$$A_v = \frac{(g_{ic}/\rho_{mw})SE_{tv}(k_1\phi c_{ic}[x] - (k_2 + R_d))}{k_1 E_{tv} + (g_{ic}/\rho_{mw})S}, \tag{8.6}$$

where k_1 is a coefficient with units ($m^3\ mol^{-1}\ Rubisco\ s^{-1}$), k_2 is a time-dependent coefficient with units (s^{-1}), and ϕ is the Bunsen coefficient, which is an alternative to Henry's Law as a means to describe the solubility of gases in solution. The Bunsen coefficient is frequently used in derivations of gas exchange processes because it has units of (m^3 gas m^{-3} solution), which resolve to a unitless term. Unlike the condition of Henry's Law, the Bunsen coefficient (ϕ) is not determined as a continuous function of p_{gj}, the partial pressure of a gas constituent j in equilibrium with the aqueous phase. Rather, ϕ is defined for a gas at standard temperature (273 K) and for a standard partial pressure of the gas of 101.3 kPa.

Using an analytical approach to solve the second derivative for $c_{ic}[x]$ on the right-hand side of Eq. (8.5), the following relation is derived:

$$c_{ic}[x] = \frac{(k_3 + C_1\exp(\alpha x) + C_2\exp(-\alpha x))}{\rho_m}, \tag{8.7}$$

where C_1 and C_2 are constants with units ($mol\ m^{-3}$) and k_3 (also with units $mol\ m^{-3}$) is defined as:

$$k_3 = \frac{k_2 + R_d}{\phi\, k_1}. \tag{8.8}$$

Equation (8.8) provides the initial balance at point x between (1) CO_2 uptake from the intercellular space at x due to the solubility of CO_2 in the liquid phase and CO_2 assimilation by Rubisco, and (2) CO_2 release into the intercellular air space at point x by mitochondrial respiration scaled to the Rubisco content. This is the "local" influence of processes in the mesophyll at point x. From this initial, "local" balance, we assume that CO_2 diffuses to point x from the substomatal cavity at either the top or bottom leaf surface, and that along the diffusive path the supply of CO_2 is mitigated by diffusive resistances and photosynthetic activity. In Eq. (8.7), modifications to the initial, "local" CO_2 balance are represented in the mathematical constants C_1 and C_2, and they reflect influences that scale with mesophyll thickness.

The term α in Eq. (8.9) provides modeled control over the balance between diffusional constraints to the transport of CO_2 from the substomatal cavity and the uptake of CO_2 by Rubisco and is defined as:

$$\alpha = \sqrt{\frac{g_{ic}\phi SE_{tv}k_1}{(p/\tau)\rho_{mw}K_{dc(w)}((E_{tv}k_1)+(Sg_{ic}/\rho_m))}}, \qquad (8.9)$$

where $K_{dc(w)}$ is the diffusivity of CO_2 in water ($m^2\,s^{-1}$). A typical value of (p/τ) for bifacial leaves can be assumed as 0.21. With this form, α in Eq. (8.9) resolves to units of m^{-1}.

In a hypostomatous leaf, where only one side of the leaf functions as a CO_2 source the terms in Eq. (8.7) with C_1 and C_2 diverge – reflecting a large dc_{ic}/dx spanning the entire thickness of the mesophyll and with $c_{ic}[x]$ decreasing as a function of distance from the substomatal cavity of the lower leaf surface. In that case, C_1 and C_2 can be modeled as:

$$C_1 = \frac{(\rho_m c_{sc} - k_3)}{1 + (\exp(2\alpha l))}, \qquad (8.10)$$

$$C_2 = \frac{(\rho_m c_{sc} - k_3)(\exp(2\alpha l))}{1 + (\exp(2\alpha l))}, \qquad (8.11)$$

where c_{sc} is the CO_2 mole fraction in the substomatal cavity and l is the mesophyll thickness (m). Thus, fundamentally C_1 and C_2 are expressions of the average CO_2 mole fraction gradient across the entire span of the mesophyll tissue and scaled to the balance between net CO_2 uptake and the resistance to CO_2 diffusion, which is reflected in α. In an amphistomatous leaf, where both surfaces of the leaf represent CO_2 sources, the terms in Equation (8.7) with C_1 and C_2 are modeled as:

$$C_1 = \frac{(\rho_m c_{sc} - k_3)(1 - \exp(-\alpha l))}{(\exp(\alpha l)) - (\exp(-\alpha l))}, \qquad (8.12)$$

$$C_2 = \frac{(\rho_m c_{sc} - k_3)(\exp(\alpha l) - 1)}{(\exp(\alpha l)) - (\exp(-\alpha l))}. \qquad (8.13)$$

Equation (8.7) provides an analytical solution to the problem of determining $c_{ic}[x]$ at a specific value for x, within a leaf. Considering the form of Eq. (8.7), it is clear that the drivers of the vertical gradient in c_{ic} through the mesophyll reflect exponential change as a function of distance from the leaf surfaces, including tradeoffs between the potential for gas- and liquid-phase conductances of CO_2 and the photosynthetic assimilation capacity as determined by Rubisco concentration.

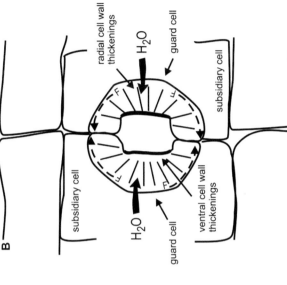

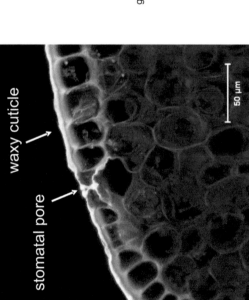

Figure 7.5 **A**. Confocal microscope image of a stomatal pore in a leaf of *Wachendorfia thyrsiflora* (red root plant). Mesophyll cells beneath the epidermis contain red-fluorescent chloroplasts. **B**. Top view of a pair of guard cells working against one another to form a stomatal opening. The guard cells possess radial thickenings of the cell wall. As water enters the cells and turgor pressure increases, the force of the turgor (F) is directed to right angles to the radial thickenings, causing the guard cell pair to press against each other at the ends of the cells. This radially directed force, coupled to longitudinal thickening along the inner (ventral) walls of the cells, causes the guard cells to work against the turgor pressures of neighboring subsidiary cells, displacing a finite volume and forming a characteristic pore. Micrograph courtesy of Professor Joerg-Peter Schnitzler (Helmholtz Zentrum München, Munich, Germany).

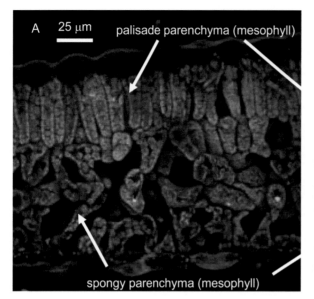

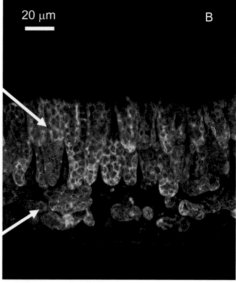

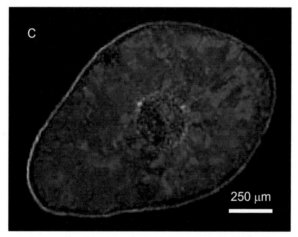

Figure 8.1 **A.** Confocal microscope picture of an oak leaf (*Quercus ilex*) showing yellow-fluorescent chloroplasts within a single layer of tightly packed palisade mesophyll cells just beneath the upper epidermis and loosely packed spongy mesophyll cells, with large, interspersed air channels, just above the lower epidermis. **B.** Two palisade mesophyll layers above spongy mesophyll tissue in a poplar leaf (*Populus trichocarpa*). **C.** Tightly packed mesophyll cells surrounding the vascular bundle in a Norway spruce needle (*Picea abies*). Micrographs courtesy of Professor Joerg-Peter Schnitzler (Helmholtz Zentrum München, Munich, Germany).

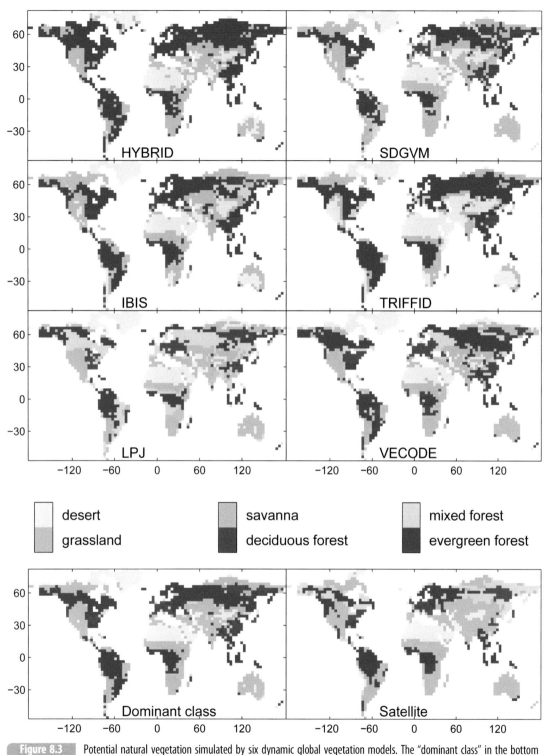

desert   savanna   mixed forest

grassland   deciduous forest   evergreen forest

Figure 8.3 Potential natural vegetation simulated by six dynamic global vegetation models. The "dominant class" in the bottom left panel is the consensus vegetation class that was predicted by the largest number of models in each grid cell. The bottom right panel is a simplified vegetation map inferred from satellite remote sensing (see Cramer *et al.* 2001). Ecophysiological properties of global vegetation are inferred from the types of correlations displayed in Figure 8.2 for different biome types and can be used to predict functional traits, such as gross primary productivity (GPP), which reflects photosynthetic potential, and net ecosystem productivity (NEP), which reflects the sum of photosynthesis and ecosystem respiration (distinguished by opposing mathematical sign). From Cramer *et al.* (2001).

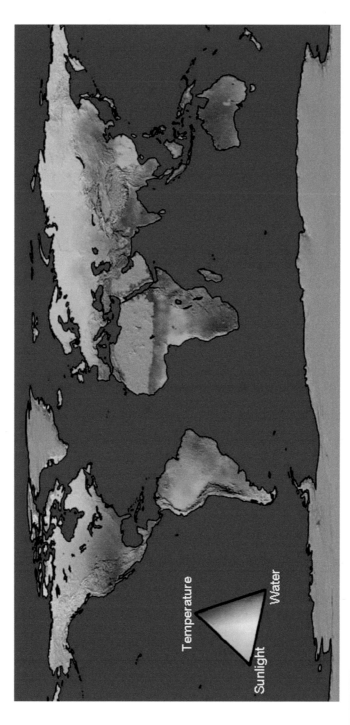

Figure 8.4 Geographic variation in climatic controls over net primary productivity (NPP) considering water, temperature, and sunlight availability. The patterns of variation in NPP were derived from satellite data on leaf area index, combined with climate data and consideration of plant functional type (PFT) differences in different biomes. Several PFTs can be found in a biome and the biomes were distinguished on the basis of their photon use efficiencies (NPP per unit of absorbed photon flux density). The biome types considered in this analysis were evergreen needleleaf forest, deciduous broadleaf forest, shrubland, savanna, grassland, and cropland. From Running *et al.* (2004).

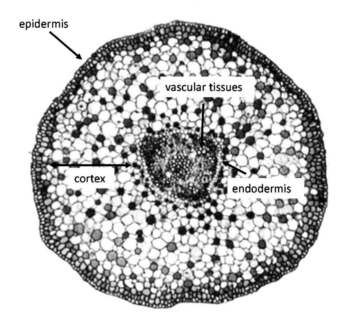

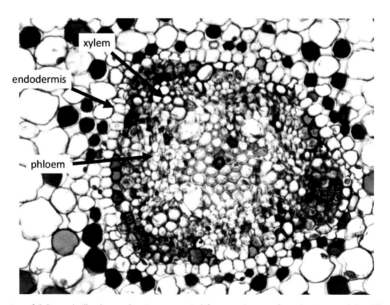

Figure 9.2 **A**. Cross-section of *Salix* sp. (willow) root showing anatomical features. See text for a description of the different anatomical features. **B**. Cross-section of a willow root at higher magnification. The central cylinder located inside the endodermis and containing the xylem and phloem tissues is known as the stele. See text for other descriptions of anatomical details. Printed with the kind permission of Alison Roberts, University of Rhode Island.

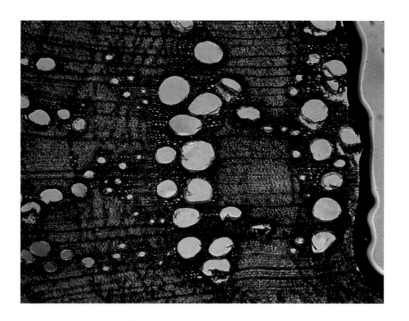

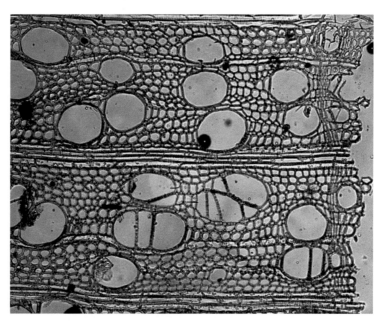

Figure 9.6 **A**. Ring-porous wood anatomy in *Quercus rubra* (red oak). **B**. Diffuse-porous wood in *Acer macrophyllum* (big-leaf maple). Note the consistency of xylem vessel diameters as a function of stem diameter in the diffuse-porous type, versus the larger diameter vessels located near the outer stem surface in the ring-porous type. Printed with the kind permission of Katherine McCulloh (Oregon State University).

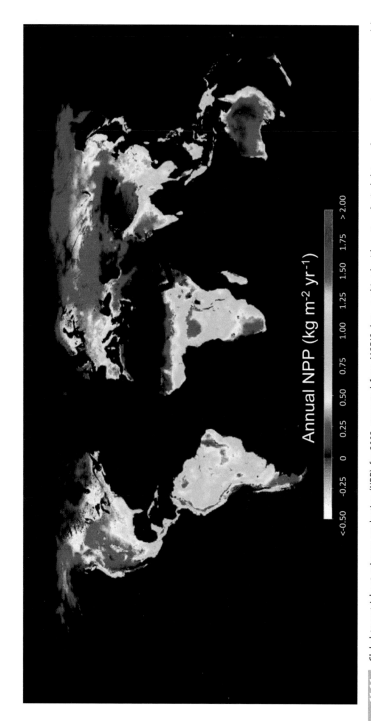

Annual NPP (kg m^{-2} yr^{-1})

<-0.50 -0.25 0 0.25 0.50 0.75 1.00 1.25 1.50 1.75 > 2.00

Figure 11.16 Global terrestrial net primary production (NPP) for 2002 computed from MODIS data combined with meteorological data and an ecosystem process model. From Running *et al.* (2004).

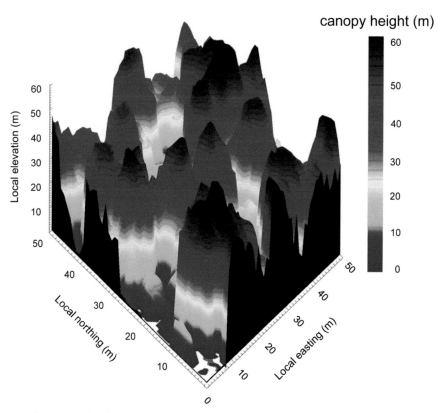

Figure B11.1 Canopy surface topography of a subsection of the Wind River Canopy Crane Research Facility in Washington State determined by discrete-return lidar. Individual lidar samples would be represented by individual points in three-dimensional space. From Lefsky *et al.* (2002).

9

Water transport within the soil-plant-atmosphere continuum

> Because water is generally free to move across the plant-soil, soil-atmosphere, and plant-atmosphere interfaces it is necessary and desirable to view the water transfer system in the three domains of soil, plant, and atmosphere as a whole . . . it must be pointed out that, as well as serving as a vehicle for water transfer, the SPAC is also a region of energy transfer.
>
> John R. Philip (1966)

Closure of the water budget for an ecosystem requires that precipitation and flows of water from neighboring ecosystems be returned to the atmosphere through evapotranspiration, transferred to storage pools, or allowed to flow out of the system. Transfer and storage of water creates capacitance in the liquid phase of the water cycle and delays the inevitable return of water vapor to the atmosphere, but a globally balanced water cycle requires that the molar equivalent of precipitated water be accounted for in the fractions stored in surface and subsurface reservoirs, plus that evaporated back to the atmosphere. Recognizing that in terrestrial ecosystems a large fraction of precipitation is returned to the atmosphere through leaf transpiration, plants occur at an important interface between the liquid and vapor phases of the water cycle. Water moves from soil into plants through viscous flow in the liquid phase, as it is "pulled" by thermodynamic forces through roots, vascular tissues, and leaf mesophyll cells, following negative pressure (tension) gradients. Tension develops in the conduction tissues as water is evaporated faster from leaves than can be replaced by flow from the soil. Physical continuity within capillary "threads" of the ascending water column is maintained by cohesive and adhesive forces that are facilitated by the electrostatic polarity of water molecules. In the vicinity of stomata, water is evaporated to the atmosphere. In the atmosphere, water is carried in the vapor phase to and from leaf and soil surfaces through diffusion near the surfaces and turbulent air motions in the well-mixed atmosphere. Given the continuous nature of these water transfer paths, and their serial relation to one another, it was recognized early in the study of plant-water relations that the "whole plant" must be considered at the center of an integrated and articulated soil-plant-atmosphere continuum, or SPAC.

As stated in the quote above from John Philip's seminal review of soil-plant-atmosphere water relations, and as we have emphasized throughout this book, the transfer of mass within the earth system is driven by free energy gradients. Thus, the concept of a SPAC rests on a thermodynamic foundation. During daylight hours, the absorption of solar radiation by soil and plant surfaces creates a free energy surplus, and the unsaturated humidity state of the atmosphere creates a free energy deficit, that together, in the case of wet or moist surfaces, drive latent heat to be transferred from the surface to the atmosphere. At night, the absorption of longwave radiation from the sky and canopy surroundings and the continued water deficit of the atmosphere replace solar radiation as primary determinants of the free energy gradients that drive latent heat loss. Ultimately, and globally, the energy and mass components of the SPAC come together in the form of an energy-driven flow of water and latent heat that replenishes the atmosphere with water vapor and balances that lost through precipitation.

The intent of this chapter is not to provide a comprehensive discussion of plant and soil water relations, and thus we will focus principally on the processes that drive water fluxes through the SPAC, rather than processes that define pools and hydrologic balances. Therefore, we will not devote significant text to describing hydrologic flows through soils, across watersheds, or from rivers to oceans. However, the transfers of water through soils, branches, and leaves to the atmosphere, or from the soil surface to the atmosphere, are critical to placing ecosystem-atmosphere water fluxes within the context of earth system processes. These transfers have been traditionally discussed within the context of several different processes and at several different scales. Rather than organize this book in a way that discusses all of these transfers in a single chapter that focuses solely on water, we have partitioned the discussion into separate chapters that focus on processes and scales. In this chapter we will discuss the *mass flow* of water through the SPAC. In the next chapter we will consider the *energy flow* of water through latent heat exchanges.

9.1 Water transport through soil

Water will flow through a soil in response to (1) the hydrostatic pressure head due to the force of gravity on the mass of a liquid (forcing water to percolate deeper into soils), (2) the electrostatic matric potential (forcing water to bind to soil particles and causing the adhesive and cohesive forces that produce capillarity), and (3) the electrostatic osmotic potential (forcing water to bind to charged ions and polar molecules in some soils, e.g., saline soils). These processes can be viewed as a combination of the positive pressure force, driven by the weight of the water column, and the negative pressure force, driven by matric and osmotic potentials (Section 2.3.1).

In the discipline of fluid dynamics, soil is treated as a *porous medium*; meaning that solid particles, which are packed together to form a compact volume, are not ideally compressed. Gaps, or pores, form at the mutual faces of soil particles, leaving spaces that can be filled with air, water, roots, fungal hyphae, or microorganisms. *Porosity* is a term used to characterize the "void volume" of soils as a fraction of total volume. *Tortuosity* is a term used to describe the "curviness" of the connected pore space in soils; most simply defined, tortuosity is the ratio of the total length of a curved, stretched pore, to the linear length from one end of the pore to the other. Soil porosity and tortuosity, through their effects on soil water potential gradients and hydraulic conductivity, exert important controls over both liquid and vapor flows.

Given equivalent mineral composition and assuming water saturation, soils with larger pores will support greater liquid flows in response to positive (downward) pressure gradients. However, soils with larger diameter pores will support weaker matric forces and thus lesser potential for adhesion to the soil matrix. Soils with smaller diameter pores will exhibit greater resistance to downward flows, but they will support greater conductivity of an upward, capillary potential as water is moved from deeper layers toward the surface. Thus, soil texture and its influence on hydraulic resistance (or conductivity) represents a fundamental state property of soils that must be considered in scaling the driving forces of free energy gradients to water fluxes.

9.1.1 The Richards equation

Early efforts to describe soil water flows through saturated soils relied on Darcy's Law (Section 6.2.3). Recalling from Chapter 6, for the case of liquid water, and in its three-dimensional form, Darcy's Law is written as:

$$F_w = -\rho_{mw} A \frac{B_k}{\mu} \nabla p_w, \qquad (9.1)$$

where F_w carries units of mol H_2O m^{-2} s^{-1}, ρ_{mw} is the molar density of liquid water (mol m^{-3}), B_k is the permeability coefficient, or the fractional cross-sectional area available to conduct porous flow, A is the cross-sectional area conducting porous flow (m^2), μ is the dynamic viscosity of water (kg m^{-1} s^{-1}), and ∇p_w is the hydrostatic pressure gradient (Pa m^{-1}). The term $(\rho_{mw} A B_k)/\mu$ can be viewed as an expression of hydraulic conductivity, facilitating liquid flow in the presence of a hydrostatic pressure gradient. As a reminder, the symbol ∇ is the gradient operator for pressure in three coordinates at a given point in the soil:

$$\nabla p_w = \frac{\partial p_{wx}}{\partial x} + \frac{\partial p_{wy}}{\partial y} + \frac{\partial p_{wz}}{\partial z}. \qquad (9.2)$$

For liquid water flows, it is more accurate on theoretical grounds to express Eq. (9.1) in terms of the vertical water potential gradient ($\nabla \psi_w$). By working with water potential, rather than pressure alone, we are expressing relations in terms of chemical potential, and thus closer to the thermodynamic root of the relation. In order to modify Eq. (9.1) to a form incorporating chemical potential, we can recognize pressure and volume as conjugate variables (Section 2.3), and we can make the analogy:

$$\psi_w = p_w \overline{V}_w, \qquad (9.3)$$

where $\overline{V}_w$ is the partial molal volume of water in m^3 mol^{-1}. Equation (9.3) allows us to rewrite Darcy's Law as:

$$F_w = -\rho_{mw} A \frac{B_k}{\overline{V}_w \mu} \nabla \psi_w, \qquad (9.4)$$

where $\nabla \psi_w$ is in J mol^{-1} m^{-1}, *the unit measure of a free energy gradient* . Often, $\nabla \psi_w$ is expressed in MPa m^{-1}, *the unit measure of a pressure gradient*. In that case, the reader should be aware that the true (and inferred) unit of expression is MPa m^{-3} (i.e, $\psi_w / \overline{V}_w$); or in other words, the volumetric water potential.

Darcy's Law was developed for the case of saturated liquid flow (with constant soil water content, θ_s). In unsaturated soils, however, the implicit assumptions that underlie Darcy's Law (e.g., absence of space- and time-dependent divergences in θ_s and independence of hydraulic conductance with regard to θ_s) are violated (Buckingham 1907). The problem of how to describe unsaturated flows challenged hydrologists for several decades in the early part of the twentieth century. Lorenzo Richards, working at Utah State University, achieved a major breakthrough on the problem in 1931, when he combined Darcy's Law with the continuity equation (Section 6.3); in essence recognizing that water transport in soil can be described in more comprehensive terms when based on volumetric mass balance, rather than hydraulic flow alone. Richards developed his theory on the basis of flow divergence, and recognized that divergence in the soil water flux, within a defined control volume, must be balanced by time-dependent changes in θ_s. Using a form similar to that shown above for Darcy's Law, we can write the *Richards equation* as:

$$\frac{\partial \theta_s}{\partial t} = -\nabla F_w = \nabla\left(A \frac{B_k}{V_{w}\mu} \nabla \psi_w \right) = \nabla\left(A \frac{B_k}{\mu} \nabla p_w \right), \tag{9.5}$$

$$\text{I} \qquad\qquad \text{II} \qquad\qquad\qquad \text{III} \qquad\qquad\qquad \text{IV}$$

where θ_s is in $m^3 \, m^{-3}$ and $\nabla \psi_w$ is in $J \, mol^{-1} \, m^{-1}$. (The negative sign disappears in Terms III and IV because flux density occurs in the direction that opposes gradients in the driving force.) In summary, the Richards equation follows from Darcy's Law, but with the added condition of *continuity and mass balance within a control volume*.

In soils, the Richards equation is often applied to the downward infiltration of H_2O through an unsaturated, porous, "turbid" medium. The presence of flow discontinuities, such as those due to animal burrows, rocks, or dead roots, which cause bias in flow paths, have the effect of uncoupling hydraulic flow from θ_s, thus invalidating the Richards equation. In a truly porous, turbid medium, the one-dimensional (vertical coordinate) form of the Richards equation can be derived in terms of a *static pressure head*; that is, the weight of a column of water on the soil system. The static pressure head is measured in units of water column depth (m), and is derived as:

$$h_w = \frac{p_w}{\rho_w g} + z, \tag{9.6}$$

where h_w is the pressure head, ρ_w is mass density ($kg \, m^{-3}$), and g is acceleration due to gravity ($m \, s^{-2}$). (The concept of static pressure head, expressed in units of height, is most often used to define flow through porous media in the fields of fluid mechanics and engineering.) In Eq. (9.6), z is expressed in units of positive meters when the water column that forces infiltration is above the surface and negative meters for vertical water infiltration below the surface. Considering the case for water flows beneath the surface, we can rearrange Eq. (9.6) to state:

$$p_w = \rho_w g \, (h_w + z), \tag{9.7}$$

which, in turn, allows us to restate the Richards equation in its most commonly used one-dimensional form:

$$\frac{\partial \theta_s}{\partial t} = -\overline{V}_w \frac{\partial F_w}{\partial z} = \frac{\partial}{\partial z}\left(\frac{\rho_w\, gAB_k}{\mu}\right)\left(\frac{\partial h_w}{\partial z} + \frac{\partial z}{\partial z}\right) = \frac{\partial}{\partial z}\left(\frac{\rho_w\, gAB_k}{\mu}\right)\left(\frac{\partial h_w}{\partial z} + 1\right).\qquad (9.8)$$
$$\quad \text{I}\qquad\quad \text{II}\qquad\qquad\qquad\quad \text{III}\qquad\qquad\qquad\qquad\qquad\quad \text{IV}$$

We can define a hydraulic conductivity (g_{hw}) expressed with units m s^{-1} as:

$$g_{hw} = \frac{\rho_w\, gAB_k}{\mu}.\qquad (9.9)$$

Once again, it is more accurate on theoretical grounds to express the flow relations of Richards equation in terms of water potential, rather than pressure head. This translation is especially important as soils dry, and matric forces become a significant contribution to the overall water potential. The pressure head ($p_w/\rho_w\, g$) can be replaced with $\left(\psi_m + \psi_g\right)$ – i.e., $\psi_m + (\rho_w g\, z\overline{V}_w)$ – in order to represent the dominant forces controlling F_w (ignoring possible contributions from the osmotic potential). Recognizing the case in which $\psi_w \sim (\psi_m + \psi_g)$, we can write a one-dimensional form of the Richards equation as:

$$\frac{\partial \theta_s}{\partial t} = -\overline{V}_w \frac{\partial F_w}{\partial z} = \frac{\partial}{\partial z}\left(\frac{AB_k}{\overline{V}_w\mu}\right)\left(\frac{\partial\left(\psi_m + (\rho_w\, gz\overline{V}_w)\right)}{\partial z}\right).\qquad (9.10)$$

As forces that control F_w, the terms $\rho_w gz\overline{V}_w$ and ψ_m oppose one another. The matric potential (ψ_m) reflects a negative pressure ("binding" force) that resists the vertical flow of water through the soil, and reduces the free energy of water, whereas $\rho_w gz\overline{V}_w$ reflects a positive pressure (weight of the water) that enhances vertical flow and increases the free energy of water; by convention, ψ_m carries a negative sign and $\rho_w gz\overline{V}_w$ carries a positive sign. According to the Richards equation, spatial divergence in the balance of one of these forces relative to the other, will in turn cause divergence in F_w and, by continuity, time-dependent storage or depletion in the soil water content (θ_s).

When soil water flows are expressed as a function of the gradient in ψ_w, as in Eq. (9.10), g_{hw} takes units of mol s kg^{-1}. We can relate hydraulic conductances when expressed with regard to the pressure head ($g_{hw}[h]$) or water potential ($g_{hw}(\psi w)$) according to:

$$g_{hw}[h] = g_{hw}(\psi_w)\left(\rho_w\, gz\overline{V}_w\right).\qquad (9.11)$$

9.1.2 The soil water retention curve

Darcy's Law and Richards equation provide relations by which to define water flux density with respect to chemical potential gradients and associated dynamics in soil water content; thereby describing seminal soil hydraulic *processes*. Of equal importance to understanding the thermodynamic nature of the soil system is the relation between soil water content and free energy density; thereby describing seminal soil hydraulic *states*. The *soil water retention curve* is an empirically derived relation between ψ_m and θ_s that reveals changes

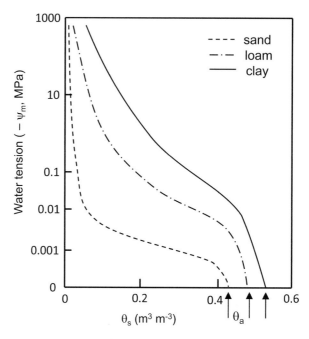

Figure 9.1 Typical soil water retention curves for three different soils that vary in mean pore size (sand > loam > clay). θ_s is soil water content, θ_a is the soil water content at which air enters the gas phase of pore volumes.

in the free energy state of the soil hydraulic system as a function of soil drying. As soil water content decreases from its maximum value, threads of liquid water in soil pore spaces (capillaries) will acquire greater tension, reflecting lesser free energy. Eventually, the tension in specific capillaries will reach a critical point at which air comes out of solution and occludes the central channel of the pores, forcing water to retract into spaces in the soil and bind as thin films on soil particles. The critical tension at which air occludes a pore is not a linear function of pore radius, and thus the soil water retention curve is non-linear in its shape. Furthermore, given the dependence of soil hydraulic conductivity on soil permeability (B_k), the loss of capillaries to air occlusion, and thus the shape of the water retention curve, causes decreases in g_{hw} as a soil dries. Typical water retention curves for a sandy soil and a clay soil are shown in Figure 9.1.

Given the hydraulic insight that can be extracted from a water retention curve, considerable effort has been devoted to describing the relation in mathematical form. In the soil sciences, empirical models that are developed from generalized observation patterns are referred to as *pedotransfer functions*. One of the pedotransfer functions used most frequently to describe the water retention curve is the Brooks and Corey equation (Brooks and Corey 1966):

$$\frac{\theta_s}{\theta_{max}} = \left(\frac{\psi_m}{\psi_{ma}}\right)^{-\lambda}, \tag{9.12}$$

where θ_s is the observed soil water content, θ_{max} is the maximum (saturated) soil water content, ψ_{ma} is the critical matric potential at which air enters the soil capillary system, ψ_m is the observed matric potential, and λ is an empirical coefficient that varies as a function of pore size distribution (approaching ∞ for homogeneous diameter distributions and approaching 0 for the most heterogeneous diameter distributions). Typical values of λ range between 0.5 and 10. The Brooks and Corey equation is simple and it provides a relatively good fit to water retention curves that resemble a "J" shape. However, it provides a poor fit to curves that resemble an "S" shape, such as those for clays. A pedotransfer function that better represents an "S" shaped curve is the van Genuchten equation (van Genuchten 1980):

$$\frac{\theta_s - \theta_{min}}{\theta_{max} - \theta_{min}} = \frac{1}{\left(1 + (a\,|\psi_m|)^n\right)^m}, \tag{9.13}$$

where θ_{min} is the residual (tightly bound) water content that remains after the soil is air dried, and a (with units 1/MPa), m (unitless), and n (unitless) are curve-fitting coefficients. Both of these functions are defined for curves that extend from the water content of initial air entry (θ_a) to θ_{min}; at $\theta_s \geq \theta_a$, the soil is assumed to be at θ_{max} and ψ_m is controlled by flow resistances (i.e., viscosity), rather than air-occluded capillary paths (i.e., B_k).

9.2 Water flow through roots

The anatomical structure of roots reflects its two primary functions – conduction and storage. A root cross-section reveals a single layer of epidermal cells at the outside boundary and a thicker tissue just inside the epidermis composed of multiple layers of cells, collectively called the *cortex* (Figure 9.2A). Cells in the cortex have a role in storing organic polymers, such as starches and proteins that can be mobilized to support future growth. At the interior boundary of the cortex lies a ring of cells called the *endodermis*. The endodermis forms the boundary between the storage function (carried out by the cortex) and the transport function (carried out by the vascular tissues – xylem and phloem) (Figure 9.2B). The vascular tissues of the root are located in a central cylinder called the *stele*. The stele contains xylem, phloem, and parenchyma cells. Xylem cells are visible in the cross-section micrographs (microscope photographs) of many roots as large, open circles, or ovals, which are hollow at maturity and when placed end-to-end, function as capillary conduits for the transport of water and solutes. Phloem cells are alive at maturity, which is evident by their color-stained appearance in micrographs. Phloem cells are specialized to conduct sugars.

The inward flow of water from the epidermis to the stele follows a *radial path* (following the radius of the root). Once it enters the stele, water follows an *axial path* as it is transported upward toward leaves. The radial flow of water occurs through two different paths. During transpiration, most water moves from the soil into the porous wall matrix surrounding epidermal cells, and then radially through the walls of cortical cells. This water does not cross membranes; it flows through the tortuous series of voids in the cell wall matrices which

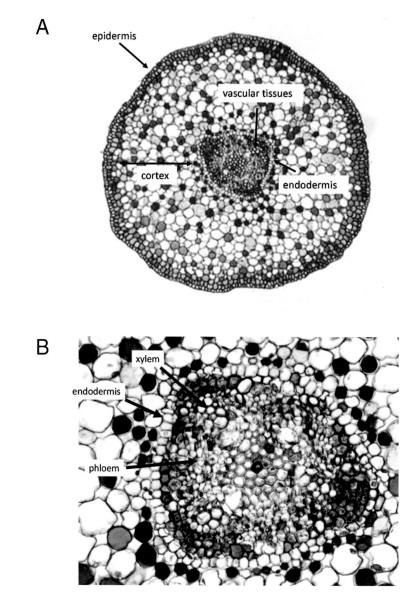

Figure 9.2   **A**. Cross-section of *Salix* sp. (willow) root showing anatomical features. See text for a description of the different anatomical features. **B**. Cross-section of a willow root at higher magnification. The central cylinder located inside the endodermis and containing the xylem and phloem tissues is known as the stele. See text for other descriptions of anatomical details. Printed with the kind permission of Alison Roberts, University of Rhode Island. See color plates section.

are formed from intertwined cellulose microfibrils. This *apoplastic flow* is driven by the negative hydrostatic pressure gradient (tension) that is propagated from leaf to root through the xylem. Apoplastic flow can be described in more common terms as water being "sucked" into the root and "pulled" to the leaves in response to the negative hydrostatic pressure gradient. Water can also cross the plasma membrane of root epidermal or cortical cells, moving from the apoplast to the symplast. This flux is forced by the osmotic potential gradient that exists between the apoplastic water of the cell walls and the symplastic water in the cell cytosol. Crossing of the membrane most likely occurs through aquaporin protein channels. Once in the cytoplasm of the cells, water can move from cell to cell through *plasmodesmata*; cytoplasmic connections that cross cell walls and link the living cytoplasm from one cell to the next. Water flow through the cytoplasm is referred to as *symplastic flow*. Apoplastic flow cannot be sustained entirely to the stele. This is because endodermal cells have thickened walls on those faces that border the cortex and vascular column, and these walls are often filled with a hydrophobic, waxy substance, called *suberin*. The suberized walls, aligned side-by-side, are called *Casparian strips*. The cytoplasm of the endodermal cells is fused to the Casparian strips, thus forcing any water and associated solutes to cross the endodermal membrane, prior to entering or exiting the stele.

Once in the xylem, water is transported vertically through a system of hollow capillary bundles. Given the lack of cytoplasm in xylem capillaries, this flow is also referred to as apoplastic flow. The continuity of the water column as it is transported through the xylem capillaries is maintained by electrostatically induced cohesion among polar water molecules. The viscosity of water (μ in Eq. (9.9)) and limited capillary potential in the apoplastic volume of the plant (AB_k in Eq. (9.9)) produces a hydraulic resistance to flow. In herbaceous plants, most hydraulic resistance to flow through the plant occurs in the radial path of the roots (Steudle and Peterson 1998); in woody species, anatomical features of xylem can cause most of the hydraulic resistance to occur along the axial path in the stem (McCulloh *et al.* 2010). It is the hydraulic resistance of the plant working against the evaporative force that pulls water upward, through a stem, that allows xylem water to be stretched and subjected to a tensile force.

9.3　Water transport through stems

The ascent of water from roots to leaves is driven by the negative pressure gradient and opposed by gravity. The thermodynamic sum of these forces is referred to as *xylem water potential* (ψ_w):

$$\psi_w = \psi_p + \psi_g \approx p_w + \rho_w gh, \qquad (9.14)$$

where ψ_p and p_w are typically negative in sign and ψ_g and $\rho_w gh$ are positive in sign. The xylem water potential represents a thermodynamic potential for water in the xylem to do work; in this case, the work that is implied is hydraulic ascent working against gravity.

Xylem water potential is affected by both long- and short-term dynamics in the hydrologic states of the soil and atmosphere. During the days or weeks between precipitation events, soil water will be progressively depleted (i.e., $\theta_s < \theta_{max}$), causing water deficits in leaves, greater tension in the xylem, and concomitantly, decreases in xylem ψ_w. This is sometimes referred to as *static xylem stress* because it develops slowly and can be treated as a nearly constant diurnal baseline (Tyree and Ewers 1991). During those periods when $\theta_s \approx \theta_{max}$, the

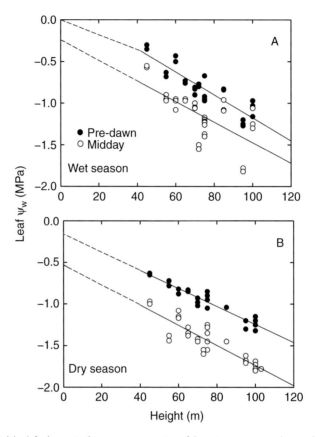

Figure 9.3   Leaf water potential (ψ_w) for leaves in the outer crown region of *Sequoia sempervirens* (sequoia) during the wet season (A) and dry season (B), and measured just prior to dawn or at midday. Redrawn from Ishii *et al.* (2008), with kind permission from Springer Science + Business Media.

soil supply of available H_2O will be adequate and static xylem stress will be minimal. However, diurnal decreases in ψ_w will occur because the flow of ascending water must continue to "work" against hydraulic resistances in the roots and stems. The impedance caused by this interaction decreases the free energy potential of xylem water. The diurnal decrease in ψ_w under these conditions is sometimes referred to as *dynamic xylem stress* because it develops in the presence of a dynamic flow of water. Dynamic xylem stress is most commonly associated with time scales of minutes-to-hours, rather than days-to-weeks.

Static and dynamic xylem water stresses, together, underlie observed patterns in ψ_w gradients. This is especially apparent in the leaf water potential gradients that develop as a function of height in tall trees (Figure 9.3). In the example of the giant sequoia, it is clear that just before dawn, following an entire night of rehydration, the value of ψ_w is significantly less than zero, even in the lowest leaves in the canopy, and a vertical gradient persists in ψ_w. This pre-dawn state of ψ_w reflects, in part, axial resistances in the conductive part of the tree that preclude equilibrium with a wet soil. In the data presented in Figure 9.3, the gradient in ψ_w increases by an almost equal amount at all heights as the day progresses and atmospheric water vapor demand increases. These diurnal patterns in ψ_w are driven predominantly by dynamic xylem stress. In comparing the gradients in ψ_w between the wetter and drier parts of the growing season, it is clear that in the presence of a drier soil the vertical gradient in ψ_w is slightly more negative and slightly steeper, than the case for a wetter soil. These seasonal patterns in ψ_w are driven predominantly by static xylem stress.

9.3.1 Wood anatomy and axial water transport through stems

Among trees, hydraulic transport varies according to anatomical characteristics of the wood (xylem tissue). Tree stem anatomy can be classified according to three fundamental types: coniferous, ring-porous, and diffuse-porous, with the latter two occurring in angiosperms. In coniferous stems, water is transported by conduits formed from overlapping *tracheids* (dead xylem cells) with numerous small openings, or pits, connecting the secondary cell wall space (the apoplasm) of an underlying tracheid to an overlying tracheid (Figure 9.4A). Thus, a continuous, porous conduit is formed for the upward transport of H_2O. In angiosperms, xylem is composed of columns of *vessels* (also dead at maturity), which attain greater diameters than tracheids and have perforated end walls (Figure 9.4B). On theoretical grounds, given the same

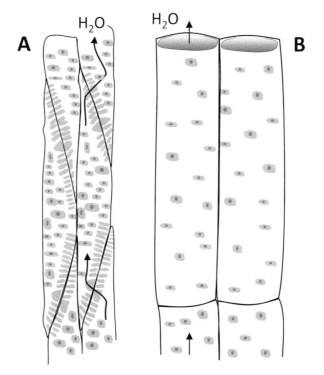

Figure 9.4 Two examples of the xylem capillary systems that exist in plants and are used for upward water transport in stems and the veins of leaves. **A**. Water transport through the perforated adjoining secondary cell walls of tracheid elements in a gymnosperm. **B**. Water transport through the stacked vessel elements in an angiosperm. Pits in the cell walls are shown as small circular or oblong features with a darker, circular membrane in the middle that controls lateral water flow among adjacent capillary columns.

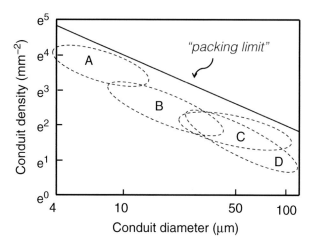

Figure 9.5 The ranges of the observed relationships between conduit diameter and conduit density (number of conduits per mm^2 of sapwood cross-sectional area) in conifer trees (A), temperate diffuse-porous trees (B), ring-porous trees (C), and tropical diffuse-porous trees (D). The "packing limit" represents the maximum possible packing of conduits of a given diameter per unit of sapwood area assuming a "square" pattern of packing. Note the tradeoff between conduit diameter and frequency. The theoretical packing limit cannot be met with conduits (vessels or tracheids) alone because of the need for fibers, parenchyma, and other non-conducting cells in the xylem to provide mechanical strength and metabolic support, respectively. Redrawn from McCulloh *et al.* (2010).

diameter, tracheids and vessels would be capable of similar hydraulic conductance; the lack of complete end wall perforations and shorter length of tracheids would be compensated by high pit densities, providing equivalent overall porosity compared to vessels (Pitterman *et al.* 2005). In reality, however, vessels and tracheids are not the same diameter at maturity; vessels have greater diameters, allowing them a hydraulic advantage over tracheids. The evolution of advantages for hydraulic conductance, however, through the emergence of wider and shorter capillaries in angiosperms, comes at a cost in terms of the structural integrity of wood. There exists a finite "packing limit" to the capillary potential of xylem. By packing wood volume with greater void space, angiosperms must accommodate lower vessel frequencies per unit of sapwood area (Figure 9.5). Given that vessels and tracheids contribute to the structural integrity of wood, as well as its capacity for hydraulic conductance, a fundamental adaptive tradeoff exists between hydraulic efficiency and resistance to mechanical stress.

   Tradeoffs can also be defined solely within the context of xylem as a hydraulic device. Ring-porous anatomy in angiosperms is characterized by wide vessels within the outermost annual ring of the sapwood (Figure 9.6A), and it is this ring that is most active in transporting water. Diffuse-porous wood has wide vessels distributed throughout the sapwood, and a larger fraction of sapwood diameter is involved in water transport (Figure 9.6B). The large diameter vessels in ring-porous wood render them more likely to be occluded by air during cold winters. As water freezes, dissolved air comes out of solution and can accumulate in vessels, forming air blockages. Within this context, it is notable that large-diameter vessels in diffuse-porous wood, which can reach the same diameters as ring-porous species,

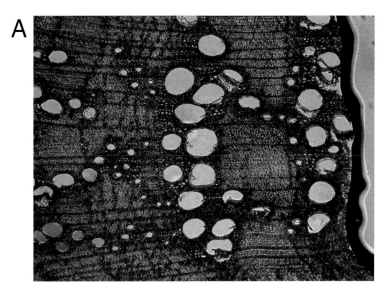

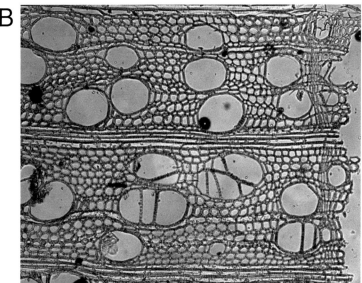

Figure 9.6   **A**. Ring-porous wood anatomy in *Quercus rubra* (red oak). **B**. Diffuse-porous wood in *Acer macrophyllum* (big-leaf maple). Note the consistency of xylem vessel diameters as a function of stem diameter in the diffuse-porous type, versus the larger diameter vessels located near the outer stem surface in the ring-porous type. Printed with the kind permission of Katherine McCulloh (Oregon State University). See color plates section.

are only observed in tropical trees (see Figure 9.5). The increased risk of air occlusion in temperate ring-porous trees is balanced by the fact that they can produce a new outermost ring of vessels to replace blocked xylem each year. The tradeoffs between utilizing wider hydraulic conduits and incurring a greater risk of air blockage have presumably imposed scaling limits on the hydraulic system – a limit that is often used to define "safety versus efficiency" tradeoffs, in this case solely within the context of xylem as a hydraulic adaptation (Sperry *et al.* 2008, Meinzer *et al.* 2010).

9.3.2 Embolism formation, repair, and detection in xylem

The air occlusions that form in xylem vessels are generally called emboli (singular = embolism). Emboli are formed when negative pressures or low temperatures are sufficient to force dissolved air from the liquid phase, into the gas phase, producing a "bubble" large enough to block capillarity. Due to the tendency for embolism formation, the negative pressures that are required to transport water upward through xylem impose a hydraulic "risk" (Tyree and Sperry 1989). Embolism events can be detected directly through acoustic techniques (one can actually detect "sound" from cavitation events), but it is more typical to use a proxy measure, such as reductions in hydraulic conductance (i.e., reductions in $A\,B_k$ in Eq. (9.9)). Emboli can be transferred to neighboring capillaries through pits located in secondary xylem cell walls in a process known as *air seeding*. As emboli spread, the water tension in residual, non-embolized elements increases, creating new risks for embolism formation – a clear case of hydraulic positive feedback. The spread of emboli can ultimately result in *catastrophic xylem failure*. Embolized, capillary conduits are non-functional until positive pressure is applied to the capillary and air from the blockage is forced back into the liquid phase. In plants that are embolized during the winter, repair often occurs the following spring when the osmotic potential of root cells draws water in from the soil, and a positive root pressure develops; the resulting root pressure "pushes" water upward into embolized conduits. In plants that are embolized during climatic drought or due to a dry atmosphere, the water for repair can come from neighboring living cells in the xylem tissue (Holbrook and Zwieniecki 1999).

Plants are capable of operating in a manner that maintains ψ_w above the critical cavitation point (Sperry *et al.* 1993). This requires that the allocation of biomass and energy to the production of xylem tissue be sufficient to accommodate the maximum (extreme) rate of transpiration that can, with relevant probability, be experienced by a plant during its lifetime in its native environment. It has been argued, on the basis of cost-benefit tradeoffs, that over-investment of resources in the construction of xylem tissue will occur at a cost to the production of other types of tissues, including those required for reproduction; whereas under-investment will leave a plant susceptible to catastrophic xylem failure. This produces a fundamental "dilemma" for plants – to persist or reproduce. While xylem is dead at maturity, and therefore of negligible consequence to longer-term maintenance costs, it is composed of lignin and cellulose, making its construction cost relatively high. Arguments have been made that plants have evolved in ways that balance investment in xylem construction against the risk of catastrophic failure (see McCulloh *et al.* 2003, Sperry 2003). In other words, the evolutionary process has, in general, effectively minimized the margin of hydraulic safety, thus maximizing the potential for reproductive fitness. Based on a broad survey of trees from different biomes, however, the margin that favors hydraulic safety appears, in general to be narrow (Choat *et al.* 2012). Most trees experience frequent periods in which they operate within 1 MPa or less, of their hydraulic safety margin. This hydraulic vulnerability makes global forests, in general, susceptible to climate warming and drying, and it demonstrates the potential for natural selection to produce phenotypes with "great benefit, but also great risk."

9.3.3 Measurement of stem hydraulic conductance

Values of g_{hw} can be determined empirically with use of a *hydraulic vulnerability curve*. The hydraulic vulnerability curve shows the rate by which g_{hw} declines from an initial state at xylem $\psi_w \approx 0$, to an ending state in which xylem ψ_w is highly negative in value and $(\Delta g_{hw}/\Delta\psi_w) \rightarrow 0$. A stem section is brought to $\psi_w \approx 0$ by rehydration from a reservoir of water under dark, humid conditions, or by forcing water into the segment with positive pressure. Once $\psi_w \approx 0$, water is allowed to flow through the stem under the force of a hydrostatic pressure head (Δp_w) created by an elevated volume of water and connected to one end of the segment. Water that flows through the segment is collected and its rate of collection is taken as F_w. The value for g_{hw} is then calculated as $F_w/(\Delta p_w/\Delta x)$, where Δx is segment length. The measure of g_{hw} can then be repeated after forcing air through the stem, or subjecting it to a high centrifugal force, thus causing progressive cavitation of xylem capillaries. In a partially embolized stem, Δp_w will have to be increased in order to reach the same F_w as was observed in the stem with $\psi_w \approx 0$; thus, the calculated g_{hw} will be reduced. After calculating g_{hw} for a range of cavitation states, a plot of g_{hw} versus $\Delta p_w/\Delta x$ provides the hydraulic vulnerability curve.

During construction of a hydraulic vulnerability curve, the progressive increase in $\Delta p_w/\Delta x$ (decrease in g_{hw}) is determined by the matric forces that interact with water within the xylem and the ratio of embolized versus intact xylem capillary columns. For porous media, in general, the relation between F_w and $\Delta\psi_w/\Delta x$ is defined by a characteristic *water retention curve*. The water retention curve has been a general feature of soil water analyses for many decades; it is a fundamental measure of the matric forces that resist pressure-driven flow through porous media. Many theoretical treatments of water flux through unsaturated porous media begin from the water retention curve. The hydraulic vulnerability curve is a type of water retention curve, and recognition of the analogy between the two has provided the basis for linking empirical and theoretical components of g_{hw}.

9.4 The hydraulic conductance of leaves and aquaporins

Studies have revealed that mechanisms exist to ensure that g_{hw} is adequate to sustain water delivery to transpiring leaves during diurnal increases in the leaf-to-air water vapor mole fraction difference (Sack and Holbrook 2006). The leaf components associated with g_{hw} are sensitive to incident PPFD and leaf temperature in ways that scale leaf xylem g_{hw} up and down in proportion to stomatal conductance (g_{sw}), providing diurnal stability in leaf ψ_w as E varies (Matzner and Comstock 2001, Guyot *et al.* 2012). A complete understanding of how this coordination is accomplished is not currently available, though it is known to involve changes in the viscosity of water in response to temperature and changes in the permeability of aquaporin channels in leaf mesophyll membranes (Cochard *et al.* 2007).

The discovery of aquaporin proteins in the cellular membranes of mammalian red blood cells in the 1990s has changed the way we think about water transport in all organisms. One of the fundamental enigmas in biology that existed for many decades, prior to this discovery, was the unavoidable paradox that H_2O, a highly polar compound, is capable of traversing the non-polar phospholipid bilayer that composes biological membranes. Most theories that were offered to explain this apparent paradox relied on the assumption that H_2O molecules were small enough to slip through "cracks" that formed in the membrane as hydrocarbon chains moved back and forth laterally. By the latter part of the 1990s, evidence emerged that water transport occurs through specialized trans-membrane proteins called *aquaporins* (Schäffner 1998, Tyerman *et al.* 1999). Aquaporins are part of an ancient set of proteins that can be found in animals, plants, and microorganisms, called major intrinsic proteins, or MIPs. The water-transport channel formed by aquaporins is composed of six helical protein chains that, together, produce a hydrophilic "pipe" across the otherwise hydrophobic hydrocarbon region of membrane bilayers. The presence of aquaporins facilitates cross-membrane transport without unfavorable polarity interactions. Aquaporins are likely to facilitate most cross-membrane water transport in roots and leaves. Substances other than H_2O can also be transported through aquaporins, including dissolved CO_2. Of particular interest, however, is the fact that protons (H^+ ions) cannot pass through aquaporins – it appears that transport through aquaporins is favored for polar compounds, but not charged ionic atoms. The aquaporin channel that crosses the membrane is relatively narrow, and it is likely that H_2O molecules cross in a single file (see review by Maurel *et al.* 2008).

9.5 Modeling the hydraulic conductance and associated effects of embolism

As a plant experiences water deficits, and xylem tension increases, the relation between g_{hw} and $\Delta\psi_w/\Delta x$ does not remain constant; rather, g_{hw} decreases non-linearly as $\Delta\psi_w/\Delta x$ increases (Figure 9.7). This non-linear dependency is principally due to the *cavitation* of capillary elements as air comes out of solution and forms emboli. Cavitation causes hydraulic dysfunction and removes conduits from the conductive path in xylem. This forces the hydraulic flux to work against a reduced capillary potential, and thus causes an increase in $\Delta\psi_w/\Delta x$. The increase in $\Delta\psi_w/\Delta x$, in turn, can lead to further cavitation. This is a clear example of positive feedback. As cavitation is propagated through the capillary system, non-linear patterns develop between g_{hw} and $\Delta\psi_w/\Delta x$.

Relying on the theoretical relations that define water retention curves, values of g_{hw} in isolated stem segments have been described according to xylem water potential and rates of cavitation (Bohrer *et al.* 2005, Janott *et al.* 2011). The water potential at which air is introduced into xylem capillaries is defined as ψ_{wa}. Thus, when $\psi_w \geq \psi_{wa}$ (i.e., ψ_w is less negative in value than ψ_{wa}), no cavitation occurs, and we can write:

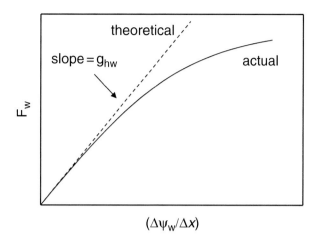

Figure 9.7 Graphical representation of stem water flow. The graph shows a *theoretical* hydraulic conductance (g_{hw}), which is constant as a function of the water potential gradient along the axial path of the stem ($\Delta\psi_w/\Delta x$), and an *actual* g_{hw}, which is variable as a function of ($\Delta\psi_w/\Delta x$). Changes in the actual g_{hw} are due to progressive cavitation of xylem capillaries due to the formation of emboli.

$$\theta_x = \left((p_s - \theta_a)\frac{\psi_a - \psi_w}{\psi_{wa}}\right) + \theta_a, \tag{9.15}$$

where θ_x is the volumetric water content of the xylem (m^3 m^{-3}), p_s is sapwood porosity (m^3 m^{-3}), θ_a is sapwood volumetric water content (m^3 m^{-3}) at the tension whereby air comes out of solution. This model assumes a linear *water retention curve* between the limits $\psi_w \approx 0$ and $\psi_w = \psi_{wa}$. Within this domain, the modeled effect between θ_x and ψ_w reflects elasticity in xylem elements that causes them to become narrower as tension increases.

Within the domain where $\psi_w < \psi_{wa}$, linear dependency of g_{hw} on ψ_w disappears and an empirical power function (after Brooks and Corey 1966, e.g., Eq. (9.12)) can be used to predict non-linear dynamics:

$$\frac{\theta_x}{\theta_{xmax}} = \left(\frac{\psi_w}{\psi_{wa}}\right)^{-\lambda}. \tag{9.16}$$

This model is validated by observations that the water retention curve changes shape at the air-entry threshold, to assume a non-linear form.

The connections of Eqs. (9.15) and (9.16) to g_{hw} are not obvious. One scheme that has been used models changes in θ and ψ_w as an effect on g_{hw} according to:

$$g_{hw} = g_{hw\,max} \begin{cases} \dfrac{\theta_a}{p_s} + \left(\left(1 - \dfrac{\theta_a}{p_s}\right)\left(\dfrac{\psi_{wa} - \psi_w}{\psi_{wa}}\right)\right)^2 & \text{if } \psi_w > \psi_{wa}, \\[2ex] \left(\dfrac{\psi_w}{\psi_{wa}}\right)^{-\lambda\eta} & \text{if } \psi_w \leq \psi_{wa}, \end{cases} \tag{9.17}$$

where $g_{hw\ max}$ is the maximum hydraulic conductance observed at $\psi_w \approx 0$ and η is a coefficient that takes the value $(2/\lambda) + 1$ (Bittner *et al.* 2012). Equation (9.17) provides a means of introducing the matric potential, the pressure potential, and the tendency for embolism formation as variables affecting g_{hw}, though the relation remains highly dependent on empirical parameterization.

9.6 Hydraulic redistribution

Prosopis tamarugo, a tree native to the Atacama desert of Chile apparently has unique water relations . . . Because of the very low water potentials of the salty surface soils, water evidently moves from the plant into the soil under certain conditions. This water may be reabsorbed by the plant and used subsequently . . . (Mooney *et al.* (1980))

The free energy gradient that drives the transport of water through the SPAC does not always occur in the direction from soil-to-plant-to-atmosphere. The most frequently observed

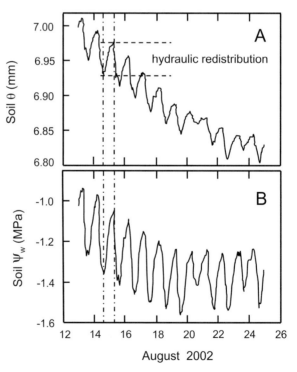

Figure 9.8 Hydraulic redistribution of water due to plant conduction in a stand of "old growth" *Pinus ponderosa* (ponderosa pine) observed at the 30 cm depth during several consecutive days in August 2002. Diurnal dynamics in soil water content (θ) is shown in panel A and diurnal dynamics in soil water potential (Ψ_w) is shown in panel B. The muted diurnal pattern between days 20 and 22 is due to cloudy weather. Redrawn from Warren *et al.* (2011), with kind permission from Springer Science + Business Media.

pattern of water transport is indeed, in this direction. However, in dry or saline soils, water can flow from plant roots to the soil. Through this mechanism water can be redistributed from one soil layer to another, using the plant as a transport conduit. These are not "counter-gradient" flows, as they always follow free-energy gradients, and are therefore fully justified in terms of thermodynamics. However, they do occur in non-conventional directions (i.e., from plant-to-soil). The importance of this phenomenon became starkly apparent during observations by Hal Mooney and co-workers in the Atacama Desert of Chile where mesquite trees appear to thrive in a saline soil and despite months without precipitation. In the extreme aridity of that ecosystem trees appear to recycle water efficiently as it exits and re-enters roots, following free energy gradients capable of direction reversal.

The process that Mooney and co-workers observed in Chile is often referred to as *hydraulic redistribution*, or in this specific case, *hydraulic lift*. Hydraulic redistribution refers to the movement of water from wetter-to-drier layers of soil, using the plant as a conduit. Hydraulic lift is a specific case of hydraulic redistribution wherein water is moved from deeper soil layers to shallower soil layers; typically at night (Figure 9.8). The multiple effects of hydraulic redistribution on ecosystem states and processes are just beginning to be clarified, though it appears that this phenomenon can affect processes ranging from bio-geochemical nutrient cycling to shifts in plant community composition (Neumann and Cardon 2012).

10

Leaf and canopy energy budgets

The plant in a field of energy, or in any considered area in which gradients exist, reacts passively to its surroundings . . . it should be clear that the sun, or even outer space, is not more removed from the plant than the next millimeter of air; it is the specific part of heat exchange being considered that determines what should be included in the plant's environment.

Klaus Raschke (1960)

The interaction of environment with a plant is through the flow of energy. There is no other way . . . All energy absorbed by a leaf must be accounted for [through storage within the leaf or loss from the leaf] and hence the energy budget for a plant leaf must balance.

David M. Gates (1968)

Energy flows from the sun to the earth where some fraction is absorbed by plant surfaces. Plants use some of that energy to drive primary production, but that is a small fraction of the total absorbed energy. Most of the absorbed energy is redistributed back to the surroundings. Following the laws of thermodynamics, energy can be transformed, transferred, or stored, but not destroyed. Thus, energy input from the surroundings must be balanced by an equal and constant output or be stored within the plant's mass. The requirement for a full accounting of energy transfers and storage, and thus reconciliation of the plant's energy budget, has been recognized for several decades as an important perspective from which to study biophysics and the evolutionary adaptation of plants to their environment. The "environment" of a plant consists of all scales between the sun and earth; though we may isolate certain components to prioritize the factors that determine energy exchange. The quotes above, by Klaus Raschke and David Gates, two of the pioneers in the field of plant-atmosphere biophysics state in very clear terms, the continuum of space that describes the "environment" with which a plant must exchange energy, and the necessity of "energy accounting" as a means to understand plant function and adaptation.

In this chapter, we develop the concept of a leaf energy budget and the models that have been developed for understanding the partitioning of absorbed energy as it is redistributed to the surroundings. Because the exchange of water between ecosystems and the atmosphere is fundamentally tied to the exchange of energy, we will also develop the biophysical concepts underlying moisture in the liquid and vapor states, and the latent energy transfers that accompany conversions from one state to the other. We will spend a large portion of the chapter developing the Penman–Monteith model for describing surface latent heat exchange at both the leaf and ecosystem scales. The Penman–Monteith model has become, arguably, one of the keystone quantitative concepts on which our estimation of surface energy budgets is based. Finally, we apply concepts from the Penman–Monteith model to understand the relative roles of plant physiology and micrometeorology in controlling latent heat exchange from leaves.

10.1 Net radiation

The solar energy that strikes a leaf surface will be reflected, absorbed, or transmitted. Given this fundamental constraint, incident solar energy can be partitioned according to:

$$R_s a_s = R_s (1 - \alpha - t), \tag{10.1}$$

where R_s is the shortwave (solar) energy flux (W m^{-2}, where $1\ W = 1\ J\ s^{-1}$); and a_s, α, and t are the fractions of R_s that are absorbed, reflected, and transmitted, respectively. Reflected radiative flux density divided by the incident radiative flux density is also called albedo. Using these relations, and our knowledge about the emission of longwave radiation (see Section 2.8), we can estimate the *net radiation balance* (R_n, W m^{-2}):

$$R_n \approx \left(R_s a_s + R_L a_L - \varepsilon_L \sigma T_s^4 \right), \tag{10.2}$$

where R_L is the longwave radiant energy (W m^{-2}) incident on the surface, a_L is the fractional absorptance of the surface for longwave radiation, T_s is the leaf surface temperature (K), ε_L is the fractional (radiant) emittance of the surface to longwave radiation, and σ is the Stefan–Boltzmann constant (5.673×10^{-8} W m^{-2} K^{-4}). In the case of the longwave components of Eq. (10.2) (the second and third terms on the right-hand side) we assume that the surface absorbs and emits radiation equally, just as in the case for a true "black-body"; thus, $a_L \approx \varepsilon_L$. For the case of a horizontal, laminar leaf the radiation balance must take account of absorption and emission of radiant energy from both upper and lower surfaces. Accepting some approximations concerning the spatial averaging of surfaces, we can write:

$$R_n \approx \left(R_s a_{su} + R_s a_{sl} \alpha_g + R_{Lu} a_L + R_{Ll} a_L - \varepsilon_L \sigma T_s^4 \right), \tag{10.3}$$

where α_g is the fractional reflectance (albedo) of the ground surface beneath the leaf; a_{su} and a_{sl} are the absorptances of solar radiation by the upper and lower leaf surfaces, respectively; R_{Lu} and R_{Ll} are the longwave radiation fluxes incident on the upper leaf surface and lower leaf surfaces, respectively; and it is assumed that absorptance of the leaf to longwave radiation is equal for the upper and lower surfaces (Figure 10.1).

Energy from the net radiation load on a unit leaf surface area must be conducted or radiated away from the surface, lost as latent heat, or stored in the heat energy and reduced entropy of the leaf's mass. After accounting for re-radiation of energy from the leaf's surfaces the principal processes leading to energy redistribution back to the atmosphere are sensible and latent heat losses. *Sensible heat* is exchanged through conduction between a leaf and the stem (or bole) of a plant or, of much greater magnitude, between a leaf's surfaces

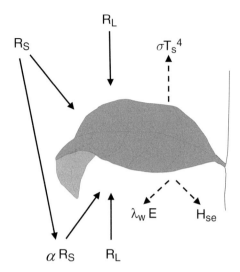

Figure 10.1   Principal energy exchanges as components of the leaf energy budget. See text for details of the symbols. Energy gains are shown as sources of absorbed energy with solid lines and energy losses are shown with dashed lines.

and the air molecules that contact those surfaces. Once transferred to the air at the leaf surface, the heat is convected away from the surface by viscous flow involving the wind. The rate of sensible heat exchange between adjoined systems of mass that exist at different temperatures is ultimately determined by the thermal conductivity of the material conducting the heat and the thermal gradient along the conductive pathway (Section 2.4). *Latent heat* is lost when the energy absorbed by a leaf is converted to an increase in the kinetic energy of water molecules, which subsequently evaporate to the atmosphere. Latent heat can also be gained when water molecules leave the vapor phase of the atmosphere and condense on a surface (e.g., as dew), releasing heat. Latent heat transfers occur within a thermodynamic system when mass undergoes a change in state without a change in temperature. In essence, latent heat represents the transfer of thermal energy (Q) into an increased entropy (S) of mass (e.g., water vapor has greater entropy than liquid water). In the case of evaporation, latent heat exchange depends on the *latent heat of vaporization* (λ_w), a coefficient that is sensitive to temperature (ranging from 2.48 MJ kg^{-1} at 10 °C to 2.41 MJ kg^{-1} at 40 °C). (A useful equation for estimating the temperature dependence of λ_w is $\lambda_w = 2.502 \times 10^6 - (2.308 \times 10^3 * T_a)$, where T_a is in °C and λ_w is estimated in J kg^{-1} water.) Thus, the rate of evaporative latent heat exchange is calculated as $\lambda_w E$, where E is evaporative flux in units of mass (or molar equivalents). Sensible and latent heat fluxes between leaves and the atmosphere are to a large extent mutually coupled. Net radiant energy that is absorbed by a leaf, but not redistributed to the atmosphere through latent heat flux, will contribute to a rise in surface temperature, causing an increase in sensible heat flux. A reduction in leaf latent heat loss will be compensated by an increase in sensible heat loss, and vice versa. Eventually, given a constant influx of energy to a moist surface and the presence of a turbulent atmosphere that is unsaturated with respect to water vapor, a steady state will be reached with regard to latent

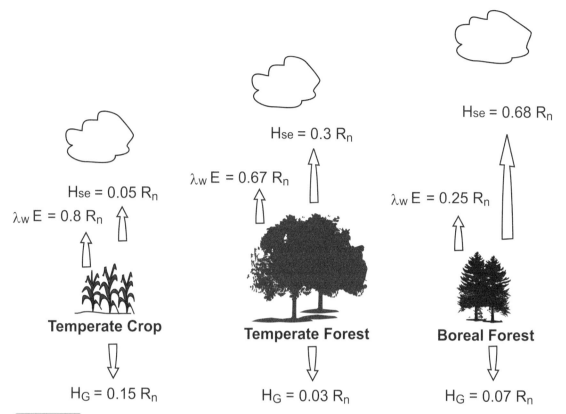

Figure 10.2 Energy losses as fractions of the net radiation gain (R_n) of three different ecosystems with variable amounts of partitioning into latent heat loss ($\lambda_w E$), sensible heat loss (H_{se}), and conduction of heat to the ground (H_G). Greater sensible heat loss rates tend to cause higher atmospheric turbulence and deeper convective planetary boundary layers.

and sensible heat losses. At that steady state, leaf temperature must be constant. Thus, in surface energy budgets we often refer to the "coupled partitioning" of absorbed energy between latent and sensible heat exchange.

Beyond the leaf scale, the same components of the energy budget can be applied to plant canopies or even entire landscapes (Figure 10.2). Latent heat loss is highest in irrigated ecosystems, such as agronomic crops, and decreases as water availability decreases. As a greater fraction of net radiation is dissipated as sensible heat loss, turbulent kinetic energy in the lower troposphere will increase, and the height of the convective planetary boundary layer will also increase. The height of the planetary boundary layer is often indicated by the height of the cumulus cloud base.

Components of the surface energy budget are affected in different ways depending on the nature of the surface being considered. For example, surface roughness becomes a more important determinant of sensible heat exchange between landscapes and the atmosphere, compared to the case for individual leaves. Greater surface roughness increases the aerodynamic conductance to sensible heat loss due to the increased potential for turbulence. In recent years it has become evident that these surface-dependent effects have important ramifications for regional-to-global climate change (Box 10.1). Of particular importance is the fact that natural succession in native ecosystems and the land-use decisions that are made for managed ecosystems can have a large impact on the way that energy is absorbed and partitioned by a landscape surface. The energy budget components that are affected by land-use variation include all of those described above, from reflective and radiative properties of a surface to the partitioning of latent and sensible heat. The influence of land-use change on future regional-to-global energy budgets will most certainly be a topic of vigorous research.

Box 10.1   **Landscape surface albedo, land-use change, and climate trends**

Human activities have caused changes to the earth's surface, many of which have the potential to influence the earth's radiation budget. Forests have a lower surface albedo than grasslands and savannas. Thus, when forests are cleared to create pasture lands or forests are removed for timber harvesting, we can expect influences on regional energy budgets.

More recently, political and economic initiatives have included strategies to expand the planting of forests as a means of enhancing atmospheric CO_2 sequestration (thus promoting climate cooling) and/or supplying the cellulose required for biofuel development (Pacala and Socolow 2004, Canadell and Raupach 2008). Such land-use changes have the potential to not only influence surface albedo, but also the atmospheric concentration of radiatively important trace gases, changes in surface roughness, and associated changes in the turbulent exchange of latent and sensible heat (Bonan 2008, Rotenberg and Yakir 2010, Anderson *et al.* 2011). (Forests tend to have greater surface roughness, larger turbulent eddies, and concomitantly lower resistances to heat exchange, compared to smoother crop and pasture surfaces.)

Many of these changes trade off against each other in complex ways, making it difficult to predict the ultimate effect on climate. For example, in tropical latitudes, the planting of forests on reclaimed pasturelands, has the potential to: (1) increase CO_2 extraction from the atmosphere (a "cooling effect"); (2) increase latent heat loss (a "cooling effect"); and (3) increase the formation of atmospheric aerosols (a "cooling effect"). In the opposite direction, however, forests will cause a lower surface albedo, greater rates of solar energy absorption, and associated higher fluxes of sensible heat and longwave radiation to the atmosphere (a "warming" effect). When considered together, the net result of these compensating effects is likely to be one of surface cooling in tropical latitudes (Anderson *et al.* 2011). The CO_2 uptake potential of forests at northern latitudes is low compared to that for tropical forests and the reduced albedo of these forests contrasts starkly with the white snow-covered surfaces of open landscapes. As regional coverage of these forests increases, the overall influence is likely to favor climate warming (Anderson *et al.* 2011). Lee *et al.* (2011) observed that the difference in mean annual surface temperature between adjacent forested and open landscapes changed by − 0.07 K per degree northward latitude (Figure B10.1A). In other words, considering albedo effects alone, forests tend to have greater warming potential as one moves northward. This effect is principally driven by the change in net radiation load across the latitudinal gradient (Figure B10.1B).

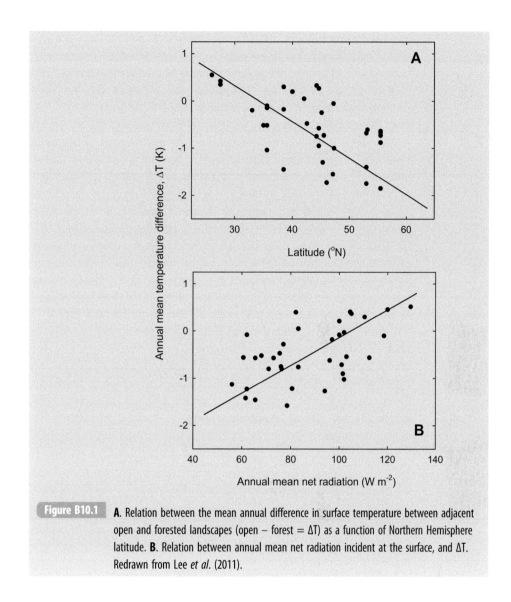

Figure B10.1 **A.** Relation between the mean annual difference in surface temperature between adjacent open and forested landscapes (open − forest = ΔT) as a function of Northern Hemisphere latitude. **B.** Relation between annual mean net radiation incident at the surface, and ΔT. Redrawn from Lee *et al.* (2011).

10.2 Sensible heat exchange between leaves and their environment

Sensible heat is transferred when there is a temperature difference between two masses that are in contact with one another and heat is conducted from the mass with greater temperature to that with lesser temperature. This process is referred to as sensible heat transfer because the flow of heat can be "sensed" either by instrumentation or, in some cases, by human touch. For example, if you step with bare feet from a carpeted surface to a tiled floor the tile will feel colder than the carpet, even though measurement with a sensor will prove them to be at the same temperature. The tiled floor "feels" colder because it has a greater capacity for heat conduction and heat is conducted from your feet to the floor at a greater rate. The tile feels colder because nerve endings in your feet detect the flux of heat and your brain interprets the associated neural input to mean that the tiled surface is "colder."

Sensible heat transfer also occurs between surfaces and a fluid in contact with the surface; once again, as long as a temperature gradient exists. Within this context, we are most interested in the transfer of heat from leaf surfaces or canopy surfaces to air. As molecules of air contact a surface, heat is exchanged through molecular conduction, either toward the surface or away from the surface, depending on the direction of the temperature gradient. The zone of molecular contact between the air and surface is thin and heat transfer from the surface up through perfectly still layers of air by molecular collision alone is slow. However, the layers of air above a surface, even in the absence of wind, are not ideally still. Layers of air will subside toward the surface as they cool and sink under the force of gravity, or they will rise as they warm and are forced upward by a pressure gradient that increases as the surface is approached. These vertical motions, driven by air density gradients, are often referred to as *free convection*, and they can carry sensible heat toward or from the surface by viscous flow. When these vertical motions are accompanied by a horizontal wind, the warmed or cooled parcels of air that rise and sink near the surface will be carried away with the forced flow, transferring sensible heat in the downwind direction through what is often called *advection*. (Convection and advection are often used interchangeably to refer to the pressure-driven viscous flows of a fluid. We will tend to use convection to indicate mean vertical flows and advection to indicate mean horizontal flows.) Building from Fourier's Law (Section 2.4), we can express surface sensible heat exchange in the vertical coordinate (z) as:

$$H_{se}[z] = -K_{dh}\, c_p\, \rho_m\, \frac{\Delta T}{\Delta z}, \qquad (10.4)$$

where $H_{se}[z]$ is the vertical sensible heat flux (J m^{-2} s^{-1}), K_{dh} is the diffusion analog coefficient for sensible heat (m^2 s^{-1}), c_p is the molar specific heat of dry air (J mol^{-1} K^{-1}), ρ_m is the molar density of dry air (mol m^{-3}), ΔT is the difference in temperature between the surface and the air adjacent to the surface (K), and Δz is the vertical distance (m) across which thermal energy is conducted.

One of the important variables in Eq. (10.4) is the molar specific heat of dry air. Molar specific heat is a measure of the thermal energy required to raise a mole of dry air one unit of Kelvin temperature. This value is not constant, but rather varies with temperature and pressure. At 273 K (0 °C) and standard pressure, the molar specific heat of dry air is 29.1 J mol^{-1} K^{-1}; this value increases to 29.2 J mol^{-1} K^{-1} at 298 K (25 °C) and standard pressure. This temperature dependence is due to changes in the way heat is channeled into the kinetic energy of air molecules as the air temperature increases. As temperature increases, slightly more heat is transferred into the internal quantum dynamics of atoms and slightly less is transferred to the kinetic energy of the molecules; this results in a slightly higher specific heat. Within the narrow temperature range considered in plant-atmosphere energy exchanges the temperature dependency of molar specific heat is ignored. The molar density of dry air (ρ_m) is also dependent on temperature and pressure. At 25 °C (298 K), ρ_m is approximately 40.87 mol m^{-3}. The dependence of ρ_m on temperature and pressure is derived from the ideal gas law ($\rho_m = P/RT$).

It is important to note that in the case of sensible heat transfer from a surface that has absorbed solar energy, the surface must increase in temperature above that of the air in order to drive a heat flux "down the temperature gradient." This is not the case in the next type of heat transfer that we will discuss, the latent heat flux. In the case of latent heat fluxes, thermal energy is extracted from the system as the entropy of water is increased without increasing the mean kinetic energy of mass. Thus, no change in surface temperature is detected.

10.3　Latent heat exchange, atmospheric humidity, and temperature

The surface-atmosphere flux of latent heat is dependent on both the amount of energy absorbed by a surface (which drives a phase change through latent heat transfer) and the water vapor concentration gradient that exists between the surface and atmosphere (which drives the flux of mass and energy away from the surface). Ultimately, latent heat exchange reflects processes that must be defined within the scope of thermodynamics, and therefore the most relevant starting point is a relation that defines the chemical potential of water vapor in air in terms of its mole fraction (referenced to the state of vapor stauration):

$$\mu_w = \mu_w{}^* + RT \ln \frac{c_{aw}}{c_{aw}{}^*}, \qquad (10.5)$$

where μ_w is the chemical potential of water vapor in a sample of air, $\mu_w{}^*$ is a standard chemical potential for water vapor, c_{aw} is the mole fraction of water vapor in the sample of air, and $c_{aw}{}^*$ is the saturated water vapor mole fraction in the same sample of air (see Section 2.3 for a discussion of chemical potential). From the form of Eq. (10.5) it is clear that as water evaporates from a wet surface, c_{aw} will increase and move closer to $c_{aw}{}^*$, and the chemical potential of water vapor in the air (μ_w) will increase; the greater the difference between c_{aw} and $c_{aw}{}^*$, the lesser the chemical potential. Within the thermodynamic context, water vapor will move spontaneously from a region of higher chemical potential (c_{aw} closer in value to $c_{aw}{}^*$) toward a region of lower chemical potential (c_{aw} further in value from $c_{aw}{}^*$). As water evaporates from the wet surface, its molecules do not contain greater kinetic energy than those at the wet surface; in fact the temperature of the wet surface and the water vapor just above the surface is the same. It is the chemical potential (i.e., potential energy) of the water vapor that changes, not the molar content of kinetic energy.

The fact that phase changes are not accompanied by temperature changes does not mean, conversely, that a temperature change in the air above a wet surface does not affect the rate of evaporation. Changes in μ_w can be caused a change in the capacity for the volume of air to hold water at saturation (i.e., changes in $c_{aw}{}^*$). The value of $c_{aw}{}^*$ is itself a function of temperature. As a wet surface absorbs energy, the air next to the surface will warm through sensible heat exchange, and its capacity to hold water vapor ($c_{aw}*$) will increase. The relation between $c_{aw}*$ and T is often written as a derived form of the *Clausius–Clapeyron relation*:

$$\frac{dc_{aw}*}{dT} \approx \frac{\lambda_w[T]c_{aw}*}{RT^2}, \tag{10.6}$$

where $\lambda_w[T]$ accounts for the temperature dependence of the latent heat of vaporization. There are several approximations inherent in Eq. (10.6), including the assumption that water vapor acts as an ideal gas and that the specific molar volume ($m^3\,mol^{-1}\,H_2O$) of water vapor is so much larger than that of liquid water that the difference between the two can be ignored. Using calculus, and making the approximation that λ_w is constant across the narrow span of temperatures considered in plant-atmosphere exchanges, we can derive the following form of the Clausius–Clapeyron relation:

$$c_{aw}{}^*[T_2] \approx c_{aw}{}^*[T_1] \exp\left(-\frac{\lambda_w}{R\,\Delta T}\right). \tag{10.7}$$

Equation (10.7) makes clear the exponential increase in $c_{aw}*$ as air temperature increases.

Equations (10.6) and (10.7) are useful for understanding the response of atmospheric humidity and changes in μ_w as T increases. As an example, we can use Eq. (10.7) to estimate that as T increases we can expect the moisture holding capacity of air to increase by approximately 7% per degree Celsius. This example illustrates the tremendous theoretical potential underlying the Eq. (10.7). However, the Clausius–Clapeyron relation is difficult to use for routine work. A more useful empirically derived relation is often used:

$$c_{aw}{}^* = \frac{a}{P}\exp\left(\frac{bT}{T+c}\right), \tag{10.8}$$

where T is expressed in units °C; P is in Pa; and a, b, and c are empirical constants with general values being $a = 0.611$ kPa, $b = 17.502$ (unitless), and $c = 240.97$ °C (Buck 1981). A plot of the relation between $c_{aw}*$ and T using Eq. (10.8) is shown as the solid trend line in Figure 10.3, demonstrating the exponential nature of the function. When integrated across finite but small spans of T, the relation between $c_{aw}*$ and T is often simplified even further through linear approximation; in which case the slope of the relation is designated as s or Δ (see Figure 10.3). A thermodynamic derivation of the Clausius–Clapeyron relation is provided in Appendix 10.1.

10.3.1 Different measures used for expressing the abundance of atmospheric water vapor

Atmospheric humidity is expressed in a variety of ways, many of which have emerged from the meteorological research community. Some of the most commonly used terms are dew point, specific humidity, relative humidity, saturation vapor deficit, and wet-bulb temperature. The *dew point temperature* (T_d) of a volume of air is the temperature to which the air must be cooled (through loss of sensible or radiated heat) in order for the water vapor mole fraction (c_{aw}) to reach the mole fraction saturation point ($c_{aw}*$); i.e., the temperature at which

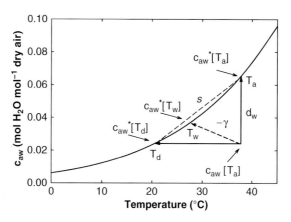

Figure 10.3 Relations among water vapor mole fraction (c_{aw}) and air temperature (T). The solid trend line represents the saturation mole fraction (c_{aw}^*) as a function of T determined using Eq. (10.8). Beginning from a parcel of air with water vapor mole fraction equal to $c_{aw}[T_a]$ at temperature T_a, a decrease in T to the dew point (represented as T_d) will intersect the saturation mole fraction trend line at $c_{aw}^*[T_d]$. An increase in c_w at the same temperature (T_a) overcomes the saturation vapor deficit (d_w), and the intercept occurs at $c_{aw}^*[T_a]$. Increasing the water vapor mole fraction through evaporation will cool the water surface, causing c_{aw} of the atmosphere to increase but the temperature of the air next to the surface to decrease. The moistened and cooled air parcel will intersect the line at $c_{aw}^*[T_a]$, where T_w is the wet-bulb temperature, s is the linearized slope of the relations between c_{aw}^* and T_a, and T_a, and γ is the psychrometric constant.

$c_{aw} = c_{aw}^*$. The dew point temperature is shown in Figure 10.3 and it is evident that c_{aw} is conserved as T is cooled to T_d. In defining the dew point, we allow for diabatic heat exchange as thermal energy is transferred from the air to the surroundings. The existence of a dew point is evident to anyone who has ventured outside in the early morning following a humid, clear night. The presence of dew on plant surfaces provides visual evidence that the surfaces cooled during the night to a temperature less than T_d. *Relative humidity* (h_r) is defined as the ratio of c_{aw} to c_{aw}^* when both are considered at the same temperature and pressure; i.e., $h_r = c_{aw}/c_{aw}^*$. At the dew point, air exists at $h_r = 1.0$ (or 100%). The *atmospheric saturation vapor deficit* (which we represent as d_w) refers to the difference between the saturation water vapor mole fraction and the actual water vapor mole fraction; i.e., $d_w = c_{aw}^* - c_{aw}$. The atmospheric saturation vapor deficit can be expressed in units of molar density (mol H_2O m^{-3}): $\rho_m c_{aw}^* - \rho_m c_{aw}$. In this case, both c_{aw} and ρ_m should be referenced to dry air, providing conserved, baseline quantities.

The *wet-bulb temperature* is the temperature a sample of air would have if it were to lose sensible heat to a wet surface, which in turn is used to evaporate water into the air, adiabatically, and cool the air to the dew point. In this case, we do not allow for diabatic heat exchange between the air and the surroundings; wet-bulb temperature is defined within the adiabatic condition. To illustrate the concept of wet-bulb temperature, consider a closed thermodynamic system consisting of a sample of air with $c_{aw} \ll c_{aw}^*$, but exposed to a surface of liquid water. Because the air is unsaturated, water will evaporate from the water surface if given an energy input. Within the adiabatic condition the energy to evaporate liquid water must come at the expense of sensible heat from the air contained within the system. We can verify this by measuring a decrease in the temperature of the air in the system; it will cool as water is evaporated from the surface. This is shown in Figure 10.3 for the case in which evaporation from a wet surface causes a parcel of air to undergo a change in state from that indicated at point A, with water vapor mole fraction defined as c_{aw}, to that indicated at point T_w, with water vapor mole fraction defined as $c_{aw}*[T_w]$. The addition of water vapor to the parcel is accompanied by a decrease in T. At T_w, the parcel will have cooled to the wet-bulb temperature (or *adiabatic dew point*). With knowledge of T_w and the original air temperature (often called the dry-bulb temperature and here indicated as T_a) we can estimate the original water vapor mole fraction of the air (c_{aw}) as:

$$c_{aw} = c_{aw}* - \gamma (T_a - T_w), \tag{10.9}$$

where γ is the *psychrometer constant* ($\gamma = c_p/\lambda_w$); thus, we only need to know the wet and dry bulb temperatures and γ in order to describe the thermodynamic state of moist air. The term "constant" is a bit of a misnomer for the case of γ, as it will vary slightly with temperature due to the temperature dependence of λ_w.

Note that the units of γ express a change in mole fraction of water vapor per unit decrease in K; the same units as those used for $dc_{aw}*/dT$, and in linear approximation, the units of γ are the same as s, the finite slope defined as $\Delta c_{aw}*/\Delta T$. The change in mole fraction of H_2O measured by γ is in fact the increase in c_{aw} required to reach $c_{aw}*$ within the adiabatic constraint. In fact, both γ and s can be used to convert a difference between T_a and T_w into a mole fraction water vapor concentration. Thus, somewhat analogous to the case for γ in Eq. (10.9), we can place s into the psychrometric context according to:

$$c_{aw}*[T_w] \approx c_{aw}*[T_a] - s(T_a - T_w), \tag{10.10}$$

where $c_{aw}*[T_w]$ refers to the saturated water vapor mole fraction at the wet-bulb temperature and $c_{aw}*[T_a]$ refers to the saturated water vapor mole fraction at the dry-bulb temperature. The approximation sign in Eq. (10.10) is required because of a simplifying convenience that we used to approximate linearity in the relation between $c_{aw}*[Ta]$ and T_a. We can relate s and γ to each other according to:

$$s \approx \frac{c_{aw}* - c_{aw}}{T_a - T_w} - \gamma. \tag{10.11}$$

10.4 Surface latent heat exchange and the combination equation

Two requirements must be met to permit continued evaporation. There must be a supply of energy to provide the latent heat of vaporization, and there must be some mechanism for removing the vapour, i.e., there must be a sink for the vapour. Analytical attacks on the problem start from one of these two points . . . (H. L. Penman (1948))

Different and (sometimes) opposing perspectives have been used historically to discuss latent heat exchange (Jarvis and McNaughton 1986). Meteorologists often approach the concept from the thermodynamic perspective, focusing on the joules of *energy* that are exchanged. Biologists often approach the concept from the hydrologic perspective, focusing on the mass of *water* that is exchanged. As described so directly in the quote above from Howard Penman, both perspectives are required for a full understanding of evaporation – energy must be absorbed by a surface in order to drive vaporization and a deficit of water must exist in the atmosphere, providing a sink to sustain the diffusive transport of mass.

On a theoretical basis, these two requirements have been combined first in the so-called *Penman equation*, or *combination equation*, and later in the more-commonly used *Penman–Monteith equation*. Thus, as we move into the discussion of this section we move from consideration of the atmospheric state of moisture and return to the fundamental processes and states that determine E, the actual evapotranspiration flux from a surface. In order to understand the combination model of evapotranspiration, we need to travel back to the early nineteenth century at which time John Dalton published his essay on evaporation at high temperature (Dalton 1802). In that essay, Dalton stated: "The evaporating force must be universally equal to that of the temperature of the water, diminished by that already existing in the atmosphere." Basically, Dalton was proposing a quantitative relation of the form:

$$\mathrm{E} = (e_s - e_a) f[\mathrm{U}], \tag{10.12}$$

where e_s and e_a are the saturation water vapor partial pressure and actual water vapor partial pressure, respectively, and $f[\mathrm{U}]$ represents a function of wind speed (recognized as a process that carries moisture away from the evaporating surface). In deriving the Penman–Monteith equation we begin from Dalton's model and state that the evaporative rate of latent heat loss from a vegetated surface can be defined in diffusive terms as:

$$\lambda_{\mathrm{w}} \mathrm{E} = \frac{\rho_{\mathrm{m}} c_p}{\gamma} \frac{(c_{\mathrm{aw}} * [\mathrm{T_s}] - c_{\mathrm{aw}}[\mathrm{T_a}])}{r_{\mathrm{tw}}}, \tag{10.13}$$

where $c_{\mathrm{aw}}*[\mathrm{T_s}]$ is the saturation mole fraction of water vapor in the air next to a wet surface and in thermal equilibrium with the surface and r_{tw} is the total surface diffusive resistance to water vapor exchange (stomatal plus boundary layer for the vegetated fraction and boundary layer alone for the non-vegetated fraction). The difference $c_{\mathrm{aw}}*[\mathrm{T_s}] - c_{\mathrm{aw}}[\mathrm{T_a}]$ is analogous to the leaf-to-air difference in the mole fraction of water vapor between the intercellular air spaces of a leaf and the boundary layer of the leaf (i.e., Δc_{w}). Now, we begin to expand the theory to accommodate the need to consider the net radiation absorbed by a surface and write:

$$\lambda_{\mathrm{w}} \mathrm{E} = \frac{s}{s + \gamma} (\mathrm{R_n} - \mathrm{H_G}). \tag{10.14}$$

Equation (10.13) reflects an Ohm's Law diffusive analog model (see Section 6.1.5), whereas Eq. (10.14) is derived from the surface energy budget. In order to combine Eqs. (10.13) and (10.14) we need to reconcile the fact that they are connected by a temperature feedback. Evaporation from a surface decreases the temperature of the surface, which in turn influences $c_{aw}*[T_s]$. We can eliminate T_s from the equations and change the point of reference to T_a through a mathematical convenience known as the Taylor Polynomial Theorem (see the mathematical concepts supplement at http://www.cambridge.org/monson). If we assume that over a narrow range of temperatures (e.g., between T_a and T_s) we can estimate s as being linear, then we can modify Eq. (10.10) to read:

$$c_{aw}^*[T_s] \approx c_{aw}^*[T_a] - s(T_a - T_s). \tag{10.15}$$

Now, we follow several steps to combine Eqs. (10.13) and (10.14): (1) we use Eq. (10.15) as a means for substitutions involving T_s and $c_{aw}*[T_s]$; (2) we recognize the relation $\gamma^* = \gamma$ (r_{tw}/r_h), where r_h is the aerodynamic resistance to sensible heat transfer; and (3) we conduct algebraic rearrangement. In the final outcome, we derive the Penman–Monteith equation as:

$$\lambda_w E = \frac{s}{s + \gamma^*}(R_n - H_G) + \frac{\rho_m c_p (c_{aw}^*[T_a] - c_{aw}[T_a])}{(s + \gamma^*)r_h}. \tag{10.16}$$

$$\text{Term} \qquad\qquad \text{I} \qquad\qquad\qquad\qquad\qquad \text{II}$$

The parameter γ^* provides linkage between $\lambda_w E$ and the transport resistances of a vegetated surface, r_{tw} reflects the combined stomatal and boundary layer resistances of the vegetated and non-vegetated surfaces, and r_h controls the boundary layer resistance to the sensible heat flux from both types of surface. This is a simple derivation of the Penman–Monteith equation, but it at least provides the sense of "combination" originally developed by Penman, as the energy (Term I) and diffusive (Term II) forcings are *combined* to describe the driving processes that determine surface evapotranspiration rate (Figure 10.4). A more complex, thermodynamic derivation of the Penman–Monteith equation is provided in Appendix 10.2, and the isothermal form of the Penman–Monteith equation, which is deployed to develop a relationship between $\lambda_w E$ and R_n, without the intervening feedback of surface cooling, is presented in Appendix 10.3.

10.4.1 Equilibrium and imposed latent heat exchange

The Penman–Monteith equation provides a useful context for evaluating surface latent heat exchange with the combined perspectives of energy budget and diffusive constraints. Depending on the surrounding environment, the rate of latent heat exchange may be more determined by its net radiation load or by the atmospheric saturation vapor deficit. On a day with relatively still air, for example, the boundary layers next to a leaf's surfaces will be relatively deep and, to some degree, they will uncouple the leaf surface from influences of the atmospheric saturation deficit. Conversely, on a windy day leaf boundary layers may be relatively shallow, in which case the influence of atmospheric saturation deficit would be imposed on processes at the leaf surface. We can explore the relative influences of these effects in quantitative terms through modification of the Penman–Monteith equation.

We start by recalling that $\gamma^* - \gamma(r_{tw}/r_{bh})$, where r_{tw} is the total resistance to water vapor exchange. The use of γ provides a means of linking latent heat loss to the thermodynamic

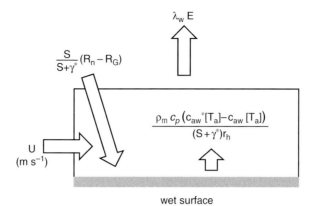

Figure 10.4 Diagrammatic representation of the principal energy budget and diffusive drivers of the surface latent heat flux (λ_wE). See text for details of the equations.

potential for surface cooling. For a hypostomatous leaf, $\gamma^* = 2\gamma(r_{tw}/r_{bh})$, and for an amphistomatous leaf $\gamma^* = \gamma(r_{tw}/r_{bh})$. Note that for a hypostomatous leaf, the resistance to heat flux (which occurs from two leaf surfaces) must be reduced by half for quantitative comparison to the resistance for transpiration (which occurs from one surface). Thus, $\gamma^*/\gamma = r_{tw}/(r_h/2)$ and $\gamma^* = 2\gamma(r_{tw}/r_h)$. For an amphistomatous leaf, the sensible and latent heat fluxes will occur from both sides of the leaf, such that $\gamma^* = \gamma(r_{tw}/r_h)$. The leaf boundary layer impedes the exchange of both latent and sensible heat, whereas the presence of stomata only impedes the exchange of latent heat. Thus, we can write $r_{tw} = r_{bw} + r_{sw}$, where r_{bw} is the boundary layer resistance to water vapor exchange and r_{sw} is the stomatal resistance to water vapor exchange. Within this context, $\gamma^* = \gamma((r_{bw} + r_{sw})/r_{bh})$.

Considering that leaf fluxes are often described in terms of conductances rather than resistances we can rewrite Eq. (10.16) to provide a form of the Penman–Monteith equation within the context of boundary layer and stomatal conductances:

$$\lambda_wE = \frac{s}{s + \gamma^*}(R_n - H_G) + \frac{\rho_m c_p(c_{wa}^*[T_a] - c_{wa}[T_a])g_{bh}}{s + \gamma^*}, \qquad (10.17)$$

where $\gamma^* = \gamma(g_{bh}/g_{tw}) = \gamma(g_{bh}(g_{bw} + g_{sw}))/g_{bw}g_{sw}$. Equation (10.17) can be evaluated at two extremes – one in which the boundary is thin and g_{bw} and $g_{bh} \rightarrow \infty$ and one in which the boundary layer is thick and g_{bw} and $g_{bh} \rightarrow 0$. We will begin with the case where g_{bw} and $g_{bh} \rightarrow 0$ due to $z_{bl} \rightarrow \infty$. In this condition, meteorological conditions at the surface are uncoupled from meteorological conditions in the atmosphere; i.e., the boundary layer acts as a form of "atmospheric insulation." We will define an additional term here, Ω_E, which represents the degree of meteorological uncoupling between the surface and atmosphere. We define perfect uncoupling as $\Omega_E = 1$ and perfect coupling as $\Omega_E = 0$; thus, when the boundary layer is thick, and g_{bw} and $g_{bh} \rightarrow 0$, we can state that $\Omega_E \rightarrow 1$. Now, consider the implications to the Penman–Monteith equation when $\Omega_E \rightarrow 1$. In that condition and assuming that $g_{bw} \approx g_{bh}$, Eq. (10.17) can be reduced to:

$$E_{eq} \approx \frac{s(R_n - H_G)}{\lambda_w (s + \gamma)} \qquad (10.18)$$

where E_{eq} represents the *equilibrium transpiration rate*. (In order to reconcile the denominator in Eq. (10.17) with the denominator in Eq. (10.18) when g_{bw} and $g_{bh} \rightarrow 0$, recall that $g_{tw} = (g_{bw} \, g_{sw})/(g_{bw} + g_{sw})$ (see Section 7.3.1), and assume that $g_{bw} \approx g_{bh}$. Solving for g_{bh}/g_{tw} within these conditions we get $g_{bh}/g_{tw} \approx 1$, and $\gamma^* \approx \gamma$.) Equation (10.18) reflects conditions in which the diabatic component of the Penman–Monteith equation exerts dominant control over latent heat exchange. The equilibrium transpiration rate is best thought of as occurring in a thick slab of homogeneous air, in contact with the leaf's surface with fixed conductances and receiving a continuous flux of net radiation. Thermodynamics requires that the energy received by the surface be partitioned to the air to an eventual point of equilibrium. Energy from absorbed radiation is transferred to the slab of air through $\lambda_w E_{eq}$. As energy is transferred to the air, its temperature must also increase driving Δc_w to increase and forcing even further increases in $\lambda_w E_{eq}$. This will continue to a point of thermodynamic equilibrium. These relations can be appreciated by rearranging Eq. (10.18) to yield $(R_n - H_G) \approx \lambda_w E_{eq}((s + \gamma)/s)$. *Thus, as the leaf boundary layer becomes thick, radiation load takes over as the primary control over E and control by stomata diminishes.* Equation (10.18) can be interpreted as reflecting a state of "meteorological control" over transpiration, as opposed to "physiological control." It is most relevant in conditions of high humidity and relatively still air. Obviously, stomata still control transpiration to the extent that water vapor diffuses through the stomatal pores into the boundary layer. The distinction implicit in the term "meteorological control" is that dynamics in stomatal conductance are not reflected in dynamics in transpiration rate, and stomatal conductance can be characterized by a negligible flux control coefficient.

Now consider the case where g_{bw} and $g_{bh} \rightarrow \infty$, because $z_{bl} \rightarrow 0$. We continue with the assumption that $g_{bh} \approx g_{bw}$, and apply some algebra, such that Eq. (10.16) reduces to:

$$E_{imp} \approx \frac{\rho_m c_p (c_{wa}{}^*[T_a] - c_{wa}[T_a]) g_{sw}}{\lambda_w \gamma} \approx \frac{\rho_m c_p d_w g_{sw}}{\lambda_w \gamma}, \qquad (10.19)$$

where E_{imp} is used to represent the *imposed transpiration rate* and d_w represents the water vapor mole fraction deficit of the atmosphere. (Note that d_w is different than Δc_w, as the former is the difference between potential and actual water vapor concentration of the air at its existing air temperature and the latter is the difference between the water vapor concentrations of a surface and the air above the surface.) During conditions in which the boundary layer is thin and g_{bw} is high, the evaporative conditions of the bulk atmosphere will be imposed on the surface. Under these conditions the partitioning of absorbed radiation to sensible thermal energy loss is so efficient that the importance of radiation load as a control over E diminishes. (Evidence that control by the diabatic component of the Penman–Monteith model is negligible in the condition of $z_{bl} \rightarrow 0$, is seen mathematically as the $g_{bh}{}^2$ term, which is implicit in the denominator of $s(R_n - H_G)/(s + \gamma^*)$, drives down the value of this relation as $g_{bh} \rightarrow \infty$.) Given that g_{sw} represents the primary constraint on E, Eq. (10.19) reflects a state of "physiological control" over $\lambda_w E$.

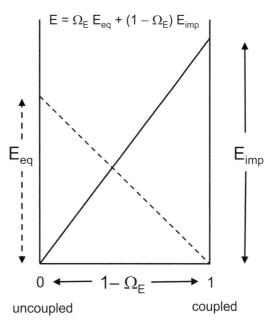

$$E = \Omega_E \, E_{eq} + (1 - \Omega_E) \, E_{imp}$$

E_{eq}

E_{imp}

$$0 \longleftarrow 1 - \Omega_E \longrightarrow 1$$

uncoupled   coupled

Figure 10.5 Equilibrium (E_{eq}) and imposed (E_{imp}) transpiration rates in relation to the degree of coupling between the air next to the leaf surface and the bulk atmosphere. The degree of uncoupling is defined as Ω_E, which varies between 0 (perfect coupling) and 1 (perfect uncoupling). The total transpiration rate from a leaf (E) can be represented as $E = \Omega_E \, E_{eq} + (1 - \Omega_E) \, E_{imp}$. As coupling between the leaf and air increases, E converges with E_{imp}, which is shown by the solid line. As coupling between the leaf and air decreases E converges with E_{eq}, which is shown by the dashed line. Redrawn from a concept presented by Jarvis and McNaughton (1986).

Leaves typically function at a state between the extremes of meteorological and physiological control, and theory has been developed to define a quantitative index for this state (Figure 10.5). The relative controls exerted by E_{eq} and E_{imp} will trade off against each other, depending on the degree of uncoupling between the surface and atmosphere. When the boundary layer of the leaf is deep, meteorological conditions at the leaf surface will be uncoupled from those in the atmosphere (the "coupling factor," Ω_E, will approach 0) and the leaf transpiration rate (E) will converge with E_{eq}. When the boundary layer is shallow, meteorological conditions at the leaf surface will be coupled to those in the atmosphere (the "coupling factor," Ω_E, will approach 1) and the leaf transpiration rate (E) will converge with E_{imp}.

Appendix 10.1 Derivation of the Clausius–Clapeyron relation

The Clausius–Clapeyron relation is derived from the initial assumption that the liquid and vapor phases of a compound are at equilibrium. We will derive the relation for water. Assuming that the internal energy of the system does not differ between the two phases (i.e., $\Delta U = 0$), we can write (see Section 2.1):

$$dG = V d p_w - S dT, \tag{10.20}$$

where G is Gibbs free energy (J mol^{-1}), V is specific molar volume (m^3 mol^{-1} H$_2$O), p$_w$ is the partial pressure of water (N m^{-2}), S is entropy (J mol^{-1} K^{-1}), and T is temperature (K), all referenced to the liquid or vapor phases. Equation (10.20) is written to resolve the difference in free energy at any transition between liquid and vapor phases. Given the condition of equilibrium, we can state that $dG = 0$ and we can write:

$$(V_1 - V_2)d p_w - (S_1 - S_2)dT = 0, \tag{10.21}$$

where V_1 and S$_1$ are state conditions for one of the phases and V_2 and S$_2$ are state conditions for the other phase. Rearranging, we obtain:

$$\frac{d p_w}{dT} = \frac{(S_1 - S_2)}{(V_1 - V_2)}. \tag{10.22}$$

We can define the latent heat of vaporization ($\lambda_w[T]$) as the temperature dependent amount of heat (J mol^{-1}) required to account for the change in entropy as liquid water evaporates. Thus, $\lambda_w = T(S_1 - S_2)$, and we can write:

$$\frac{d p_w}{dT} = \frac{\lambda_w[T]}{T(V_1 - V_2)}. \tag{10.23}$$

Noting that the mole fractions ($c_{aw}*$) and partial pressures of water in air in the saturated vapor phase are related by $c_{aw}*P = p_w$, where P is total atmospheric pressure, we can write:

$$\frac{d c_{aw}*}{dT} = \frac{\lambda_w[T]}{TP\Delta V}. \tag{10.24}$$

Equation (10.24) is the conventional form of the Clausius–Clapeyron relation, but expressed in terms of mole fraction rather than partial pressure.

Returning to Eq. (10.23), and stating some approximations, we can derive an alternative form of the Clausius–Clapeyron relation. We begin by assuming that the specific volume of the vapor phase (V_v) is so much larger than that of the liquid phase (V_l) that:

$$\frac{d p_w}{dT} \approx \frac{\lambda_w[T]}{T V_v}. \tag{10.25}$$

Assuming that water vapor in air acts as an ideal gas, we can write:

$$\frac{d c_{aw}*}{dT} \approx \frac{\lambda_w[T]}{RPT^2} p_w \approx \frac{\lambda_w[T]}{R} \frac{c_{aw}*}{T^2}, \tag{10.26}$$

and thus:

$$\frac{d c_{aw}*}{c_{aw}*} \approx \frac{\lambda_w[T]}{R} \frac{dT}{T^2}. \tag{10.27}$$

If, for the purposes of mathematical derivation we assume that λ_w is constant as a function of T, we can integrate Eq. (10.27) for two endpoints – from $c_{aw}*[T_1]$ and T_1 to $c_{aw}*[T_2]$ and T_2 – which refer to the saturation mole fractions at air temperatures T_1 and T_2, respectively. Thus:

$$\int\limits_{c_{aw}*[T_2]}^{c_{aw}*[T_1]} \frac{dc_{aw}*}{c_{aw}*} \approx \int\limits_{T_1}^{T_2} \frac{\lambda_w}{R\,T^2}\,dT, \tag{10.28}$$

which resolves to:

$$\ln\frac{c_{aw}{}^*[T_1]}{c_{aw}{}^*[T_2]} \approx -\frac{\lambda_w}{R}\left(\frac{1}{T_2}-\frac{1}{T_1}\right), \tag{10.29}$$

and can be rewritten as:

$$c_{aw}{}^*[T_2] \approx c_{aw}{}^*[T_1]\exp\left(-\frac{\lambda_w}{R\,\Delta T}\right). \tag{10.30}$$

Equation (10.30) shows the fundamental *exponential relation between saturation water vapor concentration and air temperature* that emerges from the Clausius–Clapeyron relation.

Appendix 10.2 A thermodynamic approach to derivation of the Penman–Monteith equation

We begin by defining a thermodynamic system containing a wet surface coupled to an overlying volume of air with unsaturated vapor pressure and with imaginary boundaries that distinguish the system from the surroundings. The air above the surface can be defined according to a set of state variables including pressure, volume, mass, temperature, and latent heat content. In relating these variables to one another we can draw a graph showing the potential saturation water vapor mole fraction ($c_{aw}*$) as a function of temperature (Figure 10.6), and we can define the molar sensible heat content and molar latent heat content of the air as $c_p T_a$ and $\lambda_w c_{aw}$, respectively.

Let's designate the initial state of our hypothetical control volume as point A on the graph in Figure 10.6. The disequilibrium that exists because the air exists in an unsaturated state can be defined by line AB. Line AB intersects the saturation curve at an angle because in the adiabatic state (lack of energy input from the surroundings), the energy to increase the latent heat content of the air would have to come at the expense of its sensible heat content – sensible heat must be transferred to the wet surface in order to evaporate water. *This means that the air in the control volume must cool as its humidity increases.* Thus, we can state: $\lambda_w E = -H_{se}$, and therefore the *total amount* of sensible heat that would need to be expended to evaporate water and change the air from point A to point B is $c_p(T_a - T_w)$, where c_p is the

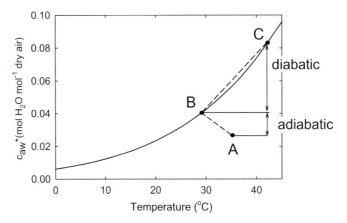

Figure 10.6   Partitioning of the increase in saturated water vapor mole fraction (c_{aw}*) as a function of air temperature into fractions due to adiabatic and diabatic energy exchange above a wet, evaporating surface. Developed from a concept presented in Monteith (1981).

specific heat of air and T_w is the wet-bulb temperature. The flux of sensible heat (H_{se}) to the wet surface is defined as:

$$H_{se} = \frac{\rho_m c_p (T_a - T_s)}{r_h} = -\lambda_w E, \tag{10.31}$$

where H_{se} is in units $J\ m^{-2}\ s^{-1}$, ρ_m is in units mol dry air m^{-3}, c_p is in $J\ mol^{-1}$ dry air K^{-1}, and r_h is the resistance to sensible heat transfer between the air and surface with units $s\ m^{-1}$. The flux for H_{se} described in Eq. (10.31) is constrained by the limit imposed when $T_a = T_w = T_s$, where T_s is the surface temperature.

Now, we can call upon Eq. (10.9) which states: $c_{aw}[T_s] = c_{aw}*[T_a] - \gamma(T_a - T_w)$, to define the latent heat transfer required to change air from point A to point B. Thus, we can write:

$$\lambda_w E = \frac{\rho_m c_p (c_{aw}*[T_s] - c_{aw}[T_a])}{\gamma\ r_{tw}}, \tag{10.32}$$

where r_{tw} is the resistance for water vapor moving between the surface and the atmosphere. Recalling that $\gamma = c_p/\lambda_w$, we can relate r_{tw} to r_h according to $\gamma* = \gamma(r_w/r_h)$. Combining Eqs. (10.31) and (10.32) we obtain:

$$-\frac{\rho_m c_p (c_{aw}*[T_s] - c_{aw}[T_a])}{\gamma\ r_w} = \frac{\rho_m c_p (T_a - T_s)}{r_h}. \tag{10.33}$$

At this point, our analysis has been complicated by the fact that we have introduced two reference potentials for temperature into the equations – the surface and the air. Complications arise because both sensible heat transfer to the surface and latent heat transfer from the surface cause T_s to change – a coupled feedback that is difficult to resolve mathematically. It would simplify matters if we could eliminate T_s and express the equations solely in terms of T_a. If we rely on the mathematical convenience provided by the Taylor Expansion Theorem to make the assumption that over the narrow range of temperatures between T_a and T_s we can estimate the slope of the relation between $c_{wa}{}^*[T_a]$ and T_a (which we represent as s) as being linear, we can write:

$$c_{aw}{}^*[T_s] \approx c_{aw}{}^*[T_a] - s(T_a - T_s), \qquad (10.34)$$

where $c_{aw}{}^*(T_s)$ is the saturated water vapor mole fraction at surface temperature. Following substitution of terms and algebraic rearrangement we can write:

$$\lambda_w E = \frac{\rho_m c_p (c_{aw}{}^*[T_a] - c_{aw}[T_a])}{(s + \gamma^*)r_h}. \qquad (10.35)$$

Equation (10.35) represents the *adiabatic component* of the coupling between latent heat exchange and surface energy budgets. Once again, the adiabatic latent heat flux is limited to the point where $T_a = T_w$. If we now relax the adiabatic assumption and allow the system to exchange energy with the surroundings, it is possible for the air to achieve temperatures higher than in the original unsaturated state and for the water vapor saturation potential to increase accordingly. In this case, the fluxes of energy from the surroundings and associated latent heat flux would be defined as *diabatic processes*. The diabatic potential can be visualized in Figure 10.6 in which the state of the air moves from point A to point C. We can describe the transition from point A to point C as the sum of adiabatic and diabatic transitions described by lines AB and BC.

Diabatic energy absorption will cause an increase in air temperature represented here as ΔT, and the change in saturated water vapor mole fraction can be represented as $\Delta c_{aw}{}^*$. Recalling Eq. (10.15), we can write: $\Delta c_{aw}{}^* = s\Delta T$. Using this relation in Eq. (10.35) we can write:

$$\lambda_w E = \frac{\rho_m c_p (s\Delta T)}{\gamma^* r_h}. \qquad (10.36)$$

Now, let us reconcile the various components of the theory. An energy balance relation for the surface (assuming no net energy storage) can be written as:

$$R_n = \lambda_w E + H_{se} + H_G. \qquad (10.37)$$

Rearranging Eq. (10.37) to solve for $\lambda_w E$, rearranging Eq. (10.36) to solve for ΔT and using this solution for substitution into Eq. (10.37) to redefine H_{se}, we obtain:

$$\lambda_w E = R_n - (H_G + H_{se}) = R_n - \left(H_G + \frac{\rho_m\, c_p (\Delta T)}{r_h} \right) = R_n - \left(H_G + \frac{\lambda_w E \gamma^*}{s} \right). \qquad (10.38)$$

After algebraic rearrangement we can write:

$$\lambda_w E = \frac{s}{s + \gamma^*} (R_n - H_G). \qquad (10.39)$$

Equation (10.39) represents the *diabatic component* of latent heat exchange. When combined with our previous expression for the adiabatic component we obtain:

$$\lambda_w E = \underbrace{\frac{s}{s + \gamma^*}(R_n - H_G)}_{\text{Term I}} + \underbrace{\frac{\rho_m c_p(c_{aw}{}^*[T_a] - c_{aw}[T_a])}{(s + \gamma^*)r_h}}_{\text{Term II}}, \qquad (10.40)$$

where Term I represents the diabatic component and Term II represents the adiabatic component. Equation (10.40) is the Penman–Monteith equation in final form, although sometimes it is written as:

$$\lambda_w E = \frac{s(R_n - H_G) + \rho_m c_p(c_{aw}{}^*[T_a] - c_{aw}[T_a])/r_h}{s + \gamma \dfrac{r_{tw}}{r_h}}, \qquad (10.41)$$

which has the advantage of specifying both the water vapor and sensible heat resistances. *The Penman–Monteith equation carries the condition of a balanced energy budget.*

Appendix 10.3 Derivation of the isothermal form of the Penman–Monteith equation

The Penman–Monteith model as written in Eqs. (10.40) and (10.41) contains an inherent feedback known as *radiative coupling*; as water is evaporated from the surface, the surface will cool. A cooler surface will radiate less longwave energy resulting in an increase in absorbed R_n, and thus an increase in $\lambda_w E$. This is a form of positive feedback on $\lambda_w E$. To eliminate the effect of radiative coupling the Penman–Monteith equation is often expressed in its *isothermal form* – i.e., defined in terms of equality between surface and air temperatures.

In deriving the isothermal form we recognize that *isothermal net radiation* ($R_n{}^*$) is defined as the net radiation that would be absorbed by a surface *if it existed at air temperature*. With respect to the surface energy budget this condition requires revision of the Stefan–Boltzmann radiant emission term ($\varepsilon_L \sigma T_s^4$), where ε_L is the fractional longwave energy emittance of the surface. The relationship between T_s and T_a is non-linear, meaning that a simple, first-order determination of T_a from T_s is not possible. Recall that we faced similar mathematical difficulty in our original derivation of the Penman–Monteith equation when we sought to remove surface temperature (T_s) as a determinant of the water vapor concentration deficit. In that case, we used the Taylor Polynomial Theorem to estimate a linear relation between the saturation water vapor mole fraction ($c_{wa}{}^*$) and air temperature (T_a). Now, we will formalize this procedure by noting the general form of a Taylor expansion series as the first-degree polynomial: $f(x) = f(c) + f'(c)(x - c)$. Letting $T_s = x$, and $T_a = c$, and recognizing from calculus that $dx^m = mx^{(m-1)}$, we can restate the relationship between surface temperature and radiant energy exchange as:

$$\varepsilon_L \sigma \, T_s^4 = \varepsilon_L \, \sigma \, T_a^4 + 4\varepsilon_L \, \sigma T_a^3 (T_s - T_a), \tag{10.42}$$

once again assuming that $T_s - T_a$ is relatively small. If $R_n{}^*$ is defined as:

$$R_n{}^* = R_S(1 - \alpha) + R_L - \varepsilon_L \sigma T_a^4. \tag{10.43}$$

then the relationship between $R_n{}^*$ and R_n can be expressed as:

$$R_n = R_n{}^* - 4\varepsilon_L \sigma T_a^3 (T_s - T_a). \tag{10.44}$$

While some simplification has been accomplished in deriving Eq. (10.44) it still contains a term for surface temperature; a fact with which we will need to reckon in the final solution.

Turning now to a path toward factoring the surface radiation-temperature feedback out of the original Penman–Monteith equation, we can write an isothermal analog of the energy budget equation ($R_n = \lambda_w E + H_{se} + H_G$) as:

$$R_n{}^* = \lambda_w E + \frac{(\rho_m c_p \Delta T \,)}{r_h} + H_G + 4\varepsilon_L \sigma T_a^3 \Delta T. \tag{10.45}$$

To simplify further algebraic manipulation, a term can be defined for the effective resistance of longwave radiation emission (r_r):

$$r_r = \frac{\rho_m c_p}{4\varepsilon_L \sigma T_a^3}. \tag{10.46}$$

The term r_r is not a resistance in the same sense as has been used to define transport resistances. However, it satisfies the same conceptual aim of a "resistance" in that it defines a mitigating factor that opposes a flux. Note that r_r carries units of s m^{-1}, which is consistent with the units for r_h. Recognizing that $1/r_h = (1/r_h + 1/r_r)$, we can write an isothermal form of the Penman–Monteith equation as:

$$\lambda_w E = \frac{s(R_n{}^* - H_G) + \rho_m c_p (c_{aw}{}^* \, [T_a] - c_{aw}[T_a])(r_h^{-1} + r_r^{-1})}{s + \gamma \dfrac{r_{tw}}{(r_h + \ r_r)}}. \tag{10.47}$$

This isothermal form of the Penman–Monteith equation retains an influence of surface cooling on H_G, the conductive flux of sensible heat to deeper layers. In cases where H_G is significant, this influence must be handled by independent accommodation of the altered temperature gradient between the surface and deeper soil layers.

11

Canopy structure and radiative transfer

A quantitative study of the architecture of both a stand and individual plant with its organs constitutes a fundamental task of phytometry.

. . . solar radiation in a plant stand is a highly complicated process dependent both on incident radiation and on the optical and geometrical properties of the vegetation.

For this reason a more specified and generalized concept of a turbid medium was proposed, according to which a stand was treated as a plate turbid anisotropic medium homogeneous in horizontal plane in terms of statistics.

Juhan Ross (1981)

The three separate excerpts quoted above from the classic book on plant canopies and their radiation regime by the biophysicist, Juhan Ross, underscore three aspects of past studies of the solar radiant flux at the canopy scale. First, description of the architecture of plant canopies is a central activity of researchers interested in controls over plant productivity and its relation to climate. Second, the relation of canopy architecture to the distribution of solar radiation within the canopy is complex. Third, in large part due to that complexity, researchers have sought ways to simplify descriptions of canopy structure and radiative transfer using statistical models. In this chapter we will use these three fundamental tenets as the context within which to explore canopy structure and its relation to the capture of solar radiation.

Leaves do not carry out their physiological functions as isolated units. Rather, they interact with the tens to hundreds of other leaves that surround them in a canopy. Much of that interaction is determined by the architectural placements of leaves on branches and, in turn, branches on stems. Each leaf absorbs, transmits, and reflects solar radiation, changes its position frequently as it is buffeted by the wind, is shaded from direct sunlight by leaves around it, and is subjected to the transfer of thermal, radiant energy from its canopy surroundings. The cumulative position and activity *of all leaves* creates a unique canopy microenvironment that feeds back to modify the biophysical activities *of each leaf*. This highly coupled system, involving primary exchanges and secondary feedbacks, results in matrices of cause-and-effect that challenge our ability to isolate controlling variables and mathematically predict precise spatiotemporal patterns of canopy-atmosphere exchange. Canopy attributes such as leaf density distribution, leaf orientation in relation to the sun, and the clumping of leaves on branches, branches within crowns, and crowns within canopies, must be appreciated before we can fully understand and predict how canopy architecture influences the exchange of mass and energy of any single canopy layer, let alone any single leaf. Studies of these attributes constitute what Juhan Ross referred to as the field of "phytometrics"; description of the placement, growth, and senescence of leaves and the influence of those dynamics on the shape, size, and geometry of the canopy.

As a means of mathematical convenience canopies are often characterized as "turbid" volumes – that is, semi-transparent volumes occupied by small, intercepting elements that absorb or redirect penetrating photons. The analogy between a turbid medium and a canopy is not ideal because leaf surfaces tend to be large compared to the path lengths at which photons are transmitted, making it difficult to apply probabilistic theories that organize a unit of canopy volume according to mean photon densities, which is possible and valid in true turbid, photometric media. With caution noted, however, this is exactly what is done – a canopy is treated as a volume analogous to a true turbid medium. By accepting the turbid medium analogy, even after noting its shortcomings, researchers have gained a conceptual foundation from which to develop the rudimentary framework for organizing the canopy radiation regime into the spatiotemporal averages required to drive models of canopy-atmosphere interactions.

In this chapter we explore the complexities of canopy structure and its influence on the canopy radiation regime. We begin with a discussion of canopy architecture; with special emphasis on the placement and orientation of leaves, the relation between leaf placement and the absorption, reflection, or transmission of solar photons, and the nature of solar radiation gradients. We then proceed to consideration of canopy radiative transfer models and the complexities of dealing with diffuse and direct radiation fields. Finally, we consider recent efforts to expand the interaction between vegetation and solar photon interception up to the global scale using satellite and aircraft remote sensing.

11.1 The structure of canopies

Leaf size, shape, and orientation, as well as the density and placement of branches and the height of trees all affect the ways that wind and light move through a canopy, and thus the ways that mass and energy are transferred. Clumps of leaves and branches, for example, create gaps in the canopy that increase photon transfer to lower layers. Canopy height will affect the effectiveness of photon capture; tall canopies trap photons more effectively than shorter canopies, so tall canopies are optically "darker." This principle underlies the differential influences of forests and grasslands that were discussed with regard to albedo and surface energy balance in the last chapter. The combined optical properties of leaves, and their distribution among vertical canopy layers, ultimately affect how much radiation is intercepted and absorbed by a canopy.

11.1.1 Leaf area index

Leaf area index (often called LAI; in this text we will also refer to it as L in the equations that we present) is the leaf area per unit of vegetated ground area (m^2 leaf area m^{-2} ground area). LAI is typically defined in two dimensions (x and y). However, it is formally derived from the three-dimensional (x, y, and z) concept of *leaf area density*, L[z], which is the amount of leaf area per unit ground area integrated with respect to vertical distance. The two-dimensional leaf area index, L, is therefore formally related to leaf area density, L[z], as:

$$L = \int_0^{z=h} L[z]dz \approx \sum_0^{z=h} L[z]\Delta z, \qquad (11.1)$$

$$\underset{\text{I}}{} \qquad \underset{\text{II}}{} \qquad \underset{\text{III}}{}$$

where $L[z]$ is expressed in m^2 (leaf area) m^{-3} canopy volume, L is expressed in m^2 (leaf area) m^{-2} ground area, and z is expressed in m. Term II provides the definition most often used in theoretical work and it is defined within the limit as $z \to 0$; this allows us to apply L to discrete, infinitesimally thin layers within the canopy. Term III provides the definition most often used in observational work wherein Δz reflects a finite height within the canopy. L is often reported for an entire canopy, in which case h is the integral or summation limit and is equal to mean canopy height. LAI is typically expressed as the *one-sided leaf area* per unit ground area. For conifers, which have cylindrical or angular needles, LAI is often expressed as the projected area of the needles, thus providing an equivalent one-sided area, or the total surface area, which requires biometric consideration needle geometry (Chen and Black 1992).

Different plant types will influence the structure of a canopy in different ways, including variable influences on the amount and distribution of LAI. Thus, a broad range of LAI values exist for ecosystems (Table 11.1). From the data presented in Table 11.1 we notice that the observed range in LAI has an upper limit; no canopy type among those surveyed exhibits a mean LAI exceeding 10 (or maximum LAI of 15 taking into account the standard deviations). In fact, most ecosystems have LAI less than 7, with the exception being tree plantations. The upper limit for LAI is controlled by tradeoffs between the requirement for positive plant carbon balance and maximum absorption of PPFD and/or between the

Table 11.1 Global survey of leaf area index of landscape classes (Asner *et al.* 2003)

Ecosystem type	Mean LAI	Standard deviation
Polar desert/alpine tundra	3.85	2.37
Moist tundra	0.82	0.47
Boreal forest woodland	3.11	2.28
Temperate savanna	1.37	0.83
Temperate evergreen broadleaved forest	5.40	2.32
Temperate mixed forest	5.26	2.88
Temperate conifer forest	6.91	5.85
Temperate deciduous forest	5.30	1.96
Temperate wetland	6.66	2.41
Temperate cropland	4.36	3.71
Temperate tree plantations	9.19	4.51
Tall medium grassland	2.03	5.79
Short grassland	2.53	0.32
Arid shrubland	1.88	0.74
Mediterranean shrubland	1.71	0.76
Tropical wetland	4.95	0.28
Tropical savanna	1.81	1.81

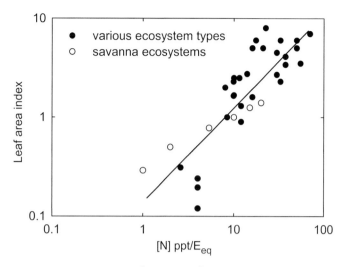

Correlation among leaf area index (expressed as m^2 leaf area m^{-2} ground area) and a dimensionless composite index of climate and leaf N concentration (obtained as the product of the mean leaf N concentration (g N g^{-1} dry mass) and annual mean precipitation (mm) divided by the annual equilibrium evaporation (mm) for a site). Redrawn from Baldocchi and Meyers (1998).

requirement for whole-plant transpiration rate to be balanced by soil water uptake. With regard to plant carbon balance, there must be a theoretical upper limit of LAI beyond which the penetration of solar energy to the lowest layers is not sufficient to sustain a positive leaf carbon budget. With regard to water balance, inequities between transpiration and soil water acquisition can cause water stress in the plant, potentially accompanied by embolized xylem columns and loss of hydraulic conductivity. Thus, the avoidance of extreme water or carbon stress requires balance between a plant's leaf area and the potential for photon or water transport to that leaf area.

A significant correlation exists between LAI and a synthetic mathematical index that combines precipitation, potential evapotranspiration, and leaf nitrogen content when assessed across a broad range of ecosystem types (Figure 11.1). The index calculated for Figure 11.1 reflects inherent physiological and biophysical correlations among LAI, net primary productivity, and climate (Monteith 1972, Budyko 1974, Sinclair *et al.* 1984, Woodward 1987, Scheffer *et al.* 2005). For example, as leaf N concentration increases, photosynthetic capacity and net primary production should increase, supporting higher LAI. The ratio ppt/E_{eq} is a reflection of the water available to drive primary productivity, with increases in the ratio potentially supporting higher LAI. The use of E_{eq} as a component of the synthetic index provides a means of connecting water availability with net radiation. Higher values of E_{eq} reflect the potential for higher net radiation loads to reduce soil moisture availability, and thus impose negative influences on potential LAI.

The LAI of a canopy is not static across time or space. It can vary over the course of a year and among different years due to variable plant phenology, plant ontogeny, environmental stress, herbivory, pathogen outbreaks, or weather. Annual species, including grasses, herbs,

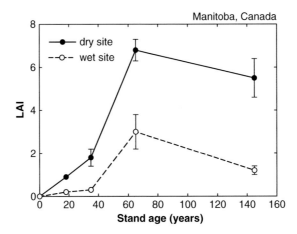

Figure 11.2   Development of leaf area index (LAI) across a chronosequence of *Picea mariana* (black spruce) boreal forest in Canada. Data are taken from dry upland sites and wet lowland sites. Redrawn from Bond-Lamberty *et al.* (2004).

and crops, must grow from a seed using limited stores of nutrient reserves. Thus, depending on growth rate annual plants may develop their leaf area more slowly compared to perennial plants. Deciduous perennial species have the potential to exhibit rapid leaf area expansion early in the growing season as they draw on larger pools of nutrients stored in stems, roots, and tree boles.

In forests, the distribution of LAI within the canopy can change as the ecosystem develops over the span of years. Observations using chronosequences are especially useful in reconstructing long-term changes in ecosystem structure. A *chronosequence* is a set of landscapes that differ in age relative to some past disturbance. With access to a chronosequence we can "trade time for space" and infer the sequential changes that would occur in an ecosystem if given enough time. In Figure 11.2, a chronosequence within the boreal biome of Canada has been used to infer increases in LAI that would occur in a single forest if it were to develop from the time of complete devastation due to fire. LAI is inferred to increase to a maximum sometime after 60 years, beyond which it decreases again. The eventual decrease in LAI is probably due to increasing tree mortality as the stand ages.

In addition to changes in total LAI, the distribution of LAI within the canopy changes over time (Figure 11.3). We can characterize the earliest successional stages of a forest as relatively open, with the canopy composed of herbaceous "pioneer" species and woody tree seedlings. There is often little vertical stratification to the LAI at this stage, and the total LAI is low compared to later stages. As time passes, established tree saplings will grow and leaves within the canopy will become progressively more light-limited; competition among neighbors will be high during these stages. Thinning rates of tree densities are high during this stage. The leaf area profile is top heavy during the stage of most intense competition and gaps in the canopy are few. Thus, the LAI of the stand often reaches a successional maximum during the late phases of the "self-thinning" stage. As the forest community and stand structure continue to develop, and tree densities begin to stabilize, the distribution

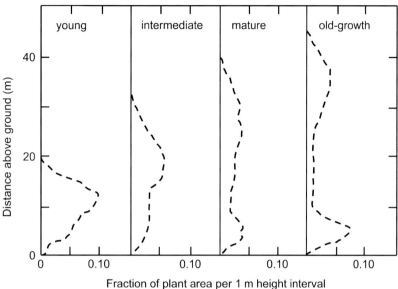

Figure 11.3 Vertical profiles of forest tree surface areas in four successional stages of a deciduous forest in the southeastern USA (Maryland). The "old-growth" forest is ultimately dominated by *Liriodendron tulipfera* (tulip poplar). Observations were made using a scanning lidar instrument (see Box 11.1). Redrawn from Harding *et al.* (2001), with permission from Elsevier B.V. Publishers.

of LAI will shift downward. This stage may also be characterized by establishment of an understory, shifting the distribution of LAI even further downward. In the final stages of forest development, species turnover may occur as "pioneer" species are not able to regenerate in the deep shade of the understory, and species with greater shade tolerance increase in importance. The vertical distribution of LAI may shift to become "bottom heavy" during this final phase, or to reflect a bifurcated distribution. Of course, this is an idealized description of the structural stages through which a forest progresses. The actual trajectory will depend on the particular species involved and the occurrence or non-occurrence of past landscape disturbances.

At a particular point in time, the distribution of LAI can also vary across space. In natural ecosystems the spatial distribution of plants is controlled by competitive relations, with greater distance among individuals occurring in ecosystems with more limited soil resources (e.g., water and/or nutrients). In some desert ecosystems the distance among plants can be large, creating "islands" of high LAI; such canopies are relatively "open" with high gap fractions. In ecosystems with limited light the horizontal spacing of plants is determined by crown diameter and architecture, but it is possible for the canopy to develop a "closed" form with a low gap fraction. In limited light, it is the vertical, rather than horizontal, distribution of LAI that is most affected. In planted crops the canopy is ordered in regular rows with bands of alternating high and low LAI. In terms of modeling the radiative regime of a canopy, the closed homogeneous form of forest canopies with low gap fractions are the simplest; greater challenges occur with spatially dispersed canopies.

11.2 The solar radiation regime of canopies

Solar photon flux density decreases as it passes through a canopy. At any given layer the photon flux is measured as the sum of two components. The *diffuse photon flux density* (I_d) is characterized by photons that have been scattered in the atmosphere or by leaves and stems. The incident diffuse photon flux from the space above the canopy is often referred to as *skylight*. The diffuse photon flux density within the canopy is the sum of the photon flux from skylight, as well as photons scattered during reflections from surfaces within the canopy. The *direct photon flux density* (I_D) is derived from solar rays that penetrate gaps in the canopy and form sun patches and sunflecks.

As long as the complete solar disc is visible from a sun patch or sunfleck, I_D is not reduced with increased depth in the canopy (Figure 11.4). The bright spots that make up sun patches and flecks are formally referred to as *numbras*. If rays of solar photons were transmitted through a canopy as truly collimated streams, the diameter of numbras should be the same as the diameter of the transmitting gaps (assuming circular gaps for simplicity). In fact, rays of solar photons are not collimated. The sun is not an ideal point source of radiation. Because of its large size the angular diameter of the sun when viewed from the earth is approximately

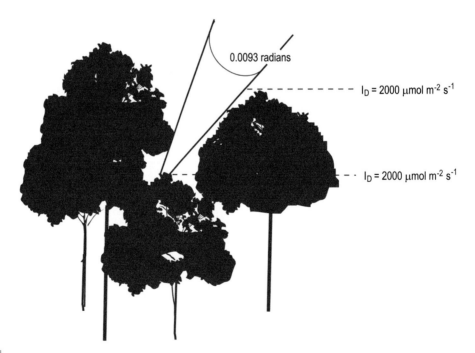

Figure 11.4   Representation of the transmission of full solar intensity through a canopy gap when the solar disc is not obstructed by canopy elements. Note that I_D does not decrease with depth through the canopy and that the rays of I_D are transmitted from the solar source with an angular diameter of 0.0093 radians; i.e., the sun is not an ideal point source and the rays of I_D are not ideally collimated.

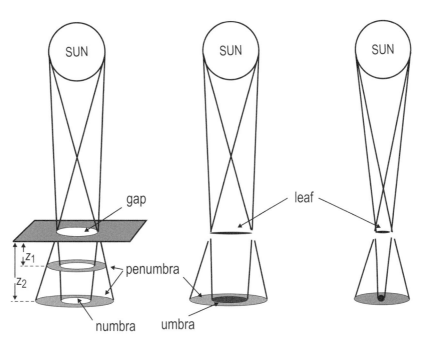

Figure 11.5 Penumbral spreading within canopies. **Left**. When solar rays pass through a canopy gap a bright spot (the numbra) is created where rays from the opposite sides of the sun overlap. At the border of the numbra a partial shadow is created (the penumbra) due to incident rays from only one side of the sun. The distance from the gap to the incident surfaces is represented as z_1 and z_2. **Middle**. When solar rays are partly blocked by a leaf, a full shadow (umbra) is created directly beneath the leaf and a penumbra is created at the border. **Right**. A smaller leaf will create a smaller umbra on an incident surface compared to a larger leaf at the same distance. The ratio of umbral to penumbral diameter will be greater for shadows created by a smaller leaf. Redrawn and adapted from Smith *et al.* (1989).

$0.53°$ (or 0.0093 radians). Consequently, the size of a numbra projected through a canopy gap will be less than the diameter of the gap and part of the photon flux will be cast as partial shade, visible as "fuzziness," at the edge of the numbra (Figure 11.5). This region of partial shade is called the *penumbra*. The diameter of the numbras will decrease and the width of the penumbras will increase as distance from a gap increases.

In the case of solar rays striking a leaf, rather than a gap, a dark shadow is cast into the canopy. The darkest part of the shadow is known as the *umbra*. The edges of umbras will also blend into regions of partial shade, or penumbras, similar to the case for numbras. Smaller leaves will cause smaller diameter umbras to be cast into the canopy, with proportionally more penumbral shade (Figure 11.5). In fact, penumbral shade is most frequently observed in canopies with small leaves (Denholm 1981).

The exact dimensions of numbras, umbras, and penumbras are difficult to define due to the gradual and continuous transition from one to another. However, penumbral width, in general, scales linearly with distance beneath a leaf (Figure 11.6). Miller and Norman (1971) developed a relation to define the half-width of penumbras as:

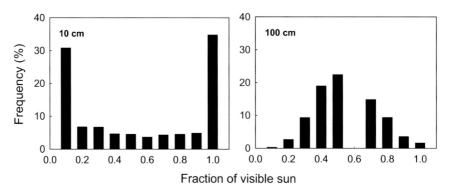

Figure 11.6 The frequency distribution of the fraction of the sun that is visible from a point 10 cm or 100 cm below a Scotch pine shoot. These data were obtained from a simulation model. The data illustrate that at a short distance below the shoot (10 cm) the sun is completely visible (resulting in a numbral sunfleck) or completely blocked (resulting in umbral shade) most of the time. At a greater distance (100 cm), the sun is only partly blocked (resulting in penumbral shade) most of the time. Redrawn from Stenberg (1995), with permission from Elsevier B.V. Publishers.

$$r_p = \frac{\alpha_{\text{sun}} z}{\cos \theta}, \tag{11.2}$$

where r_p is the radial width of the penumbral zone (cm), α_{sun} is the angular transmission of I_D (radians), z is the vertical distance between the shading leaf and the reference point where r_p is evaluated (in cm), and θ is the solar zenith angle (radians) (Figure 11.7). (The solar zenith angle is the angle of the sun relative to the highest point in the sky.) By defining r_p as a *half-width* we are referring to the length within the two indefinite penumbral boundaries wherein the PPFD is plus or minus half of its maximum penumbral value. In computer models it is not feasible to map penumbral space using such relations. Instead, the sizes and locations of penumbras are simulated using Monte Carlo models, in which a grid of points is placed within a "virtual canopy" and specific points are randomly sampled to determine their "view" of the solar disc (Oker-Blom 1984, Mõttus 2004).

Penumbral spreading of PPFD affects the photon-use efficiency of photosynthesis. Because leaf CO_2 fluxes are non-linear in their response to changing PPFD and, in the case of shade-adapted leaves, leaf CO_2 fluxes are rate-saturated at relatively low PPFD, penumbral spreading can reduce the incident PPFD on a leaf without proportional reductions in the net CO_2 flux; this results in a higher photosynthetic photon-use efficiency (i.e., a higher flux of assimilated CO_2 per unit flux of photons) (Figure 11.8). In radiation regimes dominated by the I_d penumbral spreading will have less impact on the photosynthetic CO_2 flux.

The placement and orientation of leaves, and the optical properties of leaf surfaces, determine to a large extent how photons are scattered or transmitted as solar radiation passes through a canopy. Many crop canopies are composed of randomly oriented leaves that fill most of the canopy space and reduce the potential for direct-beam penetration. In forests, leaves tend to be clumped, creating more frequent gaps, and permitting deeper penetration of I_D. Needle leaved forests tend to exhibit even more foliar clumping and tree crown

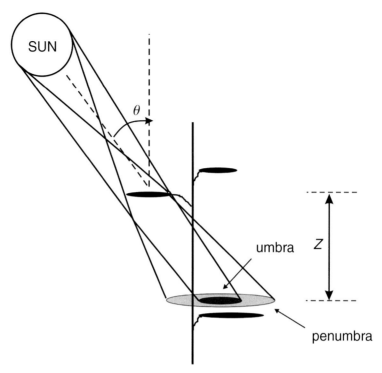

Figure 11.7 Representation of the umbra and penumbra cast as shadows from a leaf in a canopy as a function of the solar zenith angle (θ) and the distance between the occluding leaf and the reference point within the canopy (z). Redrawn and modified from Ross and Mõttus (2000), with permission from Elsevier B.V. Publishers.

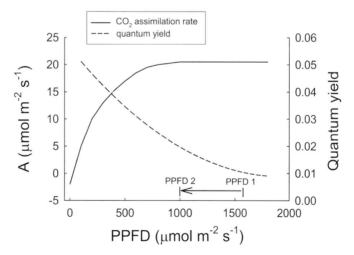

Figure 11.8 Representation of the effects of penumbral PPFD spreading on the leaf CO_2 assimilation rate (A). Note that as the PPFD is reduced through penumbral spreading from PPFD 1 to PPFD 2, there is little change in A, but a large change in the quantum yield of A (i.e., A per unit of PPFD). Thus, while A does not increase, it becomes more efficient with respect to PPFD.

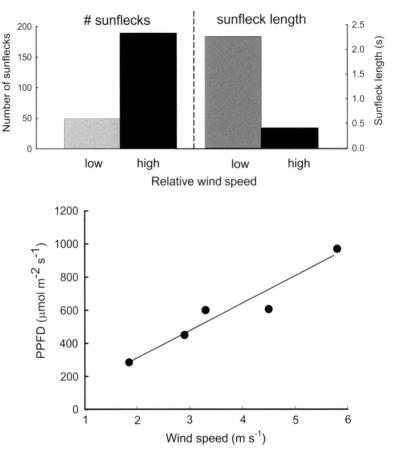

Figure 11.9   **Upper figure**. The total number of sunflecks measured in the lower portion of a cottonwood canopy on days with low wind speed (<3 m s^{-1}) or high wind speed (>3 m s^{-1}), and the average time length of sunflecks on each day. **Lower figure**. The relationship between mean wind speed and mean PPFD measured over the same 5-minute interval (1300–1305 h) each day during a 5-day period. Redrawn from Roden and Pearcy (1993), with kind permission from Springer Science + Business Media.

architecture that causes pronounced gaps between trees. The dissected nature of needle leaf canopies also offers numerous surfaces for photon reflection, which in turn results in a higher fraction of I_d, compared to broad-leaved forests or crops.

Leaf flutter can also affect the distribution of photons within a canopy. In the absence of wind the abundance of sun patches and flecks decreases from the top to the bottom of a canopy. In the presence of wind leaves move and the positions of gaps change. Leaf flutter causes an increase in the frequency of fluctuations in I_D at lower positions in the canopy (Figure 11.9). Low solar zenith angles result in reduced photon penetration causing reductions in the frequency and intensity of patches and flecks. The presence of clouds also causes more skylight and, concomitantly, increases in I_d. The amount of skylight I_d is influenced by cloud type. Thin cirrus clouds produce maximum I_d. Thick cumulus clouds cause high rates of photon scattering (increasing the ratio of I_d to I_D), but they also attenuate PPFD causing reductions in both I_d and I_D.

11.2.1 The general problem of describing photon transport through a scattering medium

> In some ways, the most fundamental problem in the theory of radiative transfer in plane-parallel atmospheres is the diffuse reflection and transmission of a parallel beam of radiation; for it will appear that the solutions of all other problems can be reduced to this one. (Chandrasekhar, S. (1950))

As solar rays strike particles and molecules in the atmosphere or plant elements in a canopy, photons are scattered. The path and fate of photons as they move from one point to another is referred to as *radiative transfer*. The key to understanding radiative transfer is an understanding of the paths taken by photons once they have interacted with a surface. The essence of this problem was captured elegantly in the quote presented above from Subrahmanyan Chandrasekhar, the influential Nobel Laureate and astrophysicist who is arguably the founder of modern radiative transfer theory. In one of his most seminal pieces of work, Chandrasekhar (1950) developed a comprehensive framework for describing photon transfer in scattering media in the form of an integro-differential transport equation:

$$\mu_{\mathrm{L}} \frac{d\mathrm{I}_t[z,\Omega]}{dz} = -E_x[z,\Omega] + \int_0^{4\pi} S_c[z,\Omega,\Omega']d\Omega' \quad , \qquad (11.3)$$

$$\text{Term} \quad \mathrm{I} \qquad\qquad \mathrm{II} \qquad\qquad\qquad \mathrm{III}$$

where I_t is the total PPFD (in mol m^{-2} s^{-1}), $\mu_{\mathrm{L}} = \cos\theta$, z defines a point within the vertical coordinate, Ω defines the geometric parameters of the incoming, direct solar photon stream according to zenith and azimuth angles, Ω' defines the geometric parameters of the incoming diffuse solar photon streams (i.e., photon streams that originate from directions other than Ω and are scattered into a solid angle defining Ω), E_x is the extinction function defining the *loss of PPFD from point z* (due to absorption and scattering) (in mol m^{-2} s^{-1}), and S_c is the scattering function defining the *gain of PPFD to point z* (from all directions defined by solid angle Ω') (in mol m^{-2} s^{-1}). The scattering (S_c) function is integrated over 4π to account for photon streams entering from the spherical volume surrounding point z. Thus, Term I is the photon flux balance with respect to height, Term II is the loss of photon flux due to absorption and backward-scattering, and Term III is the gain of photon flux due to inward-scattering from other sources. The potential role of the canopy in influencing the overall photon balance is readily recognized in Terms II and III, which would both be affected by leaf area, orientation, and surface characteristics. In its integral form, Eq. (11.3) contains implicit recognition of the conservation of photon flux in an infinitely small control volume at point z. It is the connectivity and inherent constraint among all possible scattering vectors that has proven so difficult in designing analytical forms of Eq. (11.3). Stated another way, the problem of photon scattering is a "non-local" problem requiring a difficult, "system-wide" perspective. In its mathematical essence, Eq. (11.3) is composed of an infinite system of first-order partial differential equations defining changes in the photon flux encompassing all scattering possibilities to and from point z. Using a series of partial solutions,

Chandrasekhar developed the *discrete ordinate method* in which Eq. (11.3) was recast as a series of numerical approximations of the equations specified for discrete scattering angles. Equation (11.3) is best applied to a turbid, scattering medium wherein the scattering particles are within a similar size range as the average wavelengths of the transmitted photon flux.

Radiative transfer theory based on Eq. (11.3) was originally developed for application to the atmospheric transmission of solar radiation, including attenuation due to clouds and aerosol "particles." The application of Eq. (11.3) to canopies introduces novel uncertainties. Canopies contain gaps between leaves that are orders of magnitude larger than the gaps between molecules or aerosols in the atmosphere; thus, the concept of a canopy as a truly turbid medium cannot be defended. Even if we accept as an approximation the turbid medium analogy, the application of Eq. (11.3) is difficult because of the complex and unique nature of a canopy's architecture. Ideally, we would want to solve the equation in terms of leaf transmittance and reflectance, the direct and scattered vectors of photon rays, the wavelength dependence of photon-surface interactions, and for each of these, within the context of point-to-point variation in canopy architecture. Such a solution, however, would require immense computational capabilities and detailed parameter sets given the unique structural attributes of canopies.

As an alternative, researchers have developed simpler, statistically based relations, which are capable of predicting the probability of a photon being transferred from one canopy layer to another, and which characterize the canopy in terms of mean tendencies. These relations take the form of *probability density functions* (pdfs), with the probability of photon transfer estimated from a predicted (dependent) quantity referenced to easily characterized (independent) variables associated with the canopy, e.g., LAI, leaf orientation, and solar angle. To a first approximation this approach works well, as probabilities of photon transfer can be readily converted to photon flux densities. More recently, with the advent of greater speed and computational capability, efforts are underway to fuse simpler statistically based models with the mechanistic scattering-based approaches derived from Eq. (11.3). In the sections below we will begin with a discussion of the pdf-based approaches and we will use the theory of these approaches to illustrate fundamental properties of photon flux attenuation within canopies. Following these sections, we will present a brief overview of more recent studies that bring the integro-differential transport equation back into the picture.

11.2.2 The probability of photon transfer through a canopy layer

The elementary (canopy) volume is now assumed to be constructed from a finite number of plane-parallel horizontal layers ... The flight of photons in this layered medium is analogous to piercing the vegetation canopy with long needles of infinitesimally thin radius. The relationship between the plane-parallel layers is one of statistical independence, in the sense that the realization of a random variable (contact or no contact of the needle with the canopy elements) in a layer is effectively independent of its realization in another layer. (Ranga Myneni, Juhan Ross, and Ghassem Asrar (1989))

The problem of describing photon penetration through a canopy can be viewed as a problem in probability theory (Nilson 1971, Myneni *et al.* 1989). The probability of radiative transfer from one canopy layer to another is founded on the assumption of statistical independence among successive, plane-parallel layers. In order to derive some fundamental relations between the probability of photon penetration and canopy structure, let us begin with the simplest type of canopy architecture – randomly placed, horizontal, and non-overlapping leaves in a series of stacked layers. In such a canopy, the average optical thickness (or opaque tendency) of each layer can be defined by L/N; the ratio between the cumulative canopy leaf area index (L) and the number of canopy layers (N). We will state that the area of leaves (A_L) in a single canopy layer oriented perpendicular to I_D when the sun is at its zenith, and expressed as a fraction of ground area (A_G), is A_L/A_G; thus, $L = A_L/A_G$ for a given canopy layer and $L = N(A_L/A_G)$ when summed for all layers. The ratio A_L/A_G is the *shadow ratio*; so, $(1 - A_L/A_G)$, or alternatively $(1 - L/N)$, is the *gap ratio*. When interpreted within the context of probability theory, the gap ratio represents the probability that a ray of photons passing through a layer of foliage will **not** intercept a leaf. Given the assumption of statistical independence among layers, and the condition that successive layers have the same leaf area index, the probability that a ray will pass through two layers is $(1 - A_L/A_G)(1 - A_L/A_G)$, or $(1 - A_L/A_G)^2$, and the probability (P_0) that the ray will pass through multiple, N, layers is:

$$P_0 = \left(1 - \frac{A_L}{A_G}\right)^N = \left(1 - \frac{L}{N}\right)^N. \qquad (11.4)$$

At this point we invoke a well-known principle from calculus, Bernoulli's "force of interest relation," which has been interpreted by Euler as follows: when a quantity increases at a rate proportional to its current value, and the number of events (n) leading to an increase in value is defined within the limit $n \to \infty$, the value will increase according to the exponential function:

$$\exp(x) = \lim_{n \to \infty} \left(1 + \frac{x}{n}\right)^n. \qquad (11.5)$$

Writing Eq. (11.5) as the inverse, we can state:

$$\frac{1}{\exp(x)} = \lim_{n \to \infty} \left(1 - \frac{x}{n}\right)^n = \exp(-x). \qquad (11.6)$$

Relying on calculus to define the exponential function, we can state:

$$P_0 = \exp(-L). \qquad (11.7)$$

Equation (11.7) states that for a simple canopy in which leaves are oriented perpendicular to the incoming solar rays when the sun is at its zenith, the probability of photon penetration will be equal to the inverse exponential of leaf area index. In this case, P_0 refers to the transmission probability for a layer of leaves located at vertical position (z) within the canopy and L refers to the cumulative LAI above that layer. Although Eq. (11.7) is constrained by a list of caveats, it does illustrate the conceptual simplicity by which L is the primary determinant of the statistical probability that a photon will penetrate a series of canopy layers. A formal treatment of the relation of statistical probability and canopy photon interception (or lack thereof) is provided in Appendix 11.1.

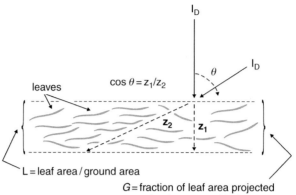

Figure 11.10 Representation of the relations among leaf area index (L), the leaf-to-sun angular orientation (G), and the solar zenith angle (θ) for solar rays penetrating a canopy. Leaves are represented as the shaded elements within the layer. L is defined according to the area of one side of the leaf. As the day progresses and the sun moves toward the horizon, i.e., as θ increases, the solar path through the layer of leaves increases, i.e., $z_2 > z_1$, thus increasing the probability of photon interception and decreasing P_0; in other words, as θ increases, the optical path length through the canopy increases and P_0 decreases. Similarly, as G increases at constant L, or as L increases at constant G, both of which translate into a higher amount of leaf area oriented perpendicular to I_D, P_0 will decrease. When combined these relations illustrate the basis for: $P_0 = \exp(-\,G\,L/\cos\theta)$.

We can develop a broader context for Eq. (11.7) by considering a solar beam coming from directions other than directly overhead and for leaves oriented in directions other than those normal to the incident beam. The effect of solar zenith angle can be appreciated by recognizing that as the sun moves away from its zenith, the path length required for photons to penetrate to a specific layer will increase, and thus P_0 will decrease. Additionally, as the fraction of leaf area in a canopy that is projected in a plane normal to the incoming solar beam increases, the probability of photon interception will increase, and once again P_0 will decrease. These relationships are shown conceptually in Figure 11.10, and they are developed more formally in terms of sun-leaf geometry in Figure 11.11.

Referring to the relations presented in Figure 11.11 we will now derive the concept of "effective LAI," which we will define as a geometrically reconciled relation between the area of a leaf that is projected at some angle other than normal and the area of its shadow on the ground. In other words, the ratio of total leaf area in a canopy or individual canopy layer to the area of intercepted I_D referenced to ground area. We start by noting that the area of leaves in a canopy layer oriented on a plane normal to the direct solar beam (A_{Ln}) relative to the total leaf area of the layer (A_L) is $A_{Ln}/A_L = \cos\alpha$. After recognizing that $\sin\beta = \cos\theta$, and using $A_{Ln} = A_L \cos\alpha$ for substitution, we can relate A_{Ln} and A_L to the area of the shadow cast by A_{Ln} on a horizontal surface (which we denote as A_{Lh}):

$$\frac{A_{Ln}}{A_{Lh}} = \sin\beta = \frac{A_L \cos\alpha}{A_{Lh}} = \cos\theta. \tag{11.8}$$

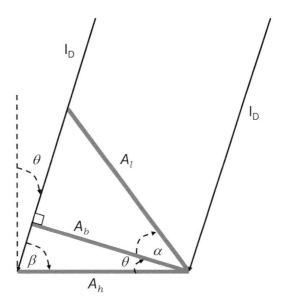

Geometric relations of the vectors defining the direct solar beam (I_D), the fraction of leaf area oriented on a plane normal to I_D (A_b), the total leaf area (A_l), the zenith angle of the sun (θ), the difference in angle between the mean orientation of the total leaf area and the leaf area oriented on a plane normal to I_D (a), and the solar elevation angle (β) (i.e., the angle of I_D above the horizontal surface). Fundamental relations within this geometrical context can be used to define G, the fraction of leaf area oriented on a plane normal to I_D (see text).

From the relations in Eq. (11.8), we can write:

$$\frac{A_{Lh}}{A_L} = \frac{\cos \alpha}{\sin \beta} = \frac{G}{\sin \beta} = \frac{G}{\cos \theta},\tag{11.9}$$

where we have introduced a new geometric coefficient, $G = \cos \alpha = A_{Ln}/A_L$. In other words, G represents the fraction of total leaf area that is oriented on a plane normal to I_D. The "effective LAI" that we set out to define is now written as L ($G/\cos \theta$). Putting these relations together, we can write an equation to define P_0 in terms of projected leaf area:

$$P_0 = \exp\left(-\frac{GL}{\cos \theta}\right) = \exp(-K_I\,L),\tag{11.10}$$

where K_I is the canopy extinction coefficient ($K_I = G/\cos \theta$). The astute reader will recognize that the form of Eq. (11.10) is similar to the well-known Beer–Lambert Law, which describes the relation between fractional photon flux attenuation and the concentration of light-absorbing molecules suspended in a solution: $I = I_0 e^{-kc}$, where I is the photon flux exiting the solution, I_0 is the photon flux entering the solution, k is the absorption coefficient of the molecule (absorption per unit concentration), and c is the concentration of the molecule (see Section 2.9). The Beer–Lambert Law is derived with the same logic as Eq. (11.10). In the case of Eq. (11.10) the relevant photon-absorbing constituent is cumulative L above a reference layer in the canopy (analogous to c in the Beer–Lambert equation), and the relevant absorption coefficient is $G/\cos\theta$ (analogous to k in the Beer–Lambert equation). For the case of leaves oriented perpendicular to the incoming solar rays, $G = 1.0$. For the case of leaves oriented parallel to the incoming solar rays, $G = 0$ (interception by the edges of the parallel leaves is ignored). An examination of Eq. (11.10) shows that as cumulative LAI increases above layer L, P_0 decreases exponentially. At the limits of the relation, P_0 goes to 1 as L or G go to 0, and P_0 goes to 0 as L goes to ∞.

In order to eventually couple radiative transfer models to scalar flux models we will need to convert the probability of photon flux penetration (P_0) to photon flux density (PPFD). That is, a conceptual "jump" is required to get from a statistical prediction that an event will occur to the magnitude of the event. The magnitude of the direct beam photon flux impinging on layer L within a canopy will be the product of I_D and the fraction of leaf area in layer L that intercepts the flux; the *sunlit fraction*. The *shaded fraction* is assumed to only receive I_d. The sunlit fraction in a canopy layer is equal to the probability of a sunfleck (P_{sf}) occurring within that layer. In order to gain a better appreciation of P_{sf}, let us refer once again to Figure 11.10. In this case P_0 is the probability of direct-beam photon transfer to the layer and therefore the probability at which I_D will strike the layer. In essence, P_0 represents the fraction of unit horizontal area that is sunlit; it is the *gap fraction*. In the limit as $\Delta L \rightarrow 0$, then $P_0 = P_{sf}$; this equality is derived formally in Appendix 11.2.

Thus, for the case of randomly arranged leaves in canopy of infinitely thin layers, $P_0 = P_{sf}$. Basically, this equality states that the probability by which a direct-beam photon is transferred through a canopy to layer L equals the probability that a unit of area in layer L will be sunlit. Once P_{sf} is determined, the shaded fraction (P_{sh}) is taken as $1 - P_{sf}$. The photon flux density measured in the sunlit fraction in each canopy layer is assumed to be the sum of both the direct and diffuse photon flux; i.e., $I_D + I_d$. From these relations the total PPFD in the sunlit and shaded fractions of a canopy layer can be determined and used as an independent variable in the modeling of canopy mass and energy fluxes.

Returning to Eq. (11.10), we will further consider the nature of the term G. Recall that we developed $G/\cos\theta$ in order to account for the "effective LAI"; the influence of leaf-sun orientation on the capacity for a unit of total leaf area to intercept the photon flux. In a hemispherical canopy with leaves exhibiting varying orientations relative to the solar zenith and azimuth angles we can define G more formally as:

$$G[z,\Omega] = \frac{1}{2\pi}\int_0^{2\pi} d\phi_L \int_0^1 g[z,\Omega]\,|\cos\,\Omega,\Omega_L|\,d\mu_L\,.$$

(11.11)

Term   I   II   III   IV

Equation (11.11) is mathematically complex, but simple in its conceptual basis. Overall, Eq. (11.11) defines G with respect to vertical distance (z) above the ground and a specific angular orientation of the direct photon stream (Ω). In Term I, ϕ_L is the fractional mean orientation of the leaf area with respect to the solar azimuth angle, in this case integrated over the hemispherical domain representing all possible orientations of the upper leaf surfaces. In Term II,

$g[z, \Omega]$ is the *leaf-area distribution function* and defines the fraction of leaf area index oriented with leaf surfaces normal to I_D. The leaf-area distribution function is defined as:

$$g[z, \Omega_L] = g'_L[z, \Omega] \ / \ L[z], \qquad (11.12)$$

where $g'_L[z, \Omega]$ represents the leaf area in layer L with surfaces normal to I_D. Looking at Term II in another way, $g[z, \Omega]$ is the probability that a given amount of leaf area within a canopy will be oriented perpendicular to I_D; the limit is 1, in which case all of the leaf area is oriented perpendicular to I_D. Term III represents a geometric function, which uses spherical geometry to define the normal between a leaf surface and the sun in terms of the solar zenith and azimuth angles according to:

$$|\cos \Omega, \Omega_L| = \cos \theta \cos \theta_L + \sin \theta \sin \theta_L \cos(\phi - \phi_L), \qquad (11.13)$$

where θ_L and ϕ_L define the mean orientation of the leaf area relative to the solar zenith and azimuth angles, respectively. In Term IV we see the differential equation used to define the integrand; in this case, $\mu_L = \cos \theta_L$. Thus, overall, the "G-function" as defined by Eq. (11.11) provides an integral expression of the leaf-sun geometry within a canopy informing us of the *projected leaf area* capable of intercepting I_D. For the cases of random, vertical, and horizontal leaf orientations within a canopy Eq. (11.11) can be solved analytically with G equal to 0.5, $2/\pi \sqrt{1 - \mu_L^2}$, and μ_L, respectively. The relationships of the G value to the solar zenith angle and different forms of crop canopies are shown in Figure 11.12, and representative analytical solutions to the relationships among G, K_I, and solar angle are provided in Table 11.2.

Several general classes of leaf inclination have been reported in past studies of canopy structure, including erectophile, plagiophile, spherical, and planophile (Lemeur and Blad 1974, Ross 1981, Norman and Campbell 1989). As the name suggests, *erectophile* canopies possess erect leaves (also called "vertical" canopies). In contrast, a *planophile* canopy possesses mostly flat leaves (also called "horizontal" canopies). The *spherical* distribution is best envisioned through analogy to the surface of a spherically shaped ball. If you were to pluck leaves and attach them to the surface of the ball while retaining their original azimuth and zenith angles, you would soon cover the surface of the ball with a spherical leaf distribution. In a *plagiophile* canopy leaves are most frequently oriented with an oblique inclination. In some crop species genetic breeding has produced variants with leaf inclination distributions that maximize productivity. For example, some corn varieties have been bred for erectophile canopy structure, increasing high photon fluxes deep into the canopy. Forest canopies do not possess constant leaf angles as a function of depth. Erect leaves are more frequently observed near the top of a canopy and horizontal leaves are more frequently observed deeper in the canopy. The transition from erect to horizontal distributions in the same canopy facilitates the penetration of I_D deeper into the canopy where it can be absorbed by lower leaves.

Using Eq. (11.10) we can explore the question: *How much cumulative LAI can a plant canopy sustain?* We can rearrange Eq. (11.10), and express it with L and K_I as:

$$L = -\ln P_0 / K_I. \qquad (11.14)$$

Table 11.2 Analytical equations defining the direction cosine, G, and the canopy extinction coefficient, K_I (Anderson 1966, Campbell and Norman 1998, Monteith and Unsworth 1990)

Leaf angle distribution	G, direction cosine	K_I, extinction coefficient
Horizontal	$\cos\theta$	1
Vertical	$2\pi\sin\theta$	$2\tan(\theta/\pi)$
Conical	$\cos\theta\cos\theta_L$	$\cos\theta_L$
Spherical or random	0.5	$1/(2\cos\theta)$

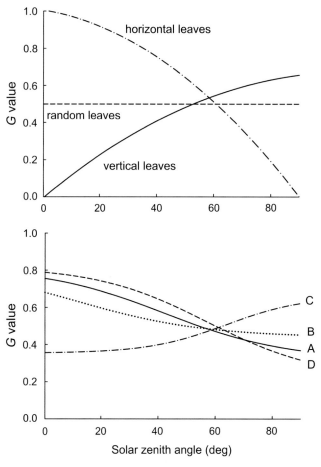

Figure 11.12 **Upper figure**. G values as a function of θ for hypothetical canopies with horizontal leaves ($G = \cos\theta$), vertical leaves ($G = 2/\pi\sqrt{1-\mu_L^2}$, or randomly oriented leaves ($G = 0.5$). Note that the G value for vertical leaves is not 1 when solar zenith angle is 90 degrees. This is due to the fact that vertically oriented leaves face various azimuthal angles (which are assumed to be random for the functions depicted in this figure). **Lower figure**. G values as a function of θ for representative crop species. A, sugar beet; B, white clover; C, rape; D, corn. Note that for most solar zenith angles that would occur during the middle of the day (i.e., 15–90°), an approximation of $G = 0.5$ is fairly representative for crop species. Redrawn from Myneni *et al.* (1989), with permission from Elsevier B.V. Publishers.

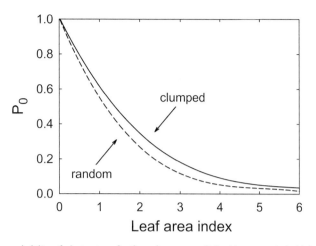

Figure 11.13 The decrease in the probability of photon transfer through a canopy (P_0) with progressively higher cumulative leaf area index for a canopy with spherical leaf orientation, but either randomly oriented leaves ($G = 0.5$) or clumped leaves ($G = 0.5$, $\lambda_0 = 0.8$). The solar zenith angle was assumed to be 45°. The theoretical limit of LAI for a canopy can be estimated at ~ 6.0, although the presence of clumped foliage can facilitate higher LAI values.

We now state a simplifying assumption: the limit below which a leaf cannot sustain a positive carbon balance is 0.05 I_0 (in other words 5% of the PPFD incident on the top of the canopy) (Jarvis and Leverenz 1983). With that assumption in place, the theoretical maximum limit for L is estimated to be 6. This value will vary depending on leaf angle distribution and degree of foliar clumping; in other words it is dependent on K_l. In fact, as we noted previously, some conifer forest canopies and tree plantations have LAI values that exceed this limit (see Table 11.1). In Figure 11.13 we present the relationship between P_0 and L, taking into account clumped or randomly distributed foliage. It is clear that in canopies with clumped foliage, the relation flattens to a near-asymptote that may explain margins of positive carbon balance to L values slightly above 6. In the next section we consider the effect of foliar clumping on PPFD penetration in more detail.

11.2.3 Foliar clumping

Natural canopies do not exhibit ideal homogeneous leaf distributions; they exhibit some degree of foliar "clumping" (e.g., Baldocchi *et al.* 1985, Chen 1996). Foliar clumping is the result of both genetic controls over a plant's architectural form, and acclimation of that form to the prevailing environment. The combined result of these influences produces greater clumping in the upper, sunlit layers, and lesser foliar clumping in the lower shaded layers. Greater clumping of leaves in the most sunlit parts of a canopy results in high photosynthetic rates per unit of canopy volume and increases the probability of transfer of I_D to deeper layers. Lesser foliar clumping and reduced overlap of leaves in the lower, shaded layers of a canopy, increases the leaf area exposed to penetrating rays of I_D, as well as enhancing the capture of I_d from the hemisphere above the leaves.

Foliar clumping poses challenges for the mathematical description of the canopy photon regime. Descriptions of clumped architecture begin with consideration of individual shoots or branches and end with consideration of the entire canopy. The effective LAI for a shoot is described according to the planar shadow area it projects (essentially A_h) when illuminated with a photon source of known zenith and azimuth origin. Thus, we can write:

$$A_{Lh} = \frac{1}{\pi} \int\limits_{0}^{2\pi} d\phi \int\limits_{0}^{\pi/2} A_{Ln}[\phi, \theta] \cos\theta \, d\theta, \tag{11.15}$$

where ϕ and θ are the azimuth and zenith angles of the photon source relative to the main axis of shoot orientation, respectively (Chen 1996). Clumping in the shoot (γ) can be expressed as $\gamma = 0.5 \, A_L/A_{Lh}$. Various approaches have been designed to measure the silhouette area of individual branches in the laboratory (Fassnacht *et al.* 1994), in the field (Smolander and Stenberg 2001), and using computer simulations (Smolander and Stenberg 2003).

As we move from consideration of clumping at the shoot scale to clumping at the canopy scale, two crucial pieces of information are required: gap size and gap fraction (Chen *et al.* 1997). Gap size refers to the two-dimensional breadth or area of canopy gaps and gap fraction refers to the fraction of ground area beneath a canopy that has an open view of the sky. These co-dependent variables can be combined to obtain an estimate of the degree to which foliar and crown clumping facilitate the penetration of I_D through the canopy. Various methods have been devised to quantify these variables, including automated track-mounted and hand-held sensors that move through the forest and quantify PPFD along linear transects (Gower and Norman 1990, Stenberg 1996, Baldocchi *et al.* 1997, Chen *et al.* 1997). Using data on the extent of sun and shade patches detected by the moveable sensors, gap size can be calculated and summed to provide a spatially integrated estimate of gap fraction. Using Eq. (11.8) as a foundation, and measurements of the gap fraction, the effective LAI (L_e) of the canopy can be calculated as:

$$L_e = 2 \int\limits_{0}^{\pi/2} \ln\left(\frac{1}{P_0[\theta]}\right) \cos\theta \, \sin\theta \, d\theta, \tag{11.16}$$

where $P_0[\theta]$ is the observed canopy gap fraction at a specific zenith angle θ. A more robust estimate of the canopy gap fraction can be determined after observing $P_0[\theta]$ across a full range of zenith angles. The "full-zenith" gap fraction can then be used to estimate the canopy clumping index (λ) (see Chen 1996), which is analogous to γ, but scaled to the entire canopy. The overall LAI for the canopy with clumped foliage is then estimated as:

$$L = (1 - f_\alpha) \, L_e \lambda, \tag{11.17}$$

where f_α is the fraction of the total canopy surface area represented by woody tissue. In some cases, λ is partitioned into separate functions for crown and foliar clumping; this is especially relevant for canopies with needles or small leaves, in which penumbral effects make it difficult to quantify λ using optical measurements.

Now, with an understanding of how foliar clumping is quantified and defined, we can consider the topic of radiative transfer in canopies with clumped foliage. Clumping of foliage, in both broad-leaved and coniferous canopies, facilitates the transmission of I_D (Figure 11.13). Using this logic, Eq. (11.10) can be modified to permit less extinction of I_D as a function of canopy depth. The modified form of Eq. (11.10) is often referred to as the *Markov model* (Nilson 1971) and it is justified according to the statistical theory of Markov chains (see Appendix 11.1). In practice the model takes the form:

$$P_0 = \exp\left(\frac{-\lambda_0 GL}{\cos\theta}\right), \tag{11.18}$$

where λ_0 is the clumping coefficient that calibrates the probability that a photon will be transmitted through a canopy layer. In the Markovian sense, λ_0 links the probability of photon transfer in a given canopy layer to the probability of photon transfer in the layer immediately above; i.e., "*future probability is a function of the present state.*" Clumping is interpreted as producing a Markovian dependency among canopy layers whereby the probabilities of photon transfer through successive canopy layers are connected. In practice, the term λ_0 is defined as 1 if leaves are randomly distributed, less than 1 if leaves are clumped and greater than 1 if leaves are uniformly distributed. Thus, when leaves are clumped the probability of I_D transmission through a layer is increased. When leaves are randomly distributed, Eq. (11.18) simplifies to Eq. (11.10). In addition to the Markov model, other mathematical functions have been used to represent the clumped nature of foliage. We have listed past models that have been used for various canopies, and the relevant clumping parameters in Table 11.3.

11.2.4 The non-Gaussian nature of canopy photon transmission

Canopies with clumped foliage promote heterogeneity with regard to transmission of the incident photon stream. The gaps through which I_D and I_d are transmitted, and the foliar surfaces that intercept the beams, are distributed non-randomly within the canopy. Let us assume that we can move a photon sensor back and forth across a single layer of leaves. The non-random transmission of photons above that layer will cause "bright spots" and "gray spots" with regard to measured PPFD. In other words, the probability of measuring any given PPFD in that layer will be described by a non-Gaussian density function. At spots immediately below an occluding leaf or branch the observed PPFD will reflect only I_d. At spots below gaps both I_D and I_d will be received. Thus, the time-averaged PPFD for each spot can be predicted with a Gaussian probability distribution, but the probability of a given PPFD occurring at a particular instant in time across the spatial extent of the layer will be distributed in non-Gaussian form (Figure 11.14). Time-averaged PPFDs within a layer will approach a Gaussian distribution due to the natural averaging that occurs as the sun moves across the sky and shifts specific points between full sun and shade; this will average out the "bright" and "dark" spots that cause the skewed nature of instantaneous observations. Within the vertical profile of the canopy the nature of skewed probability density functions will change as the frequency of unobstructed gaps changes and as the degree of foliar clumping changes. In the upper canopy the probability density function will be negatively

Table 11.3 A representative listing of the types of radiative transfer models used for different canopies to predict P_0. For the negative binomial and Markov models leaf clumping is represented by a clumping coefficient (ΔL in the case of the negative binomial model and λ_0 in the case of the Markov model). (From: Baldocchi and Collineau 1994)

Crop	Spatial distribution	Clumping coefficient
Maize, sparse	Random: Poisson	
Maize, dense	Random: Poisson	
Maize, rows	Clumped: Markov	$\lambda_0 = 0.4$–0.9
Sunflower	Random: Poisson	
Sorghum	Random: Poisson	
Cotton	Random: Poisson	
Oranges	Clumped: Negative binomial	$\Delta L = 1.0$
Deciduous forest	Clumped: Negative binomial or Markov	$\Delta L = 2.65$
		$\lambda_0 = 0.53$–0.67
Rice	Random: Poisson	
Wheat	Random: Poisson	
Pine forest	Clumped: Markov	$\lambda_0 = 0.6$–1.0
Douglas fir	Clumped: Markov	$\lambda_0 = 0.36$–0.45
Sorghum	Clumped: Markov	$\lambda_0 = 0.8$
Cotton	Clumped: Markov	$\lambda_0 = 0.65$

skewed, reflecting sunflecks and sun patches, penumbra, and some shading. Deeper in the canopy the probability density function will be positively skewed reflecting the predominance of diffuse light and fewer sunflecks and penumbra.

Recognition of the non-Gaussian distribution of PPFDs in plant canopies, along with the non-linear response of photosynthesis and stomatal conductance to variation in PPFD, has caused a reassessment of how canopy photon transfer is treated in canopy flux models (Sinclair *et al.* 1976, dePury and Farquhar 1997). Rather than using a single value and function that describes the exponential decay in PPFD as a function of canopy depth (e.g., the K_l value we derived in Eq. (11.10), it is now routine to consider two independent sample populations for the PPFD at any given canopy layer – one for the direct beam and one for the diffuse beam. (The penumbral beam is still ignored in most models.) This type of model is often called the *sun/shade* model when applied to radiative transfer alone, or the *sun/shade big leaf model* when applied to predictions of leaf fluxes.

11.2.5 Process-based radiative transfer models

As an alternative to the statistical schemes described above several models have emerged that are derived from the first principles of photon scattering (Myneni *et al.* 1989, Disney *et al.* 2000, Jensen 2001). While these models are still probabilistic in their final solution, they at least begin with assumptions about the photon scattering process rather than the pdf of sampled PPFDs. Thus, we will refer to these models as *process-based radiative transfer models*. These types of models re-introduce some of the intellectual foundations laid out in

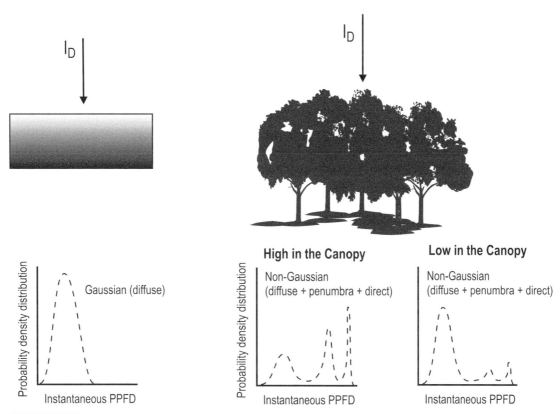

Figure 11.14 Hypothetical comparison between profiles of PPFD after penetration of the direct component of solar radiation in a single leaf (left) or a canopy (right). In the leaf, the radiation is diffused as it reflects off the numerous cellular surfaces, and exits the leaf predominantly as diffuse radiation distributing a Gaussian distribution with respect to the total PPFD. In the canopy, the radiation penetrates as several fractions, direct (sun patches and sunflecks), penumbral, and diffuse, exhibiting Gaussian distributions within each fraction, but exhibiting a non-Gaussian, or skewed, distribution with respect to the total PPFD.

the work of Chandrasekhar (1950). Process-based models have advantages in that they can be applied in three dimensions, thus using more of the available information on canopy structure. The improved three-dimensional resolution of process-based models has made them especially useful in remote-sensing applications. One disadvantage of process-based models is that they are computationally expensive. The high computational costs arise from their need to track multiple scattering events at multiple points in the canopy.

Process-based radiative transfer models can be assigned to two groups: *geometric-optic models* and *photon-scattering models*. Geometric-optic models emphasize the effects of canopy architecture on beam penetration, whereas photon-scattering models emphasize the effects of individual canopy elements (leaves, stems, etc.) on photon reflection. The geometric-optic approach is especially useful for canopies with large gap fractions, where plants are widely spaced, and for predicting the PPFD that will exist at any specific point in the canopy.

The photon-scattering approach is useful for bidirectional reflectance models that are used to retrieve information on canopy structure from spectral patterns of surface reflection.

Geometric-optic models are based on the general principles of geometric-optic physics, which describe photon propagation in terms of transmitted *rays*. In the simplest application of geometric-optic principles a ray of photons is allowed to penetrate a canopy "envelope," be reflected once by the soil or leaves, and exit the canopy without further interaction; these are the so-called *single-scattering models*. Some geometric-optic schemes have been developed to introduce multiple scattering (Li and Strahler 1995, Chen and LeBlanc 2001). In geometric-optic models the canopy is represented as a group of three-dimensional plant crowns of specified size and shape (the geometric component). Various shapes have been used to produce abstracted plant crowns. The abstract envelopes that surround each crown shape provide photon transmission boundaries and volumes for radiant energy balance; canopy envelopes have been designed using different crown shapes, including those with rectangular or triangular cross sections, as well as arrays of ellipsoids, cones, and cylinders (Figure 11.15). A careful analysis is required in order to find the best shape to represent a crown as the most intuitive choices are not always the best choices. For example, the assumed choice of a cone to best represent coniferous crowns for pine and spruce trees in Finland was not as good as the choice of ellipsoids (Rautiainen *et al.* 2008).

The geometric-optic approach is applied to landscapes constructed in cyberspace. Data from real canopies, including crown size and shape, foliage distribution and orientation, and plant spacing are used to construct a virtual canopy. Once established, the transfer of photons to specific points within the virtual canopy can be explored, either from the perspective of an "observer" located within the canopy (the *procedural approach*) or from the perspective of a photon ray traveling through the canopy (the *Monte Carlo approach*). In the procedural approach, the shadow projections of leaves are computed to determine if an "observer" located at a specific point in the canopy is in sun or shade; the computation of these projections depends on the leaf angle and angle of incoming sunlight. In the Monte Carlo approach, statistical theory is used to describe the transfer of an ensemble of photons "fired" from a single point of origin. Typically, the transfer of 10^6 photons is followed to obtain reliable results. The trajectories or rays of these fired photons are traced through sub-volumes of the virtual canopy; a process called *ray tracing*.

In geometric-optic modeling, the *landscape scene* is characterized by light and dark patches that reflect structural attributes. The simplest geometric-optic approaches consider the shadows cast on the landscape by opaque shapes. More complicated models calculate the probability of beam penetration through each abstracted envelope. One of the most comprehensive geometric-optic model frameworks has been developed by a large international group, including Myneni and co-workers in the United States (Myneni 1991, Myneni *et al.* 1995), Cescatti in Italy (Cescatti 1997), and Sinoquet in France (Sinoquet *et al.* 2007). Various iterations of their models have evolved toward a common approach in which the differential equation for photon transport is solved using the simplified discrete ordinates scheme; the angular coordinates of photon transport are discretized into a small subset of specified angular coordinates. An operational limitation of geometric-optic models, and one of the primary motivating factors driving the development of photon-scattering models, is

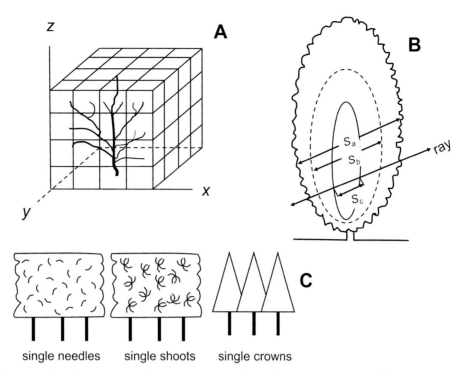

Figure 11.15 Representations of various canopy abstraction methods used to develop radiative transfer models. **A**. Division of canopy into a grid of sub-volumes, each of which is subjected to analysis of photon transmission probabilities (see Norman and Welles 1983); **B**. Representation of individual plant crowns in geometric-optic models in which embedded ellipses are used to define photon travel path lengths (S_x), which is then used along with leaf area density and leaf orientation functions to predict the probability of penetration by a ray of photons (see Wang and Jarvis 1990, Cescatti 1997); **C**. General representation of the different scales used in organizing predictions of the probability of photon interception by individual needles located in specified canopy layers, individual shoots with clumped needles, and individual crowns, as in geometric-optic modeling. Redrawn from Baldocchi and Collineau (1994).

the poor representation of the diffuse radiation environment; partially shaded areas are poorly represented in most landscape scenes.

In photon-scattering models, a principal aim is to describe the nature of diffuse radiation within a canopy. Leaves have high absorption coefficients for photons in the visible part of the spectrum, so the number of scattering reflections that occur once a photon gets into the canopy is small. Photons in the near-infrared part of the spectrum, however, are not readily absorbed by leaves, so the number of scattering reflections is large and the optical properties of leaf surfaces exert an important control over near-infrared scattering dynamics. A principal assumption in photon-scattering models is that the probability that a reflected or scattered photon will recollide with a canopy element is a simple function of the photon's associated wavelength of transmission and the structural complexity of the canopy. This so-called *recollision probability* can be estimated from remotely sensed observations of the reflected photon flux or it can be simulated using Monte Carlo models. In the case of a remotely sensed, reflected photon flux the ratio of absorbed (I_a) to intercepted (I_i) photon fluxes can be approximately related to the mean scatter coefficient of leaves (ω_L) and the recollision probability (p_r) according to:

$$\frac{I_a}{I_i} = \frac{1 - \omega_L}{1 - p_r\,\omega_L}. \qquad (11.19)$$

In this case, $I_i = I_a + I_s$, where I_s is the total scattered photon flux and it is assumed that p_r is the same for all photon scattering events. The term ω_L refers to the fraction of PPFD incident on a leaf that is transmitted or reflected (i.e., scattered) and $(1 - \omega_L)$ is the fraction of PPFD that is absorbed. Derivation of Eq. (11.19) requires the assumption that the ground beneath the canopy has albedo equal to zero. Monte Carlo models have been used to estimate the recollision probability. Virtual photons are "fired" from specific geo-referenced points inside or outside the canopy and their trajectories are traced in cyberspace. Based on the scattering behavior of the photons, probabilities of interaction with canopy elements and the probability that a photon will re-emerge from the upper surface of a canopy can be estimated (Mõttus 2007). The fates of the fired photons (e.g., absorption, transmission, or reflection) are determined by a multiple-state, Markovian chain model. The probability of a photon intercepting a canopy element, as opposed to passing through the canopy, is calculated using the same binomial assumptions described above for the statistical models (see Appendix 11.1). The true innovation found in photon-scattering models is the method by which they handle interactions beyond the initial contact, which enables calculation of the recollision probability. An analytical approximation of the recollision probability has been formulated as a function of LAI and canopy gap fraction (Stenberg 2007). In the past few years, hybrid models have also been produced, bringing together the advantages of both geometric-optic and photon-scattering models. In the hybrids, the principles of geometric-optic modeling are used to resolve the entry of photon rays into canopy envelopes and interaction between photons and canopy elements during the first-order scattering event. Then, higher-order scattering events are described by photon-scattering schemes (Li and Strahler 1995, Chen and Leblanc 2001).

11.2.6 Modeling surface albedo

Canopy albedo is a key parameter of biosphere-atmosphere exchange models, in part determining the potential for absorption of solar radiation and thus the energy available for sensible and latent heat exchange. As a first approximation we expect the albedo of a canopy to be close in value to the mean reflectivity of leaves. However, albedo is a scale-dependent property; measured canopy albedos are often half, or less, the mean value of albedo for leaves in the canopy. Scattering and subsequent light trapping by leaves in successively deeper layers of a canopy reduce the whole canopy albedo relative to mean leaf reflectances. The successive trapping of the photon flux explains the inverse relationship between albedo and canopy height that has been previously noted (Dickinson 1983).

The albedo of a canopy reflects the sum of two opposing streams of photons that are initially intercepted by the canopy; *upwelling and downwelling*. The upwelling stream represents the scattering of photons out of the canopy, whereas the downwelling stream represents photons scattered deeper into the canopy. This framework forms the basis of the class of reflectance models known as *two-stream models* (see Myneni *et al.* 1989). Models that further divide the upward and downward hemispheres into quartile sectors with scattered photons have been produced (e.g., four-stream models; see Li and Dobbie 1998) and improve the accuracy of the albedo prediction, especially at low solar angles. For general purposes, an analytical equation capable of simulating canopy albedo (α_c) with the assumption of one scattering event (such as would occur near the outer surface of a canopy) and semi-infinite depth can be derived from the two-stream approximation as (Dickinson 1983):

$$\alpha_c = \frac{\omega_L}{\left(1 + (1 - \omega_L)^{1/2}\right)^2}. \tag{11.20}$$

Equation (11.20) is for the assumption of random leaf orientation within the canopy. For the extreme case where all the leaves are uniformly layered in a canopy we can derive a modified form of Eq. (11.20) as:

$$\alpha_c = \frac{\omega_L}{1 + (1 - \omega_L^2)^{1/2}}. \tag{11.21}$$

From Eqs. (11.20) and (11.21) we can predict that the albedo of a canopy composed of randomly distributed leaves is approximately half as large as for a canopy of uniformly distributed leaves; i.e., when ω_L is small. When considered together, both equations show mathematically that the tendency for a canopy of randomly distributed leaves to trap scattered photons is greater than for a canopy of uniformly distributed leaves.

General values for albedo for different earth surfaces can be found in many textbooks concerning biophysics. Values for surface albedo, however, continue to evolve as new technological approaches, particularly those involving the use of satellite spectrometers become available. We have presented a range of values for earth surfaces garnered from the recent literature in Table 11.4. Many of these values have been derived from various solar wavebands, but they show convergence around some common trends. Coniferous, needleleaf forests tend to have some of the lowest albedo values among terrestrial ecosystems. Glacial surfaces have relatively high albedo when covered with fresh snow, but these change to lower values as bare ice is exposed. Among terrestrial ecosystems, herbaceous crops exhibit higher albedo than forests, burned forests exhibit lower albedo than burned tundra, and dry vegetated surfaces exhibit a higher albedo than wet vegetated surfaces. The ocean exhibits the lowest albedo, generally being less than 10%.

Studies of canopy albedo for forests (Ollinger *et al.* 2008) and a broad range of ecosystem types (Hollinger *et al.* 2010) have uncovered a positive correlation between albedo and mean leaf N concentration, up to a limit. Recognition of such a correlation provides a new tool for coupling surface albedo to functional aspects of ecosystems in land surface models, as processes such as net photosynthesis rate, stomatal conductance, and the emission of certain volatile organic compounds (such as isoprene) are also positively correlated with leaf N content. The cause(s) of these correlations may be due to covariance between leaf structure and function. Recall that the net photosynthesis rate of leaves is positively

Table 11.4 Albedo estimates of earth surfaces

Surface type	Albedo (%)	Reference
General global vegetation classes		Alton (2009)
Broadleaf forest	10	
Needleleaf forest	7	
C_3 grassland	14	
C_4 grassland	12	
Shrubland	7	
North American forests		Romàn et al. (2009)
Needleleaf (Arizona)		
Managed	6–10[1]	
Unmanaged	7–11	
Deciduous broadleaf (Tennessee)	12–16	
Needleleaf (burned) (Canada)		
1850	5–8	
1930	7–10	
1964	7–11	
1981	10–13	
Mediterranean ecosystems		Fernández et al. (2010)
Forest/shrubland	13	
Agricultural crops	20	
Marshes	12	
Sandy dunes	33	
Sahel agricultural ecosystems (millet)		Ramier et al. (2009)
fallow field	20–23 (wet)	
	32 (dry)	
crop field	26–29 (wet)	
	34 (dry)	
Arctic tundra (Alaska)		Rocha and Shaver (2009)
Unburned	17	
Burned	3–6	
Tropical forest (Amazonia)	12.5	Culf et al. (1995)
Tropical pasture (Amazonia)	17	Culf et al. (1995)
Greenland (coastal snow)	93	Kokhanovsky and Schreier (2009)
Greenland (bare ice during melt period)	<25	
Bare rock (aridland)	30	Zhou et al. (2003)
Sand (aridland)	40	Zhou et al. (2003)
Ocean ($\cos \theta = 0.6$–1.0)[2]	3–5	Jin et al. (2004)

[1] Variation is due to season and measurement approach.
[2] θ represents the solar zenith angle and ocean albedo is reported as a broadband value 0.25–4 μm.

correlated with leaf N concentration because of the requirement for high concentrations of the enzyme Rubisco in leaves with high photosynthesis rates (see Section 4.2). In leaves with a high biochemical capacity for photosynthesis the internal structure of the leaf, particularly within the mesophyll tissues, must facilitate an effective diffusive flux from the stomatal pores to the chloroplasts (see Section 8.4). High photosynthetic fluxes in leaves are typically correlated with high ratios of mesophyll cell surface area and effective exposure of that surface area to the intercellular air spaces of the leaf. The numerous interfaces between mesophyll cells and the intercellular air spaces in leaves with high photosynthetic capacities would tend to increase intra-leaf photon scattering and potentially increase the upwelling photon stream that represents leaf albedo (see Section 8.3).

11.3 Remote sensing of vegetation structure and function

It is beyond the scope of this book to provide a detailed treatment of remote sensing approaches and methods. However, a general treatment is warranted because remote sensing is important in the scaling of fluxes to landscapes and in defining canopy structure and function. Waveband-specific surface albedo is often measured with satellite and aircraft sensors in order to discern vegetation features such as LAI and even discern species-specific differences in leaf chemical composition and metabolic activity. Satellite remote sensing of surface vegetation provides broad spatial coverage at relatively high resolution (1 km) with a high rate of repetition (once per day to once every eight days depending on the signals needed). Older studies relied on the Advanced Very High Resolution Radiometer (AVHRR), which was carried on various satellites and provided relatively broad spectral differentiation. Reflectance in the visible part of the spectrum, using two channels of the instrument (detecting radiation at 0.58–0.68 μm and 0.73–1.10 μm for Channels 1 and 2, respectively), was used to construct the *normalized difference of vegetation index* (NDVI). The NDVI ratio of reflectance detected in Channel 1 (Ch_1, red) to Channel 2 (Ch_2, near-infrared) is calculated as:

$$NDVI = \frac{(Ch_1 - Ch_2)}{(Ch_1 + Ch_2)},\qquad(11.22)$$

and NDVI has been shown to correlate with LAI (Cihlar *et al.* 1991). Given the inherent correlations of NDVI to LAI, and the use of LAI as a basis for normalizing ecosystem fluxes, researchers have used NDVI in combination with various models to infer surface physiological dynamics, including transpiration and photosynthesis (Running and Nemani 1988). A second parameter that has been used in combination with remotely sensed surface reflectance is the *fraction of absorbed photosynthetically active radiation* (fPAR):

$$fPAR = \frac{aPAR}{PAR_0}$$
$$aPAR = (PAR_0 + PAR_s) - (PAR_c + PAR_t),\qquad(11.23)$$

where aPAR is the absorbed PAR; PAR_0 is the downwelling PPFD incident at the top of the canopy; PAR_s and PAR_c are the PAR fluxes reflected by the soil and canopy, respectively; and PAR_t is the PAR flux transmitted through the canopy to the soil (Asrar *et al.* 1989). Because the absorption of PAR is principally related to photosynthetic activity, aPAR is a measure of physiological function in a vegetated surface and $aPAR/PAR_0$ is approximately equal to NDVI. The value of aPAR is often determined using upward and downward looking sensors located above and below the canopy, and it can be correlated with satellite-measured vegetation indices (e.g., NDVI) for the same landscape scene; this provides a means of coupling vegetation photosynthetic activity with spectral reflectance.

More recently, remotely sensed data have been obtained with the Moderate Resolution Imaging Spectrophotometer (MODIS), which was originally launched with the TERRA satellite in 1999 (Myneni *et al.* 2002). MODIS provides multi-angular reflectance signals that reveal more insight into the three-dimensional structure of vegetation compared to AVHRR. MODIS takes advantage of the fact that reflected radiation is anisotropic. Structural attributes of the vegetation uniquely modify the paths and density of reflected photon streams. MODIS uses its multi-angular detection system, coupled with a three-dimensional canopy radiative transfer model and advanced algorithms, to directly estimate LAI and aPAR. This is essentially an inverse modeling problem – i.e., given certain properties of scattered radiation, can we describe the structural and absorptive features of the canopy that caused those properties? One of the key algorithms used in the MODIS analysis is the *bidirectional reflectance distribution function* (BRDF). The BRDF contains mathematical descriptions of how photons are scattered by the vegetated surface including explicit description of viewing geometry, reflectance from ground versus canopy, mutual shadowing by canopy elements, multiple-scatter dynamics, and corrections for atmospheric transmission. With insight generated from the BRDF, MODIS data can be used to calculate surface albedo across broad geographic regions.

The use of MODIS has provided new capabilities for resolving ecosystem-atmosphere CO_2 and H_2O fluxes at the global scale (Running *et al.* 2004, Mu *et al.* 2007). One of the products of the MODIS data set is the Enhanced Vegetation Index (EVI), an improvement over NDVI because it is not as sensitive to saturation of the reflectance signal from densely vegetated surfaces. MODIS products also include a modeled estimate of CO_2 fluxes, such as that associated with gross primary productivity (GPP). Using MODIS data GPP has been estimated as a function of aPAR according to the relation:

$$GPP = \varepsilon \, aPAR, \qquad (11.24)$$

where ε is radiation-use efficiency with units g C MJ^{-1} and aPAR carries units of MJ m^{-2} day^{-1}. The value of ε is "tuned" according to surface temperature and atmospheric vapor pressure deficit. Once GPP is estimated it can be combined with ecosystem process models to estimate net primary productivity (NPP) as $NPP = GPP - R_e$, where R_e is ecosystem respiration rate. This process has resulted in the development of global maps of NPP (Figure 11.16). It should be emphasized that the GPP and NPP estimates derived as MODIS products are *modeled values* conditioned on remotely sensed estimates of aPAR and LAI. There are significant sources of error in MODIS-modeled productivity estimates, including radiometric measurements and conversion algorithms (e.g., for LAI and aPAR), spatially

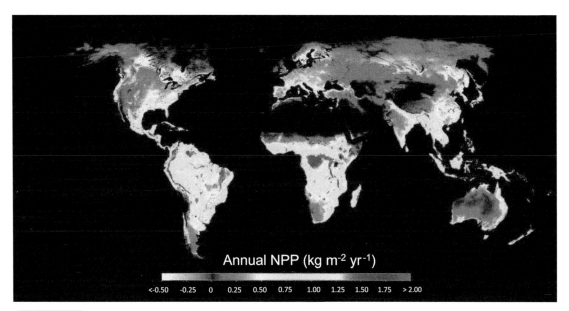

Annual NPP (kg m^{-2} yr^{-1})

<0.50   -0.25   0   0.25   0.50   0.75   1.00   1.25   1.50   1.75   >2.00

Figure 11.16 Global terrestrial net primary production (NPP) for 2002 computed from MODIS data combined with meteorological data and an ecosystem process model. From Running *et al.* (2004). See color plates section.

explicit meteorologic inputs (e.g., for temperature and vapor pressure), and the derivation of ecologic parameters and processes (e.g., for plant and ecosystem physiology) (see Zhao *et al.* 2005, 2006, Heinsch *et al.* 2006, Turner *et al.* 2006). One of the most recent developments in the use of remotely sensed data, especially in spectrally resolved form, involves the use of aircraft-deployed lidar (light detection and ranging) systems (Box 11.1).

Appendix 11.1 Reconciling the concepts of statistical probability and canopy photon interception

We have cast the prediction of PPFD profiles in terms of statistical probability in deriving Eq. (11.7). Here, we explore a more formal connection between the concepts of probability and photon transfer. The problem to be solved is to predict the probability that a photon will contact a leaf as it passes through successive canopy layers. In statistical terms this is a *binomial problem*; there are only two possible outcomes, contact or no contact. Stated mathematically:

$$P[x = n|N] = \frac{N!}{n!(N-n)!} \; (1-p)^{N-n} \; p^n. \qquad (11.25)$$

Equation (11.25) is the *binomial probability function*, and is read as "the probability (P) that the number of successes (contacts) x will equal n in N random trials is . . .." In this case, p is the probability of each individual success. Thus, if a series of random trials are conducted

Box 11.1	Lidar and canopy imaging

The use of *airborne lidar* (light detection and ranging) systems has provided new opportunities for characterizing canopy structure, particularly in its three-dimensional details (Lefsky *et al.* 2002, Wagner *et al.* 2008). Surface detection with lidar involves the time lapse between emission and return of a laser pulse aimed at the surface. Thus, the use of lidar is restricted to lower altitude deployment and most useful for characterizing small surface footprints with high spatial resolution (0.5–2 m). Terrestrial lidar deployment involves use of a laser beam in the near-infrared part of the electromagnetic spectrum, where reflectance from vegetation is high. The reflected radiation returned to the lidar will come as a mix of energy pulses from the reflectance footprint. The energy reflected from the more distant surface features will arrive later in time. Thus, by processing the time versus energy relation, insight into three-dimensional surface topography can be discerned (Figure B11.1). Lidar is typically deployed with an associated global positioning system (GPS) to provide reference coordinates for the reflectance scene. Lidar systems come in two different types, each using a different method of signal processing: *discrete-return lidar* systems measure the heights of selected energy peaks and are capable of processing a higher density of returned signals, whereas *full-waveform systems* continuously measure all of the returned energy. Both systems have been used in recent analyses, with discrete-return systems being favored for high precision tree mapping (Hudak *et al.* 2008), and full-waveform systems favored for broader-scale canopy structure (Asner *et al.* 2008a). Recently, airborne systems have been deployed with combined full-waveform lidar and a spectral sensor (Asner *et al.* 2008b), providing resolution of canopy structure across broad areas at as fine a scale as 1–2 meter square-grid resolution

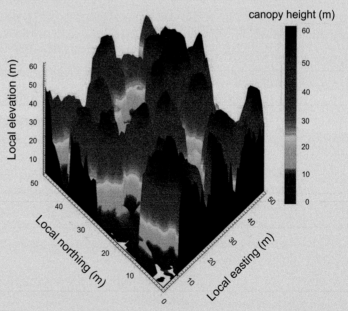

Figure B11.1	Canopy surface topography of a subsection of the Wind River Canopy Crane Research Facility in Washington State determined by discrete-return lidar. Individual lidar samples would be represented by individual points in three-dimensional space. From Lefsky *et al.* (2002). See color plates section.

(Asner *et al.* 2011). To the extent that individual species provide unique spectral reflectance patterns, this hybrid system provides the combined advantages of lidar for mapping canopy structure and the advantages of spectral imaging for mapping the presence or absence of certain species in the canopy. This type of high-resolution lidar sensing will become even more valuable as an ecosystem management tool in the future, providing detailed maps of carbon storage or carbon removal across a landscape.

the expected frequency with which one would observe zero contacts or one contact or two contacts, can be calculated. The pattern of *expected* frequencies for the different numbers of contacts would follow a *binomial distribution*.

In certain binomial situations, the number of random trials is large and the probability of success is small. In these cases, the frequency with which success is observed approaches a characteristic distribution pattern referred to as the *Poisson distribution*. The probability density function that describes a Poisson event is stated as:

$$P[x = n] = \frac{(m_x)^n \exp(-m_x)}{n!},$$ (11.26)

where m_x is the mean of a population of measurements of variable x. Equation (11.26) is read "the probability that the random variable x takes the value n is equal to ...." The Poisson distribution is derived as a limiting case of the binomial distribution as $N \to \infty$.

The Poisson probability predicts the likelihood of random and rare events. Given the infrequent occurrence of canopy light gaps, the penetration of I_D is indeed a random and rare event. Accordingly, the probability by which I_D can be expected to penetrate to a specific layer in the canopy can be predicted by a form of Eq. (11.26). The cumulative L of a canopy can be divided into a large number of horizontal layers (N). Progressing from the upper surface through each successive layer the cumulative L will change by ΔL, where $\Delta L = L/N$. Within each layer, the algebraic probability (p) of a photon making contact with a leaf is assumed to be equal to the mean number of contacts per layer (i.e., $m_x = m_c = G \Delta L/\cos \theta$). As $\Delta L \to 0$ and $N \to \infty$, the Poisson probability of observing no contacts (P_0) takes the form:

$$P_0 = \exp(-m_x) = \exp\left(\frac{-GL}{\cos \theta}\right).$$ (11.27)

P_0 is the probability that I_D will penetrate to a specific canopy layer. One important assumption within the Poisson model is that all leaves in the canopy are arranged randomly in space; anything other than random orientation would create a conditional dependence for the penetration of I_D from one layer to the next, violating the condition of independent trials that is required for Poisson probabilities.

In order to accommodate canopies in which a conditional dependence of photon penetration from one layer to the next exists, photon penetration models have been developed from the principle of Markovian chains. Markov theory predicts that if one canopy layer registers few photon contacts, then the canopy layer immediately below will also register few contacts. One simple way to add conditional probability to a model is to include a conditional multiplier. Starting with the Poisson model we can add the condition that the probability of observing a contact between a photon and a leaf is "$\lambda G \Delta L / \cos \theta$" for the case where the canopy layer above the leaf has registered a contact, and "$\lambda_0 G \Delta L / \cos \theta$" for the case where the layer above has not registered a contact (Nilson 1971). The conditional probabilities that no contact will occur within a canopy layer would be "$1 - \lambda G \Delta L / \cos \theta$" and "$1 - \lambda_0 G \Delta L / \cos \theta$." Thus, λ and λ_0 are Markov coefficients that calibrate the probability that a leaf will intercept a photon to the frequency of interceptions in the layer above. The maximum degree of independence among layers occurs when $\lambda = \lambda_0 = 1$. With λ and λ_0 defined as unity, the probability of intercepting a photon becomes unconditional and solely dependent on "$G \Delta L / \cos \theta$," thus reducing to the Poisson function. For the case when $\lambda_0 \neq 1$, and considering $N \to \infty$ (i.e., when $\Delta L \to 0$), the Markov probability of observing no contacts takes the form:

$$P_0 = \exp\left(\frac{-\lambda_0 G L}{\cos \theta}\right). \tag{11.28}$$

The Markov model is most useful for describing photon penetration through canopies with clumped foliage.

Appendix 11.2 The theoretical linkage between the probability of photon flux penetration (P_0) and the probability of a sunfleck (P_{sf}) at a specific canopy layer

Mathematically, we can derive the equality $P_0 = P_{sf}$ from an initial abstraction of the canopy as a volume of leaves divided into horizontal (plane) layers with progressively greater cumulative LAI (L) as we move from the top to the bottom of the volume (see Gutschick 1991). Recognizing that as L increases an inherent negative relation exists between L and P_0, we can state:

$$P_{sf} = f\left(-\frac{dP_0}{dL}\right), \tag{11.29}$$

where f represents "function of." Equation (11.11) is required to adhere to the principle of photon flux conservation – any change in the probability of direct beam transmission through a canopy volume should be reflected in the probability of direct beam interception as a sunfleck. In this case, we are working within the limit, as $\Delta L \to 0$. The higher the negative slope represented by dP_0/dL (in other words the steeper the decrease in P_0 as L increases), the less likely a sunfleck will occur at canopy layer L. To develop the relation further we recognize that the average direct beam photon flux density on a leaf in layer L ($\bar{I}_D$) can be defined as:

$$\bar{I}_D = I_D(\cos \theta)\left(-\frac{dP_0}{dL}\right). \tag{11.30}$$

Further, we recognize that the average I_D on a leaf equals the probability of a sunfleck multiplied by the fraction of leaf area in the canopy layer projected in a plane normal to I_D (i.e., the G-function):

$$\bar{I}_D = I_D G P_{sf}. \tag{11.31}$$

Combining Eqs. (11.30) and (11.31) results in:

$$I_D G P_{sf} = I_D (\cos \theta) \left(-\frac{dP_0}{dL} \right). \tag{11.32}$$

Rearranging Eq. (11.32) and substituting from Eq. (11.10) for P_0, an expression is obtained for P_{sf}:

$$P_{sf} = \frac{\cos \theta}{G} \left(-\frac{dP_0}{dL} \right) = \frac{\cos \theta}{G} \left(\frac{d \exp(-GL/\cos \theta)}{dL} \right). \tag{11.33}$$

Differentiating Eq. (11.33) using $c = -G/\cos \theta$, and $u = cx$, and taking the chain rule of differential calculus as:

$$\frac{de^{cx}}{dx} = \frac{de^u}{du} \frac{du}{dx} = e^u c = c e^{cx}, \tag{11.34}$$

and recognizing that:

$$\frac{de^{-x}}{dx} = (-e^{-x}), \tag{11.35}$$

the relationship between P_{sf} and P_0 can be expressed as:

$$P_{sf} = \frac{\cos \theta}{G} \left(-\frac{G}{\cos \theta} \right) \left(-\exp \frac{-GL}{\cos \theta} \right) = \exp \frac{-GL}{\cos \theta} = P_0. \tag{11.36}$$

12

Vertical structure and mixing of the atmosphere

A partir d'une hauter variable avec la situation atmospherique (de 8 km à 12 km) commence une zone caractérisée par lá très faible décroissance de température ou même par une croissance légère avec des alternatives de refroidissement et d'echauffement. Nous ne pouvans préciser l'épaisseur de cette zone; mais, d'après les observations actuelles, elle pataît atteindre au moins plusieurs kilometers.

[At some variable height in the atmosphere (between 8 km and 12 km) there begins a characteristic decay of the low temperature trend, or even a slight increase in temperature with alternating heating and cooling. We can specify the thickness of this zone and from the current observations it appears to be at least several kilometers.]

Leon Philippe Teisserenc de Bort (1902)

Prior to 1900 most meteorologists recognized that atmospheric temperature decreased with height and they assumed that this decrease was continuous. Late in the nineteenth century, however, the French meteorologist Leon Philippe Teisserenc de Bort used hydrogen balloons with precisely calibrated thermometers to demonstrate that while temperature did indeed decrease with height to approximately 8–12 km, above that height the temperature remained constant, or even increased slightly. Teisserenc de Bort referred to the atmosphere above 8–12 km as the "isothermal zone." Later, studies by the German meteorologist Richard Assmann confirmed the observations of Teisserenc de Bort, and even extended them to note that the so-called "isothermal zone" was actually a zone with consistent temperature increase as a function of height. This condition of warmer air above cooler air is now referred to as an "inversion," thus distinguishing it from the more commonly observed pattern of decreased temperature with height. In later writings, during the early twentieth century, Teisserenc de Bort postulated the existence of two atmospheric layers separated by the inversion – the *troposphere*, literally translated from the Greek word "tropein," which means to turnover, and the *stratosphere*, literally translated from the Latin word "stratificationem," which means to form into layers. Since his writings on this subject, the temperature inversion observed by Teisserenc de Bort has become known as the *tropopause*, the region separating the troposphere from the stratosphere. The discoveries of Teisserenc de Bort were not only important for our understanding of the physical arrangement of the atmosphere, but by focusing on atmospheric turnover in the troposphere, they laid the foundation for our current understanding of surface-atmosphere transport and even more general aspects of atmospheric physics.

In this chapter we explore structural and functional attributes of the atmosphere and how these attributes interact to cause vertical mixing. We begin with a discussion of tropospheric structure and its response to diurnal changes in solar forcing at the earth's surface. Solar energy sustains atmospheric buoyancy gradients that develop during the day and act as the "engine" that drives vertical mixing. Solar-driven mixing is a principal cause of surface-atmosphere exchange, especially the deep convective transport that spans the entire troposphere and extends to the tropopause. As we develop the concept of tropospheric mixing, and continue our emphasis on thermodynamics that was introduced in earlier chapters, we will develop concepts that explain the causes of atmospheric stability, or lack thereof. Within that context we will establish a framework for understanding the vertical differentiation of those driving and resistive forces that create atmospheric turbulence and thus those forces that determine the potential for turbulent transport. We will consider atmospheric turbulence as a dynamic process in the next chapter. Here, we will focus on spatial characterization of the atmospheric domains of shear- and buoyancy-generated forcings that create turbulence.

12.1 Structure of the atmosphere

Many of the principles by which we understand and predict atmospheric dynamics are based on the theories of fluid mechanics. Atmospheric fluid motions are potentially complex, and in some ways enigmatic. They include aspects that are highly *organized* (as evidenced in the rotational motions of whirlwinds, tornados, and cyclones), as well as *chaotic* (as evidenced in the intermittency and stochasticity by which gusts of wind can be detected on "blustery" days). At the largest scales, such as those associated with regional circulation cells, the properties of turbulent eddies converge with what we characterize as "weather," a concept that accommodates the stationarity we apply to forecasts of daily mean temperature and pressure. At smaller, local scales, the states of turbulent eddies change at the scale of seconds to minutes, providing the kinetic condition for ecosystem-scale exchanges of mass and energy with the atmosphere. In both cases, the atmospheric motions can be formally defined as turbulent in that their state properties represent departures from mean values; they only differ in the timeframe within which those departures occur.

Turbulence in the troposphere in and of itself implies time-dependent changes in the properties of the atmosphere. Turbulence characterizes a dynamic atmosphere. However, it is not so dynamic that structural organization does not exist. While taking on the properties of a diffuse compressible fluid, the atmosphere is not ideally mixed, and layers, such as those described by Teisserenc de Bort can be observed. That is, spatial domains with similar state properties can be distinguished from other spatial domains with different state properties. The boundaries between atmospheric layers are dynamic, changing frequently according to diurnal and seasonal cycles. Nonetheless, the mean properties of the layers are reliable enough to allow for some degree of classification.

12.1.1 The troposphere and its associated layers

The lowermost region of the atmosphere, with an average depth of 11 km, is the *troposphere*. The upper portion of the troposphere is defined as the *free troposphere*, which is characterized by strong horizontal winds that approach idealized *geostrophic flow* (i.e., exhibiting little variation in speed as a function of height). Geostrophic flow is driven by regional pressure gradients that are balanced by the Coriolis force. (The Coriolis "force" emerges from application of Newton's laws of motion to a rotating frame of reference. When combined with the centrifugal force caused by a rotating spherical earth, the Coriolis force explains the apparent angular trajectories of atmospheric parcels that move along N–S meridian lines in the free troposphere.) Horizontal transport in the free troposphere is rapid and trace gas molecules can be moved long distances before being mixed back toward the surface. The lowermost 10–25% of the troposphere is called the *planetary boundary layer* (PBL). Winds in the PBL respond to changes in surface conditions at time scales of an hour or less. The surface imposes a drag force on PBL winds, which in turn causes high rates of kinetic energy dissipation and induces high rates of shear near the surface. During the day, with high solar energy fluxes, energy is conducted from the surface to the atmosphere, warming the lowest part of the PBL and creating vertical buoyancy gradients. Higher above the surface, rising parcels of air begin to organize into convective updrafts that drive deep vertical mixing. A distinct boundary identified by a relatively thin region of temperature inversion (different from the inversion observed by Teisserenc de Bort) separates the PBL from the free troposphere and is known as the *capping inversion*. The capping inversion acts as a "lid" to the PBL and forms a natural barrier between the turbulent flows of the PBL and the laminar flows of the free troposphere. The PBL is the most relevant atmospheric layer with regard to surface-atmosphere mass, energy, and momentum exchanges.

Insight into the nature of the PBL can be gained by studying diurnal dynamics in its structure and stability (Figure 12.1). To illustrate this we start on the left-hand side of Figure 12.1, at noon on a relatively clear, warm day. In such conditions, the PBL is already in a state of high turbulence and can be described according to three layers – *the surface layer*, *the convective mixed layer*, and *the entrainment zone*. (Often, the surface layer is considered part of the convective mixed layer during the day and the nocturnal stable layer at night.

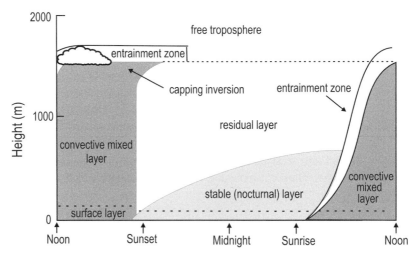

Figure 12.1 The diurnal cycle in the formation of the convective mixed layer during the daytime period, and its conversion to the nighttime stable layer and overlying residual layer. This cycle is typical of clear, warm days and clear nights. Redrawn from Stull (1988).

However, we will discuss it separately in order to make clear its unique role in atmospheric processes.) These layers are most accurately defined in terms of wind flow characteristics and gradients (or lack thereof) in state variables such as temperature, density, and moisture content. However, they can also be distinguished with regard to vertical position. The surface layer represents the lowermost portion of the PBL, where the atmosphere interacts directly with the surface and where most shear-generated turbulence is generated. The convective mixed layer represents the bulk of the PBL above the surface layer (at least during the midday period of warm, sunny days) and where convective turbulence begins to organize according to buoyancy gradients into relatively strong updrafts, which are in turn balanced, with regard to transported mass, by cooler, denser downdrafts. The entrainment zone lies at the top of the convective mixed layer, in the vicinity of the capping inversion. Air is exchanged between the free troposphere and PBL in localized areas of perturbation within the entrainment zone.

Considering the transition from midday to sunset, and moving from left to right along the timeline of Figure 12.1, surface heating weakens and turbulence within the convective mixed layer decreases in intensity. As the surface loses heat through longwave radiation to the nighttime sky, the air in the surface layer will cool and settle into a stable state known as the *stable nocturnal layer*. Above the stable nocturnal layer lie the remains of the convective mixed layer, which is often referred to as the *residual layer*; referring to the fact that it contains residual properties from the previous day's convective mixed layer. As surface cooling continues through the night the nocturnal stable layer will grow upward and consume the residual layer.

As we continue even further along the timeline of Figure 12.1, in the morning, with the onset of another day of surface warming, the convective mixed layer will develop once again from the surface up; the layer of stable air that remains from the previous night, and lies above the developing mixed layer, will become thinner and eventually disappear as the convective mixed layer grows upward, re-establishing itself beneath the capping inversion. In the following short sections we will elaborate on the nature of these dynamic atmospheric layers.

12.1.2 The surface layer

The surface layer is often approximated as the lowermost 10% of the PBL (Figure 12.2). Most turbulence in the surface layer is generated by the drag force that the earth's surface imposes on the wind and the Kelvin–Helmholtz instabilities associated with wind shear (see Section 7.2). The surface layer can be divided into two sublayers. The *roughness sublayer* extends from the ground to a distance 2–3 times the height of a vegetated canopy. The thickness of the roughness sublayer is proportional to the degree of "roughness" and structural heterogeneity at the canopy surface. The complex nature of wind flows in the roughness sublayer makes it difficult to reliably measure or model scalar fluxes; measurements or predictions at one point may vary from those a few meters away due to surface heterogeneity. The measurement of ecosystem-atmosphere exchange is typically made from a platform that is elevated above the roughness sublayer, allowing time-dependent statistical characterization of the flow and associated turbulent exchanges to approximate a state of stationarity.

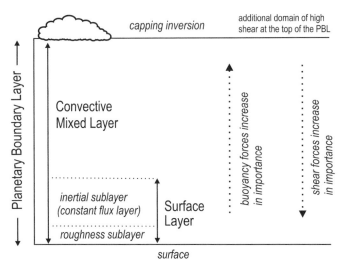

Figure 12.2 Schematic showing the surface layer in relation to the convective mixed layer and the opposing trends in shear-generated turbulence near the surface and buoyancy-generated turbulence that becomes organized higher above the surface. The thickness of the various layers is not drawn to scale.

Above the roughness sublayer is the *inertial sublayer.* The inertial sublayer represents the top portion of the surface layer and often extends from the top of the roughness sublayer to a few hundred meters above the surface; although there are no clear physical demarcations for either of these sublayers. In the inertial sublayer, shear forces give way to buoyancy forces as dominant controls over vertical scalar transport, especially during periods of high surface sensible heat flux, which tends to enhance atmospheric instability (Dupont and Patton 2012). The momentum flux from the horizontal wind to the earth's surface is approximately constant through the inertial sublayer (approximately 90% of the momentum flux at the surface, on average). That is, there is no significant vertical divergence in the amount of shear within the inertial sublayer and turbulent fluxes are approximately conserved as a function of height. Certain theoretical derivations require normalization by constant scalar fluxes. So, in the interest of theoretical convenience the inertial sublayer is often defined as that part of the lower atmosphere where fluxes are approximately constant with height (i.e., less than 10% variance). In fact, the inertial sublayer is sometimes called the *constant flux layer.*

12.1.3 The convective mixed layer

A primary cause of vertical motion and turnover in the convective mixed layer is the forcing due to negative pressure (buoyancy) gradients. As heat and moisture are transferred from the earth's surface to the overlying atmosphere, *Rayleigh–Taylor instabilities* develop. These instabilities are similar to Kelvin–Helmholtz instabilities in that they both result in fluid mixing, but the manner by which they cause mixing is fundamentally different. Rayleigh–Taylor instabilities form at the interface of two fluid layers with different densities, in this case under the influence of the gravitational force. As air parcels near the surface are

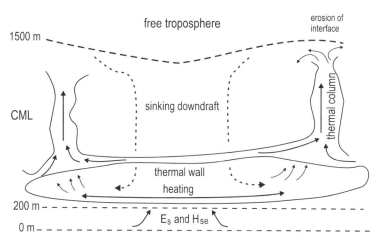

Figure 12.3 Structure in the convective mixed layer (CML) due to rising updrafts and sinking downdrafts. Sensible heat flux (H_{se}) and evaporative moisture flux (E_s) at the surface causes increased buoyancy of the air which rises to form a thermal wall a few hundred meters above the surface. Warmed air from the thermal wall flows laterally to feed updraft vertices (thermals). Slower, sinking downdrafts balance the updrafts. Typical scales for this type of atmospheric structure are approximately 1.5–3.0 times the height of the CML between thermal columns (Lothon *et al.* 2009). This figure was re-drawn from a concept presented in McNaughton (1989).

warmed, they rise and displace sinking parcels of cooler air from above. Thus, the interface between the fluid layers is unstable and the potential energy of this instability is converted to the kinetic energy of vertical mixing.

Once they are set in motion, warmed parcels of rising air will coalesce into slabs that are up to several kilometers in breadth and remain intact as they rise to a few hundred meters above the surface (Figure 12.3). Through this process, horizontal and vertical structure develops in the convective mixed layer. Thermal columns form at the vertices of the slabs causing updrafts that extend to the top of the PBL where their momentum is eventually stopped as the air cools and reaches buoyant equilibrium with its surroundings. Between the columns of rising air, downdrafts from the top of the PBL sink toward the surface, thus maintaining mass balance. The location and spacing of the updrafts and downdrafts are affected by surface properties, as well as natural scale-sensitive atmospheric instabilities in the convective system. There are measurable differences in vertical velocity of air in the updrafts and downdrafts, with the velocity being higher in the narrower expanse of updrafts (Figure 12.4). The alternation of neighboring domains of updrafts and downdrafts creates mixing through large convective eddies that increase in size and speed as the intensity of solar radiation increases. Vertical mixing in the convective mixed layer is so thorough that gradients in scalar concentration and potential temperature are minimal.

12.1.4 The entrainment zone

Thermal updrafts will develop greater velocity as the surface warms during the day, and associated air parcels will impact the capping inversion with progressively greater force.

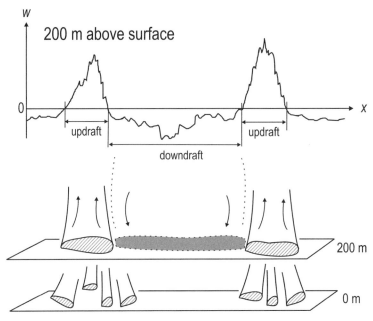

Figure 12.4 The coalescence of updrafts with increasing height above the surface. The cross-sectional area of updrafts is illustrated in the lower part of the figure. A hypothetical plot of vertical wind speed (*w*) as a function of horizontal distance (*x*) is shown for 200 m above the surface in the upper part of the figure. Note the broader spatial expanse and lower wind speed of downdrafts compared to updrafts. Redrawn from Hunt *et al.* (1988).

Localized regions of the inversion are eroded as the rising parcels strike its lower surface. The distortions caused by collision of the rising thermals disrupt laminar flow at the base of the inversion, and create billowing "puffs" of cumulus clouds that penetrate into the inversion. As thermal updrafts penetrate the inversion, localized regions of mixing, or entrainment, can occur, disbursing mass and energy that originated near the surface into the free troposphere and conversely mixing mass and energy from the free troposphere downward into the PBL.

Cumulus clouds form above the inversion boundary because convective updrafts are often propelled by sufficient velocity to cross the *lifting* (or *lifted*) *condensation level*, the height at which rising air parcels will cool to the point of water condensation on aerosols, dust, or other particles that act as *cloud condensation nuclei*. The formation of cumulus clouds is most frequently observed following clear, warm mornings with high solar energy loads on the surface. In conditions with sufficient soil moisture, updrafts will be laden with water vapor and the formation of cumulus clouds is more likely to be followed by precipitation events. In conditions with insufficient soil moisture, clouds formed at the tops of thermal columns will be less dense and precipitation will be less likely. Thus, convective rain triggers are controlled to a large extent by local soil moisture conditions and the potential for vegetation to drive evapotranspiration from a heated surface. If regional soil moisture and evaporative potential decrease below a critical level, convective rain triggers are likely to also decrease in frequency, requiring that moist air and accompanying precipitation be advected into the system from neighboring regions before evapotranspiration-precipitation cycles are re-established (Siqueira *et al.* 2009).

12.1.5 The stable nocturnal layer and residual layer

At dusk, surface heating ceases and mixing in the PBL decreases. The air nearest the surface will cool and settle, forming the stable nocturnal layer. Tracers released into this layer exhibit little vertical dispersion. Stability in the nocturnal layer is proportional to the rate and magnitude of surface cooling; evenings with clear skies and weak winds exhibit the most stable layers. The lower intensities of turbulence and weakened vertical transport make nocturnal turbulent flux measurements near the surface difficult. Above the stable nocturnal layer lies the residual layer. During the early evening some turbulence is retained in the residual layer as temperature gradients from the previous day linger. As surface cooling continues through the night the residual layer will dissipate, leaving only the nocturnal stable layer.

12.2 Atmospheric buoyancy, potential temperature, and the equation of state

The buoyant properties of air are defined according to *potential temperature* and *virtual potential temperature*. A parcel of air can be characterized by an actual temperature and a potential temperature. Potential temperature (θ_t) is the temperature that the parcel would have if it were dry and if it were adiabatically moved from its actual position in the atmospheric column (at pressure P) to a reference (standard) position (at pressure P_0, typically 100 kPa). A formal derivation of θ_t is provided in Appendix 12.1. The advantage to using θ_t, rather than T_a, is that θ_t is independent of changes in the pressure or volume of an atmospheric parcel as it moves through the atmosphere. That is, it provides a means by which to evaluate changes in the heat content of parcels as they change vertical position, without the complication of concomitant changes in pressure and volume. Potential temperature will not be the same for every air parcel because each parcel will contain different amounts of heat when normalized to the reference position (i.e., the initial condition defined by T_a will be potentially different for different parcels of air).

Potential temperature is formally defined according to a relation known as Poisson's equation ($\theta_t = T_a(P_0/P)^{0.286}$), but it can be approximated as:

$$\theta_t \approx T_a + z\left(g/c_p\right), \tag{12.1}$$

where T_a is absolute temperature (units of K), z is height above or below the 100 kPa reference level (m), g is the gravitational acceleration constant (m s^{-2}), and c_p is the specific heat of dry air at constant pressure (J kg^{-1} K^{-1}). The quantity g/c_p represents the dry-air adiabatic lapse rate. According to Eq. (12.1), potential temperature will increase with height; this is the state of a stable atmosphere – cooler air at the surface and warmer air above. Vertical convection occurs when the surface is warmed and the vertical gradient in actual air temperature opposes the stable gradient in potential temperature. An air parcel that achieves positive buoyancy and rises, will adiabatically expand and cool in its actual temperature, but assuming no transfer of heat from its surroundings, it will retain the same potential temperature.

Virtual potential temperature is defined as the temperature that a parcel of dry air must have in order to equal the density of moist air that exists at a specific reference height. Thus, virtual potential temperature accounts for the combined effects of heat, moisture, and altitude on air density. Virtual potential temperature (θ_{vt}) for moist but unsaturated air is often calculated as:

$$\theta_{vt} \approx \theta_t(1 + 0.61\,c_w), \tag{12.2}$$

where c_w is the water vapor mole fraction and 0.61 is derived from the approximated ratio of gas constants for dry and moist air.

Increases in potential temperature and virtual potential temperature occur when heat, and heat plus moisture, respectively, are transferred to air through exchange with the surroundings. At the earth's surface, transfer of sensible and latent heat cause θ_t and θ_{vt} to increase, thus establishing a change in buoyancy. Vertical mixing due to buoyancy gradients, and associated convection, has the ability to mix the lower atmosphere to considerably greater heights compared to the vertical motions caused by shear stress at the earth's surface. A good starting point for understanding the dynamic motions of buoyantly driven convection is to define it in terms of hydrostatic disequilibrium. In the vertical coordinate there are two opposing forces that interact to determine the state of the atmosphere. The gravitational force acts on the mass of the atmospheric column to produce a downward directed force and the vertical pressure gradient, increasing toward the earth's surface, facilitates an upward directed (buoyancy) force. To understand how the pressure gradient translates to an upward force imagine a large balloon filled with air of relatively low density (either warm air or helium). The balloon will rise when released because the pressure of the atmosphere is higher on the bottom surface of the balloon than on its top; this pushes the balloon upward, or in other words gives it buoyancy. For a homogeneous atmosphere (an assumption we make for the purposes of simplification), in which atmospheric mass density (ρ_a) is constant, we can describe the state of *hydrostatic equilibrium* as:

$$-\Delta P = \rho_a g \Delta z, \tag{12.3}$$

where ΔP represents the vertical pressure difference between points z_1 (closer to surface) and z_2 (further from surface), $\Delta z = (z_2 - z_1)$, and g is the acceleration due to gravity (m s^{-2}). Thus, the pressure (P) at any height is due to the weight of the atmosphere above that point. Disequilibrium in the hydrostatic state of the atmosphere, and the creation of buoyant motions, is caused by an imbalance between the pressure and gravitational forcings that influence a parcel of air.

As an exercise to better understand the concept of hydrostatic disequilibrium imagine a hypothetical atmospheric parcel that gains sensible heat (H_{se}) at the earth's surface. We might ask: how high can the parcel rise in response to its heat gain? To start with the simple case, let us assume that this parcel is not allowed to expand as the air inside is warmed.

Under such conditions all of the added heat (ΔH_{se}; in units J kg^{-1}) will go into increasing the temperature (ΔT) of air in the parcel according to:

$$\Delta H_{se} = c_v \Delta T, \tag{12.4}$$

where c_v is the *specific heat of dry air at a constant volume* (J kg^{-1} K^{-1}). Given the constraint that the parcel cannot expand, the equation of state ($PV = nRT$) requires that the increase in temperature be accompanied by a proportional increase in pressure. (Many treatments of atmospheric physics will use a normalized expression of atmospheric density, the specific volume (V), which represents the volume occupied by a unit mass (m^3 kg^{-1}), and is equal to $1/\rho_a$. Expressed in terms of specific volume, the equation of state becomes $PV = RT$.)

Now let us relax the assumption that the air parcel cannot expand. In this case, some of the absorbed heat will go toward increasing air temperature and some will drive volumetric expansion. Expansion requires that work be done against the surrounding atmosphere. Taking the definition of work as force times distance, the amount of work that must be done to expand the parcel is equal to the force required to overcome ambient pressure (P) multiplied by the expanded specific volume (ΔV). Expressed formally:

$$\Delta H_{se} = c_v \Delta T + P(\Delta V). \tag{12.5}$$

Equation (12.5) is analogous to the First Law of Thermodynamics defining the conservation of energy as $\delta Q = d U + \delta W = 0$, where Q is thermal energy (J), U is internal energy (J), and W is the work (J) done by a closed thermodynamic system on its surroundings. In meteorological studies, ΔV is not typically observed. Instead, volumetric dynamics are expressed in terms of their associated pressure changes, revealing:

$$\Delta H_{se} = c_p \Delta T - V(\Delta P), \tag{12.6}$$

where c_p is the *specific heat of dry air at a constant pressure*. A derivation of Eq. (12.6) is provided in Appendix 12.1.

Heat exchange between parcels of air and their surroundings is particularly relevant at certain atmospheric boundaries. As discussed above, at the ground surface sensible heat is conducted to overlying air molecules. At the inversion that caps the top of the PBL heat is conducted to the free troposphere. In the atmospheric column between these boundaries heat exchange between parcels of air and the surrounding atmosphere is negligible at the time scales associated with buoyant ascent and descent. Thus, as parcels of air rise and fall within the convective mixed layer their state properties can be predicted with good approximation by the equation of state, and we can state $\Delta H \approx 0$; in other words, the rising parcels exhibit adiabatic ascent (Section 2.6).

To continue with our analysis and derive an expression that predicts the distance that a parcel can rise given a finite amount of surface heat uptake, and assuming adiabatic ascent, we now need to reconsider the hydrostatic equation. Using Eq. (12.6) as a means to substitute for ΔP in Eq. (12.3), we obtain:

$$\Delta H_{se} - = c_p \Delta T + V \rho_a g \Delta z. \tag{12.7}$$

Equation (12.7) provides us with the relation we need to quantify the adiabatic ascent of an air parcel (Δz) given a finite amount of sensible heat uptake (ΔH_{se}).

As air parcels rise adiabatically they will do work against the surrounding atmosphere and cool. This provides the thermodynamic explanation for the decrease in temperature as a function of height that was observed in Teisserenc de Bort's balloon experiments. In addition to using Eq. (12.7) to quantify the distance of adiabatic ascent, we can use it to quantify the amount of associated adiabatic cooling. In this case we are not interested in translation of ΔH_{se} into a dynamic property of ascent; rather, we are assuming that a unit of H_{se} has been exchanged at some time in the past and has set a parcel in motion. Now, we want to determine the amount by which that parcel will cool (ΔT) as a function of the height of ascent (Δz). The value ($\Delta T/\Delta z$) will be independent of ΔH_{se}, as long as no further exchange of heat with the surroundings is allowed. Recognizing that $V\rho_a = 1$ and assuming that $\Delta H = 0$ during the ascent, we can write:

$$-\frac{\Delta T}{\Delta z} = \frac{g}{c_p}. \tag{12.8}$$

The term $-\Delta T/\Delta z$ is known as the *adiabatic lapse rate*, and it resolves to -9.8 K km^{-1} for ideally dry air. Moist air cools more slowly as it rises due to the release of heat as moisture condenses to form aerosol and cloud droplets. Observed lapse rates in natural conditions are typically in the range -5 K km^{-1} to -6 K km^{-1}.

12.3 Atmospheric stability

Stability within a fluid refers to the potential for turbulence. *Dynamic stability* refers to the potential for shear-generated turbulence. *Static stability* refers to the potential for buoyancy-generated turbulence. We will consider the causes and patterns of dynamic stability in detail in the next chapter. To further illustrate the concept of static instability, consider the case of an air parcel that is held in place at some height in the atmosphere. If, upon release, the parcel sinks or rises, because it is more or less dense, respectively, than its surroundings, the atmosphere is characterized as statically unstable at that height. If the parcel is at buoyant equilibrium with its surroundings and neither sinks nor rises upon release, and this is the case no matter which height the parcel is released, the atmosphere is characterized as statically neutral. If the parcel moves to a new position and then stops when released from different positions in the atmosphere, the atmosphere is characterized as statically stable. In a stable atmosphere, air below the imaginary parcel exists at higher density and air above exists at lower density. During diurnal transitions between the unstable convective mixed state and the stable nocturnal state the PBL will pass through the "*neutral state.*" (Buoyancy gradients in the atmosphere can be caused by relatively small temperature fluctuations (e.g., as small as 0.03 K). Thus, an ideally neutral atmosphere is rare to negligible in its frequency of occurrence. More commonly, reference to a "neutral atmosphere" or "neutral state" is intended to convey atmospheric domains in which shear, rather than buoyancy, is the dominant cause of turbulence.)

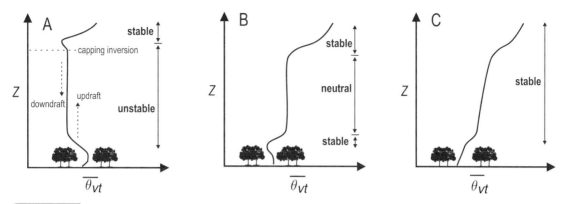

Figure 12.5   **A.** The relationship between mean virtual potential temperature $(\bar{\theta}_{vt})$ and height showing statically unstable conditions that support turbulent updrafts and downdrafts. This relationship might occur at midday during clear weather. **B.** The relationship between $(\bar{\theta}_{vt})$ and height showing statically stable and neutral conditions, such as might occur in the late afternoon or early evening as a result of cooling at the surface. Because surface cooling is often initiated from the top of the canopy, unstable conditions may persist within the closed canopy airspace, causing some local within-canopy turbulence. **C.** The relationship between $(\bar{\theta}_{vt})$ and height showing statically stable conditions, such as might occur in the early morning after a clear night, or above an open canopy. Drawn from a concept presented in Stull (1988).

Static stability is often characterized through gradients in mean virtual potential temperature (Figure 12.5). A negative gradient in $\bar{\theta}_{vt}$ ($\partial\bar{\theta}_{vt}/\partial z < 0$) indicates static instability, a positive gradient ($\partial\bar{\theta}_{vt}/\partial z > 0$) indicates stability and the absence of a detectable gradient ($\partial\bar{\theta}_{vt}/\partial z = 0$) indicates neutrality. The use of $\partial\bar{\theta}_{vt}/\partial z$ to infer stability is valid as long as the entire vertical profile is known. When the entire profile is not known, it is common for $\partial\bar{\theta}_{vt}/\partial z$ to be characterized for a portion of the PBL, providing what is known as the *local lapse rate*. Use of local lapse rates to infer static stability for the entire atmosphere is not recommended. Take the case of local neutrality (i.e., $\partial\bar{\theta}_{vt}/\partial z = 0$), which is adiabatic. In other words, the actual temperature cools as a function of height but not the potential temperature. The presence of an adiabatic column of air does not preclude upward buoyant fluxes. Buoyant fluxes are not driven by the lapse rate, but by a difference in density between a parcel of air and its surroundings. Thus, a parcel can rise through a local, adiabatic column of air if it has obtained heat beyond the lower boundary of the locally defined column and dissipates its kinetic energy beyond the upper boundary of the column.

The *Richardson number* (*Ri*), links the relative magnitudes of vertical gradients in $\bar{\theta}_{vt}$ (scaling with buoyancy-generated turbulence) and $\bar{U}$ (scaling with shear-generated turbulence):

$$Ri = \frac{(g/\bar{\theta}_{vt})(\partial\bar{\theta}_{vt}/\partial z)}{(u_*/kz)\,(\partial\bar{U}/\partial z)}, \qquad (12.9)$$

where u_* is the surface friction velocity with units of m s^{-1} (discussed in more detail in the next chapter), k is the von Karman constant (a proportionality constant that can be assumed as 0.4 for most surface layer applications), and $\bar{U}$ is the observed mean wind speed (m s^{-1}).

The Richardson number is dimensionless. It carries a value that is 0 for neutral static stability, positive for stable static stability, and negative for unstable static stability. The sign of Ri is determined by the direction of the $\bar{\theta}_{vt}$ gradient; i.e., when $\bar{\theta}_{vt}$ decreases with distance from the surface (reflecting unstable conditions), the gradient is negative; when $\bar{\theta}_{vt}$ increases with distance from the surface (reflecting stable conditions), the gradient is positive. This is the same sign convention used for diffusive gradients. In an atmosphere with $Ri \gg 1$, the vertical gradient in $\bar{\theta}_{vt}$ strongly favors static stability. In an atmosphere with $Ri \ll 1$, the vertical gradient in $\bar{\theta}_{vt}$ strongly favors static instability. The usefulness of Ri is restricted to the surface layer where gradients in $\bar{\theta}_{vt}$ and $\bar{U}$ are significant. During diurnal progressions from unstable to stable atmospheric regimes the Ri will increase and change sign from negative to positive.

Laboratory studies have shown that when drag is the dominant destabilizing force in a fluid, laminar flow converts to turbulent flow at a threshold value for Ri. This value is called the *critical Richardson number* (Ri_c). In general, Ri_c in the range 0.21–0.25 is valid for the surface layer. As is evident from Eq. (12.9), below Ri_c the convective forcing represented in the numerator is small compared to the advective forcing represented in the denominator. Another commonly used variation of Ri is the so-called *bulk Richardson number* (Ri_b). The calculus of Eq. (12.9) is stated with regard to the limit as $\Delta z \rightarrow 0$. This definition of Ri works well for theoretical formulations. However, in the real atmosphere, gradients in $\bar{\theta}_{vt}$ and $\bar{U}$ are measured across finite space. For this reason, the definition of Ri is often relaxed from ∂z to Δz; Ri_b is then derived for a discrete vertical expanse of the atmosphere. One commonly used form of the relationship defining Ri_b is:

$$Ri_b = \frac{(g/\bar{\theta}_{vt})(\Delta\bar{\theta}_{vt}/\Delta z)}{(\Delta\bar{U}/\Delta z)^2}. \qquad (12.10)$$

An alternate expression that is useful in characterizing the static state of the atmosphere is the ratio of height (z) to the Obukhov length (L). The *Obukhov length* represents the height at which gradients in buoyancy replace shear as the dominant control over turbulence:

$$L = -\frac{u_*^3\,\bar{\theta}_{vt}}{gk\left(\overline{w'\theta_{vt}'}\right)_0}, \qquad (12.11)$$

where $\left(\overline{w'\theta_{vt}'}\right)_0$ is the flux of virtual potential temperature at the surface. The Obukhov length carries units of meters when $\left(\overline{w'\theta_{vt}'}\right)_0$ carries units of m K s^{-1}. A greater understanding of the Obukhov length is achieved when it is recognized that u_*^3/kz is an expression of the turbulence kinetic energy generated by shear stress at the surface. The relative control over turbulence by shear decreases at a greater rate as a function of height, compared to that for buoyancy; the Obukhov length represents the height where these controls crossover and above which u_* diminishes as a governing parameter. As the potential for shear-generated turbulence at the surface increases, the two terms will converge at a higher height, and so the Obukhov length will be longer. Conversely, as the potential for buoyancy-generated turbulence increases, relative to shear-generated turbulence, the two terms converge at a lower height, and so the Obukhov length will be shorter. Static stability is often defined as the ratio of z to L. The advantage of using z/L, compared to Ri, is that L is constant with height through the surface layer whereas Ri may or may not be constant. (If the vertical gradients in $\bar{\theta}_{vt}$ and $\bar{u}$ are linear, then Ri will be constant with height, otherwise not.) Like the Richardson number, z/L carries a value of 0 for neutral conditions, a negative sign for statically unstable conditions, and a positive sign for statically stable conditions; the higher the value of a negative z/L, the more unstable the atmosphere, and the higher the value of a positive z/L, the more stable the atmosphere. In this case, mathematical sign is determined by the nature of $(\overline{w'\theta_{vt}'})_0$. A virtual potential temperature flux toward the atmosphere (i.e., during unstable conditions) is defined as a positive flux, whereas a virtual potential temperature flux toward the surface (i.e., during stable conditions) is defined as a negative flux.

The potential to develop large vertical buoyancy gradients from high surface fluxes of sensible and latent heat into the atmosphere, *and* to diminish those gradients during surface cooling at night, is highest when wind shear at the surface is lowest (Figure 12.6). At high $|z/L|$ in both stable and unstable regimes, the presence or absence of buoyancy gradients is influenced by the amount of shear at the surface. Regimes with high shear-generated mixing near the surface develop weak gradients in θ_{vt}, and are thus characterized with larger L, and concomitantly smaller z/L, in both stable and unstable regimes. It should be noted that analyses such as those shown in Figure 12.6 are biased by the fact that u_* appears in the terms defining both axes (note that u_*^3 is found in the numerator of Eq. (12.11)), creating autocorrelation. Nonetheless, the relationship is informative in demonstrating the trend toward a greater role for shear in atmospheric regimes tending toward neutrality, and lesser role for shear in regimes tending toward the unstable or stable states.

The interaction between shear and buoyancy controls the nature and extent of vertical structure in the daytime PBL (Figure 12.7). In the shear-dominated condition, with weak heat exchange at the surface, the surface layer is often overtopped by a near-neutral upper

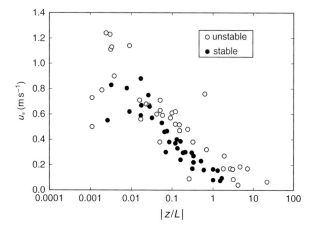

Figure 12.6 The relationship between atmospheric friction velocity (u_*) and the absolute value of atmospheric stability (z/L) above a boreal forest in northern Finland. The absolute value of z/L was used to facilitate comparison of stable and unstable regimes. Friction velocities are highest at near-neutral conditions or in regimes of low static instability or stability. Redrawn from Joffre *et al.* (2001), with kind permission from Springer Science + Business Media.

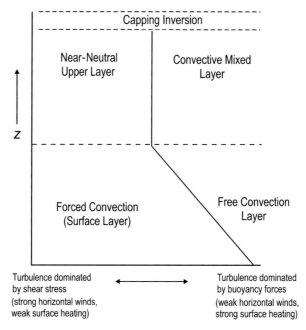

Figure 12.7   Vertical structure of the planetary boundary layer as a function of shear- and buoyancy-induced turbulence. z = height above the surface. Redrawn from Holtslag and Nieuwstadt (1986), with kind permission from Springer Science + Business Media.

layer. In the condition where surface heat exchange is high a free convection layer can form near the surface with strong thermal updrafts that feed the overlying convective mixed layer. At intermediate levels of surface heating, the free convection layer can be displaced from the surface and forms between the surface layer and convective mixed layer.

Appendix 12.1 Derivation of potential temperature and conversion from volume to pressure in the conservation of energy equation

We begin with a mathematical statement of the First Law of Thermodynamics: $d\mathrm{U} = \delta\mathrm{Q} - \delta\mathrm{W}$, where U is internal energy, Q is thermal energy, and W is work, all in J kg^{-1}. We define a change in enthalpy (H) as $d\mathrm{H} = d\mathrm{U} + d\mathrm{P}V = d\mathrm{U} + Vd\mathrm{P} + \mathrm{P}dV$, where H and U are in J kg^{-1}, P is in N m^{-2}, and V is in m^3 kg^{-1}. We state that adiabatic work is conducted through a change in the specific volume of an atmospheric parcel at any given pressure (i.e., $d\mathrm{W} = \mathrm{P}dV$), and that entropy and heat are translatable according to $d\mathrm{S} = \delta\mathrm{Q}/\mathrm{T}$ (Section 2.2). Combining all of these relations we can write:

$$d\mathrm{H} = d\mathrm{U} + (Vd\mathrm{P} + \mathrm{P}dV), \tag{12.12}$$

$$d\mathrm{H} = (\delta\mathrm{Q} - \mathrm{P}dV) + (Vd\mathrm{P} + \mathrm{P}dV), \tag{12.13}$$

$$d\mathrm{H} = (\delta\mathrm{Q} + Vd\mathrm{P}) = \mathrm{T}d\mathrm{S} + Vd\mathrm{P}. \tag{12.14}$$

Within the adiabatic condition, $TdS = \delta Q/T = 0$ and $dH = VdP$. Now, we introduce a new relation: $dH = c_p\,dT$, where c_p is the specific heat of air at constant pressure (J kg K^{-1}). Using this relation for substitution of dH, and $V = RT/P$ as substitution for V, dividing both sides by PV, and conducting algebraic rearrangement, we can state:

$$\frac{dP}{P} = \frac{c_p}{R}\frac{dT}{T}. \tag{12.15}$$

Integration with respect to pressure and temperature yields:

$$\left(\frac{P}{P_0}\right)^{R/c_p} = \frac{T}{T_0}. \tag{12.16}$$

Taking T_0 as θ_t, T as T_a (the temperature of an air parcel at its original height with pressure (P)), and P_0 as the pressure at a reference height, and calculating $R/c_p = 0.286$, we derive:

$$\theta_t = T_a\left(\frac{P_0}{P}\right)^{0.286}. \tag{12.17}$$

Now, let us take a closer look at the basis for converting between volume and pressure in parcels of air undergoing adiabatic ascent. To convert from a volumetric basis to a pressure basis, we assign specific state properties to the parcel (after it has exchanged heat with the surface) with its pressure being equal to $P + \Delta P$, its temperature being equal to $T + \Delta T$, and its volume being equal to $V + \Delta V$. Applying the equation of state, we obtain:

$$(P + \Delta P)(V + \Delta V) = R(T + \Delta T). \tag{12.18}$$

Expanding Eq. (12.18), assuming that the product between ΔP and ΔV is negligibly small compared to the other quantities, and taking $PV = RT$, we obtain:

$$P\Delta V + V\Delta P = R\Delta T. \tag{12.19}$$

Rearranging Eq. (12.19) to isolate $P\Delta V$, and substituting into Eq. (12.5) we obtain:

$$\Delta H_{se} = (c_v + R)\Delta T - V(\Delta P), \tag{12.20}$$

where c_v is the specific heat of dry air at constant volume (J kg K^{-1}). We now define $c_p = c_v + R$. This equality can be appreciated by examination of Eq. (12.20) with the caveat that pressure is kept constant (i.e., $\Delta P = 0$). In that case, $\Delta H/\Delta T$, which is the definition of specific heat, is equal to $c_v + R$. Expressing Eq. (12.20) in terms of c_p reveals:

$$\Delta H_{se} = c_p\Delta T - \Delta P. \tag{12.21}$$

13 Wind and turbulence

> There are two great unexplained mysteries in our understanding of the universe. One is the nature of a unified generalized theory to explain both gravity and electromagnetism. The other is an understanding of the nature of turbulence. After I die, I expect God to clarify the general field theory to me. I have no such hope for turbulence.
>
> Theodore von Kármán (unpublished)

Processes in the atmosphere have long had the power to capture human imagination and determine the fate of major historical events. Wind and atmospheric transport produced the "pure and white clouds" described in the poetry of Keats and are responsible for the dust storms that blinded armies during the Napoleonic wars. In our own lives we experience and depend on wind every day. It affects how airplanes fly, the efficiency of automobile travel, and our ability to predict weather. *Wind is air with velocity. Given that mass multiplied by velocity is a measure of momentum, wind can also be referred to as air with momentum.* Through its velocity the wind is coupled to "forces" that drive or resist its flow, such as pressure and friction, respectively. Instabilities in the atmospheric flow develop as these forces work against one another, causing gustiness, or more formally, turbulence. Turbulence reflects departures in the velocity vectors of the wind from their mean values. Turbulence represents a variance about the mean velocity. Turbulent departures from the mean wind flow are complex and not currently subject to precise mathematical description, as stated in the humorous quip reprinted above from the renowned atmospheric physicist, Theodore von Kármán. The transport of mass, energy, and momentum in the atmosphere occur through both the mean and turbulent components of the wind. Thus, in order to understand atmospheric transport, we must develop an understanding of the wind.

In this chapter we establish the context of wind and turbulence as atmospheric processes. We begin with the fundamental driving and resistive forces that produce wind, and we pay special attention to how those forces create turbulent instabilities that form near the earth's surface. We will establish a thermodynamic framework for these instabilities by introducing the concept of turbulence kinetic energy and discussing patterns of redistribution in this energy as turbulence forms in the various time and space domains of the atmosphere. We will then spend considerable time gaining a conceptual appreciation of wind eddies, the physical manifestation of atmospheric turbulence. Finally, we will establish two alternative mathematical frameworks, Eulerian and Lagrangian, within which to describe turbulence. It is from these frameworks that will emerge many of the models that are introduced in the following two chapters.

13.1 The general nature of wind

Wind is the flow of air. Wind has speed and direction. Wind is ultimately a process defined by physics; it is mass moved by a force. Wind can therefore be explained according to Newton's Laws of Motion. Stated simply, velocity in a mass is related to an external force applied to that mass. The external force, in turn, is capable of accelerating the mass according to $F = ma$, where F is force, m is mass, and a is acceleration. Furthermore, the application of force to a mass that collides with a second mass causes an equal and opposing force from the second mass. In the case of wind, the relevant, but opposing, forces are pressure and friction. A difference in pressure from one point on the earth's surface to another creates a force "differential" that accelerates atmospheric mass from the point of higher pressure toward the point of lower pressure. As accelerated molecules in the atmospheric mass collide with slower moving molecules within the mass itself and at the atmosphere-surface boundary, resistance to flow is encountered. Consistent with the theme of thermodynamics, which we have used as a basis for understanding fluxes throughout this book, we can state that wind, and the resistance it encounters, is a result of work being performed on the atmosphere. Starting from Newton's Second Law expressed as $a = F/m$ we can state:

$$a = -\left(\frac{\Delta P}{\rho_a \Delta x} + \frac{F_D}{m}\right), \qquad (13.1)$$

where a is in units of m s^{-2}, ΔP is the pressure difference between two points (N m^{-2}), ρ_a is atmospheric mass density (kg m^{-3}), Δx is the distance across which the pressure difference occurs (m), and F_D is the drag force (resistance) that causes deceleration caused by the resistance (g m s^{-2}), and m is mass (g). The drag force is caused at the molecular scale by friction and pressure due to collisions at molecular surfaces (see Section 7.2), a combination of forces that result in fluid viscosity. Molecular viscosity is likely important at fluid-surface boundaries, but in turbulent flows, higher-scale interactions will often emerge to become important determinants of shear stresses within the fluid itself. In the case of wind eddies, for example, parcels of air considerably larger than molecules can be transferred between adjacent, interacting layers, transferring momentum and creating shear; this process is often used to justify the concept of *eddy viscosity* as an emergent frictional force within the turbulent fluid that opposes the inertial forces propelling the flow. The sign convention on the right-hand side of Eq. (13.1) requires that positive acceleration occurs in a direction opposite to that of the pressure gradient. In other words, the pressure gradient increases from low-to-high pressure (positive ΔP), but acceleration of the air mass increases from high-to-low pressure. The drag force, F_D, works in the direction opposite that of the acceleration, but similar to that of the pressure gradient; i.e., the drag force vector is parallel to that from low-to-high pressure. Given the condition that during the course of a day ΔP tends to change from one state to another, wind speed will also change according to the product of acceleration and time (t):

$$U(t_1) = U(t_0) + a\Delta t, \qquad (13.2)$$

where $U(t_1)$ is the wind speed (m s^{-1}) at time t_1 characterized by a new state of ΔP, $U(t_0)$ is the wind speed at time t_0 characterized by the original state of ΔP, and $t_0 + t_1 = \Delta t$.

Using the Cartesian coordinate system we define three orthogonal, vector components that are summed to provide the "composite" wind speed observed at any instant in time. The longitudinal wind velocity is u; this is the vector describing the horizontal wind along coordinate x. The lateral wind velocity is v; this is the "cross-wind" vector along coordinate y. The vertical wind velocity is defined as w; this is the wind vector along coordinate z. (We will follow the convention of designating the overall, observed wind speed as U, and its Cartesian vector components as u, v, and w.) The relative magnitudes of the vector components will vary through time and thus in their relative contributions to U. Variations in these vector quantities can exhibit regularity in their time-dependent trends, such as those characterized by sinusoidal periodicity, as well as stochastic variances. Thus, a time series for U can be decomposed to the sum of the *mean wind velocity*, a *periodic wave velocity*, and a *random fluctuating velocity* (Figure 13.1), all of which contain varying contributions from u, v, and w.

The mean wind velocity reflects "quasi-constant" components of the driving forces behind atmospheric circulation. These forces include: (1) regional pressure gradients associated with synoptic weather dynamics and (2) local, near-surface pressure gradients associated with sloped topography (i.e., nighttime downslope drainage flows). The mean wind velocity reflects advection of energy and mass along these gradients. (The term "quasi-constant" is intended to convey the trend that these pressure gradients do not change appreciably in their magnitude or distribution during the span of time typically used to derive the mean wind velocity.) The periodic wave velocity reflects time-dependent variance around the mean wind velocity with a periodicity on the scale of a few hours. Wave velocities are not always present in a wind speed time series, and are often not explicitly considered in analyses of turbulent transport. Waves tend to occur at night under stable thermal stratification. If the waves are regular in their periodicity, the phases of the waves will offset one another resulting in no significant net, mean directional transport. The random fluctuating velocity is associated with turbulence; motions that occur on shorter time scales (e.g., fractions of a second-to-minutes) than those reflected in the mean wind or waves. Such short-term dynamics can be caused by instabilities in the flow that reach critical values and trigger sudden shifts in the local pressure field. The random fluctuating component of observed wind speed is responsible for the turbulent transport of heat, energy, and mass between ecosystems and the atmosphere.

13.2 Turbulent wind eddies

Atmospheric turbulence contains motions that vary across a broad spectrum of spatiotemporal scales (called *eddies*). The size of eddies increases as height above the ground increases. As large eddies move through the atmosphere their inherent instabilities cause them to decompose into smaller eddies creating an *inertial cascade* as energy is transferred from the largest to smallest motions. The cascade of eddies to progressively smaller sizes

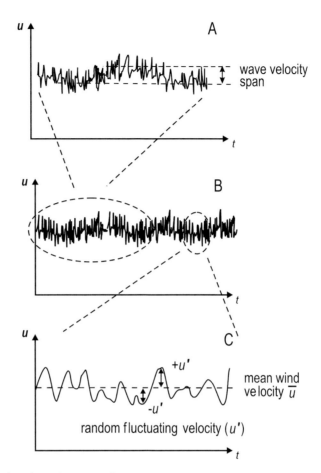

Figure 13.1 The components of an observed time series for longitudinal wind velocity (u). The wind velocity recorded as a time series is presented in panel B. The time record is composed of embedded patterns that exhibit wave-like repetition (panel A) as well as random fluctuations (panel C). For the analysis of turbulent processes that occur within a short time domain (e.g., minutes-to-an-hour) the wave-like repetition is typically ignored and Reynolds averaging is applied to the mean and random fluctuating components alone.

will continue until they are so small that all of their energy is consumed by work against viscosity, and their kinetic energy is converted to heat.

It is difficult to physically characterize eddies because their position changes as they are pushed along by the longitudinal wind. Eddies do not sit in one spot, rotating as ideal, vertically oriented vortices. Instead, they are composites of vertical bursts, gusting upward and downward, as they are advected horizontally. The dynamic nature of eddies forces an observer to measure their properties as passing "structures" as they move past a fixed observation point. Thus, eddies are characterized using *time-series measurements* combined with the assumption that they do not change in their essential features as they are advected past the observer; this assumption is referred to as *Taylor's frozen turbulence hypothesis* (Figure 13.2). In deriving the hypothesis, Taylor (1938) wrote: "If the velocity of the air

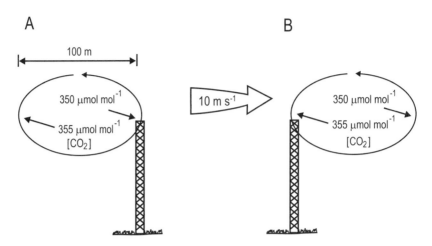

Figure 13.2 Representation of Taylor's frozen turbulence hypothesis. A hypothetical eddy that is 100 m in diameter is shown to be advected from left to right at the rate of 10 m s^{-1} and carrying a difference in CO_2 mole fraction of 5 μmol mol^{-1} from the leading edge to the trailing edge. Taylor's hypothesis states that the difference in CO_2 mole fraction can be detected using time series measurements separated by a 10-second interval. Redrawn from Stull (1988), with kind permission from Springer Science + Business Media.

stream which carries the eddies is very much greater than the turbulent velocity, one may assume that the sequence of changes (in the wind speed) at the fixed point is simply due to the passage of an unchanging pattern of turbulent motion over the point." In other words, we use time series measurements to "trade space for time."

If we ignore periodic wave velocities as causes of variance about the mean wind velocity (as discussed above these wave velocities are relatively rare), then we can focus on the random, fluctuating velocity as a measure of turbulent eddies. That is, we can assume that the pattern and magnitude of the observed variances contain insight into the structure and turbulent speed of eddies. Various statistical methods are deployed to extract that insight, including chaos theory, Fourier transformation, conditional sampling, and wavelet analysis. The most fundamental characterization of turbulence is obtained through *Reynolds averaging* (Appendix 13.1). In Reynolds averaging we assume that a fluid with turbulence can be described at any instant as the sum of its mean state and any fluctuation, or departure (represented with a "prime"), from the mean state: $x = \bar{x} + x'$. Fluctuations from the mean are defined statistically as variances (i.e., $(x - \bar{x})^2$), and the total unbiased variance in the time series ($\sigma[x]^2$) is represented as:

$$\sigma^2[x] = \frac{\sum_{i=1}^{n}(x - \bar{x})^2}{n - 1}, \tag{13.3}$$

where n represents the number of observations in the time series. We can use the total variance to characterize *turbulence intensity* (e.g., i_u) as $i_u = \sigma_u / \bar{u}$, where in this case, σ_u is the standard deviation (square root of the variance) of the longitudinal wind velocity normalized to the mean longitudinal wind velocity, $\bar{u}$. This same measure of turbulence intensity can be applied to the cross-wind (i_v) and vertical wind (i_w) velocities.

We can extend the concept of Reynolds averaging to two variables in a time series. The product of two variables is the product of their individual means plus a *covariance*. As an example, we can calculate the product between longitudinal wind velocity (u with units m s^{-1}) and a scalar concentration (c with units g m^{-3}), which represents the longitudinal flux density for the scalar entity in units of g m^{-2} s^{-1}. From Reynolds averaging we can then derive the instantaneous flux density as the sum of the mean flux (Term I) and covariance flux (Term II):

$$\overline{u\mathrm{c}} = \overline{u}\,\overline{\mathrm{c}} + \overline{u'\mathrm{c}'}.$$

$$\quad\;\; \mathrm{I} \qquad \mathrm{II} \tag{13.4}$$

The covariance flux is taken as the *turbulent flux*. The overbar in Term II indicates time-averaging of the covariance. The time-averaged, turbulent flux (Term II) is referred to as the *eddy covariance*, or *eddy flux*. This term can be applied to vertical, longitudinal, or cross-wind covariances (i.e., $\overline{w'\mathrm{c}'}$, $\overline{u'\mathrm{c}'}$, or $\overline{v'\mathrm{c}'}$, respectively).

13.3 Shear, momentum flux, and the wind profile near the surface

The interactions among inertial and frictional forces, and the transfer of momentum, that combine to create shear in air flows across leaf surfaces (Section 7.2), can be applied in analogous manner to the formation of turbulent wind eddies in the surface layer of the PBL. Shear is evident in the atmosphere as a decrease in wind speed as the earth's surface is approached (Figure 13.3). According to conservation laws, the near-surface decrease in wind speed must be accompanied by a transfer of momentum toward the surface (Figure 13.4); momentum must be conserved. Thus, there are fundamental relations among the concepts of shear, wind speed, and momentum transfer that should be amenable to mathematical description. Let us explore these relations by partitioning a change in the overall wind speed, as the surface is approached, into its fundamental components. We can write an expression that is derived from the Navier–Stokes equations (see Section 2.7):

$$\frac{d\mathrm{U}[t, x, y, z]}{dt} = \frac{\partial \mathrm{U}}{\partial t} + u\frac{\partial \mathrm{U}}{\partial x} + v\frac{\partial \mathrm{U}}{\partial y} + w\frac{\partial \mathrm{U}}{\partial z} \approx -\frac{1}{\rho_a}\frac{\partial \mathrm{p}}{\partial x} + \frac{\mu}{\rho_a}\left(\frac{\partial^2 \mathrm{U}}{\partial x^2} + \frac{\partial^2 \mathrm{U}}{\partial y^2} + \frac{\partial^2 \mathrm{U}}{\partial z^2}\right), \tag{13.5}$$

$$\text{Term I} \qquad\qquad\quad \mathrm{II} \qquad\qquad\qquad\qquad \mathrm{III}$$

where p is pressure (in units N m^{-2}), μ is dynamic viscosity (units of kg m^{-1} s^{-1}), and ρ_a is the dry air mass density (kg m^{-3}), and we assume the condition of an incompressible atmosphere. As we work from left to right in Eq. (13.5) we can see the time-dependent derivative moving into the pressure and viscosity terms and space-dependent derivatives moving into the second-order terms that define divergence in the wind speed gradient in three dimensions. Equation (13.5) is written in the form of a conserved budget and it reflects

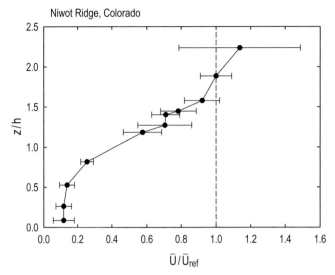

Niwot Ridge, Colorado

Figure 13.3 The vertical profile of mean wind speed ($\overline{U}$) above and within a subalpine coniferous forest in Colorado. The height above the surface (z) is expressed relative to the canopy height (h), which is approximately 11.4 m. The mean wind speed is expressed relative to the mean wind speed at a reference height ($\overline{U}_{ref}$), in this case at $z = 1.89$ h. Redrawn from Turnipseed *et al.* (2003), with kind permmision from Springer Science + Business Media.

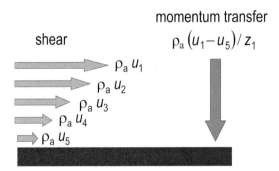

momentum transfer

shear

$\rho_a \left(u_1 - u_5\right)/z_1$

$\rho_a\, u_1$

$\rho_a\, u_2$

$\rho_a\, u_3$

$\rho_a\, u_4$

$\rho_a\, u_5$

Figure 13.4 Shear is represented as a vertical profile in the product between longitudinal wind velocity (u) and atmospheric mass density (ρ_a) at five heights above a surface. The decrease in $\rho_a u$ as the surface is approached indicates the deposition of momentum from the wind to the surface.

continuity constraints (see Section 6.3); that is, divergence in the space-dependent compo-nents of flow must be balanced by divergence in the time-dependent components of flow. The second moment derivatives that compose the right-hand side of Term III represent the frictional forces per unit mass in the directions of the three Cartesian coordinates, *x*, *y*, and *z*. (Statistical moments, in general, are described further in the mathematical concepts supple-ment provided online at: http:\\www.cambrige.org/monson) Thus, Term III in Eq. (13.5) reflects the condition that divergence in the horizontal pressure gradient must be balanced by divergence in the resistive frictional forces.

We refer to frictional forces within the fluid, in general, as *viscosity*; more formally, absolute viscosity (μ) is the proportionality constant that links the *shearing stress* (τ), with units of N m^{-2}, to the vertical velocity gradient ($\partial U/\partial z$):

$$\tau = -\mu \frac{\partial U}{\partial z}. \tag{13.6}$$

The negative sign on the right-hand side of the equation indicates that the amount of stress in the fluid *increases* vertically as the wind speed *decreases*. Now, let us develop a deeper understanding of τ by considering the forces that create mechanical stress in a fluid. Consider a unit volume of fluid – a "fluid particle" or "control volume" – located in the zone of shear stress, near a stationary surface, and subjected to differential transfer of momentum on its upper and lower faces. The spatially averaged viscous force ($\overline{F}_v$) per unit of atmospheric mass (m_a) on the control volume in the longitudinal (x) coordinate can be represented as:

$$\frac{\overline{F}_v}{m_a} = \left(\frac{\tau(\text{upper}) - \tau(\text{lower})}{\Delta z} \right) \frac{\Delta x \Delta y}{\rho_a \Delta x \Delta y}, \tag{13.7}$$

where $m_a = \rho_a \Delta x \Delta y \Delta z$ is the atmospheric mass contained within the volume of the particle, and where we have taken account of the fact that the viscous forces are distributed across the two-dimensional area of the upper and lower faces of the particle ($\Delta x \Delta y$). Now, differentiating Eq. (13.7) as $\Delta z \to 0$ and using Eq. (13.6) for substitution, we obtain:

$$\frac{F_v}{m_a} = \frac{1}{\rho_a} \frac{\partial \tau}{\partial z} = \frac{1}{\rho_a} \frac{\partial}{\partial z} \left(\mu \frac{\partial U}{\partial z} \right) = \frac{1}{\rho_a} \frac{\partial^2 U}{\partial z^2}. \tag{13.8}$$

Simplifying, we can state:

$$\frac{\partial \tau}{\partial z} = -\mu \frac{\partial^2 U}{\partial z^2}. \tag{13.9}$$

To this point, we have used the concept of absolute, or dynamic, viscosity (μ) in our derivations. In some applications, especially those involving the interaction of moving fluid particles, it is desirable to define viscosity in terms of the kinetics of motion, rather than force. In those cases, we can derive the *kinematic viscosity* (v) as: $v = \mu/\rho_a$. Thus, kinematic viscosity is the dynamic viscosity normalized to fluid density. From the derivation of Eq. (13.9), the mathematical concept of viscosity and its relation to shear stress and momentum transfer can be more clearly appreciated. From this exercise, we can begin to understand the causes of the decrease in wind speed as the earth's surface is approached and its relation to inertial and resistant forces. The stress that develops within a fluid due to momentum transfer within particles of the fluid itself, and eventually to the stationary surface, is often referred to as *viscous shear stress*. Viscous shear stress is ultimately due to the differential forces that influence the upper and lower surfaces of fluid particles as they move in the near-surface flow.

A second type of stress that can be transferred through a fluid is due to the presence of turbulent eddies that transfer momentum to fluid particles, and eventually to the surface. This type of momentum transfer also leads to stress within the fluid, in this case involving "eddy viscosity," and it is referred to as *Reynolds shear stress*. Mathematically, Reynolds shear stress emerges from the momentum equation following the application of Reynolds averaging. We begin with Eq. (13.5), apply Reynolds averaging, and simplify:

$$\frac{\partial \bar{u}}{\partial t} \approx -\frac{1}{\bar{\rho}_a}\left(\frac{\partial \bar{p}}{\partial x}\right) + \frac{1}{\bar{\rho}_a}\frac{\partial}{\partial z}\left(\mu\frac{\partial \bar{u}}{\partial z} - \bar{\rho}_a\overline{u'w'}\right). \tag{13.10}$$

This new budget equation reveals the Reynolds shear stress on the far right-hand side of the equation; in the term $(-\bar{\rho}_a\overline{u'w'})$, where $\bar{\rho}_a$ is the mass density of dry air at the time-averaged air temperature and pressure for the same measurement period as that for the turbulent velocity covariance $\left(\overline{u'w'}\right)$.

A derivation of Eq. (13.10) is presented in Appendix 13.2. As reflected in Eq. (13.10), Reynolds shear stress represents the statistical covariance between two turbulent wind vectors, one that moves parallel to the surface (u') and one that moves perpendicular to the surface (w'). Physically, this covariance represents the work of turbulent eddies as they transfer momentum through the atmosphere toward the surface. Recall that the statistical covariance between two turbulent quantities represents a flux density (Section 13.2). The momentum flux at the surface, due to Reynolds shear stress, can be represented as $(-\bar{\rho}_a\overline{u'w'})_0$. By convention, the negative sign indicates that the direction of the flux is from the atmosphere toward the surface. We can write a general equation for the Reynolds shear stress as:

$$\tau = -\bar{\rho}_a\overline{u'w'}. \tag{13.11}$$

In the turbulent planetary boundary layer, Reynolds shear stress is a greater source of momentum transfer than viscous shear stress. With the concepts of mechanical stress, shear, and momentum transfer in hand, we can appreciate a more fundamental

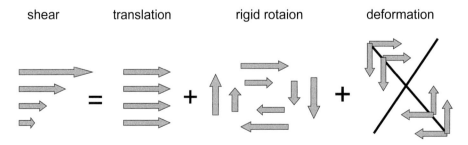

shear   translation   rigid rotaion   deformation

Figure 13.5 Representation of shear as the sum of translation, rigid rotation, and pure deformation. Redrawn from a concept presented in Blackadar (1997), with kind permission from Springer Science + Business Media.

recognition of shear as the vector sum of three component processes: translation, rigid rotation, and deformation, all of which reflect the transfer of momentum (Figure 13.5).

13.3.1 Surface friction velocity

A scaling parameter that is useful in normalizing atmospheric processes driven by shear-generated turbulence is the *friction velocity* (u_*), defined as $u_* = \sqrt{\overline{|u'w'|}_0}$. The units of u_* are m s^{-1}, the same as wind speed. The friction velocity provides insight into the inherent tendency for wind to transfer momentum to the surface, create shear stress, and produce turbulence. Friction velocity is the square root of the Reynolds shear stress. Thus, following from Eq. (13.11), we can write:

$$\tau = -\rho_a\, u_*^2. \tag{13.12}$$

We can derive a relation linking wind shear (expressed as the vertical gradient in wind speed) and u_*:

$$\frac{\partial U}{\partial z} = \frac{u_*}{kz}, \tag{13.13}$$

where k is the von Karman proportionality constant, which is dimensionless and typically takes a value of 0.4 (Högstrom 1988, Frenzen and Vogel 1992). Integration of Eq. (13.13) with respect to z leads to the logarithmic wind profile equation:

$$U[z] = \frac{u_*}{k}\, \ln\!\left(\frac{z}{z_0}\right). \tag{13.14}$$

Inherent in Eq. (13.14) is the assumption that in a neutral atmosphere, surface roughness will interact with the wind to cause U = 0 at some height above the surface; the rougher the surface, the higher the point where U = 0. The theoretical height at which U extrapolates to zero, when expressed as a function of ln z, is known as the *aerodynamic roughness length* (z_0). In Eq. (13.14) z_0 represents the constant of integration. It is clear from Eq. (13.14) that z_0/z scales positively and exponentially as a function of the ratio $u_*/U[z]$. In other words, the greater the potential for the surface to extract momentum from the atmosphere, and induce shear stress, the greater the value of z_0; that is, the "rougher" the surface (in aerodynamic terms). The value of z_0 is fundamentally a measure of surface drag. A derivation of Eqs. (13.13) and (13.14) is provided in Appendix 13.3.

As a real, canopied surface is approached, reductions in U[z] do not follow the idealized relation defined in Eq. (13.14). In fact, U[z] will, on average, reach a value of zero at some height above that predicted by z_0. The reality of an upwardly displaced surface, as distinct from z_0, is distinguished as the *zero plane displacement height* (d_H); a term that we will simplify and refer to as *displacement height*. The difference between d_H and z_0 is proportional to surface roughness (Figure 13.6).

Theoretically, d_H should represent the mean height within a canopy where Reynolds shear stress diminishes to zero (Thom *et al.* 1975, Raupach and Thom 1981). Given a canopy with

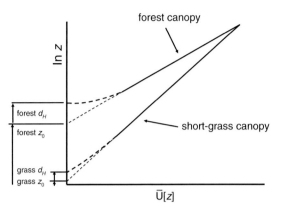

Figure 13.6 Theoretical figure showing relationship between the aerodynamic roughness length (z_0) and zero plane displacement height (d_H) for hypothetical forest and short-grass canopies. The presence of the canopy causes the relation between the time-averaged wind speed at height z ($\overline{U}[z]$) and ln z to bend and extrapolate to zero at height greater than z_0. The relative slopes of the two solid lines show that $\overline{U}[z]$ decreases more for a unit change in height, in the case of the taller forest canopy, compared to the shorter grass canopy, illustrating the greater capacity for the taller canopy to extract momentum from the wind.

d_H, the most relevant vertical dimension for defining turbulent transport is $(z - d_H)$, rather than z alone. Thus, Eq. (13.14) can be rewritten as:

$$\overline{U}[z] = \frac{u_*}{k} \ln\left(\frac{(z - d_H)}{z_0} \right) \qquad (13.15)$$

where $\overline{U}[z]$ is the time-averaged wind speed as a function of height (z). Direct determination of d_H is difficult, especially for forests. Difficulties arise because of physical causes (e.g., canopy patches create heterogeneity in aerodynamic roughness), and theoretical causes (e.g., curvature in the wind profile as the surface is approached complicates extrapolation to d_H) (Shaw and Pereira 1982, Wieringa 1993). One method of estimating d_H involves use of Eq. (13.14) along with the observed wind speed profile well above the canopy to determine u_*. We will refer to this as the "theoretically derived" value of u_*. The theoretically derived value of u_* can then be compared to that obtained from a sonic anemometer located near the top of the canopy using the relation: $u_* = \sqrt{|\overline{u'w'}|}$. We will refer to this as the "empirically derived" value of u_*. Assumed values of d_H in Eq. (13.15) can then be varied iteratively to determine the value at which the theoretically and empirically derived estimates of u_* converge (Foken and Nappo 2008).

An examination of classic textbooks on atmospheric dynamics will yield some general "rules of thumb" for estimating d_H. In general, d_H is 60 to 80% of canopy height (Oke 1987, Monteith and Unsworth 1990). These values, however, are biased toward measurements above crops; in forests d_H has been estimated to be in the higher portion of this range. The factor that most influences d_H is the vertical distribution of leaf area and porosity (openness) of the canopy. Shaw and Pereira (1982) used a higher-order turbulence closure model to examine the relations among d_H, z_0, LAI, and surface drag. In general, it was predicted that: (1) as canopy closure increases (i.e., porosity decreases) the surface becomes aerodynamically smoother and z_0 decreases; (2) values of d_H/h, where h is canopy height, can range from 0.4 to 0.9; and (3) the highest values of d_H/h are associated with canopies that have high LAI and where the zone of maximum LAI is positioned higher above the ground. For application in weather and climate models, Raupach (1994) developed analytical equations for expressing z_0 and d_H as functions of LAI (here represented as L) and h:

$$1 - \frac{d_H}{h} = \frac{1 - \exp(-\sqrt{aL})}{\sqrt{aL}} \quad \text{and} \quad \frac{z_0}{h} = \left(1 - \frac{d_H}{h}\right)\exp\left(-k\frac{u_h}{u_*}\right), \quad (13.16)$$

where a and k are empirical coefficients.

13.4 Turbulence kinetic energy (TKE)

Turbulence in the PBL is powered by kinetic energy (referred to as *turbulence kinetic energy*, or TKE). Kinetic energy (KE) is defined as KE = 0.5 mv^2, where m is mass and v is velocity. Using this relation, we can define the kinetic energy contained in a fluid flow as KE = 0.5 mU^2 = 0.5 m($u^2 + v^2 + w^2$). From Reynolds averaging, we can partition the wind's overall kinetic energy into mean kinetic energy (MKE) and time-averaged turbulence kinetic energy (TKE), when averaged across an interval of time and expressed per unit of atmospheric volume, as:

$$\text{MKE} = 0.5\,\bar{\rho}_a(\bar{u}^2 + \bar{v}^2 + \bar{w}^2), \quad (13.17)$$

$$\text{TKE} = 0.5\,\bar{\rho}_a(\overline{u'^2} + \overline{v'^2} + \overline{w'^2}), \quad (13.18)$$

where overbars represent time averaging and recalling that $\bar{\rho}_a$ is the time-averaged mass density of air and has units kg m^{-3}. The statistical and physical foundations upon which we define TKE are made clear by the squared terms in Eq. (13.18), which refer to wind speed variances (the random fluctuating velocity vectors of wind speed), and thus reflect velocity in the eddy motions of turbulence. Once produced, TKE can be passed from larger eddies to smaller eddies, be redistributed to higher or lower levels in the atmosphere, or be dissipated by conversion to heat. We can write a budget that balances the production of TKE against these possible fates. Assuming horizontal homogeneity, and considering turbulence in the vertical coordinate (i.e., w') as the only spatially explicit divergent term in the budget, the time-dependent production and loss of TKE for a unit volume of atmosphere can be defined as:

$$\frac{\partial \text{TKE}/\bar{\rho}_a}{\partial t} = \left(\frac{g}{\bar{\theta}_{vt}}\left(\overline{w'\theta_{vt}'}\right)\right) - \left(\overline{u'w'}\frac{\partial \bar{u}}{\partial z}\right) - \left(\frac{\partial\left(\overline{w'(\text{TKE}/\rho_a)}\right)}{\partial z}\right) - \left(\frac{1}{\bar{\rho}_a}\frac{\partial\left(\overline{w'p'}\right)}{\partial z}\right) - \varepsilon,$$

$$\qquad\qquad\quad \text{I} \qquad\qquad\quad \text{II} \qquad\qquad\quad \text{III} \qquad\qquad\quad \text{IV} \qquad\quad \text{V}$$

$$(13.19)$$

where g is acceleration due to gravity, p′ is the turbulent pressure fluctuation, and ε is the rate of heat production due to viscous forces. In Eq. (13.19), the time derivative of the time-averaged TKE is equal to the production (or loss) of energy from the buoyancy gradient (Term I), production due to the momentum deposited by wind shear (Term II) (in this case momentum flux, $\bar{\rho}_a \overline{u'w'}$, where $\bar{\rho}_a$ is the time-averaged mass density of dry air, is deposited to the ground and takes a negative sign), loss (or gain) by vertical redistribution of energy as turbulent eddies are transported to higher or lower levels in the atmosphere (Term III), loss (or gain) due to vertical redistribution of energy through pressure perturbations (e.g., by gravity waves) (Term IV), and loss due to the conversion of kinetic energy to heat (Term V). Any residual energy within the balance of the various terms on the right-hand side of Eq. (13.19) will be retained within the unit volume of atmosphere and contribute to the time-dependent change in TKE reflected on the left-hand side of the equation. The *dissipative* nature of turbulence is made clear in Eq. (13.19). Without the continuous production of TKE through Terms I and II, turbulence will eventually subside as its energy is consumed through Term V. The vertical domain of the daytime convective mixed layer, where the TKE budget is most relevant, and the relative production or loss rates for each TKE term are provided in Figure 13.7.

In order to gain a deeper understanding of the mechanisms that control TKE, each term of the TKE budget will be considered in more detail. On clear days with high levels of surface heating TKE is produced at a high rate through an increase in near-surface buoyancy according to Term 1. Radiant energy is converted into buoyant pressure, which in turn is converted into kinetic energy as air parcels rise. Buoyant production of TKE is associated with relatively large eddies, or updrafts (Leclerc *et al.* 1990, Meyers and Baldocchi 1991). High in the convective boundary layer, near the capping inversion, buoyant TKE is consumed as rising air parcels converge toward density equilibrium with their surroundings. During the early evening, as the ground cools due to radiative heat loss, buoyant TKE is consumed near the surface causing turbulence to be suppressed.

Production of TKE through Term I is especially important in tropical latitudes during *deep convection* events. During deep convection, air that is heated and moistened at the surface can rise rapidly causing extreme instability. Condensation of moisture at the top of deep convection cells forms dense cumulus clouds and releases heat which even further strengthens the buoyancy gradient. Updrafts in deep convection cells can, in extreme cases, penetrate the entire free troposphere and reach the tropopause (up to 18 km in the tropics). In the latter case, heat and mass from surface fluxes can be transported directly into the stratosphere. At the outer margins of the convective cell the air cools and forms downdrafts with the potential for heavy precipitation, often associated with seasonal monsoons.

Production of TKE by wind shear is most important at the two boundaries of the PBL – near the suface and at the top of the capping inversion where shear is produced by interactions with the winds of the free troposphere. As height increases, shear production tapers off, and the local TKE budget reflects consumption of shear-induced TKE as energy is passed down the turbulence frequency spectrum to smaller eddies. Term II reflects the production of shear-induced TKE. The momentum flux carries a negative sign (deposition to ground), which makes the overall contribution of this term to the TKE budget positive when multiplied by the negative sign relating Term I to Term II.

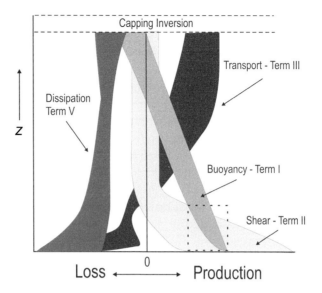

Figure 13.7 The generalized, and relative, distribution of the primary terms of the normalized TKE budget in the daytime convective mixed layer, and their relative magnitude in terms of TKE production or loss. The pressure redistribution term (Term IV) is not presented because it tends to be small with high uncertainties. The dashed box represents the domain where L, the Obukhov length, is determined as the height of the crossover between dominance by shear versus buoyancy forcings on TKE. Redrawn from Stull (1988), with kind permission from Springer Science + Business Media.

Term III is often referred to as *turbulence transport*. This term reflects the fact that some TKE is transported to greater heights (during daytime warming) or lesser heights (during nighttime cooling) as larger eddies that move across broad distances break up into smaller, local eddies. Through this process there is a spatial redistribution of TKE; that is, TKE is transported and then deposited, with turbulence itself being the transport agent. In general, during the day, turbulence transport leads to a net loss of TKE from the lower half and a net gain by the upper half of the PBL. We emphasize that transport of TKE does not create or destroy turbulence, *it only moves it*. During the day, Term III is particularly influenced by the presence of large, buoyancy-powered eddies capable of moving TKE long distances before decomposition to smaller eddies.

Pressure within the PBL is not homogeneous. Motions associated with turbulence and the mean wind can produce a dynamic local pressure environment and there are small fluctuations in atmospheric pressure due to point-to-point differences in the mass of the air column. Thus, the time-averaged covariance between w' and $p'(\overline{w'p'})$, in this case the net vertical flux of pressure fluctuations, is a factor that potentially affects the TKE budget and is shown in Term IV. In reality, atmospheric pressure fluctuations that are relevant to the TKE budget are small (typically less than 0.01% of the total, local TKE) and difficult to measure. Of potentially greater importance are pressure fluctuations associated with wave propagation. As thermal updrafts strike the capping inversion, at the top of the PBL, they cause localized impressions, or "dents," that subsequently rebound downward under the influence of the gravitational force. The vibrational displacement caused by these impact events is transmitted through the PBL as gravity waves. The propagation of such waves can create turbulence, as well as additional covariance between w' and p', both of which can transport or produce TKE.

Molecular dissipation, as shown in Term V, occurs as TKE is passed down a cascade from large to small eddies, eventually reaching the scale where the energy contained in each eddy is insufficient to overcome molecular viscosity. Regimes that foster small, high-frequency eddies will also foster the greatest rates of TKE dissipation. This would include turbulence regimes close to the surface, especially those within the air space of canopies where wake and wave formation around leaves and stems enhance the redistribution of TKE from larger to smaller eddies. In 1941, A. N. Kolmogorov derived a length scale to characterize that part of the eddy spectrum that is most involved in energy dissipation. Reasoning that dissipation must be proportional to viscosity (in this case kinematic viscosity, v), Kolmogorov derived the relation $\eta = \left(v^3/\varepsilon\right)^{1/4}$, where η is the *Kolmogorov microscale*, the turbulent length scale most effective in dissipating kinetic energy to heat. In most surface layer regimes, η is approximated as 1 mm, further demonstrating that it is only eddies with the smallest length scales that contribute to energy dissipation.

13.5 Turbulence spectra and spectral analysis

As noted previously, wind speed varies in complex ways, but within that complexity we can find evidence of cyclic organization. For example, when wind speed is observed as a time series, updrafts eventually transition to downdrafts and forward surges transition to backward surges, creating cyclical patterns through time. In order to accommodate inference about the cyclic nature of wind speed we have formulated the "rotating eddy" as a physical analog, and we rely on mathematical transformations that allow us to move from *digital data* (that characterize time-series measurements) to *phase-frequency data* (that characterize the periodicity of cyclic processes). One of the most widely used methods for converting digital data into phase-frequency data is the *fast Fourier transform* (or FFT), in which a digitized time series can be defined by a composite of the phases and frequencies of sine and cosine waves (Figure 13.8). Using Fourier analysis we can ask: what cyclic frequencies and wave functions provide the most parsimonious description of the velocity variance in a digitized time series? Stated another way: what is the *power* of each eddy frequency as a contributor to the TKE contained in a sample of turbulent wind? The evaluation of *frequency density* across the entire spectrum of represented frequencies is known as *spectral analysis* (Lumley and Panofsky 1964). The essence of spectral analysis is the recognition that the mean square variance in a set of observed wind velocities (σ_x^2, where x represents u, v, or w wind vectors), is equal to the integral of the individual variances across all observation frequencies. Thus, with regard to the spectral analysis of u, or the product of u and c, we can write:

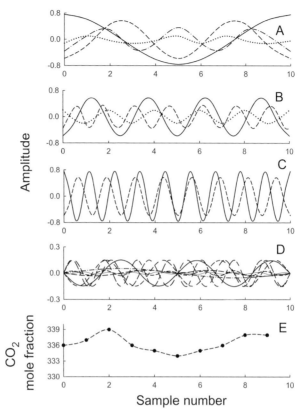

Figure 13.8 The phase-frequency components of a digital time series of CO_2 mixing ratio measurements presented as a series of superimposed sine and cosine waves. Using Fourier analysis, discrete time series data are translated into complex numbers composed of real (cosine component) and imaginary (sine component) parts, which can be expressed in phase space as cyclic frequencies. **A.** The cosine components of the time series data expressed as cyclic frequencies (in cycles per total time of measurements, n) for n = 1, 2, 3, and 4. **B.** The cosine components of cyclic frequencies for n = 5, 6, and 7. **C.** The cosine components of cyclic frequencies for n = 8 and 9. **D.** The sine components of cyclic frequencies for n = 1, 2, 3, 4, 6, 7, 8, and 9. **E.** The original time series showing ten evenly spaced measurements of CO_2 mixing ratio. The broken line in panel E is the mean of the CO_2 measurements and can be thought of as an offset (or anomaly) upon which the sum of the cyclic amplitudes can be superimposed to reconstruct the original time series.

$$\sigma_u^2 = \int\limits_0^\infty S_u[f]df; \quad \sigma_{uc}^2 = \int\limits_0^\infty S_{uc}[f]df, \tag{13.20}$$

where $S_u[f]$ is the spectral density of the variance at frequency f and $S_{uc}[f]$ is the spectral density of the covariance of u and c at frequency f. In this application, spectral density is the contribution of each wavenumber to the TKE of the atmosphere; it is a measure of "turbulent power" and has units J s^{-2} (or watts per unit frequency). The spectral representation of the covariance, or flux, is called the *co-spectrum*.

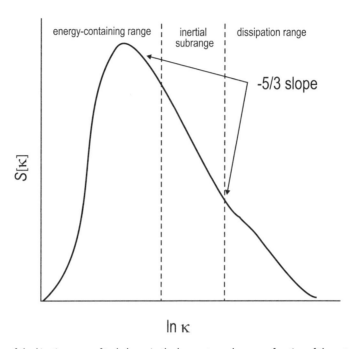

Figure 13.9 Representation of the kinetic energy of turbulence in the lower atmosphere as a function of the natural log of wavenumber (κ). The parameter $S[\kappa]$ is the spectral density, which represents the contribution of each wavenumber to the TKE of the atmosphere. It is clear that the greatest contribution of turbulent energy is provided by eddies with relatively low wavenumbers, and thus low cyclic frequencies (i.e., large eddies). Note the $-5/3$ slope characteristic of wavenumbers near, and within, a region of the spectrum known as the inertial subrange and for the case of spatially homogeneous turbulence. Redrawn from Kaimal and Finnigan (1994), with permission from Oxford University Press.

An analysis of spectral density reveals characteristic features and distributions across eddies of TKE, or turbulence "power." As shown in Figure 13.9, spectral analysis reveals: (1) a distinct *energy-containing range* of frequencies, where the kinetic energy of turbulence is produced; (2) an *inertial subrange* of frequencies, where energy is neither produced nor consumed, but rather passed from larger to smaller eddies; and (3) a *dissipation range* of frequencies, in which TKE is consumed by atmospheric viscosity. The transport of scalar entities occurs principally within eddies of the energy-containing range, although eddies from the other ranges can also contribute. In his original theoretical treatment of turbulence, Kolmogorov (1941) concluded that at high Reynolds numbers, turbulence within the inertial subrange, especially at the higher frequency bands, exhibits motions in which the velocity vectors are, on average, equal in all directions; or *isotropic*. Thus, while turbulence within the inertial subrange has sufficient energy to drive transport, it has no directional bias, which would be required to cause net changes in the distribution of scalar entities. *In reality*, this ideal "isotropy" does not occur; boundary conditions always impose some degree of bias on the directional distribution of wind vectors, and therefore some directional transport is possible

Box 13.1	General spectral scaling in turbulent transport

There are a few universal relationships that have emerged from spectral and co-spectral analyses of turbulence and turbulent fluxes. These relationships are often used as diagnostic metrics with which to evaluate the accuracy and pattern of variation in observations of turbulence. For example, the homogeneous inertial subrange of the turbulence frequency spectrum can be identified as the band where spectral density, $S_x[\kappa]$, scales with the $-5/3$ power of angular wavenumber (κ; where $\kappa = 2\pi f/U$, where U is wind speed in units of m s^{-1} and f is frequency in units of s^{-1}) (Figure 13.9). (Wavenumber is expressed as the reciprocal of wavelength, and thus carries units of m^{-1}.) Thus, we can write:

$$S_x[\kappa] = a_x \varepsilon^{2/3} \kappa^{-5/3}, \qquad (B13.1.1)$$

where x refers in generic fashion to any one of the wind velocity vectors, u, v, or w; a_x is the Kolmogorov constant (with values typically between 0.5 and 0.6); and ε is the turbulence kinetic energy dissipation rate. Co-spectral analysis also reveals a characteristic slope for the relation between a normalized flux and κ, and with a power value of $-7/3$. When applied to different atmospheric situations the shape of the spectra and co-spectra for wind velocity vectors can vary quite a bit, although these power relations in the inertial subrange are fairly well conserved.

  The original set of equations for computing turbulence power spectra and co-spectra were developed for fewer than 20 hours of turbulence observations above a limited number of sites with "ideal" (flat, homogeneous) terrain (Kaimal *et al.* 1972, Wyngaard and Cote 1972). These general equations, based on an admittedly limited set of measurements, have been the foundation for most of the modeling of turbulence and turbulent transport over the past several decades. More recently, the form of the equations has been re-evaluated by Su *et al.* (2004) using 40 000 hours of turbulence data collected above two different mixed hardwood canopies. These more recent studies confirmed many of the previous relations reported from the earlier campaigns, though with some inclusion of modified model parameters. The fundamental scaling relations reported originally for turbulence and turbulent transport appear to be valid for a broader scope of surfaces, with a broader range of roughness characteristics, than originally anticipated.

within the inertial subrange. Moving to even smaller eddies, beyond the scale of the inertial subrange, there is no significant transport associated with motions smaller than the Kolmogorov microscale, or within the energy dissipation range, since TKE within this range is converted to heat. In general, we can state that spectral and co-spectral densities peak in the energy-containing region of the spectrum and decrease as TKE is redistributed from the energy-containing range toward the dissipation range. Spectral and co-spectral analysis of the turbulent atmosphere is often used to derive scaling relations that can be applied to the development of turbulent transport models and that can be used as diagnostic tools to evaluate the accuracy of observations above natural ecosystems (Box 13.1).

13.6 Dimensionless relationships: the Reynolds number and drag coefficient

Sometimes in quantitative modeling it is useful to have terms that express unitless relationships among controlling variables; thus, removing them from dimensional bounds and facilitating their use as general scaling coefficients. These *dimensionless numbers* are composites of variables that reflect natural, interacting processes, and thus retain fundamental relationships, though within a dimensionless framework. (Technically, dimensionless numbers do indeed have a dimension; they are scaled to the value of 1. By convention, however, we refer to unitless numbers as dimensionless numbers.)

The most useful dimensionless numbers are derived in terms of easily quantified measures, e.g., time, length, mass. Ironically, dimensionless numbers are derived through a process known as *dimensional analysis*. Within the study of fluid mechanics, one of the most common types of dimensional analysis is *Buckingham Pi theory*. The Buckingham Pi theorem states that *a quantitative relationship that describes a physical process with m variables can be rewritten in terms of its fundamental dimensions as $(m - n)$ dimensionless relationships, where n is the number of fundamental dimensions*. The mathematical constraint revealed in the Buckingham Pi theorem emerges from matrix algebra.

In order to provide an example of dimensional analysis, let us consider development of a dimensionless relationship for wind shear. Theoretically, wind shear can be described with five dimensional variables: wind speed (U, with units m s^{-1}, and referenced to a specific height above the ground, z), kinematic viscosity (v, with units m^2 s^{-1}), mass density of air (ρ_a, with units g m^{-3}), the distance, or fetch, over which the wind contacts the surface (d, with units m), and the drag force (F_D, with units g m s^{-2}). An analysis of the units for these five variables reveals that they can be grouped around three fundamental dimensions: length, time, and mass. From the Buckingham Pi theorem we can deduce that the five variables can be organized into two dimensionless relationships. In practice, we define these relationships as the *Reynolds number* (Re) and the *drag coefficient* (C_D), whereby:

$$Re = \frac{U[z]d}{v},\tag{13.21}$$

$$C_D = \frac{F_D}{\rho_a U^2[z]d^2}.\tag{13.22}$$

In both of these relations we have indicated the wind speed as U[z] to make clear that the derived dimensionless values are dependent on height (z). The Reynolds number (Re) describes the ratio of inertial forces (reflected in U) to viscous (resistive) forces (reflected in v). A low Re means that the inertial forces are small compared to the viscous forces and the streamlines of the fluid flow are relatively more stable. A high Re means that the inertial forces are much larger than the viscous forces, forcing the streamlines to be reorganized as turbulent flow. The drag coefficient, C_D, defines the tendency for the inertial force to work against friction. In formal terms, C_D is the aerodynamic force on a surface per unit kinetic energy and surface area. A high value of C_D means that the potential for the fluid to deposit momentum to the surface, and thus create a drag force in the direction of the flow, is also high. It is clear that a relation should exist between C_D and z_0, the aerodynamic roughness length; a rougher surface should induce greater drag (see Raupach 1992, and Wieringa 1993, for descriptions of the theoretical linkages between C_D and z_0). It should also be clear that a rougher surface, with its increased drag, will have greater potential to induce turbulence by enhancing shear stress in the fluid, and thus inducing greater potential to transfer heat and mass from a surface into the well-mixed air above the turbulent boundary layer.

Intuitively, these two dimensionless numbers should scale in proportion with each other. In other words, turbulence (reflected in Re) should be determined to some extent by the drag force (reflected in C_D). In fact, such a relation exists. We begin by stating the relation in general terms as:

$$C_D = f(Re), \tag{13.23}$$

where f is used to indicate "function of." The utility of Eq. (13.23) can be appreciated after some rearrangement of Eqs. (13.21) and (13.22), which results in:

$$F_D = \left[\rho_a U^2(z) d^2\right] f(Re). \tag{13.24}$$

The threshold for the onset of turbulence is a value for Re of approximately 2000. Most atmospheric flows are turbulent ($Re \gg 2000$) as v is on the order of 10^{-5} m^2 s^{-1}, d is on the order of meters, and u is on the order of 0.1 to 10 m s^{-1}. For example, the Re value for a 3 m s^{-1} wind speed over a 3 m wide strip of corn field would be 90 000. In contrast, the Re value for wind moving at a speed of 0.25 m s^{-1} over a 10 cm wide leaf is 250. In general terms, the Re value and turbulence intensity are high when the characteristic length scale of the flow is much larger than the thickness of the laminar boundary layer.

13.7 The aerodynamic canopy resistance

The transfer of heat and mass from the upper surface of a canopy to the turbulent PBL is mitigated by an *aerodynamic resistance* (r_a), which is *inversely proportional* to F_D. The inverse relationship between r_a and F_D can be understood when it is recognized that a turbulent flux is enabled as momentum from the horizontal wind is deposited to the canopy surface, causing a drag force and resulting in shear stress that redirects momentum into the vertical turbulent wind, thus facilitating the formation of eddies that carry mass and energy to or from the surface. The greater the drag force, the larger the eddies and the greater the potential for turbulent exchange.

Recall that the covariance between the horizontal turbulent wind and vertical turbulent wind is defined as the friction velocity (u_*). Thus, r_a scales inversely with u_*. We can express the aerodynamic resistance for sensible heat flux (r_{ah}), as:

$$r_{ah} \approx \frac{\ln\left((z - d_H)/z_0\right)}{ku^*}.$$ (13.25)

We have used an approximation for this relation because the roughness height (z_0) for momentum is slightly different than that for sensible heat due to differences in the vertical distributions of the sinks for momentum and sources for sensible heat within the canopy.

13.8 Eulerian and Lagrangian perspectives of turbulent motions

There are two perspectives from which to evaluate the temporal and spatial properties of turbulence – Eulerian and Lagrangian. The *Eulerian* perspective refers to the analysis of a fluid flow as it moves past a fixed observation point. The *Lagrangian* perspective refers to the tracking of individual fluid "parcels" as they move within the flow. We can develop formal concepts of Lagrangian and Eulerian perspectives with the use of differential calculus. The total derivative of one variable with respect to another (e.g., dy/dx) is often defined for functions in which independent variables are not truly independent from each other, but can be related to a single, truly independent variable. As an example, consider the case in which the concentration of a constituent (c) increases with time in a parcel of air that is moving through the atmosphere contacting various sources and sinks. Further, assume that the sole relation that interests us is how c changes as a function of time (dc/dt). Now, let us assume that a "chart" exists of all sources and sinks that affect c and the position of these sources and sinks are provided in terms of Cartesian coordinates. In order to resolve changes in c as a function of time, we will need to trace the path of the parcel as it interacts with all of the charted sources and sinks. The solution to dc/dt will depend on four independent variables (t, x, y, and z), only one of which (t) is truly independent. The relationship of the other three to dc/dt is, to some degree, dependent on t since it takes time for the parcel to move along any of the three coordinates. Thus, we can write:

$$\frac{dc}{dt} = \frac{\partial c}{\partial t}\frac{dt}{dt} + \frac{\partial c}{\partial x}\frac{dx}{dt} + \frac{\partial c}{\partial y}\frac{dy}{dt} + \frac{\partial c}{\partial z}\frac{dz}{dt}.$$ (13.26)

In Eq. (13.26) we have expressed the overall change in c as a function of all four variables using the chain rule of calculus and knowledge that the three spatial coordinates are all dependent on t. Recognizing that $dt/dt = 1$ and expressing the spatial derivatives in terms of wind velocity, we can restate Eq. (13.26) as:

$$\frac{dc}{dt} = \frac{\partial c}{\partial t} + u\frac{\partial c}{\partial x} + v\frac{\partial c}{\partial y} + w\frac{\partial c}{\partial z}.$$ (13.27)

The total derivative, expressed on the left-hand side of Eqs. (13.26) and (13.27), is some-times called the *Lagrangian derivative*. Within the Lagrangian perspective the observer travels with the parcel of fluid as it moves through the coordinate domain on a flow defined by vectors u, v, and w, contacting sources and sinks.

In contrast, consider a time series of observations made from a fixed location immersed in the fluid; that is, from an Eulerian perspective. In this case, numerous fluid parcels arrive at the observation point, each one originating from coordinates defined in the x, y, and z directions, and thus carrying the signature of a set of potentially unique interactions with sources and sinks. We can rearrange Eq. (13.27) to provide the Eulerian derivative:

$$\frac{\partial c}{\partial t} = \frac{dc}{dt} - \left(u\frac{\partial c}{\partial x} + v\frac{\partial c}{\partial y} + w\frac{\partial c}{\partial z} \right). \tag{13.28}$$

The Eulerian perspective takes into account the fact that the observer is only seeing a parcel at one point in time, and thus is not able to discern the total changes that have occurred in c as it traveled to the observation point. Rather, he or she is only able to discern a *partial* change in c with respect to time ($\partial c/\partial t$). Additionally, in the Eulerian perspective the observer is seeing the composite effect of numerous parcels arriving from multiple directions. It is the collective nature of these advected parcels that cause the observer to see something different in c compared to the Lagrangian derivative. If you could take the Lagrangian (complete) perspective, and subtract from it the divergence in c that occurs as parcels travel along the various flow vectors, then you would be able to predict the incomplete (partial) perspective at one point in time. Thus, the Eulerian derivative can be expressed as the Lagrangian derivative minus any divergent effects.

The Eulerian and Lagrangian perspectives can be used to develop scales of length and time with regard to fluid motions. Consider a time series of data collected from the Eulerian perspective. Once again, in the Eulerian condition, we are viewing samples of air without knowledge of all the changes that have occurred as they have traveled to the observation point. If we assume frozen turbulence according to Taylor's hypothesis, we can convert our time series of wind speed measurements into a series of measurements across the spatial extent of a turbulent eddy. Let us assume that we collect a time series on the longitudinal vector of wind speed (u). We can use the time series to define the *Eulerian length scale* (Λ_E) with respect to u as:

$$\Lambda_{Eu} = \overline{u}t_{Eu} = \overline{u}\int_{t=0}^{\infty} r_u[\Delta t] = \overline{u}\int_{t=0}^{\infty} \frac{\overline{u'[t]\ u'[t+\Delta t]}}{\sigma_u^2}dt, \tag{13.29}$$

where $\overline{u}$ refers to the time-averaged longitudinal wind velocity; u' is the turbulent, or variant, longitudinal wind velocity; $r_u[\Delta t]$ refers to an autocorrelation function; and σ_u^2 refers to the mean square variance of u' (i.e., $\overline{u'^2}$). The parameter, t_{Eu} represents the *Eulerian time scale*, which is defined by the autocorrelation function. The autocorrelation function describes the average covariance of the turbulent wind speed when evaluated across Δt, normalized by the total variance and integrated across time between 0 and ∞. (Clearly, it is not possible to integrate the autocorrelation function to $t = \infty$. The integral,

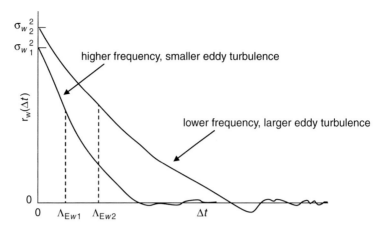

Figure 13.10 Schematic drawing of the autocorrelation functions ($r_w\Delta t$) for vertical wind velocity (w) as a function of progressively larger time intervals between measurements (Δt) in two different turbulence regimes. The Eulerian length scale is indicated at Λ_{Ew}. It is clear that complete autocorrelation equals the variance of w (indicated as σ_w^2 at the y-intercept), while a complete lack of autocorrelation occurs as the function approaches 0. Redrawn from Lenschow (1995).

as written, sets a theoretical boundary on the autocorrelation. In real use, the time integral will need to be set beyond the point where the covariance between time intervals approaches 0.) The autocorrelation function is an expression of the degree to which two sequential measurements of wind speed are correlated and thus part of the same coherent motion that we define as an eddy. In essence, using the autocorrelation function we derive the time and length scales for the dominant, independent eddies in a sample of observed wind. Using Eq. (13.29) it can be shown that the degree of correlation between measured, successive wind speeds decreases exponentially as the time between the measurements increases (Figure 13.10).

The *Lagrangian length scale* (Λ_L) is defined in much the same way as that for the Eulerian length scale except for the nature of the velocity observations contained in the autocorrelation function. Recall that in the Lagrangian condition we are able to define the entire path taken by a fluid parcel as it moves in the flow. This perspective must be built into the autocorrelation function. The *Lagrangian time scale* (t_L, also denoted as τ_L in some past treatments) reflects the shortest travel time interval within which Lagrangian wind speeds are autocorrelated. Lagrangian wind speeds are defined from the travel time and distance between the initial position of a hypothetical buoyant parcel and its height at the time of observation. It is difficult to characterize individual eddies, and thus Λ_L, within the Lagrangian perspective. Tracing the entire travel paths of parcels is not possible. One of the means of determining Lagrangian length and time scales is to use wind data collected at a site, and then use time trends in those data to trace the paths of hypothetical parcels of air moving along flow paths in cyberspace.

In order to accommodate valid statistical description in either the Eulerian or Lagrangian perspectives, observations of wind speeds must exist during a period of *stationarity*, i.e., a period in which calculated statistical parameters have reached stable (with respect to time) values. During a period of stationarity one string of time series data looks statistically the same as the next. This is known as the *ergodic condition*. In the ergodic condition time-averages will equal ensemble averages (see Appendix 13.1). The estimation of characteristic length and time scales is required to ensure that the assumption of ergodicity can be met. The ergodic condition will only occur if statistical averaging is carried beyond the autocorrelation limits of the Eulerian and Lagrangian scales. Otherwise, statistical entities will not be based on truly independent measures and statistical stability cannot be assured. Fluxes calculated from time series data in a non-stationary state are sensitive to the choice of averaging time; fluxes calculated from truly stationary data are not.

13.9 Waves, nocturnal jets, and katabatic flows

As discussed above in Section 13.1, in addition to turbulent transport, deviations from the mean wind velocity can be caused by periodic waves, which thus have the potential to influence scalar transport. There are several potential causes of atmospheric waves. Topographic features such as hills and mountains can cause pressure gradients that permit air flows to oscillate back and forth in wave-like fashion, creating *regional waves*. As the wind meets the topographic irregularity of a hill, the pressure force pushes it up and over, causing acceleration in the wind's speed. The pressure force is relaxed on the lee side of the hill, allowing the wind to decelerate. Vertical oscillations can occur in the leeward flow as the forces of buoyancy and gravity create upward and downward momentum that causes layers of air to "overshoot" and "undershoot" the original flow vector. The oscillation is eventually dampened as these instabilities weaken with distance in the leeward flow. These gravity-induced waves tend to be propagated in conditions of static stability (Finnigan *et al.* 1984, Sun *et al.* 2004); in unstable atmospheres the natural tendency is for air parcels to continue rising, removing the potential for gravity to restore an upwardly forced parcel.

Canopy waves are often observed above tall canopies. Canopy waves tend to have periodicities at slightly longer time scales than those of local turbulence (Figures 13.11 and 13.12), and thus spectral analysis will reveal their presence (Fitzjarrald and Moore 1990, Lee *et al.* 1997, Turnipseed *et al.* 2004). The causes of canopy waves are not clear, although it has been proposed that they develop from shear-induced instabilities at the top of a canopy or from fluctuating pressure fields that occur as atmospheric subsidence occurs after sunset. Whatever the mechanism, co-spectral analysis has revealed that the potential for canopy waves to transport mass and energy can be significant and, therefore, important for understanding nighttime surface-atmosphere fluxes. In addition to the direct effects of moving scalars up and down at the front of a pulsating wave, larger waves can create localized turbulence, which in turn can drive limited turbulent transport in the otherwise stable atmosphere.

Another phenomenon that often occurs in the nighttime stable atmosphere and has the potential to drive transport is the *nocturnal jet*. During the day, turbulent mixing in the

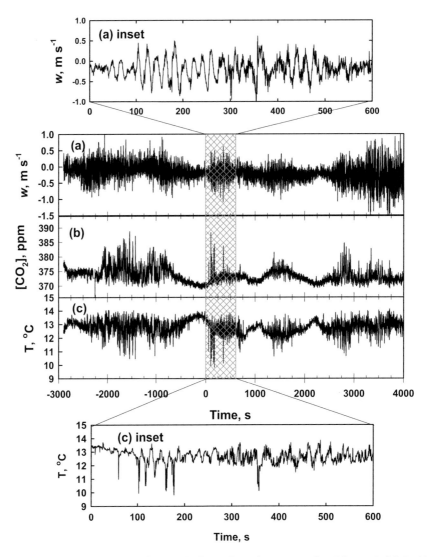

Figure 13.11 A plot of high-frequency time series data for vertical velocity, CO_2, and temperature for a 2-hour period during highly stable nighttime conditions above a subalpine coniferous forest at Niwot Ridge, Colorado. Variance in the vertical wind speed (w) appears to cycle with a periodicity of approximately 20 minutes. Concurrent with cycling in w, larger fluctuations in temperature and CO_2 are evident. It appears that a low frequency, periodic fluctuation was observed, which can be interpreted as a mesoscale, mountain gravity wave propagated down past the measurement site. Within the larger scale of the mountain wave, subperiods can be identified which contain canopy waves (shown in the insets) each with periods of $\sim$ 15–25 s. Redrawn from Turnipseed *et al.* (2004).

convective mixed layer resists the potential for winds from the free troposphere to penetrate the PBL. As the sun sets, turbulence in the mixed layer is dissipated and air next to the ground begins to stratify and form the stable nocturnal layer. Reduced turbulence in the early nighttime mixed layer weakens any resistance to penetration of winds from aloft. As a result,

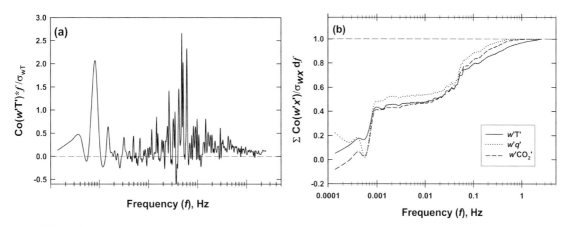

Figure 13.12 **(a)** The dual presence of mid-frequency canopy waves superimposed on the lower frequency mountain wave is clearly shown when looking at co-spectra for the entire period depicted in Figure 13.11. Large spectral peaks were observed at frequencies of 0.05 and 0.0008 Hz, corresponding to 20 s and 21 min, respectively. These frequencies reflect the local canopy and mesoscale mountain waves, respectively. **(b)** Integration of the co-spectra indicated that between 25–50% of the total scalar (temperature, moisture, and CO_2) flux, was accounted for within the low frequency wave (note the large jump in the cumulative flux at the 0.0008 Hz frequency); indicating prominent influences of the mesoscale waves on local turbulent fluxes. Redrawn from Turnipseed *et al.* (2004).

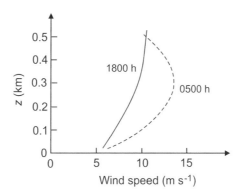

Figure 13.13 The vertical profile in wind speed at two different times of the day: late in the day, when turbulent mixing is high and the wind profile is "softened," and early in the morning, when the nocturnal jet has developed a few hundred meters above the ground. Redrawn from data collected near Wangara, Australia, as reported in Malcher and Kraus (1983).

a sharp gradient in wind speed can develop immediately above the stable nocturnal layer, producing the nocturnal jet a few hundred meters above the ground (Blackadar 1957); the resulting jet can reach velocities of 10–30 m s^{-1} (Figure 13.13). Near the lower boundary of the jet, shear between layers of air with different velocities can produce Kelvin–Helmholtz waves. Studies using model systems have suggested that turbulence in Kelvin–Helmholtz waves can be amplified if differences in density occur between the interacting air layers. In

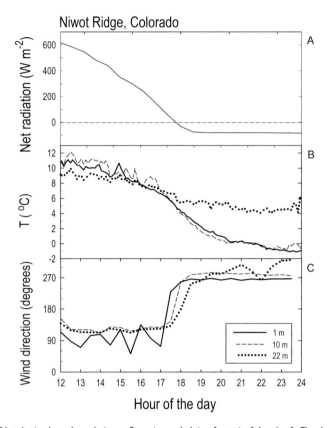

Figure 13.14 The formation of katabatic, downslope drainage flows in a subalpine forest in Colorado. **A.** The decrease in net radiation measured at 22 m during the afternoon and early evening. **B.** The decrease in air temperature at three heights during the afternoon and early evening. Note that the air cools down first near the top of the canopy (10 m), followed by cooling near the surface (1 m), and above the canopy (22 m). **C.** The change in wind direction from easterly, upslope flow (~ 90º) to westerly, katabatic drainage flow (~ 270º) as the surface cools. Note that the drainage flow develops first near the ground, then progresses upward during the night. Canopy height is 11.4 m.

that case, static instability can develop within the waves, causing even more turbulence. These local effects can potentially affect nighttime turbulent transport, although actual observations of processes involving Kelvin–Helmholtz waves are few (Banta *et al.* 2002, Cheng *et al.* 2005, Cooper *et al.* 2006).

The suppression of turbulent mixing in the nocturnal atmosphere sets the condition for momentum gradients and advective fluxes to occur near the surface. Stability in the nocturnal layer is maintained by gravity, which forces the coldest, most dense air to the surface; the settling of stable air tends to quench turbulence in the nocturnal layer. The establishment of this stable state often causes layers of air near the surface with diverging densities to become decoupled from one another. On sloped terrain, *drainage flows* (or *katabatic flows*) can develop as layers of near-surface air increase in density and are forced downslope by gravity (Figure 13.14). (The word *katabatic* is literally derived from Greek, whereby "katabatikos" means "going downhill.") Depending on the nocturnal pattern of radiative heat loss, drainage flows can develop as "slabs" of descending air that range from a few centimeters to several meters thick (Mahrt *et al.* 2001a, Turnipseed *et al.* 2003). In the presence of horizontal concentration gradients, these slabs of air can carry mass and energy downslope.

Appendix 13.1 Rules of averaging with extended reference to Reynolds averaging

It is the nature of turbulent wind speeds and surface-atmosphere fluxes to vary from one instant to the next. Characterization of variable systems is often achieved through statistical averaging accompanied by estimation of the random error. Here, we consider three types of averaging. *Time averaging* is applied to sequential measurements made over a time interval and is calculated as the measurement sum divided by Δt. *Spatial averaging* is applied to a set of measurements made at different points at the same instant in time across a defined spatial domain (often represented as a spatial area), with the measurement sum divided by the areal extent of the domain. *Ensemble averaging* is applied to a set of independent realizations (or replications) of the same measurement, with the measurement sum divided by the number of realizations. In studies of the atmosphere, a single realization is all that is normally possible. Thus, the ensemble average is impractical as a statistical tool in studies of the atmosphere. However, it is the most fundamental of averages when it comes to understanding the mean tendencies of atmospheric states and processes. Accordingly, approaches have been developed to approximate the ensemble average by making serial observations under similar sets of conditions.

Each of these averages (time, space, and ensemble) can be broken down into further statistical components using a technique called *Reynolds averaging*. In Reynolds averaging, any single measurement can be split into mean and turbulent (or fluctuating) components. The mean component refers to the tendency of the observation to cluster around the values for other observations of the same type. The turbulent component refers to the realization that due to variability in natural systems, individual observations are separated from this mean tendency to some quantitative degree. Following the rules of Reynolds averaging:

$$u = \bar{u} + u'; \quad v = \bar{v} + v'; \quad w = \bar{w} + w'; \quad c = \bar{c} + c', \tag{13.30}$$

where u, v, w, and c are instantaneous values of the wind velocity vectors and concentration, respectively; $\bar{u}$, $\bar{v}$, $\bar{w}$, and $\bar{c}$ are mean values for some time-averaged interval containing the instantaneous measurements; and u', v', w', and c' are fluctuations from the mean at the instant of the measurement.

Certain rules can be applied to the mean and turbulent components of averaged variables (Table 13.1). One of the more important rules is that $\bar{u'} = 0$, which is required by the definition of a mean – i.e., the positive and negative deviations from a mean must sum to 0. When the time-averaged product between the mean and fluctuating components of two different variables is evaluated, we can conclude

Table 13.1 Some of the most commonly used rules of Reynolds averaging

(1) $\overline{(\overline{u})} = \overline{u}$

(2) $\overline{(u + c)} = \overline{u} + \overline{c}$

(3) $\overline{(\overline{u}c)} = \overline{u}\,\overline{c}$

(4) $\overline{(du/dt)} = d\overline{u}/dt$

(5) $u = \overline{u} + u'$

(6) $\overline{u'} = 0$

(7) $\overline{(\overline{u}c')} = \overline{u}\,\overline{c}' = \overline{u}\cdot 0 = 0$

(8) $\overline{(uc)} = \overline{(\overline{u} + u')(\overline{c} + c')} = \overline{(\overline{u}\,\overline{c})} + \overline{(u'\overline{c})} + \overline{(\overline{u}c')} + \overline{(u'c')}$

$\qquad\qquad\qquad = \overline{u}\,\overline{c} + 0 + 0 + \overline{u'c'} = \overline{u}\,\overline{c} + \overline{u'c'}$

$\overline{(uv')} = \overline{u}\,\overline{v}' = \overline{u}\cdot 0 = 0$. As an example, consider the derivation of Rule 8 of Table 13.1. The middle two terms of the factorial drop out, leaving the product of the mean components as one term and the product of the fluctuating components as the other. The latter product cannot be assumed as 0. It is non-linear and must be evaluated as a higher-order statistical moment.

Appendix 13.2 Derivation of the Reynolds shear stress

For a viscous, incompressible, Newtonian fluid, we can express the condition of continuity as:

$$\rho_a \frac{d\mathrm{U}}{dt} \approx \frac{\partial \mathrm{U}}{\partial t} + u\frac{\partial \mathrm{U}}{\partial x} + w\frac{\partial \mathrm{U}}{\partial z} \approx -\left(\frac{\partial p}{\partial x}\right) + \mu\left(\frac{\partial^2 \mathrm{U}}{\partial x^2} + \frac{\partial^2 \mathrm{U}}{\partial z^2}\right). \tag{13.31}$$

Applying Reynolds averaging to Eq. (13.31), we obtain:

$$\begin{aligned}\rho_a \frac{d\mathrm{U}}{dt} &\approx \frac{\partial(\overline{u} + u')}{\partial t} + (\overline{u} + u')\frac{\partial(\overline{u} + u')}{\partial x} + (\overline{w} + w')\frac{\partial(\overline{u} + u')}{\partial z} \\ &\approx -\left(\frac{\partial(\overline{p} + p')}{\partial x}\right) + \mu\left(\frac{\partial^2(\overline{u} + u')}{\partial x^2} + \frac{\partial^2(\overline{u} + u')}{\partial z^2}\right)\end{aligned}. \tag{13.32}$$

Using continuity and recognizing that $\partial(\overline{u} + u')/\partial x = 0$ (see Table 13.1), we can simplify to:

$$\rho_a \frac{d\mathrm{U}}{dt} \approx \frac{\partial(\overline{u} + u')}{\partial t} + (\overline{w} + w')\frac{\partial(\overline{u} + u')}{\partial z} \approx -\left(\frac{\partial(\overline{p} + p')}{\partial x}\right) + \mu\left(\frac{\partial^2(\overline{u} + u')}{\partial z^2}\right). \tag{13.33}$$

Applying time averaging, and respecting the outcome from Reynolds averaging that $\overline{x'} = 0$, we can write:

$$\overline{\rho}_a \frac{d\overline{\mathrm{U}}}{dt} \approx \frac{\partial \overline{u}}{\partial t} + \frac{\partial \overline{w}\,\overline{u}}{\partial z} + \frac{\partial \overline{w'u'}}{\partial z} \approx -\left(\frac{\partial \overline{p}}{\partial x}\right) + \mu\left(\frac{\partial^2 \overline{u}}{\partial z^2}\right). \tag{13.34}$$

Assuming lack of divergence in the mean flow vectors we can write:

$$\bar{\rho}_a \frac{d\overline{U}}{dt} \approx \frac{\partial \bar{u}}{\partial t} + \frac{\partial \overline{w'u'}}{\partial z} \approx -\left(\frac{\partial \bar{p}}{\partial x}\right) + \mu \left(\frac{\partial^2 \bar{u}}{\partial z^2}\right). \tag{13.35}$$

Using algebra for rearrangement, we can write:

$$\frac{\partial \bar{u}}{\partial t} \approx -\frac{1}{\bar{\rho}_a}\left(\frac{\partial \bar{p}}{\partial x}\right) + \frac{1}{\bar{\rho}_a}\frac{\partial}{\partial z}\left(\mu \frac{\partial \bar{u}}{\partial z} - \bar{\rho}_a \overline{u'w'}\right), \tag{13.36}$$

where the Reynolds shear stress $\left(-\bar{\rho}_a \, \overline{u'w'}\right)$ emerges on the right-hand side of Eq. (13.36) as a term grouped with the viscous shear stress; with both represented as components of the frictional force that resists the pressure force.

Appendix 13.3 Derivation of the logarithmic wind profile

The logarithmic dependence of mean wind speed on height in a near-neutral atmosphere has been observed in numerous past studies. However, we might also ask: is logarithmic dependence expected on the basis of theory? We can derive a theoretical expectation of the wind profile using *Prandtl's mixing layer theory* (Prandtl 1925). We begin by assuming that parcels of air moving along the x-coordinate within a flow with mean speed $(\bar{u})$ can be distinguished as a function of each parcel's height (z). Specifically, we assume that the deviation from $\bar{u}$ (represented as u') increases (due to greater shear) as we move closer to the surface. Thus, we can write:

$$u' = \bar{u}[z \pm l] - \bar{u}[z] = \pm l \frac{\partial \bar{u}}{\partial z}, \tag{13.37}$$

where l is the *mixing length*; in this case reflecting the minimum distance at which we can distinguish the flow's velocity gradient. In the case of a wind flow, the mixing length is the distance at which parcels of air are mixed into the mean wind gradient.

We can reasonably make the assumption that the nature of turbulence is such that $u' = \pm(Cw')$. Given that u' is proportional to $\partial \bar{u}/\partial z$ as defined within Eq. (13.37), our assumption derives from the corollary that $\partial \bar{u}/\partial z$ is proportional to w' (where C is the proportionality constant). Additionally, we can state that the mixing length is linearly proportional to z according to the von Karman proportionality constant (k); thus, $l = kz$. From this, it is seen that both mixing length and mean eddy size scale linearly with height above the surface.

Taking all of these relationships and assumptions into account, we can write an expression for the momentum flux $(\overline{w'u'})$ as:

$$\overline{w'u'} \approx l^2 \left(\frac{\partial \bar{u}}{\partial z}\right)^2 \approx (kz)^2 \left(\frac{\partial \bar{u}}{\partial z}\right)^2, \tag{13.38}$$

$$\frac{\overline{|w'u'|}^{0.5}}{kz} = \frac{u_*}{kz} = \frac{\partial \overline{u}}{\partial z}. \tag{13.39}$$

Equation (13.39) can be integrated with respect to the vertical distance between z_0 and z, where z_0 is the aerodynamic roughness length:

$$\int_{z0}^{z} \frac{\partial \overline{u}}{\partial z} dz = \int_{z0}^{z} \frac{u_*}{kz} dz. \tag{13.40}$$

Considering each side separately we can integrate as:

$$\int_{z0}^{z} d\overline{u} = \overline{u}[z], \tag{13.41}$$

$$\int_{z0}^{z} \frac{u_*}{kz} dz = \int_{z0}^{z} \frac{u_*}{k} \frac{dz}{z} = \frac{u_*}{k} (\ln z - \ln z_0) = \frac{u_*}{k} \ln \left(\frac{z}{z_0} \right). \tag{13.42}$$

Resulting in the final form of the logarithmic dependence of $\overline{u}$ on height (z):

$$\overline{u}[z] = \frac{u_*}{k} \ln \frac{z}{z_0}. \tag{13.43}$$

14

Observations of turbulent fluxes

In the summer of 1968, scientists at the Air Force Cambridge Laboratories in Bedford, Massachusetts, conducted a surface-layer experiment over a very flat, uniform site in southwestern Kansas ... The Kansas experiment saw the first systematic application of tower-based sonic anemometry and computer-controlled data acquisition in a field experiment ... The statistical relationships developed from these measurements, both spectral and time-averaged, have since been tested and compared by many experimenters, and possible inconsistencies have been discussed at length in the literature, but the basic structure revealed by the two experiments remains substantially unchanged.

Kaimal and Wyngaard (1990)

Most of the net transport of mass and energy between an ecosystem and the atmosphere occurs through the turbulent wind. As discussed in the last two chapters, turbulence varies greatly in its properties depending on where and when it occurs, but it also exhibits coherency and organization in its motions. Both the variable and coherent aspects of turbulence are induced by features of the earth's surface; especially by topographic features, such as the location of hills, mountains, and valleys, and canopy characteristics such as depth and roughness. From a theoretical perspective, it had been reasoned since the groundbreaking work in 1954 by Andrei Monin and Alexander Obukhov that near-surface shear forces, in the neutral surface layer above a horizontally homogeneous land surface, could not sustain vertical divergence in the momentum flux. This led to the inevitable conclusion that surface fluxes must be conserved as a function of height in the surface layer, which in turn led to the concept of a "constant flux layer." In the late 1950s and early 1960s several research groups took up the aim of validating these theoretical predictions – an aim that required careful observations above flat, simple terrain. The initial experiments conducted in Kansas (and later in Minnesota) in the late 1960s did indeed validate the predictions and this validation has withstood numerous subsequent tests and debates. As referenced in the paper cited above from Kaimal and Wyngaard, after more than 30 years of testing, the tenets of these early experiments are still widely accepted and used.

In this chapter we will begin with consideration of the atmospheric surface layer. We will discuss those properties of the layer that are crucial to an understanding of transport processes, at least with regard to statistical characterization. Following consideration of surface layer processes, we will focus on the zone of high shear that occurs near the "rough" surface of a canopy. Surface roughness enhances momentum transfer, alters the structure of wind fields, and influences the partitioning of kinetic energy among the various scales that define turbulent motions. The chaotic nature of near-surface turbulence is born from shear at the canopy-atmosphere interface. However, canopy architecture also has the potential to impose spatial organization on the wind. Canopies represent porous exchange volumes that allow the wind to penetrate at certain points and vent at other points; while in the canopy, the wind can transfer momentum to leaves and stems, as well as bend and swirl around them, creating new turbulent motions defined by the size and placement of canopy elements.

In order to illustrate an interaction between the wind and canopy, let's adopt the perspective of a fluid particle and follow it as it is transported on various turbulent eddies from high in the atmospheric surface layer, through the roughness sublayer, through a forest canopy, and ultimately to the ground. Let's assume that the particle's "journey" occurs on a day with near-neutral static stability. As the particle moves through the atmospheric surface layer, it will be buffeted by turbulent pressure pulses, nudging it back and forth along random trajectories. If through these chaotic motions, the particle moves toward the surface, it will encounter a progressively sharper gradient in shear stress and rows of swirling vortices moving with the longitudinal wind flow. Downward-oriented gusts will buffet it at a higher frequency and with more velocity than upward-oriented gusts. As the particle is pushed into the canopy it will encounter fine-scale vortices of swirls and wakes next to leaves and branches. Deeper in the canopy, vortical motions will decrease as the density of canopy elements also decreases and the particle will lose momentum to drag forces as it interacts with the boundary layers of leaves and the soil. Eventually the particle will come to rest on the ground.

Clearly, the presence of surface roughness has the potential to alter both the magnitude and direction of forces that determine turbulent exchanges between ecosystems and the atmosphere. Two challenges that we face are how to make sense of observations made in the presence of these roughness effects and how to use the insight gained from observations to design models that predict transport dynamics. In this chapter we will address the first challenge: how do we make sense of observations of turbulent transport made near the surface? In the next chapter we will take up the issue of turbulent transport models.

14.1 Turbulent fluxes in the atmospheric surface layer

In addition to studying the general characteristics of turbulence, an aim of researchers in the Kansas and Minnesota campaigns was to derive a general relation between fluxes and concentration gradients; for many atmospheric constituents fluxes are more difficult to measure than concentration gradients. Building on the conceptual foundation of diffusion, it was thought that a general theory of atmospheric transport that related gradients to fluxes, through an *eddy diffusivity coefficient* (K_D), could expand what began as a theory at the molecular scale. (We will designate the eddy diffusivity coefficient as K_D to distinguish it from the molecular diffusivity coefficient, which we will retain as K_d.) The theory used to justify the relation between gradients and fluxes in the atmospheric surface layer derives from a form of dimensional analysis known as *Monin–Obukhov Similarity Theory* (MOST; Monin and Obukov 1954). Starting from the assumption that the time-averaged vertical turbulent flux density ($\overline{F}_{vt}$; mol m^{-2} s^{-1}) measured in the neutral surface layer is conserved with respect to flux density at the ground surface, we can state: $\overline{F}_{vt} \approx \left(\overline{\rho_m} \, \overline{w'c'} \right)_0$, where $\left(\overline{w'c'} \right)_0$ is the time-averaged eddy covariance measured at the ground surface and $\overline{\rho_m}$ is the molar density of dry air at the time-averaged air temperature and pressure for the same measurement period as that used for the eddy covariance observations. Using this relation as a foundation, we can state:

$$\overline{F}_{vt} \approx \left(\overline{\rho_m} \overline{w'c'}\right)_0 \approx -\overline{\rho_m} \frac{kz\overline{u}_*}{\phi} \frac{\partial \overline{c}}{\partial z} \approx -K_D \overline{\rho_m} \frac{\partial \overline{c}}{\partial z} , \tag{14.1}$$

where we have introduced a negative sign in those relations between flux and mean concentration gradients and where ϕ is a dimensionless MOST scaling parameter defined as:

$$\phi = \frac{kz}{f_*} \frac{\partial \overline{c}}{\partial z} , \tag{14.2}$$

and f_* is the surface turbulent flux density normalized to friction velocity $\left[\left(\overline{w'c'}\right)_0 / \overline{u}_*\right]$, with units that resolve to mole fraction. (The symbol f_* may be written in more specific terms, such as T_*, c_{c*}, or c_{w*} for the fluxes associated with sensible heat, carbon dioxide, and water vapor, respectively.) Using Eq. (14.1), turbulent fluxes can be estimated on the basis of mean concentration gradients; this recognition is used to derive a form of MOST known as *K theory* (where "K" refers to the eddy diffusivity). Referring to Eq. (14.1), K_D can be defined in aerodynamic terms as:

$$K_D = \frac{u_*}{\phi} [kz], \tag{14.3}$$

where k is von Karman's constant (a dimensionless number that can be estimated as 0.4 in the atmospheric surface layer). The first term (u_*/ϕ) reflects "turbulent potential," derived from shear-generated turbulence and the second term (kz) reflects the *mixing length*. The mixing length is the average distance a parcel of air moves before being mixed into the concentration gradient; it is a reflection of mean eddy size. As z increases, u_* decreases; that is, the turbulent potential is greatest at the surface where shear is highest. The decrease in u_* as a function of z is accompanied by an increase in mean eddy size, and thus an increase in mixing length. The decrease in u_* and increase in mixing length are approximately proportioned as z increases in the neutral surface layer, meaning that $\overline{F}_{vt}$ remains approximately constant, as long as there is no divergence in $\partial \overline{c}/\partial z$. This is why the neutral surface layer is sometimes called the constant flux layer.

    In neutral conditions, by convention, ϕ is taken as unity and K_D reduces to u_*kz. In the presence of unstable thermal stratification, a new length scale must be introduced (L, called the Obukhov length), and K_D must be modified in a way that scales u_*/ϕ_j to an index of static instability, e.g., z/L. It is currently not possible to do this on sound theoretical terms, so it is done through observation and the design of empirical relationships between ϕ and z/L (Kaimal *et al.* 1972, Kaimal and Finnigan 1994). With u_* in units of m s^{-1} and z in units of m, K_D will have units of m^2 s^{-1}; the same units as the molecular diffusivity coefficient (K_d). One powerful outcome of MOST is that ϕ, and as a result K_D, remain constant for all scalars in a neutral atmosphere. Thus, the flux of any one scalar can be defined in terms of another. At its core, MOST leads to the approximation that, *in the case of shear-generated turbulence, all scalars are transported by eddies in the same way*. This represents a powerful realization for our understanding of surface layer fluxes.

Accepting its role in the theoretical underpinnings of flux-gradient relations, we want to emphasize that there are important caveats and limitations to the application of MOST. MOST is only valid in the surface layer, where fluxes are, to a first approximation, constant with height. Even in the surface layer, however, MOST can break down in the face of high rates of vertical convection. Under such conditions, large eddies can create "chimney"-type updrafts that transport mass and energy in a way that uncouples their transport from the influence of local concentration gradients. In highly convective conditions, K_D becomes dependent on z_i/L, rather than z/L, where z_i is the depth of the convective boundary layer (Johansson *et al.* 2001). In the lower portion of the atmospheric surface layer, especially above tall forest canopies, the coupling between fluxes and gradients can also break down, invalidating MOST. In this domain, high shear stress can create large eddies that propel mass and energy across distances far greater than those represented by the mean free path within their respective concentration gradients.

14.2 The effect of a plant canopy on atmospheric turbulence

As we move our focus to the atmosphere just above a canopy, in the roughness sublayer, the universal Monin–Obukhov functions that describe flows in the surface layer become invalid. How then, do we make sense of turbulence in the vicinity of canopy roughness elements? In early theoretical treatments the turbulence regime at the canopy surface was approximated as the turbulence regime of the surface layer, but with energetic accommodation for small, high frequency vortices in the vicinity of leaves and branches (reviewed in Raupach and Thom 1981). We now realize that this view was much too simple. Canopies not only modify the turbulence field in the higher frequencies of the eddy spectrum, but also in the lower frequencies. It is not only small, local vortices created near leaves and branches that make up the canopy's turbulence regime, but also large, sweeping eddies that traverse the entire canopy. Influences on both the high and low frequency components of turbulence must be reconciled in order to balance the momentum budget in the vicinity of canopies. Furthermore, canopy elements are seldom arranged with random placement, and therefore turbulence is created with the systematic "memory" or "imprint" of a canopy's structure. Structural aspects of a canopy can alter wind direction and local rates of energy dissipation. It is clear that unlike the situation in the surface layer, the characteristics of turbulence fields in canopies will be more dependent on unique aspects of each canopy, making the formulation of general theories and universal scaling relations even more challenging.

14.2.1 Observations of turbulence at the canopy surface

One of the most characteristic features of near-neutral wind flows in the vicinity of a canopy is the sharp inflection in velocity at the canopy surface (Figure 14.1). The inflection has been observed at the top of many different types of canopies, both natural

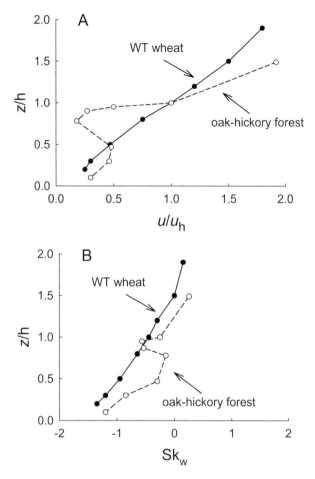

Figure 14.1 **A**. The relationship between normalized height of measurement (z/h) and normalized longitudinal wind velocity (u), both relative to the top of the canopy at height (h). The data are for two canopies – a wheat canopy measured in a wind tunnel experiment (WT wheat) and an oak-hickory forest canopy in Tennessee. **B**. The relationship between z/h and skewness (Sk_w) (the third-order statistical moment) of the vertical wind velocity measured above, and within, the two canopies. The negative skew within the two canopies indicates a preponderance of downward-moving (negative) wind gusts. The wheat data were redrawn from Raupach *et al.* (1996) and the forest data were redrawn from Baldocchi and Meyers (1988), with kind permission from Springer Science + Business Media.

and artificial, covering a broad range of heights and surface roughness features (Raupach *et al.* 1996). Thus, it appears to be a general feature of near-canopy flows. The inflection is caused by the drag force encountered by the wind at the canopy surface; rougher surfaces induce higher drag, which consumes momentum from the wind at higher rates and causes greater inflection in the velocity profile. The canopy drag force is congruent with high rates of Reynolds shear stress in the turbulent regime near the canopy surface (see Section 13.3). Variation in canopy roughness is, in general, correlated with variation in the size and placement of leaves; larger leaves, clustered near the tops of trees tend to

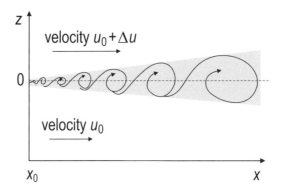

Figure 14.2   The formation of turbulent mixing at the interface of two plane mixing layers flowing in the same direction at different velocities. In this case, we assume that one layer (with velocity $u_0 + \Delta u$) flows above the point $z = 0$ and one layer (with velocity u_0) flows below the point. The layers are allowed to come into contact with each other at point x_0, where the turbulent layer begins to form. As the flows continue in the x direction turbulence at the interface grows. Redrawn from Raupach *et al.* (1996), with kind permission from Springer Science + Business Media.

produce a rougher canopy surface. The point of greatest inflection occurs at $z = h$, where h is canopy height. The steep decline in the streamwise mean wind speed ($\overline{u}$) through the canopy profile, and the fact that momentum flux at the ground is nearly 0 for many different types of canopies, is evidence of the efficacy with which canopies extract horizontal momentum from the wind.

Within the context of fluid mechanics, high Reynolds shear stress near the canopy surface is explained on the basis of plane mixing layers, which are common features at fluid-surface interfaces (Figure 14.2, see Acton 1976). Plane mixing layers are distinguished from classical ("rough-wall") boundary layers in that turbulent vortices form between layers of the moving fluid, rather than being generated solely at the fluid-surface boundary. As instabilities develop between plane mixing layers they have the potential to produce turbulent eddies traveling in directions above and below the zone of instability. Theory and observations from artificial canopies in wind flume experiments have provided insight into how bidirectional gusts are produced at the interface of plane mixing layers. Observations of vertical velocity fluctuations above canopies have confirmed the presence of waves that exhibit coherency and evolve to regularly spaced vortex rolls in the stream-wise direction (Raupach *et al.* 1996). Spacing of the rolls is defined by an Eulerian length scale (Λ_w) which, in turn, is correlated with a shear length scale (L_s, in units of m) defined empirically as (Finnigan 2000):

$$L_s = \frac{\overline{U}[h]}{\partial \overline{U}[h]/\partial z} , \tag{14.4}$$

where $\overline{U}[h]$ is the mean wind speed (m s^{-1}) at the top of the canopy and $\partial \overline{U}[h]/\partial z$ is the vertical wind shear. Across a broad range of canopies, the relation $\Lambda_w \approx 8L_s$ is fairly well conserved (Raupach *et al.* 1996, Brunet and Irvine 2000; although see Thomas and Foken 2007 for examples of departure from the common pattern). The process of roll formation is likely initiated when eddies intrude into the region of near-canopy stress from higher in the atmosphere (Raupach *et al.* 1996, Finnigan 2000). The intrusions raise near-surface shear stress above a critical threshold, causing the emergence of wave-like instabilities that grow in size and organize into roll and braid-like structures (Figure 14.3). The hairpin, bent rolls shown in Figure 14.3 bring counter-rotating vortices next to each other causing the formation of swiftly flowing, alternating downgusts and upgusts. As the turbulent structures continue to grow they become unstable and eventually break apart to form the classic, less-organized eddy structures associated with classic, "rough-wall" boundary layers.

14.2.2 The types and frequencies of eddies in canopies

In the near-neutral atmospheric surface layer the probability of observing an upward-oriented gust is approximately equal to that of observing a downward-oriented gust. In the vicinity of canopies, however, downward-oriented gusts, often called *sweeps*, are observed as short, narrow bursts with high speeds, whereas upward-oriented gusts, called *ejections*, occur at lower speeds and have larger cross-sectional areas. Thus, the probability of observing a sweep at any point within a time series of wind speed data is less than that of observing an ejection (Figure 14.4). Together, sweeps and ejections account for most of the venting that occurs in the canopy airspace. Canopy sweeps originate from the combined force of gusts generated by neighboring vortices, as discussed above (see Figure 14.3). Canopy ejections are the result of the upwardly oriented pressure fields that develop during the time between sweeps within the canopy airspace.

A more complete picture of the distribution of turbulence within a canopy can be attained using a technique called *conditional analysis*. In conditional analysis we plot the joint probability distribution in x–y graphical space of designated events, in this case the occurrence of wind gusts. The joint probability distribution provides insight into the most likely combinations of correlated events. Taking the example of observed longitudinal (u') and vertical (w') turbulent wind speeds we can construct a two-dimensional plot with four quadrants (Figure 14.5), distinguished as follows:

Quadrant 1 $u' > 0$; $w' > 0$ outward interaction;
Quadrant 2 $u' > 0$; $w' < 0$ sweep;
Quadrant 3 $u' < 0$; $w' < 0$ inward interaction;
Quadrant 4 $u' < 0$; $w' > 0$ ejection.

The interactions in Quadrants 1 and 3 are rare in most canopies. Correlations in Quadrant 2 are the result of covariance between a positive u' and negative w', reflecting a sweeping downward-oriented gust moving in the same direction as the streamwise flow. The opposing motion, in which the gust rotates opposite the streamwise flow (i.e., a negative u'), is most often observed with an upward-oriented gust (i.e., a positive w'). The results shown in Figure 14.5 are typical of many forest canopies.

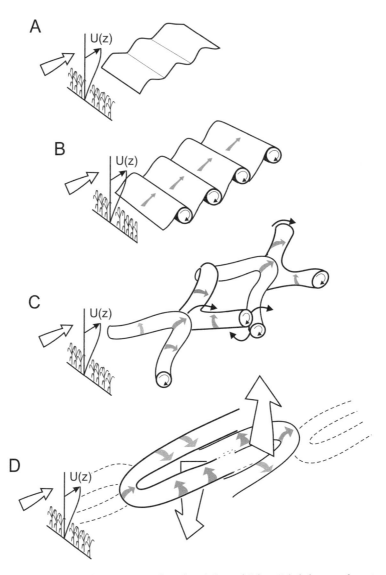

Figure 14.3 Visualizations of the time-dependent progression from: (panel A) initial Kelvin–Helmholtz wave formation forming near the top of a canopy within the region of shear stress (indicated by the profile of wind speed, U(z)), to (panel B) forward rolling of the vortices, to (panel C) the formation of fully organized vortex rolls that bend under the force of surrounding turbulence and form hairpin pairs due to mutual induction of their neighboring vortex fields, to (panel D) the conversion to fully developed paired hairpin structures that generate upward and downward directed gusts due to the mutual activity of neighboring vortices. Redrawn from Finnigan *et al.* (2009).

Spectral analysis of turbulence within canopies has revealed that in general there is a shift toward eddies at higher frequencies and a steeper slope to the inertial subrange, compared to the atmospheric surface layer (Baldocchi and Meyers 1988, Amiro 1990, Gardiner 1994, Liu *et al.* 2001, Su *et al.* 2008). Both of these effects can be explained as the wind working

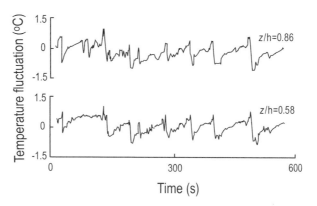

Figure 14.4   Temperature fluctuations in the air of a mixed forest canopy in Canada. Measurements were taken at two heights within the canopy (where z/h is the height of measurement (z) relative to the canopy height (h)). The pattern of fluctuation reveals evidence of canopy "sweeps" (downward-moving cooler air of high velocity originating above the canopy) and canopy "ejections" (upward ramps of warmer, slower-moving air originating within the canopy), as indicated by the coordinated decreases and increases in temperature, respectively, at the two heights. Temperature is measured as deviation from the starting point, which is taken as 0. Redrawn from Gao *et al.* (1989), with kind permission from Springer Science + Business Media.

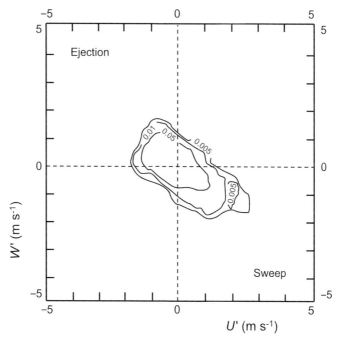

Figure 14.5   Joint probability distribution of the normalized streamwise turbulent wind speed (u') and the vertical turbulent wind speed (w') measured in a plantation spruce forest and presented in conditional analysis format. $U' = u'/\sigma_u$ and $W' = w'/\sigma_w$, where σ_u and σ_w are the standard deviations of the observed streamwise and vertical wind speeds, respectively. The probability values indicate the range of joint U' and W' values distributed across the four quadrants after excluding extreme tails of the distribution at P = 0.005, 0.001, and 0.05, respectively. Redrawn from Gardiner (1994), with kind permission from Springer Science + Business Media.

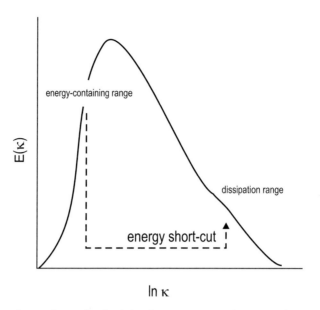

Figure 14.6   A schematic drawing showing the transfer of turbulent kinetic energy in a plant canopy from larger eddies in the energy-containing range of the spectrum to smaller (higher wavenumber) eddies associated with wake and wave motions closer to the TKE dissipation range. Redrawn from Finnigan (2000).

against the drag of leaves and other canopy elements, which creates small wakes, diverting the energy of larger eddies into smaller eddies without going through the more orderly progression of energy transfer characteristic of the inertial cascade; thus, a *spectral short-cut* is created (Figure 14.6). One consequence of these interactions is the prevention of the isotropy that would theoretically occur within the inertial subrange (Kolmogorov 1941, Finnigan 2000). Prevention of isotropy provides a means for net directional transport *within canopies* at those inertial subrange frequencies that would not normally contribute to transport *within the atmospheric surface layer.*

Laboratory studies have provided an opportunity to observe directly the fine-scale structure of wakes and swirls produced by wind on the lee side of canopy elements (Poggi and Katul 2006). Turbulent wakes take the form of small vortices that line up in succession in the streamwise direction behind the canopy element, similar to *von Karman streets*, a commonly observed structure in fluid wakes (Taneda 1965). Measurements of turbulence spectra in the wakes on the lee side of tree boles are consistent with the predicted vortices of von Karman streets (Cava and Katul 2008). The average frequency of rotation ($\overline{f}[z]$, with units s^{-1}) in wake vortices as a function of height can be estimated as a function of mean tree bole diameter ($\overline{d}[z]$) and streamwise wind speed ($\overline{U}[z]$), both referenced to height (z) within the canopy, according to:

$$\overline{f}[z] = \frac{0.217\overline{U}[z]}{\overline{d}[z]} \, , \tag{14.5}$$

where 0.217 is a dimensionless number known as the Strouhal number. The existence of fine-scale shear in the wake of leaves and branches enhances within-canopy dissipation of turbulent kinetic energy.

In summarizing the effects of canopies on turbulence we can identify three spatial domains within the atmospheric surface layer that can be differentiated on the basis of controlling variables (Poggi *et al.* 2004), and that take us back to the journey of the hypothetical atmospheric particle that we referred to in the introductory paragraphs of this chapter. Well above the canopy, in that domain where Monin–Obukhov theory is dependable, turbulence length scales are determined solely as a function of height above the surface (z) and the spectral distribution of turbulence kinetic energy follows patterns expected of an orderly transfer cascade from larger to smaller eddy motions. As the canopy is approached, in the roughness sublayer, turbulence is generated by plane mixing layers working against the sharp gradient in shear stress. In this domain, turbulence length scales are determined by both wind speed and shear, and skewed patterns begin to emerge in the spectral probability distribution of turbulence kinetic energy. Finally, within the canopy airspace, eddies are re-sized and re-oriented according to unique aspects of canopy structure. In this domain, turbulence kinetic energy is transferred directly to wake vortices and leaf or branch waving motions; turbulence length scales are highly influenced by quantitative aspects of the local canopy structure and can be described according to the characteristic dimensions of canopy elements.

14.3 Turbulent fluxes above canopies

By the principles of conservation of mass and energy and assuming horizontal homogeneity, we can state:

$$\int_0^{z_r} \frac{\partial \overline{F_v}}{\partial z} dz = \pm \int_0^{z_r} \frac{\partial \overline{S}}{\partial z} dz, \tag{14.6}$$

where $\overline{F_v}$ is the time-averaged vertical flux density, z_r is a reference height above the canopy, and $\overline{S}$ represents the time-averaged, net source (positive relation) or sink (negative relation) diffusive flux for the scalar or vector of interest within the canopy below z_r. Using Reynolds averaging to partition F_v into its turbulent and mean components (F_{vt} and F_{vm}, respectively), and recognizing the vertical flux as a statistical covariance, we can write:

$$\overline{F_v} = \overline{\rho}_m (\overline{wc}) = \overline{F_{vm}} + \overline{F_{vt}} = \overline{\rho}_m (\overline{w}\overline{c}) + \overline{\rho}_m \left(\overline{w'c'} \right),$$
$$\text{Term I} \qquad\qquad \text{II} \qquad\qquad\qquad \text{III} \tag{14.7}$$

where the time-averaged vertical flux density is equal to the time-averaged correlation between vertical wind speed (w) and scalar mole fraction (c) (Term I), which is equal to the sum of the Reynolds averaged mean and turbulent flux densities (Term II), which is in turn equal to the sum of the Reynolds averaged mean and turbulent covariances between velocity and concentration. (The molar density of dry air (ρ_m) is included in each product because we are using mole fractions to define c, and F_v is defined as a flux density.) Combining Eqs. (14.6) and (14.7), and integrating across vertical space to z_r, we can write:

$$\int_0^{z_r} \frac{\partial \overline{F_v}}{\partial z} dz = \overline{\rho}_m \int_0^{z_r} \frac{\partial \overline{wc} + \overline{w'c'}}{\partial z} dz = \pm \int_0^{z_r} \frac{\partial \overline{S}}{\partial z} dz, \qquad (14.8)$$

where the diffusive source or sink flux ($S[z]$) is assumed to be at steady state. Assuming that $\overline{w} = 0$, we can write:

$$\overline{\rho}_m \int_0^{z_r} \left(\overline{w'c'}\right) dz = \pm \overline{\rho}_m \left(\overline{w'c'}\right)_0 \pm \int_0^{z_r} L[z] \frac{\overline{\rho}_m \left(c_a[z] - c_i\right)}{r_b + r_s} dz,$$

$$\text{Term I} \qquad\qquad\qquad \text{II} \qquad\qquad\qquad \text{III} \qquad\qquad\qquad (14.9)$$

where $\overline{\rho}_m\left(\overline{w'c'}\right)_0$ is the turbulent flux density to or from the soil surface (mol m^{-2} s^{-1}), $L[z]$ is leaf area density (m^2 leaf area m^{-3} canopy volume), $c_a[z]$ is the ambient mole fraction of a constituent at height z, c_i is the average leaf intercellular mole fraction of the constituent, and r_b and r_s are the boundary layer and stomatal resistances to the diffusive flux (s m^{-1}), respectively. Term I is often called the *eddy flux*. Equation (14.9) contains the implicit assumption that $F_{vm} = 0$; that is, that the mean vertical flux does not contribute to the total vertical flux. This assumption will be discussed further below, as its validity can be challenged. However, for the moment, and for convenience, we will assume that it is valid. Formal derivation of the eddy flux and its relation to the vertically integrated diffusive flux is presented in Appendix 14.2.

We introduced the concept of spectral density as a means of characterizing the total covariance across all eddy frequencies, and therefore the total flux, in Chapter 13 (Section 13.5). Now, we can connect that theory with the eddy flux. The eddy flux reflects the integral sum of the co-spectrum of turbulent transport:

$$\overline{\rho}_m \int_0^{z_r} \overline{w'c'} dz = \int_0^{z_r} \int_{\omega=0}^{\infty} S_{wc}[\omega] d\omega dz, \qquad (14.10)$$

where S_{wc} is the co-spectral density (mol m^{-2} radian^{-1}) and ω is the angular frequency (with units radians s^{-1} and related to natural frequency, f, as $\omega = 2\pi f$). It is a challenge to validate spectral theory through observations as some of the observed eddies exist at frequencies that exceed the limits of statistical stationarity; thus, the contributions of eddies to the observed flux density may be influenced by differences in the diurnal environment. We often relax the definition of the eddy flux to include only that portion of the spectral density contained in eddies sampled in a specified measurement period; typically 30 minutes. The eddy flux measurement has been applied to hundreds of sites across the globe in the interest of obtaining direct observations of the CO_2 and water fluxes from different ecosystems and providing observational constraint on models of the global carbon and water cycles (Baldocchi *et al.* 2001). A brief discussion of the approaches required to measure the eddy flux in a variety of ecosystems and to evaluate its accuracy and uncertainties is provided in Box 14.1.

Prior to 1960, most researchers interested in quantifying the rate of mass exchange between ecosystems and the atmosphere did so by sampling known areas of the surface using enclosure techniques followed by spatial aggregation and averaging. Mathematical aggregation, however, carries the problem of compounded errors and inadequate representation of non-linear scaling relations. Micrometeorologists of the 1960s began exploring new approaches, including the use of near-surface winds to mix, or physically integrate, flux signals from larger areas of the surface. The underlying concept was simple: *if time-dependent variance in the velocity vectors of the wind and concentrations of scalar quantities could be resolved, then an integrated measurement of the flux could also be resolved.* This concept recognized the surface flux as a simple covariance between velocity and concentration. These early efforts led to eventual development of the eddy flux approach.

The eddy flux approach depends on atmospheric turbulence to carry scalar entities past measurement sensors and to thoroughly mix the air so that the scalar of interest does not accumulate in the canopy airspace. Without thorough mixing the observed eddy flux is not effectively coupled to the cumulative activity of sources and sinks within the canopy. This causes the eddy flux to underestimate the true surface flux. Inadequate mixing is especially common at night when atmospheric stability increases. In these circumstances the true surface flux must be determined as the sum of the eddy flux and the time-dependent change in scalar concentration beneath the flux sensors:

$$
F_{vt(z_r)} = \bar{\rho}_m \, \overline{w'c'}_{(z_r)} + \bar{\rho}_m \int_0^{z_r} \frac{d\bar{c}}{dt} \, dz.
$$

$$\text{Term I} \qquad\qquad \text{Term II} \qquad\qquad\qquad\qquad \text{(B14.1.1)}$$

In addition to eddy flux instrumentation, measurement towers typically include a sampling system that measures the vertical profile in mean scalar concentration and thus permits an approximation of Term II.

The eddy flux method is the only method currently available for directly measuring turbulent fluxes. How do we know that eddy flux measurements are accurate? There are several independent approaches that can be used to address this question, including analysis of co-spectral density, comparison of measured fluxes against observations of plant growth or balanced scalar budgets, and evaluation of energy budget closure. Co-spectral analysis is often used to assess how well observations of turbulent transport follow theoretical and observational norms that have been established during the many past campaigns that have been conducted on ideal terrain (e.g., the Kansas and Minnesota campaigns of the 1960s). From those studies, we know the behavior to expect when normalized fluxes are plotted against a spectral scale of turbulence; deviations from the expected patterns can be used to flag uncertainties in the flux time series. The biometric approach to eddy flux validation has been applied to the CO_2 flux. Tower-based observations of net CO_2 fluxes are converted to a measure of net primary productivity (NPP), often through models that are conditioned on the observations. Measurements of the biomass increment of plants in the tower footprint can then be compared to estimated NPP to determine the accuracy of the eddy flux measurement. Energy-budget closure relies on the assumption that if the eddy flux measurements at a site are accurate, then closure of the local energy budget, which depends on two turbulent fluxes (sensible and latent heat exchange), should be complete. By similarity, we assume that if we are making accurate measurements of the turbulent fluxes for sensible and latent heat then we are also making accurate measurements of the turbulent fluxes for other scalars, such as CO_2. The eddy flux approach was originally developed for very simple terrain and for scalar fluxes for which instrumentation was available to measure turbulent deviations at a high frequency. As this approach has been moved into natural ecosystems with complex terrain, and applied to scalars for which quantification is difficult, new research efforts have emerged to accommodate the theory and intensify efforts toward accurate flux validation. More details about the eddy flux measurement can be found in Kaimal and Finnigan (1994), Baldocchi (2003), and Foken (2008).

At this point, it is worth returning to the flux-gradient relation (K-theory) that we presented in Section 14.1 and to now consider its relation to the eddy flux. If we assume that the eddy flux can be related to a diffusive-analog flux density, we can state:

$$\int_0^{z_r} \overline{w'c'} dz = -K_D \int_0^{z_r} \frac{\partial c}{\partial z} dz. \tag{14.11}$$

It is important to realize that the forces that underlie molecular diffusion are different than those that underlie the eddy flux. In the case of molecular diffusion, scalar transport is forced by the thermal motion of diffusion particles; in the case of the eddy flux, scalar transport is forced by the velocity of the fluid. To establish the theoretical analogy stated in Eq. (14.11) we must assume that the length scale for transport in both cases is smaller than, or at least similar to, the length scale of the within-canopy concentration gradient (Figure 14.7). This assumption holds true for molecular diffusion. The mean free path for molecular diffusion is similar to the effective length scale of the relevant concentration gradient. This assumption was originally thought to be also valid for wind eddies transporting scalars through the air space of canopies. Now, we realize that this is not the case. Rather, canopy turbulence often occurs as large eddies at the same length scale as canopy depth, and thus single eddies are capable of crossing, in one sweep, the entire span of concentration gradients. This has the potential to uncouple turbulent transport from the local concentration gradient. As a result, the equality expressed in Eq. (14.11) may be without foundation when applied to many canopies; especially tall, rough forest canopies with the high shear stresses that produce large eddies (Denmead and Bradley 1985).

Empirical evidence for uncoupling between turbulent transport and concentration gradients above forest canopies is found in past measurements of sensible heat, latent heat, and CO_2 fluxes (Figure 14.8). In the atmosphere immediately above the pine forest depicted in Figure 14.8 net fluxes and concentration gradients for all three scalars have opposite signs, which is consistent with the flux-gradient relationship. However, within the canopy, net fluxes for sensible heat and CO_2 have the same sign as the concentration gradient, demonstrating *counter-gradient transport*. In the case of latent heat, a significant net flux occurs despite the absence of a measurable gradient. Mathematically, this would imply an infinitely large K_D.

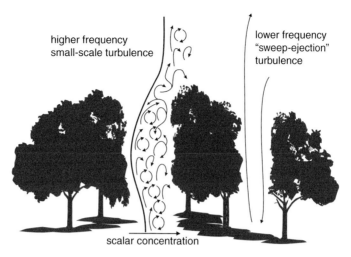

higher frequency
small-scale turbulence

lower frequency
"sweep-ejection"
turbulence

scalar concentration

Figure 14.7 Theoretical justification of the flux-gradient approach at the canopy scale requires that turbulence within the canopy be composed of relatively small eddies, such that the mixing length is comparable in scale to the scalar concentration gradient. In reality, turbulence in many canopies is composed of relatively large eddies that traverse the length of the canopy. In the left figure, the vertical profile of the scalar concentration gradient is provided as the heavy solid line.

14.3.1 Air density and the eddy flux

Now, we return to one of the simplifying assumptions used to derive the equality expressed in Eq. (14.9). When this relation is assessed over ideal, flat terrain it is commonly assumed that $\overline{w}$ equals 0. This assumption is required to justify equality between the sum of source-sink activity within the ecosystem and the turbulent flux divergence, alone. The assumption is based on the fact that there is no source or sink for dry air at the surface, and assuming no significant change in atmospheric pressure, the mass of air at the surface must remain constant; thus, the mean flux density of rising air, $(\overline{wc}_{air})_{up}$, must be balanced by the mean flux density of sinking air, $(\overline{wc}_{air})_{dn}$, where c_{air} refers to the mass density (g m^{-3}) of moist air. A careful theoretical analysis (Webb *et al.* 1980), however, has revealed that the assumption $(\overline{wc}_{air})_{up} = (\overline{wc}_{air})_{dn}$ can only be reconciled if we accurately account for the influences of sensible and latent heat exchange on the mass density of air at the surface. To illustrate the point, imagine a surface that is exchanging sensible and latent heat with the overlying atmosphere. Air parcels in contact with the surface will expand as they take in heat and moisture. As parcels of lower density rise, they will be replaced by sinking parcels with cooler and drier air. This exchange of parcels with different mass densities will result in a net gain of mass at the surface. Given the requirement for conservation of mass, a mean, upward flux of dry air that compensates for the mass imbalance must, in theory, exist. In the presence of combined sensible and latent heat fluxes at the surface, we can state:

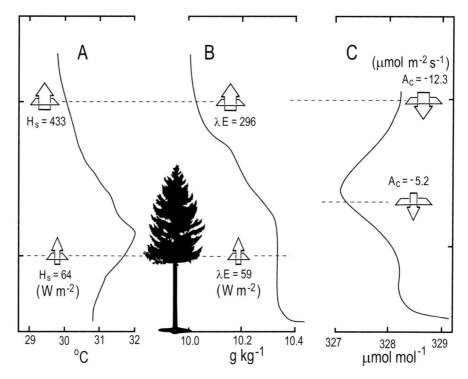

Figure 14.8 Representative fluxes of sensible heat (H$_s$), latent heat (λE), and canopy-scale CO$_2$ (A$_c$) within a pine forest in relation to gradients in temperature, water vapor concentration, and CO$_2$ concentration. Note that above the canopy, fluxes of all three scalars occur from regions of high concentration to regions of low concentration. However, in the space below the canopy, fluxes for sensible heat and CO$_2$ occur in a direction from low concentration to high concentration, and the flux for latent heat is significant in magnitude despite no apparent gradient in water vapor concentration. Redrawn from Denmead and Bradley (1985), with kind permission from Springer Science + Business Media.

$$F_{vj} \approx \overline{w'c_j'} + a\,\frac{\overline{c}_w}{\overline{c}_a}\,\overline{w'c_w'} + \overline{c}_j\left(1 + a\,\frac{\overline{c}_w}{\overline{c}_a}\right)\frac{\overline{w'\theta_t'}}{\overline{\theta}_t},$$

$$\text{Term I} \qquad\qquad \text{II} \qquad\qquad\qquad \text{III} \qquad\qquad\qquad\qquad (14.12)$$

where F_{vj} is the vertical flux density of scalar j (with units g m^{-2} s^{-1}); $\overline{w'c_w'}$ and $\overline{w'\theta_t'}$ are the turbulent fluxes of water vapor and sensible heat, respectively, a = m$_a$/m$_w$ (the ratio of the molecular weight of dry air to water vapor), $\overline{c}_j$ is the mean density concentration (g m^{-3}) of scalar j, $\overline{c}_a$ is the mean density (g m^{-3}) of dry air, and $\overline{c}_w$ is the mean density (g m^{-3}) of water vapor (for derivation see Webb *et al.* 1980, Wyngaard 1990). Term I is the time-averaged eddy flux for scalar j, and Terms II and III together account for the required theoretical, mean vertical flux of scalar j due to gradients in mass density. Terms II and III take their mathematical form from the equation of state. With Terms II and III in place, the total mass flux can be correctly estimated as the covariance turbulent flux. The correction represented by Terms II and III is often called the *Webb–Pearman–Leuning (or WPL) correction*. In one sense, the WPL relation is not a "correction," but rather the application of proper physics. The most conserved covariance flux is found with mass quantities expressed in mole fractions or mixing ratios, not mass densities, which are "sensed" by most mass spectrometers. It is because of our reliance on observed mass densities that errors in the calculated covariance fluxes can arise because of vertical gradients in atmospheric density. Failure to apply the WPL correction can cause a "dead," inert surface to appear as though it is exchanging CO_2 and H_2O. This was demonstrated by Ham and Heilman (2003) who used the eddy covariance technique without the WPL correction, and observed apparent photosynthesis and transpiration from an asphalt parking lot! However, when the correction was applied, the fluxes were correctly predicted to be zero. *We want to emphasize that the WPL correction is only relevant when concentration is observed or expressed in terms of mass density. If concentration is expressed as the mole fraction or mixing ratio with respect to dry air, then the WPL correction is not needed.*

14.4 Mesoscale fluxes

Variation in the surface properties of neighboring landscapes can cause local-to-regional atmospheric circulation cells capable of transporting mass and energy across broad areas (Figures 14.9 and 14.10). The transport of mass and energy within these large, regional cells is often referred to as a *mesoscale flux* (Pielke *et al.* 1991, 1998). An example of mesoscale circulation is found in the sea breezes and land breezes that occur in coastal areas. During the morning, coastal inland areas heat up faster than the sea, causing air above the land to rise and air from the sea to flow inland; this causes a regional circulation known as a *sea breeze* to observers on the coast. At night, the land cools faster than the sea and the circulation is reversed, giving rise to a *land breeze*. Similar effects can be predicted for mid-continental regions. Differential sensible and latent heat fluxes from land patches can induce the formation of large eddies; sometimes spanning several times the length of the convective mixed layer (Baidya and Avissar 2000).

Large-scale atmospheric circulations have become an important topic in studies of the localized convective activity that accompanies regional monsoons. In the wet tropics, deep convection, characterized by rapidly rising moist air is driven by broad regional heating and is typically associated with the rising branch of Hadley Cell circulation. This is an equatorial phenomenon with hemispheric meteorological ramifications. In coastal regions with dry, often elevated terrain, monsoon convection can occur because of steep land-sea thermal contrasts. Rapid heating of upland coastal terrain during the day, and associated high rates of upward convection can cause near-surface low pressure cells to develop. If cooler, ocean air is capable of flowing into this cell, monsoonal precipitation can develop. More recently, it has become clear that convective precipitation cells, similar to those observed in monsoon systems, can form in certain soil moisture regimes in inland areas that experience rapid heating (Wang *et al.* 2007, Siqueira *et al.* 2009). Working in dry regions of Sahel Africa, for example, Taylor *et al.* (2011) observed that convective activity, which ultimately led to

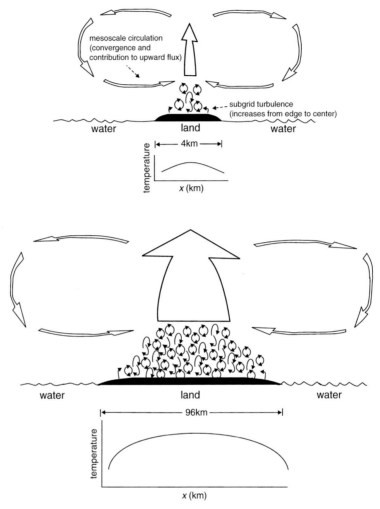

Figure 14.9 Mesoscale circulation cells and local turbulence developing over an island as the result of differential heating between the land and sea. A larger island is capable of developing larger temperature differentials and, as a result, larger mesoscale circulation cells.

precipitation, was triggered most frequently above soils with *spatial moisture gradients*, rather than those that were homogeneously "dry" or "wet" (Figure 14.11). It was the combination of a dry patch grading into a wet patch across the space of 10–40 km that most frequently produced convective triggers. Apparently such gradients provided the correct spatial alignment of sensible heat loss and updraft activity at the dry end of the gradient and near-surface moisture flow from the moist end of the gradient, to optimally trigger moist, convective flow.

Mesoscale circulations can also be caused by shear-induced interactions as wind passes over terrain characterized by high relief. For example, as wind passes over topographic features, such as hills and mountains, atmospheric "wave" and "rotor" motions can form on

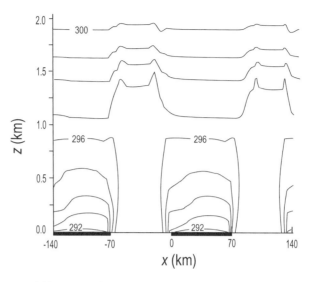

Figure 14.10 The potential temperature field generated from numerical simulations using the Regional Atmosphere Modeling System (RAMS). Strips of water (indicated by heavy black lines) alternated with strips of land, each with width 70 km. Mean wind speed was 1 m s^{-1} and contour interval is 1 K. Note the development of deep columns of warm air caused by thermal updrafts above the strips of land, potentially drawing cool air at lower levels from adjacent strips of water, generating mesoscale circulations between the land and water strips. Redrawn from Zeng and Pielke (1995a).

the lee side (Turnipseed *et al.* 2004). In another example, the advection of cool air over warmer lakes can induce the formation of rolling vortices; the cooler air with its higher density tends to "tumble" as it passes over the lake, inducing the formation of an elongated, rolling vortex (Sun *et al.* 1997). All of these mesoscale circulations are capable of trans- porting mass, energy, and momentum, *in addition to the "local" turbulent fluxes.*

The relation between mesoscale and "local" turbulent fluxes has been studied using regional weather prediction models capable of large eddy simulation. Local turbulent fluxes are high near the surface, whereas mesoscale fluxes are high aloft (Figure 14.12). Mesoscale eddies that form as a result of regional circulation will become organized at length scales greater than those that affect local fluxes. Thus, mesoscale and local turbulence regimes occur in distinct frequency domains and exert influences in different spatial domains of the atmosphere. In the upper region of the PBL the dominant influence of large, organized eddies is obvious. Even above the PBL, in the free troposphere where local fluxes becomes diffuse, mesoscale eddies can penetrate and cause circulatory transport as air from aloft sinks toward the ground (Zeng and Pielke 1995b).

Despite their tendencies toward distinct frequency domains, mesoscale fluxes can influ- ence local fluxes. Mesoscale motions can induce shifts in the local wind direction and they can cause low-frequency contributions to the local flux. Taking into account the potential for interactions between mesoscale and local fluxes, we can describe the overall turbulent vertical flux as:

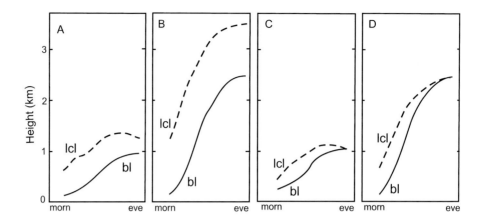

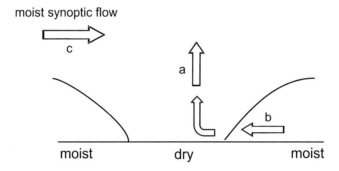

Figure 14.11 In the upper panels are shown the results of a simulation study conducted by Siqueira *et al.* (2009), in which the diurnal growth of the convective boundary layer (bl) and the lifted condensation level (lcl) are shown for the atmosphere above a wet soil without synoptic flow of moisture into the system (panel A), a dry soil without synoptic flow into the system (panel B), a wet soil with moist synoptic flow into the system (panel C), and a dry soil with moist synoptic flow into the system (panel D). The intersection of the bl and lcl indicate the time of a potential convective trigger event, leading to a convective rain storm. Growth of the bl is greatest above dry soils with high sensible heat exchange (panel B). However, without the introduction of moist air through synoptic flow, a convective trigger is unlikely. The probability of a convective trigger is increased when moist air is delivered into the region by synoptic flow above the boundary layer and entrained into the growing bl. In studies in the arid regions of Sahel Africa, a second control over growth of the bl and the triggering of convective rain storms is shown in the lower figure (redrawn from Taylor *et al.* 2011), where heterogeneous patches of dry and wet soil lead to the near-surface flow of moist air (b) into a dry patch where convective lift is high (a). The probability of triggering a convective rain event is increased if moist synoptic air is carried into the region from above the bl (c), and entrained into the growing bl above the dry patch.

$$\overline{w\,c} = \overline{w'\,c'} + \left(\overline{w'\,c''} + \overline{w''\,c'}\right) + \overline{w''\,c''},$$

$$\text{Term}\ \ \text{I} \qquad \text{II} \qquad \text{III} \qquad \text{IV} \qquad \text{V} \qquad\qquad (14.13)$$

where the total flux (Term I) is composed of the sums of the local turbulent flux (Term II), the interactions between the mesoscale and components (Terms III and IV), and the mesoscale

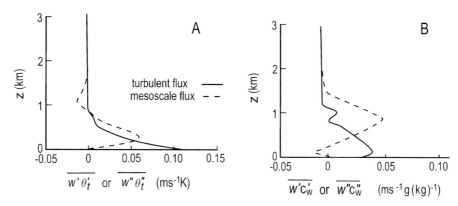

Figure 14.12 Turbulent and mesoscale sensible heat (panel A) or moisture (panel B) fluxes as a function of height (z), generated from the Regional Atmosphere Modeling System (RAMS), a numerical mesoscale model used here in the two-dimensional mode. For the simulations, mesoscale motions were simulated across alternating strips of land and water with width of 40 km and a mean wind speed of 3 m s^{-1}. PBL height is approximately 1 km. Redrawn from Zeng and Pielke (1995b).

flux (Term V). Given the most often used averaging time (30 min) for determining the turbulent eddy flux above terrestrial ecosystems, mesoscale contributions to the total flux typically emerge in the mean (advective) flux terms.

Appendix 14.1   Derivation of Monin–Obukhov similarity relationships

In their theoretical work in the 1950s, Monin and Obukhov developed a form of *similarity theory*, reasoning that the behavior of shear-induced turbulence in the neutral surface layer is predictable, at least with regard to statistical averaging, and that universal scaling relations can be derived. Similarity theory emerges from dimensional analysis, which is used in a variety of scientific disciplines to evaluate quantitative relations after normalization across time and space (see Section 12.6). The derived equations are often referred to as *similarity relations*, because they reflect relations of similar form, both mathematically and graphically. Monin–Obukhov similarity theory (MOST) was originally developed (Monin and Obukhov 1954) as a means to use four fundamental parameters – u_*, z, $g/\overline{\theta}_t$, and $\left(\overline{w'\theta_{vt}'}\right)_0$ – which reflect three fundamental dimensions – length, time, and temperature – to describe flow behavior in the surface layer. These parameters can be combined to produce a series of similarity relations. As an example, consider the MOST relation that provides a non-dimensional expression of the variance in vertical wind speed (σ_w):

$$\phi_w = \sigma_w/u_*. \tag{14.14}$$

In this case, u_* is used as a characteristic wind speed scalar to normalize σ_w (both u_* and σ_w have units of m s^{-1}). Normalization of scalar concentration gradients and fluxes can be achieved by relating them to one of the fundamental scaling parameters:

$$T_* = \frac{-(\overline{w'\theta_{vt}'})_0}{u_*}, \tag{14.15}$$

$$c_{w*} = \frac{-(\overline{w'c_w'})_0}{u_*}. \tag{14.16}$$

The utility of these relations is that common terms appear in all expressions. This facilitates algebraic substitution and provides the basis for evaluating one parameter in terms of another. Some of the most useful similarity relations are those that normalize scalar mean concentration gradients. For example:

$$\phi_M = \frac{\partial \overline{u}}{\partial z} \frac{k(z-d_H)}{u_*}, \tag{14.17}$$

$$\phi_H = \frac{\partial \overline{\theta}_{vt}}{\partial z} \frac{k(z-d_H)}{T_*}, \tag{14.18}$$

$$\phi_{cw} = \frac{\partial \overline{c_w}}{\partial z} \frac{k(z-d_H)}{c_{w*}}. \tag{14.19}$$

where ϕ_M, ϕ_H, and ϕ_{cw} are the MOST parameters reflecting normalized mean profiles in horizontal wind speed, temperature, and humidity, respectively. Using these relations, we can state the cases for normalized momentum (τ), sensible heat (H_{se}), and moisture fluxes (E_s) as:

$$\tau = u_*^2 = \frac{\partial \overline{u}}{\partial z} \frac{ku_*(z-d_H)}{\phi_M}, \tag{14.20}$$

$$H_{se} = -u_* T_* = -\frac{\partial \overline{\theta}_{vt}}{\partial z} \frac{ku_*(z-d_H)}{\phi_H}, \tag{14.21}$$

$$E_s = -u_* c_{w*} = -\frac{\partial \overline{c_w}}{\partial z} \frac{ku_*(z-d_H)}{\phi_{cw}}. \tag{14.22}$$

In statically neutral conditions, empirical observations have revealed that ϕ_M, ϕ_H, and ϕ_{cw} approach a constant value, which by convention is taken as unity (see Kaimal and Finnigan 1994). Thus, one can define the eddy diffusivity coefficients for the turbulent fluxes of momentum (K_M), sensible heat (K_H), and moisture (K_{cw}) as equal to $ku_*(z - d_H)$, and it can be assumed that $K_M = K_H = K_{cw}$. This is the basis for so-called K-theory, or the flux-gradient relationship, which has been used in the past to infer fluxes from concentration gradients.

Appendix 14.2 Derivation of the conservation equation for canopy flux

The flux of a scalar into and out of a defined volume is subject to the laws of conservation. Consider a small isothermal volume. Now, allow the volume to shrink such that its dimensions approach that of a single, fixed point (i.e., Δx, Δy, and $\Delta z \to 0$). We refer to this as the *control volume*. We will consider the mass balance for a single scalar (j) that is present within the control volume at mole fraction c_j. Respecting continuity, we can state:

$$\frac{d\rho_m c_j [t, x, y, z]}{dt} = \frac{\partial \rho_m c_j}{\partial t} + \frac{dx}{dt}\frac{\partial \rho_m c_j}{\partial x} + \frac{dy}{dt}\frac{\partial \rho_m c_j}{\partial y} + \frac{dz}{dt}\frac{\partial \rho_m c_j}{\partial z} \pm \frac{\partial S_{bj}}{\partial x} \pm \frac{\partial S_{bj}}{\partial y} \pm \frac{\partial S_{bj}}{\partial z}$$
$$\pm \frac{\partial S_{cj}}{\partial x} \pm \frac{\partial S_{cj}}{\partial y} \pm \frac{\partial S_{cj}}{\partial z}.$$

$$(14.23)$$

The right-hand side of the equation contains the partial derivatives that refer to the local (in time and space) components of the total derivative, S_{bj} represents the flux density of all biological sources or sinks (units of mol m^{-2} s^{-1}), and S_{cj} represents the flux density of all chemically reactive sources or sinks (units of mol m^{-2} s^{-1}). The derivatives with respect to time (dx/dt, dy/dt, dz/dt) represent the Cartesian components of wind speed (units of m s^{-1}), with u, v, w being the streamwise, lateral, and vertical velocity components, respectively:

$$\frac{dx}{dt}\frac{\partial \rho_m c_j}{\partial x} = u\frac{\partial \rho_m c_j}{\partial x}; \quad \frac{dy}{dt}\frac{\partial \rho_m c_j}{\partial y} = v\frac{\partial \rho_m c_j}{\partial y}; \quad \frac{dz}{dt}\frac{\partial \rho_m c_j}{\partial z} = w\frac{\partial \rho_m c_j}{\partial z}. \quad (14.24)$$

Assuming horizontal homogeneity in the distribution of biological and chemical sources and sinks, and using Eq. (14.24) for substitution, we can write:

$$\frac{d\rho_m c_j}{dt} = \frac{\partial \rho_m c_j}{\partial t} + u\frac{\partial \rho_m c_j}{\partial x} + v\frac{\partial \rho_m c_j}{\partial y} + w\frac{\partial \rho_m c_j}{\partial z} \pm \frac{\partial S_{bj}}{\partial z} \pm \frac{\partial S_{cj}}{\partial z}. \quad (14.25)$$

Partitioning all terms into mean and turbulent components using Reynolds averaging, and time-averaging all variances and source or sink fluxes, we obtain:

$$\frac{d\overline{\rho}_m \overline{c}_j}{dt} = \frac{\partial \overline{\rho}_m \overline{c}_j}{\partial t} + \left(\overline{u}\frac{\partial \overline{\rho}_m \overline{c}_j}{\partial x} + \overline{\rho}_m \overline{c}_j\frac{\partial \overline{u}}{\partial x} + \overline{v}\frac{\partial \overline{\rho}_m \overline{c}_j}{\partial y} + \overline{\rho}_m \overline{c}_j\frac{\partial \overline{v}}{\partial y} + \overline{w}\frac{\partial \overline{\rho}_m \overline{c}_j}{\partial z} + \overline{\rho}_m \overline{c}_j\frac{\partial \overline{w}}{\partial z} \right)$$
$$+ \left(\frac{\partial \overline{\rho_m u' c_j'}}{\partial x} + \frac{\partial \overline{\rho_m v' c_j'}}{\partial y} + \frac{\partial \overline{\rho_m w' c_j'}}{\partial z} \right) \pm \frac{\partial S_{bj}}{\partial z} \pm \frac{\partial S_{cj}}{\partial z}. \quad (14.26)$$

From this point we can further simplify the derivation. First, we assume that the mean fluxes in the streamwise and lateral coordinates are zero due to horizontal homogeneity. In other words, we assume that if a volume of air is advected into the upwind side of the control volume along the x-dimension it will displace an equivalent volume of air out of the downwind side of the control volume, carrying equal quantities of scalar j. Additionally, we assume that the streamlines contributing to u are ideally parallel to the surface, such that $\overline{w}$ is approximated as 0. Finally, we assume that scalar j is chemically inert. With these simplifying assumptions, Eq. (14.26) can be written:

$$
\frac{d\left(\overline{\rho_m c_j}\right)}{dt} = \frac{\partial\left(\overline{\rho_m c_j}\right)}{\partial t} + \frac{\partial\left(\overline{\rho_m u' c_j'}\right)}{\partial x} + \frac{\partial\left(\overline{\rho_m v' c_j'}\right)}{\partial y} + \frac{\partial\left(\overline{\rho_m w' c_j'}\right)}{\partial z} \pm \frac{\partial \overline{S}_{bj}}{\partial z}. \tag{14.27}
$$

To simplify the derivation even further we assume that the horizontal turbulent flux divergences ($\partial\left(\overline{\rho_m u' c_j'}\right)/\partial x$ and $\partial\left(\overline{\rho_m v' c_j'}\right)/\partial y$) are small and negligible compared to the vertical turbulent flux divergence ($\partial\left(\overline{\rho_m w' c_j'}\right)/\partial z$). This is justified because turbulence intensities in the x- and y-coordinates are small compared to those in the z-coordinate. Thus, we can write:

$$
\frac{d\left(\overline{\rho_m c_j}\right)}{dt} = \frac{\partial\left(\overline{\rho_m c_j}\right)}{\partial t} + \frac{\partial\left(\overline{\rho_m w' c_j'}\right)}{\partial z} \pm \frac{\partial \overline{S}_{bj}}{\partial z}. \tag{14.28}
$$

Of course, real landscapes rarely adhere to all of these simplifying assumptions. Thus, a major research topic in the field of micrometeorology is evaluation of the terms that we have eliminated. This topic is particularly important in landscapes that occupy hilly or mountainous terrain, where the wind streamlines can bend in various ways and horizontal gradients in c_j can be significant due to horizontal heterogeneity in vegetation structure and species composition.

The time-averaged covariance between w' and c_j' (i.e., $\overline{\rho_m w' c_j'}$) is referred to as the eddy flux. The eddy flux is measured at a specific "reference" height (z_r), in which case, we can consider the vertically integrated form of Eq. (14.28) as:

$$
\int_0^{z_r} \frac{\partial S_{bj}}{\partial z} dz + \int_0^{z_r} \frac{d\left(\overline{\rho}_m \overline{c}_j\right)}{dt} dz = \int_0^{z_r} \frac{\partial\left(\overline{\rho}_m \overline{c}_j\right)}{dt} dz \pm \int_0^{z_r} \frac{\partial\left(\overline{\rho_m w' c_j}\right)}{\partial z} dz \approx \int_0^{z_r} \frac{\partial\left(\overline{\rho}_m \overline{c}_j\right)}{\partial t} dz \\
+ \left[\left(\overline{\rho_m w' c_j'}\right)_{(z_r)} \pm \overline{F}_{j0}\right],
$$

$$\tag{14.29}$$

where $\overline{F}_{j0}$ is the time-averaged molar source or sink flux of constituent j at the ground surface. Equation (14.29) conserves the relations inherent in the continuity equation. Any space-dependent divergence in the time-averaged scalar flux must be balanced against a time-dependent divergence in the spatially averaged concentration of a control volume.

One final issue concerns the use of units in the eddy flux relation. In this derivation we have expressed the scalar concentration in units of dry air mole fraction. With w' expressed as m s^{-1}, the eddy flux density would be expressed in units of mol m^{-2} s^{-1}.

If scalar concentration is expressed as a mass density (e.g., $g\,m^{-3}$), then the molar density of dry air (ρ_m) is not needed in the eddy flux expression (i.e., $\overline{w'c'_j}$ is sufficient to resolve an eddy flux density with units $g\,m^{-2}\,s^{-1}$). In that case, however, influences of temperature and pressure gradients on mass density must be reconciled either empirically (by standardization of time-series measurements to a constant temperature and pressure) or theoretically (through application of the Webb–Pearman–Leuning relation). If the scalar concentration is expressed relative to moist air, the influence of moisture variation through the measured time series must be reconciled to the standard case for dry air.

15

Modeling of fluxes at the canopy and landscape scales

Single-layer models of evaporation from plant canopies are incorrect but useful, whereas multilayer models are correct but useless.

Raupach and Finnigan (1988)

This quote from a general discussion paper by Raupach and Finnigan is actually the title of the paper, and it succinctly summarizes one of the underlying compromises that must be made when developing models of earth system processes – *convenience versus accuracy*. Models are necessary for allowing us to proceed from observations to generalizations, and therefore from descriptions to projections. However, a model, by necessity, is an imperfect representation of a process. In earlier chapters we addressed the nature of models that have been developed at the biochemical and leaf scales. Here, we address models describing the turbulent exchanges between canopies or landscapes and the atmosphere.

If models are accurate in their depiction of transport processes, the area-specific turbulent flux that is predicted at the scale of a whole canopy should converge with the integral spanning all leaf diffusive fluxes determined within the canopy. Given this constraint, one could argue that knowledge of turbulence and its representation in a model is not necessary to determine surface-atmosphere fluxes; one need only determine the integral sum of diffusive fluxes in the underlying leaves. This type of bottom-up model, however, in which diffusive fluxes are simply summed to give a canopy flux, is incapable of describing how those fluxes distribute their respective scalar entities across the space within and above the canopy, and thus how variance in scalar concentration gradients can affect diffusive source or sink activity. In essence, modeled fluxes are uncoupled from dynamics in their associated concentration gradients. In order to dynamically link fluxes to gradients, atmospheric transport *must be considered*; mass or energy must be dispersed to or from the immediate vicinity of sources and sinks. One of the primary challenges facing modelers seeking to describe dynamic flux systems at the canopy scale is the inclusion of accurate scalar dispersion algorithms.

In this chapter, we begin with a discussion of the general structure of source-sink models and their use to predict mass and energy exchange in response to driving variables. We then pick up the topic of Eulerian versus Lagrangian perspectives on turbulent transport, which was introduced in Chapter 12, and we discuss past efforts to incorporate those perspectives into canopy and ecosystem models as a means of linking source and sink activity to atmospheric concentration gradients. It is the calculus of variations as revealed in Eulerian and Lagrangian approaches to describing time-dependent changes in flux systems that allows us to track dispersion within fixed coordinate frameworks.

15.1 Modeling canopy fluxes

Canopy flux models are constructed from one of two perspectives: the forward or inverse directions (see Box 1.2 in Chapter 1). In inverse modeling, observations of gradients in either environmental drivers of the flux, or concentrations of the scalar entity transported by the flux, are used to predict the spatial and temporal distributions of the coupled fluxes. In forward modeling, direct observations of the flux distribution are used to predict coupled gradients in environmental drivers or scalar concentrations.

15.1.1 Big-leaf and multiple-layer canopy flux models

One of the more common types of models used for predicting canopy fluxes is the *big-leaf model* (also called the *single-layer model*). In big-leaf modeling, microclimatic gradients within the canopy are averaged to provide spatiotemporal means for each driving variable, and these averaged values are then used in an inverse manner to predict whole-canopy fluxes (Norman 1981, Amthor *et al.* 1994). The use of averaged environmental drivers and the assumption that the canopy flux is coupled to averaged drivers, forces this form of modeling to carry the assumption of linear scaling between averaged fluxes and drivers. The assumption that a canopy reflects a mean entity that can be described by a single set of driving variables and a single set of response functions allows users to ignore unique distributions of sources and sinks; this mitigates the need for complex numerical schemes to resolve changes in flux along microenvironmental gradients. Thus, big-leaf modeling can be viewed as the simplest type of inverse modeling – detailed transport schemes that would normally be used to link observed gradients to flux distributions are not used. With their simplified averaging schemes, however, big-leaf models have the potential to carry the same non-linear averaging errors that were discussed for individual leaves (Section 8.6). Averaging errors arise when the distribution of a driving variable is heterogeneous and the response function, linking a flux to the variable, is non-linear in its form.

Recall from Chapter 8 that in the case of individual leaves averaging errors tend to resolve themselves, to some degree, because across the vertical expanse of a leaf the distribution of fluxes is linearly correlated with the distribution of driving variables. For example, we discussed how the distributions of temporally averaged PPFD and c_{ic} across leaves scale in approximately linear fashion with the maximum net CO_2 assimilation rate. This scaling provides a natural "convenience" that resolves averaging errors. Similar patterns can be found in the distributional covariance of PPFD and net CO_2 assimilation rate (A_n) among leaves within a canopy (Figure 15.1; also see Lloyd *et al.* 1995, Sellers *et al.* 1996). In a manner analogous to the case for individual leaves, this linear covariance is facilitated by the differential allocation of leaf nitrogen through the vertical profile of the canopy (Figure 15.2). Through evolutionary selection, plants have attained N allocation patterns that balance the concentrations of photosynthetic enzymes with the mean availability of PPFD and CO_2. Thus, the highest biochemical potential for A_n, and highest leaf N concentration, occurs in the highest layers of the canopy, where the time-averaged PPFD is also highest, and where stomatal conductances tend to be high, facilitating high

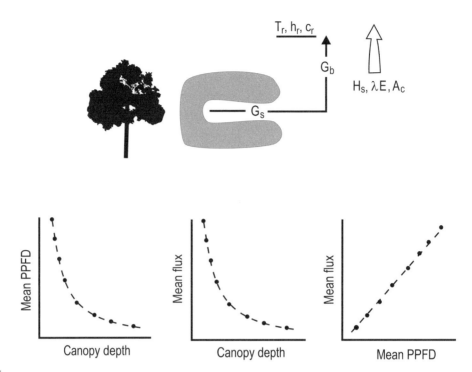

Figure 15.1 Assumptions that underlie "big-leaf" models of canopy flux density. In the upper part of the figure whole-canopy stomatal conductance (G_s) and boundary layer conductance (G_b) are represented as single, canopy-integrated values. Fluxes of sensible heat (H_s), latent heat (λE), and CO_2 (A_c) are calculated relative to the mean differences in temperature (T_r), humidity (h_r), and CO_2 mole fraction (c_r) between the air within the canopy and at a reference height above the canopy. In the lower part of the figure, the assumption of common exponential decreases in the mean PPFD and mean flux with increased canopy depth is represented, which forms the basis for the resultant assumption of linearity between mean PPFD and mean flux. The upper part of the figure was adapted from Raupach and Finnigan (1988). The overall concept of linear, big-leaf, canopy-level scaling is developed in Sellers *et al.* (1992).

rates of CO_2 diffusion into the leaf and high rates of Rubisco activity. This pattern of distribution is often described as "optimal" or "coordinated" and it produces advantageous (from the plant's perspective) plant nitrogen-use efficiencies (Field 1983, Hirose and Werger 1987, Chen *et al.* 1993). The linear relation between A_n and PPFD provides theoretical justification for the big-leaf modeling approach. The relation is not always ideal and there are likely yet-to-be-discovered limitations and imposed stresses that can modify presumed patterns of optimality (Anten 2005, Lloyd *et al.* 2010). In fact, several studies have noted that the vertical gradient of leaf concentration in many canopies is shallower than the vertical gradient in time-averaged PPFD (Meier *et al.* 2002, Niinemets 2007, Dewar *et al.* 2012). Nonetheless, it is a fair approximation to assume that linear relations exist in these variables across broad groups of plants and ecosystem types (Reich *et al.* 1997, 2003), and to apply the linear approximation as a convenience in justifying

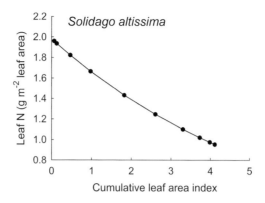

Figure 15.2   Leaf N concentration in *Solidago altissima* as a function of canopy position using cumulative leaf area index to denote vertical canopy depth. The line represents an exponential best fit to the data. Redrawn from Hirose and Werger (1987), with kind permission from Springer Science + Business Media.

big-leaf canopy modeling. As an example of big-leaf modeling, in Box 15.1, we present a description and application of one of the more widely used models; the "big-leaf" form of the Penman–Monteith equation, which is often used to predict canopy-scale H_2O and energy fluxes. Once again, we emphasize that big-leaf modeling exemplifies the inverse approach because fluxes are inferred from gradients (in this case a spatially and temporally averaged gradient).

Having stated that it is a fair approximation and convenience to assume linear scaling in the application of big-leaf models, we now note that there is a different, more subtle flaw in the deployment of such models. In assuming exponential extinction of PPFD through the vertical profile of a canopy, and thus linear scaling with exponential declines in maximum A_n, big-leaf models contain the inherent assumption that each canopy layer exists as a diffuse (well-mixed) radiative medium, much like vertical tissue layers in an idealized leaf. Horizontal discontinuities in PPFD are, in essence, converted into averaged, homogeneous gray layers. In a leaf, photons are scattered in the uppermost tissue layers, producing a diffuse photon flux that decreases exponentially as a function of distance. At each point within the leaf's profile, from top to bottom, the PPFD can be defined with a progressively lower time-averaged mean value defined by time-dependent variance distributed in Gaussian manner at values greater or lesser than the mean. That is, the total PPFD available to drive the CO_2 flux at any layer within the leaf can be defined by a single probability density function. In optical terms, the diffuse radiative nature of each tissue layer means that they exist as a gray, turbid system.

The optical system, and its effect on radiative transfer, is fundamentally different for canopies. Canopies possess clumped leaves, and leaves oriented at different angles, which allow the direct photon flux to penetrate to deeper canopy layers. Thus, the PPFD available to drive CO_2 fluxes at any layer within the canopy will be defined by two probability density functions, one for the diffuse fraction (I_d) and one for the direct fraction (I_D). (Penumbral

Box 15.1   **The Penman–Monteith (PM) model of ecosystem evapotranspiration**

One of the most commonly used big-leaf models is the Penman–Monteith model, which can be applied to either single leaves or entire canopies (see Section 10.4). The Penman–Monteith equation takes the form:

$$\lambda_w E = \frac{s(R_n - H_G) + \rho_a c_p (c_{wa^*}[T_a] - c_{wa}[T_a])/r_{ah}}{s + \gamma \dfrac{r_{ew}}{r_{ah}}}. \tag{B15.1.1}$$

In applying Eq. (B15.1.1) to the canopy scale, r_{ah} represents aerodynamic resistance to sensible heat transfer, which is dependent on physical features of the atmosphere and canopy including wind speed, atmospheric stability and surface roughness, and r_{ew} represents the canopy conductance to H_2O flux, which is dependent on the stomatal conductance of all leaves and the surface conductance of the underlying ground (Raupach and Finnigan 1988, Schulze *et al.* 1994). Equation (B15.1.1) can be rearranged to solve for g_{ew} (where $g_{ew(c)} = 1/r_{ew}$), and details of canopy structure can be used to partition g_{ew} into its canopy ($g_{ew(c)}$) and ground ($g_{ew(g)}$) components. Using the "big-leaf" Penman–Monteith equation, Schulze *et al.* (1994) analyzed the relations among $g_{ew(c)}$, $g_{ew(g)}$, and leaf area index (LAI, or L in equations) (Figure B15.1). Depending on L, the total ecosystem conductance for H_2O flux reflects a primary control over soil evaporation (for canopies with low L) or canopy transpiration (for canopies with high L) (Figure B15.1). The area between the lines reflects the contribution of ground evaporation. At L > 4, the conductances converge due to the thick canopy and low penetration of R_n to the ground; at L < 4, $R_{n(g)}$ and $g_{ew(g)}$ reflect significant controls over ecosystem $\lambda_w E$.

The big-leaf form of the PM model also has been useful in probing the interplay between conductance and absorbed energy as alternative controls over the rate of ecosystem evapotranspiration. In a manner similar to what we discussed for individual leaves, we can identify the *equilibrium evapotranspiration rate* (E_{eq}) at one extreme whereby the canopy boundary layer conductance is small and evapotranspiration is controlled by meteorological conditions, and the *imposed evapotranspiration rate* (E_{imp}) at the other extreme,

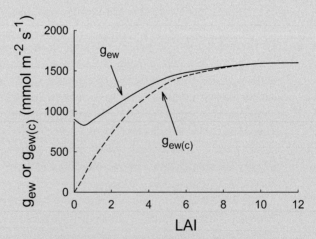

Figure B15.1   Relation between modeled canopy-only portion of the ecosystem conductance to H_2O ($g_{ew(c)}$) or canopy plus ground (total) ecosystem conductance to H_2O (g_{ew}) versus canopy leaf area index (LAI). Redrawn from Schulze *et al.* (1994).

whereby the canopy boundary layer conductance is large and evapotranspiration is controlled by stomatal conductance. The relation between E_{eq} or E_{imp} is made clear through a plot of the ratio E/E_{eq} (sometimes called the Priestley and Taylor coefficient; after Priestley and Taylor 1972) as a function of the canopy conductance ($g_{ew(c)}$) (Figure B15.2). When $g_{ew(c)}$ is low, stomatal conductance is low relative to the canopy boundary layer conductance, and diffusive flux at the leaf level is the primary constraint to canopy transpiration rate. Under these conditions E decreases below its equilibrium value. As $g_{ew(c)}$ increases, canopy boundary layer conductance and the amount of absorbed radiant energy exert greater control over E. This shift in control is seen as E/E_{eq} approaches unity. At some high, critical value for $g_{ew(c)}$, E/E_{eq} approaches an asymptote between 1.1 and 1.4. (The asymptote exceeds unity because of atmospheric mixing that brings dry air from high in the atmosphere down to the canopy, increasing E beyond the predicted equilibrium point.) It is clear from this example that the big-leaf application of the PM model can provide a perspective on the surface evapotranspiration rate that integrates the physiological and meteorological controls.

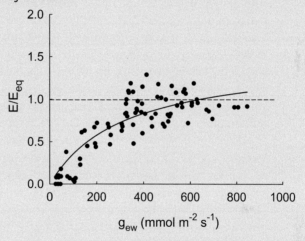

Figure B15.2   The ratio of measured surface evaporation (E) to equilibrium evaporation (E_{eq}) as a function of ecosystem conductance to H_2O (g_{ew}) for an aspen (*Populus tremuloides*) canopy in the southern boreal forest of Canada. The "best-fit" non-linear relationship reveals that E/E_{eq} reaches an asymptote at 1.18. Redrawn from Blanken *et al.* (1997).

PPFD represents a third fraction that could be included in describing the non-Gaussian distribution of PPFD, but it will be ignored here for the purpose of simplicity.) Accordingly, canopy layers are not gray, turbid systems, but rather a heterogeneous patchwork of bright and gray spots. I_D is not typically reduced as direct solar rays penetrate a canopy, and thus the A_n of sunlit leaves will approach light saturation at all canopy layers. In contrast, I_d decreases with depth, and the A_n for shaded leaves will be light limited and determined by the low-PPFD region of the A_n:PPFD relation. For each canopy layer, a fraction of the leaf area will exhibit zero-order dependence on changes in PPFD and another fraction will exhibit approximate first-order dependence. In the face of this non-linearity, determination of an average A_n for each canopy layer as a function of a single time-averaged, Gaussian PPFD will result in unavoidable errors. Big-leaf canopy models will propagate these errors. Through tuning of the relation between A_n and PPFD the errors can be minimized and big-leaf models can be forced to portray nearly accurate A_n:PPFD responses. Such tuning, however, does not reflect the true biochemical or physiological response of A_n to the PPFD gradient.

In a study by Sinclair *et al.* (1976), which was later refined by dePury and Farquhar (1997), it was demonstrated that the use of linear averaging is better justified when the total canopy flux is solved as the sum of the independently derived fluxes of sun and shade leaves, rather than the average response of a single leaf; in other words, if the A_n:PPFD functions are applied *prior to* determining the canopy integral flux. With this approach, the canopy is considered as the analog of two big leaves (one receiving I_d and one receiving I_D) (Figure 15.3). This is an assumption that recognizes, at least in part, the non-Gaussian nature of the distribution of photon flux densities within canopy layers. The predictions that result from the modified model (which can be called a *sun/shade big-leaf model*) are similar to those of more complex models and differ from those of conventional big-leaf models (Figure 15.4). In Figure 15.4, the sun/shade big-leaf and multiple-layer models are similar in their predicted outputs, but the outputs of both of these models differ from that for the simplest big-leaf model. In the left figure the differences are greatest at high LAI where the peaks in I_D and I_d are similar in magnitude. In the right figure, the differences are greatest during periods when I_d is a large fraction of PPFD (e.g., days with high, diffuse clouds and high "skylight" fluxes). The differences in the right figure are due to greater spreading of PPFD over the available leaf area when photons are incident as I_d, rather than as I_D. The multiple-layer and sun/shade models accommodate the effects of photon spreading at high diffuse fractions, whereas the simplest big-leaf model does not. *The sun/shade big-leaf model is a compromise between the simple, but inaccurate nature of the traditional big-leaf models, and the complex, and accurate nature of the traditional multiple-layer models.*

The most common type of forward-directed canopy model is the *multiple-layer model*. In multiple-layer models the vertical distribution of the canopy flux is resolved directly (Figure 15.5). In such models it is not necessary to assume linearity between the mean distributions of forcing variables and fluxes. In fact, no a priori assumptions need be made concerning the distribution of fluxes within the canopy. Thus, one advantage to the use of multiple-layer models is that the effects of canopy structure and resource allocation on scalar flux can be examined. For example, consider the question of how N allocation pattern influences the canopy-integrated CO_2 flux (A_c). In other words, to what degree would a uniform vertical distribution of leaf N concentration affect A_c compared to a pattern whereby upper canopy leaves are provisioned with more N than lower canopy leaves? It would not be possible to address such a question with big-leaf models since they depend on the a priori assumption that canopy N distribution scales linearly with mean PPFD. In a multiple-layer model, however, hypothetical adjustments can be made to the distribution of N and the resultant effects on A_c can be determined. Past studies of this type

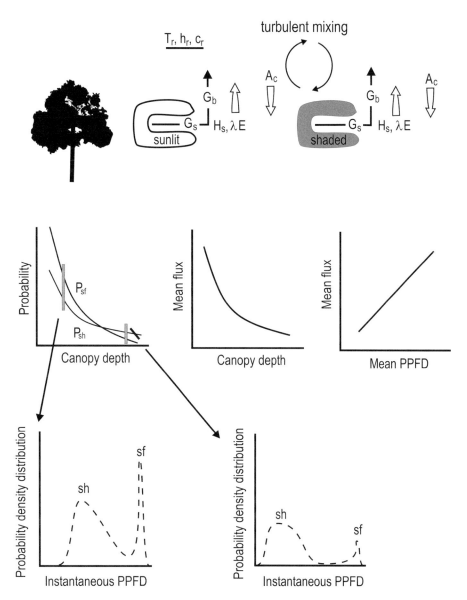

Figure 15.3 Schematic representation of a sun/shade big-leaf model. In the upper part of the figure, a modeling approach is represented by which the fluxes for sunlit and shaded leaves are determined separately then summed to obtain the total canopy flux. In the middle figures, the vertical extinction of I_b and I_d is represented as exponential decreases in the statistical probability of observing each photon flux (P_{sf} and P_{sh}, respectively) as a function of canopy depth. In the lower figures it is shown that the probability distribution function for PPFD in each vertical layer is separate for I_b and I_d; here represented as sunfleck (sf) or shaded (sh) fractions. The upper part of the figure was adapted from Raupach and Finnigan (1988).

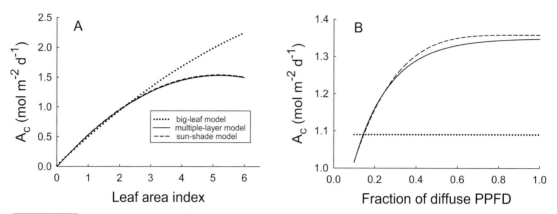

Figure 15.4   Predictions of canopy-level CO_2 assimilation rate (A_c) as a function of LAI (panel A) and fraction of diffuse PPFD incident on the top of the canopy (panel B). A_c was predicted using the biochemical models of Farquhar and von Caemmerer (1982) coupled to three different types of canopy flux models; a big-leaf model, a multiple-layer model, and a sun/shade big-leaf model. Redrawn from dePury and Farquhar (1997).

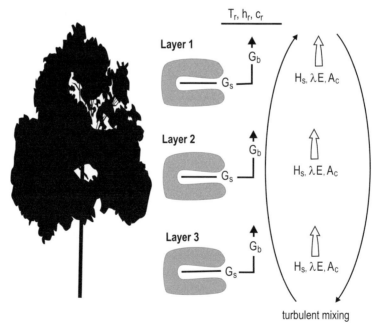

Figure 15.5   Schematic representation of multiple-layer canopy models. Fluxes for each canopy layer are calculated independently and summed to provide the total canopy flux. In this case, fluxes for sensible heat (H_s), latent heat (λE), and canopy CO_2 assimilation rate (A_c) are calculated relative to temperature (T_r), humidity (h_r), and CO_2 mole fraction (c_r) at a single reference height above the canopy.

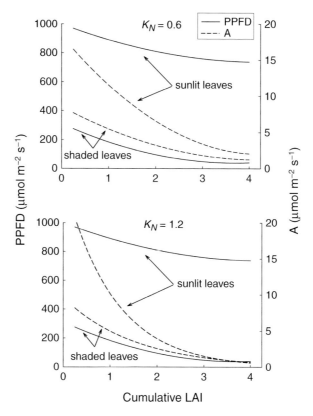

Figure 15.6 **Upper figure**. Incident PPFD and simulated net leaf CO_2 assimilation rate (A) for the sunlit and shaded leaf fractions of a hypothetical canopy as a function of cumulative leaf area index (LAI). The nitrogen allocation coefficient (k_N) has been set at 0.6 (on a scale where a K_N of 0 indicates uniform N distribution and K_N values greater than 0 indicate progressively higher leaf N concentration in upper canopy layers). **Lower figure**. The same as the upper figure except that K_N has been set at 1.2. The total canopy N was kept constant for both simulations. The multiple-layer model and parameterization that was used for these simulations is according to Leuning *et al.* (1995). The total canopy-integrated daily CO_2 assimilation rate (A_c) for the upper figure is 0.732 mol m^{-2} day^{-1} and for the lower figure is 1.230 mol m^{-2} day^{-1}. Redrawn from Leuning *et al.* (1995).

have shown that different N allocation patterns do indeed cause differences in A_c (Figure 15.6). The lower value for A_c at $K_N = 0.6$ in Figure 15.6, compared to $K_N = 1.2$, suggests a benefit to canopy CO_2 assimilation rate when N is distributed in non-uniform fashion.

Modeling canopy processes is fraught with complexities and uncertainties. Modelers are constantly trying to sharpen the physiological and meteorological processes represented in their models and remove interdependencies among those processes. One of the key challenges is equifinality – the potential for multiple causes to lead to the same modeled outcome. In the case of canopy CO_2 exchange models, equifinality often emerges in efforts to accurately and independently simulate radiative transfer within the canopy and the vertical distribution of leaf photosynthetic capacity (Box 15.2).

Box 15.2	Equifinality and physiological tradeoffs in canopy flux models

Canopy models that couple fluxes to environmental drivers have evolved to the point where detailed physiological processes are connected through direct interactions and/or feedbacks to gradients in environmental drivers. Some of the earliest models simulated canopy processes with a big-leaf format, and processes were driven by mean climate parameters. As knowledge about the averaging errors in big-leaf models emerged, canopy radiative transfer schemes were modified to account separately for physiological processes in sunlit and shaded canopy fractions. With the advent of large-scale tower flux networks, providing canopy flux measurements across a range of ecosystems, it became possible to evaluate the error between models and observations. When observations were compared to modeled predictions of CO_2 fluxes using the canopy scheme of one of the most commonly used land surface models, the Community Land Model (CLM), it was found that the model projected nearly universal overestimates of canopy net CO_2 assimilation rate (A_c) (Bonan *et al.* 2011). Evaluation of model processes revealed two possible reasons for this error: (1) the modeled PPFD through canopies, to the lowest layers, was too high; and/or (2) the modeled vertical distribution of Rubisco in canopies was too high. Either of these options would cause canopy photosynthesis rates to be too high. The model could be modified to bring the estimates of A_c back into balance with observations by either decreasing N availability (thus decreasing Rubisco concentrations, especially at the lower canopy layers), or decreasing the canopy gap fraction (thus decreasing PPFD at the lower canopy layers). This analysis brings to light two challenges that modelers must reconcile. First, is the issue of equifinality. There are multiple compensations and tradeoffs in model processes that can lead to the same projected outcome, but for different reasons – too great a photon flux or too great a Rubisco concentration can lead to the same model error. Second, as the inadequacies of models become apparent, there is a tendency for modelers to add more processes with more parameters in order to provide more points of adjustment – this results in "overparameterization" of the model. In this case, parameters could be added to account for steeper vertical gradients in the allocation of N to Rubisco, or steeper gradients in extinction of the incident PPFD. As the parameterization of a model increases, so too does its potential for equifinality. Modelers often seek the ideal balance between having too much versus too little detail in a model.

15.2 Mass balance, dynamic box models, and surface fluxes

Taking into account the principles of mass balance and continuity it is possible to develop a theoretical framework using the entire convective mixed layer of the atmosphere as a control volume. This provides a means for estimating entire landscape fluxes. Imagine the convective mixed layer as a box that rests on the ground and encloses a landscape, much like a chamber enclosing a leaf, though in this case with a "lid" that increases in height through the day. Assuming horizontal homogeneity, the time-dependent change in mean scalar concentration within the box will equal the sum of the landscape flux introduced at the lower boundary and the entrainment flux introduced at the upper boundary. The entrainment flux is due to vertical growth of the mixed layer. As the capping inversion of the mixed layer moves upward during the day, air from aloft will become entrained into the volume; in essence, producing a vertical *entrainment flux* at the upper boundary. When, for example, air in the free troposphere above the capping inversion contains a greater CO_2 concentration than air in the convective mixed layer (such as occurs on a clear, sunny day), the entrainment flux will represent a positive CO_2 flux that opposes photosynthetic CO_2 assimilation by the landscape. The difference between the surface (landscape) flux and entrainment flux will have to be balanced by the time-dependent change in the spatially averaged CO_2 concentration of the mixed layer. If we assume effective turbulent mixing within the box, then we can invoke the continuity equation to create a *dynamic box model*, from which we can estimate the landscape flux (Denmead *et al.* 1996). We begin with the following finite form of the continuity equation written with respect to hypothetical scalar j (see Section 6.3):

$$-\frac{\Delta \overline{F}_j}{\Delta z} = \overline{\rho}_m \frac{\Delta \overline{c}_j}{\Delta t}. \tag{15.1}$$

If we assume that $\Delta \overline{F}_j$ represents the time-averaged difference between the turbulent landscape flux (F_{vtj}) at the lower boundary and the entrainment flux at the upper boundary, we can derive the following expression:

$$\frac{d \rho_m c_{ij}}{dt} = \frac{\overline{F}_{vtj}}{z_i} - \overline{\rho}_m \left(\frac{\overline{c}_{oj} - \overline{c}_{ij}}{z_i} \right) \left(\frac{dz_i}{dt} \right), \tag{15.2}$$

$$\text{Term} \qquad \text{I} \qquad \text{II} \qquad \qquad \text{III}$$

where $\overline{c}_{oj}$ is the spatially-averaged mean mole fraction of j in the air above the capping inversion, $\overline{c}_{ij}$ is the spatially-averaged mean mole fraction of j in the air of the mixed layer, z_i is the height of the mixed layer, and Term III represents the entrainment flux assuming that the mean vertical velocity ($\overline{w}$) is zero. Thus, using measurements of $\overline{c}_{oj}$, $\overline{c}_{ij}$, and dz_i/dt, taken from the well-mixed planetary boundary layer, we can estimate the value of $\overline{F}_{vtj}$. A derivation of Eq. (15.2) is provided in Appendix 15.1.

Let us take a moment to examine one application of Eq. (15.2). We will consider the surface and entrainment fluxes of heat and moisture and their influences on mean concentrations in the atmosphere above a boreal forest in Canada. Potential temperature (θ_t) fluxes tend to decrease linearly with height across an unstable convective mixed layer (i.e., a negative vertical flux divergence) (Figure 15.7). This reflects the sum of an upward flux as sensible heat is transferred from the surface, and a downward flux as sensible heat from above the capping inversion is entrained into the growing mixed layer. Overall, the flux tends to be positive in sign (net upward flux) in the lower 80% of the mixed layer, and negative in sign (net downward flux) in the upper 20%. (Recall that the sign of the flux is opposite that of the gradient. In this case, the gradient in θ_t is negative from the surface upward so the flux in that direction is positive; the gradient in θ_t is positive moving upward across the capping inversion and the flux is negative.) The turbulent flux of moisture exhibits positive divergence in the vertical coordinate. This reflects the sum of a "wet" upward flux due to surface evapotranspiration and a "dry" downward flux due to the mixing of dry air from aloft, both of which are positive in sign (downward mixing of dry air into a column of moist air is equivalent to the net upward flux of moist air). Continuity requires that a negative

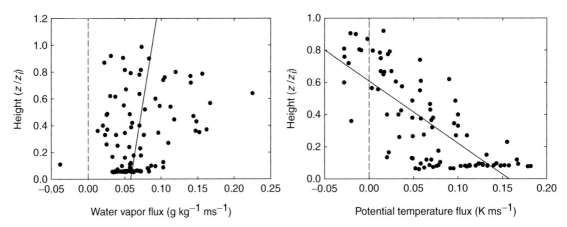

Figure 15.7 Height-dependent profile of water vapor flux and potential temperature flux above a boreal forest ecosystem in Canada. The change in water vapor flux with increasing height reflects a positive flux divergence, whereas the change in potential temperature flux with height reflects a negative flux divergence. Height is represented as the ratio of convective mixed layer height (z_i) to reference height (z). Redrawn from Davis et al. (1997).

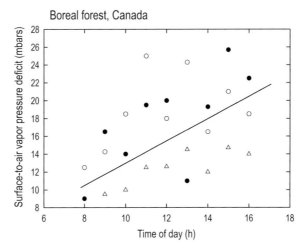

Figure 15.8 Diurnal change in the surface-to-air vapor pressure deficit in the convective mixed layer above a boreal forest in Canada. The data are from three different days (represented by different symbols). Redrawn from Davis et al. (1997).

divergence in heat flux and a positive divergence in moisture flux be balanced by warming and drying of the convective mixed layer as the day progresses. Measurements of vertical profiles in heat and moisture using aircraft sampling show that indeed the average surface-to-atmosphere water vapor pressure difference (VPD) increases during the day (Figure 15.8). The VPD is analogous to the leaf-to-air water vapor mole fraction difference (Δc_w), which, when increased, has the physiological consequence of causing partial stomatal closure in vegetation. Thus, as the convective mixed layer grows during the day, and entrains warm, dry air from above, we can predict a direct, physiologically induced decrease in landscape transpiration rate.

15.3 Eulerian perspectives in canopy flux models

Eulerian and Lagrangian logic are used in both big-leaf and multiple-layer models to link turbulent flows and fluxes to within-canopy mean concentration gradients. Recall that with an Eulerian perspective, the observer (or sensor) remains at a single point in the flow field and records observations of flow velocity and scalar density as "parcels" of air move past the observation point; with a Lagrangian perspective, the observer moves with individual parcels of air, exchanging scalar entities with point sources and sinks located throughout the flow domain (Section 13.8) (Figure 15.9). The Eulerian perspective is clearly represented in flux models that begin from a statistical covariance. To derive an Eulerian relation describing the covariance flux, we begin from the calculus of continuity. Thus, for the vertical coordinate, we can state:

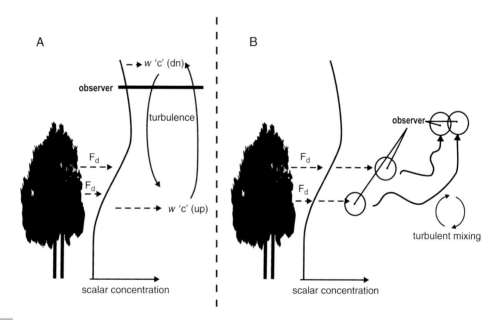

Figure 15.9   **A**. Schematic demonstration of the principle of Eulerian transport applied to canopy flux models. The cumulative effect of scalar diffusive flux (F_d) across leaf surfaces within the canopy creates a scalar concentration gradient (in this case the leaves are a net sink for scalar, creating a gradient with higher concentration above the canopy and lower concentration within the canopy). Scalar flux between the canopy and the atmosphere is observed from a stationary point above the canopy. **B**. Schematic demonstration of the principle of Lagrangian scalar transport applied to canopy flux models. Parcels (or particles) of air are allowed to interact with the concentration gradient to pick up scalar and transport it to various points within and above the canopy. In this case, the observer is within the parcels and moves with them to the final point of transport.

$$\rho_m \frac{\partial c_j}{\partial t} = -\frac{\partial F_j}{\partial z}, \tag{15.3}$$

where $\bar{c}_j$ is in units of mole fraction, ρ_m is the molar density of air (mol air m^{-3}), and F_j is the total flux density for scalar j with units mol m^{-2} s^{-1}. Recognizing that F_j can be represented by the statistical covariance between vertical wind speed (w) and c_j (i.e., $F_j = \rho_m w c_j$), multiplying both sides of Eq. (15.3) by the turbulent component of the vertical wind speed (w'), and applying Reynolds averaging, leads to:

$$w' \frac{\partial(\bar{c}_j + c_j')}{\partial t} = -w'(\bar{w} + w')\frac{\partial(\bar{c}_j + c_j')}{\partial z}. \tag{15.4}$$

Expanding Eq. (15.4), dropping those terms on the right-hand side containing $\bar{w}$ (with the assumption that $\bar{w} = 0$), dropping $\partial w'\bar{c}_j/\partial t$ from the left-hand side due to the Reynolds averaging outcome that terms with mixed fluctuating and mean components resolve to 0 (see Table 13.1), assuming near-neutral stability, and applying time-averaging, we obtain:

$$\frac{\partial \overline{w'c_j'}}{\partial t} = -\overline{w'^2}\frac{\partial \bar{c}_j}{\partial z} - \frac{\partial \overline{w'w'c_j'}}{\partial z} - \overline{c_j'}\frac{\partial \overline{p'}}{\partial z}. \tag{15.5}$$

$$\text{Term} \qquad \text{I} \qquad\qquad \text{II} \qquad\quad \text{III} \qquad\quad \text{IV}$$

Equation (15.5) is a continuity expression of the vertical eddy flux. Using Eq. (15.5) we gain the perspective that within an infinitesimally small control volume at height z, the time-dependent change in $\overline{w'c_j'}$ is dependent on the degree to which the vertical concentration gradient and vertical eddy flux gradient co-vary with w' (turbulence in the vertical coordinate) and the influence of the vertical pressure gradient. In essence, Eq. (15.5) relates time and space dependencies in the turbulent flux of scalar j. The context of the model is Eulerian because it resolves the turbulent flux divergence across a fixed point (z). A common parameterization that has been used to simplify Eq. (15.5) is:

$$\overline{c_j'}\frac{\partial \overline{p'}}{\partial z} \approx \frac{\overline{w'c_j'}}{t_{Ew}}, \tag{15.6}$$

where t_{Ew} is the Eulerian time scale for vertical wind speed (Wyngaard 1992). Using this approximation and assuming that $\partial w'c_j'/\partial t = 0$ (i.e., stationarity) we can write:

$$\overline{w'c_j'} \approx -\overline{w'^2}t_{Ew}\frac{\partial \bar{c}_j}{\partial z} - t_{Ew}\frac{\partial \overline{w'w'c_j'}}{\partial z}. \tag{15.7}$$

$$\text{Term} \qquad \text{I} \qquad\qquad \text{II} \qquad\qquad \text{III}$$

Equation (15.7) provides deeper insight into determinants of the eddy flux. Term II makes it clear that the eddy flux can, in theory, be modeled as the statistical variance in vertical turbulent wind speed operating across a scalar concentration gradient. Term III is known as a *triple correlation* and it represents the third statistical moment defining divergence in the eddy flux. The third moment defines *skewness* in the vertical distribution of the eddy flux;

skewness that is caused by an interaction of the eddy flux with the distribution of w'. (Statistical moments are described further in the Supplement on Mathematics provided in the electronic materials that support this book: http\www.cambridge.org/monson.)

If the correlation between the eddy flux and w' were to be Gaussian, Term III would resolve to zero and Eq. (15.7) would provide a valid means for modeling the eddy flux within the context of the flux-gradient relationship. In that case, we could be confident in the approximation $K_D \approx \overline{(w'^2)} t_{Ew}$. This is often the case in the atmospheric surface layer where fluxes are approximately constant as a function of height. However, the turbulence regime of the roughness sublayer, especially above tall canopies, tends to be non-Gaussian with regard to distributions in w'; the organized sweep and ejection motions that are common within the canopy and near the canopy surface cause the spatiotemporal probability density distributions of w' to be skewed. (The skewed nature of the flux above tall canopies can be seen in the conditional analysis described in the previous chapter in Section 14.2.2.) Thus, flux budgets for the roughness sublayer are often incapable of closure using only the second statistical moment (Term II). In order for Eq. (15.7) to be fully resolved, we would need to define the non-zero nature of Term III. Unfortunately, efforts to close Term III create a non-zero, fourth statistical moment – called *kurtosis*. Closure of the fourth moment will create even higher-order moments in need of closure, and so on. This pattern brings to light one of the fundamental limitations to Eulerian modeling; it is not readily amenable to mathematical closure. As non-Gaussian probability density functions are defined by higher-order statistical moments, there will always be more mathematical unknowns than equations. Ultimately, processes driven by turbulence are not amenable to description by Gaussian statistics.

The closure problem does not mean that Eulerian modeling of the eddy flux is *intractable*, but it does mean that it is *inexact*. In some cases, models have been developed to circumvent the closure problem and derive higher-order moments by *approximation*. In the case of Eq. (15.7) one commonly used approach minimizes Term III with the approximation that skewness in the vertical distribution of the eddy flux is small enough to be negligible, and Term II can be approximated as:

$$\overline{w'^2} t_{Ew} \frac{\partial \overline{c}_j}{\partial z} \approx K_D \frac{\partial \overline{c}_j}{\partial z}. \tag{15.8}$$

The approximation expressed in Eq. (15.8) may be defensible for some short canopies. However, it can be challenged in the case for tall forest canopies.

Eulerian models are most accurate when closure approximations are applied to statistical moments several orders removed from the term of interest. Closure is typically accomplished by approximating the highest-order moment as amenable to Gaussian description. Alternative approaches have been developed in which large eddy simulation (LES) models are used to compute turbulence statistics (Shaw and Schumann 1992, Dwyer *et al.* 1997, Shen and Leclerc 1997, Albertson *et al.* 2001, Shaw and Patton 2003); and these approaches show considerable promise. In LES models, the equation of motion is applied directly to the largest eddies of a specific spatial domain, and transport through the smallest eddies is parameterized with fixed values. LES models are computationally intensive but are able to simulate, in high temporal and spatial resolution, many of the characteristic features of turbulence.

15.4 Lagrangian perspectives in canopy flux models

> We have seen that the intermittency of scalar transport in plant canopies and its large scale appear to preclude the use of gradient diffusion descriptions in space ... (O. T. Denmead and E. F. Bradley (1987))

> A better abstraction is to regard the canopy as a collection of a vast number of scalar source elements, or tiny chimneys, puffing out plumes of scalar material. To explore this view it is necessary to consider the process of dispersion from each source element by using Lagrangian principles: that is, by finding statistics of the motion of the wandering fluid elements that carry the scalar. (M. R. Raupach (1989a))

The modeling of turbulent exchanges between ecosystems and the atmosphere is difficult in large part because of the lack of mathematical frameworks capable of describing the motions of turbulence, and therefore the process of time-dependent dispersion. As a first approximation, in the past, researchers used diffusion to model atmospheric dispersion. However, turbulent transport often occurs across length scales that do not match well with the mean free path used to characterize scalar concentration gradients. In more recent years, researchers have begun to develop new mathematical frameworks within which to describe the dispersive aspects of ecosystem-atmosphere exchange.

A key concept that is better captured in Lagrangian models, compared to Eulerian models, is *turbulent persistence*. Turbulent motions have the potential to delay the dispersion of scalars, causing them to accumulate locally within the canopy; this type of persistence cannot be predicted by diffusion models. Imagine a small piece of wood floating downstream and entering a swirling, whirlpool of water – its downstream journey will be delayed. Lagrangian persistence is modeled from knowledge about the distributions of diffusive sources or sinks within a canopy, combined with a probability function that permits prediction as to where and when ensembles of fluid parcels moving in a turbulent fluid will interact with those sources and sinks, and after a specified interval of time, where the parcels will reside. This is expressed mathematically as:

$$\rho_{\mathrm{m}} c_j[z, t] = \frac{1}{z} \left(\int\limits_{z_0}^{z} \int\limits_{t_0}^{t} S_j[z_0, t_0] \, P_{\mathrm{t}} \, [z, t | z_0, t_0] dz \, dt \right), \qquad (15.9)$$

where $c_j[z, t]$ represents the scalar mole fraction of scalar entity j referenced to position z and time t, S_j represents the volumetric dispersive source or sink strength for scalar j (with units mol j m^{-3} s^{-1}) referenced to position z_0 and time t_0, and P_{t} is the transition probability function describing the probability that a parcel of air that starts from z_0:t_0 will end up at z:t, with that probability normalized by distance z.

The core Lagrangian component of Eq. (15.9) is P_{t}, the *transition probability function*. Conceptually, P_{t} provides a statistical prediction of the distribution of parcels after release from a point source, in this case with the point defined by z_0 and t_0 (Figure 15.10). In finite time and space, parcels can take numerous possible paths from the source, with all paths summing to a

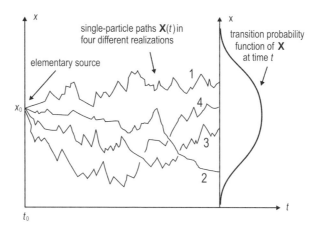

Figure 15.10 The transition probability function is determined for different times after release of a parcel of air from a point source where a scalar has been added or subtracted. The upper case, bold notation for **X** represents vector notation, reflecting the fact that the space traversed is defined in three Cartesian dimensions. Redrawn from Raupach (1989b).

probability density function that defines the change in position Δz (where $\Delta z = z_0 - z$) after time Δt (where $\Delta t = t_0 - t$). In practice, P_t is derived from the statistical properties of observed wind fields. Contained within P_t is the potential to represent turbulent persistence. In this case, persistence is represented by a modeled velocity time series through specified values of Δt (and thus Δz) that permit patterns of *autocorrelation* to be defined among time series elements. Within the interval of autocorrelation, the dispersed parcels are said to be in the *near field* relative to their source, and dispersion can be expressed as $\sigma_z = \sigma_w \Delta t$, where σ_w is the standard deviation of the parcel's vertical wind speed. The near field provides the domain of turbulent persistence. With a small value for σ_w, dispersion is slow and the positions of parcels tend to remain correlated with one another as a function of time. As σ_w is increased, dispersion will be greater, the position of parcels with respect to one another will be less correlated as a function of time, and the tendency toward persistence will decrease. As σ_w changes from smaller to larger values, the mean position of the parcel cloud will always be zero, but the width of the cloud will vary as a function of $\sigma_w \Delta t$. Given a non-zero σ_w, beyond a critical time, autocorrelation will break down completely and parcel velocities can be expected to change in random fashion. At this point, the parcels will have left the near field and entered the *far field* (Figure 15.11). Given its random nature, far field dispersion resembles diffusion, and thus for large travel times, Lagrangian dispersion models will converge toward an equivalent, diffusion-based Eulerian model. In both Lagrangian far-field dispersion and Eulerian diffusion, the depth of the cloud of parcels will increase as a function of $t^{1/2}$, rather than t; i.e., $\sigma_z = \sqrt{2\sigma_w t_L t}$, where t_L is the Lagrangian time interval, the time required for transition between near-field and far-field dispersion (Raupach 1989b, Rodean 1996).

For most time intervals, the dispersion of a scalar will be defined as an interpolation between the near-field and far-field limits. The interpolation can be approximated by defining the probability of autocorrelation as an exponential function of t. With knowledge about S_j and P_t, Eq. (15.9) can be integrated with respect to space and time, across all sources

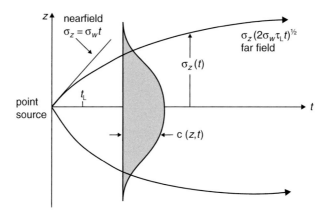

Figure 15.11 Representation of the spread of parcels of air carrying a scalar released from a point source as a function of time (t). In the near field, the spread of the scalar can be described as a linear function of t. In the far field, the spread of the scalar can be described as a function of $t^{1/2}$. The integral time scale is t_L. The near field is defined at $t \ll t_L$, where sequentially measured parcel velocities are autocorrelated. The far field is defined at $t \gg t_L$, where parcels spread randomly according to the laws of diffusion. Within the limits of parcel spread, the concentration of scalar ($c(z,t)$) can be described by the transition probability function, and can be approximated as Gaussian.

or sinks within any one canopy layer, and solved to provide the mean mole fraction of a scalar. In this case, the mean concentration represents an ensemble average (c_j), though with the assumption of a steady state, it is equal to the time average ($\bar{c}_j$). By summing to obtain the total scalar contained in parcels arriving at the top of the canopy, with specified Δt, from all sources within a unit of ground area, and determining instantaneous deviations from $\bar{c}_j$, turbulent fluxes can be determined.

The main challenge in resolving S_j and P_t is to computationally move air parcels through a virtual canopy volume in a manner that accurately reflects wind flows. The most commonly used approach depends on a Markovian "random-walk" function. This function computes the trajectory of a large number of fluid parcels released into a *dispersion matrix* in cyberspace; the dispersion matrix is parameterized with observed wind data. The dispersion matrix (D_{ik}) has units of resistance (s m^{-1}) and is related to the transport of the scalar according to:

$$D_{ik} = \rho_m \left(\frac{c_{ji} - c_{jr}}{S_j \Delta z_{ik}} \right), \tag{15.10}$$

where i is the transport destination, or receptor point; k is the initial location, or source point; c_{ji} and c_{jr} are mole fractions of the scalar of interest at the receptor point (i) and a reference point (r) above the canopy, respectively; S_j represents the source or sink strength at level k; and Δz_{ik} represents the distance between points i and k. The dispersion matrix can be computed by uniformly releasing fluid parcels into cyberspace at a known rate from different points in the canopy, following their trajectories with the random-walk model, and computing the local concentration field (Baldocchi 1992); analytical models have also been designed to define parcel flow probabilities (Raupach 1989b, Vandenhurk and McNaughton 1995, Massman and Weil 1999, Warland and Thurtell 2000).

For the simplest case of one-dimensional diffusion the vertical displacement of fluid parcels (Δz) can be computed as the product of the Lagrangian vertical velocity (w_L) and time step (Δt). Incremental changes in vertical velocity must be computed with the Langevin equation, a stochastic differential equation that is comprised of a deterministic forcing (a memory term) and a random forcing (Thomson 1987, Raupach 1988):

$$\frac{dw}{dt} = -aw + b\xi[t], \tag{15.11}$$

where the dispersion coefficients a and b provide empirically derived connections (with units s^{-1} and m s^{-1}, respectively), to the behavior of the conditional transition probability function, and $\xi[t]$ is a random fractional time-dependent increment with a mean of zero and a variance of one. The dispersion coefficients are near zero for well-mixed flows. In essence, they act as resistors that account for counter-gradient transfer.

Lagrangian models based on the Langevin equation will fail inside a plant canopy where the variance in w is skewed with a tendency to increase with height. A net downward drift occurs because descending fluid parcels enter domains with reduced w. Thus, in the dispersion matrix of cyberspace an artificial accumulation of mass occurs near the ground. In nature, intermittent wind gusts disproportionately transfer mass and maintain vertical mixing within the canopy volume. Heuristic approaches have been used in models to remove the unrealistic accumulation of mass that would otherwise not occur (Legg and Raupach 1982, Wilson *et al.* 1981, Leclerc *et al.* 1988).

Appendix 15.1 Derivation of the model for planetary boundary layer (PBL) scalar budgets in the face of entrainment

The model that is derived by the following approach is described in greater detail in McNaughton (1989), Raupach (1991), and Denmead *et al.* (1996). The convective mixed layer is capped at the upper limit by a temperature inversion and at the lower limit by the ground's surface. A vegetated landscape is assumed to exist as a source or sink for mass exchange within the column of air, which is better conceived as a control volume within which mass balance must occur. A downward entrainment flux occurs throughout the day as the capping inversion moves upward and air is entrained into the mixed layer below. This entrainment flux can be represented mathematically if we consider that the mean scalar mole fraction inside the well-mixed convective mixed layer ($\overline{c_i}$) of height z_i is changed by the addition of a thin layer of air from above (Δz_i) characterized by a different mean scalar mole fraction ($\overline{c_o}$) derived from the free troposphere outside the convective mixed layer. The change in mole fraction of the convective mixed layer is then represented by:

$$\overline{c_i} + \Delta\overline{c_i} = \left(\frac{\overline{c_i} z_i + \overline{c_o}(\Delta z_i)}{z_i + \Delta z_i} \right). \tag{15.12}$$

Equation (15.12) reflects a linear, two-member mixing model in which one end-member (the thin layer of air from above) is mixed into the other end-member (the convective mixed layer). Moving $\overline{c_i}$ from the left-hand side to the right-hand side, multiplying by unity as $(z_i + \Delta z_i)/(z_i + \Delta z_i)$, dividing both sides by Δt in order to derive the rate of change in scalar concentration, and rearranging through algebra results in:

$$\frac{\Delta \overline{c_i}}{\Delta t} = \frac{(\overline{c_o} - \overline{c_i})}{(z_i + \Delta z_i)} \frac{\Delta z_i}{\Delta t}. \tag{15.13}$$

Evaluating Eq. (15.13) within the limit as Δz_i and Δt approach 0, and including the landscape flux from the lower boundary, we obtain the following form of the continuity relation:

$$\frac{d(\rho_m c_i)}{dt} = \frac{\overline{F_t}}{z_i} - \overline{\rho}_m \left(\frac{\overline{c_o} - \overline{c_i}}{z_i} \right) \left(\frac{dz_i}{dt} \right). \tag{15.14}$$

Using Eq. (15.14) along with atmospheric measurements of $\overline{c_i}$, $\overline{c_o}$, and dz_i/dt, we can estimate changes in $\overline{F}_{vt}$ the time-averaged vertical landscape flux (mol m^{-2} s^{-1}).

In those cases where the mean vertical wind speed ($\overline{w}$) is not zero, it can be included as a variable in the scalar budget. A non-zero $\overline{w}$ will work against the upward growth of the capping inversion in the case of atmospheric subsidence (a negative $\overline{w}$) and in the same direction of upward growth of the capping inversion in the case of atmospheric expansion (a positive $\overline{w}$):

$$\frac{d(\rho_m c_i)}{dt} = \frac{\overline{F}_{vt}}{z_i} - \overline{\rho}_m \left(\frac{\overline{c_o} - \overline{c_i}}{z_i} \right) \left(\frac{dz_i}{dt} - \overline{w} \right). \tag{15.15}$$

16

Soil fluxes of CO$_2$, CH$_4$, and NO$_x$

Soil is a special natural body, distinct from other rocks . . .

V. V. Dokuchayev (quoted by V. Vernadsky 1938)

In one of his influential essays on the nature of the biosphere, which subsequently laid the foundation for modern biogeochemistry, Vladimir Vernadsky quoted his geology teacher, Vasily Dokuchayev, in a manner that made clear Dokuchayev's view that soil is uniquely influenced by the organisms that live within; it lies at the interface between geology and biology. Roots and microorganisms exchange material with the soil and in doing so influence its chemical composition and physical structure. The interactions between the biological and geological components of soil ultimately determine how it interacts with the atmosphere. Soil is the medium through which terrestrial plants gain access to most mineral elements of the geosphere (carbon being the principal exception), and to liquid water from the hydrosphere, both of which exert major controls over the capacity for plants to exchange CO$_2$, H$_2$O, and energy with the atmosphere. Microbial activities in soil carry out the recycling of organic matter, returning carbon to the atmosphere, thus closing the terrestrial carbon cycle. Certain types of bacteria, in symbiotic relations with plant roots, or free-living in soil and water, facilitate the fixation of N$_2$ from the atmosphere. Chemolithotrophic microorganisms in soil oxidize inorganic compounds to generate energy that is subsequently used to drive the autotrophic assimilation of carbon and at the same time produce volatile trace gases (including nitrogen oxides and N$_2$) that are emitted to the atmosphere. Any concentrated consideration of ecosystem-atmosphere exchange must include soil processes in order to fully comprehend the relevant mass and energy fluxes.

We begin this chapter with an introduction to the decomposition of soil organic matter – a principal source of soil respiration and the process responsible for recycling nutrients between dead and living biomass. We then take up the topic of transport mechanisms for the exchange of mass between soils and the atmosphere, once again focusing on the flux of CO$_2$ from soil respiration. Finally, in a series of sections we consider recent observations and models that focus on biological controls over the exchanges of CH$_4$ and NO$_x$ between soils and the atmosphere. Cumulatively, these trace gases represent major contributions to the terrestrial carbon and nitrogen budgets, as well as to the reactive photochemistry and energy partitioning that occurs in the atmosphere.

16.1 The decomposition of soil organic matter

The vertical nature of soil structure is often described by a taxonomy based on *horizons*, or layers, classified according to depth, as well as physical, geological, and biological attributes

(Bridges 1990, FitzPatrick 1993). Horizons are not static through time with regard to thickness, density, or chemical composition; organic material is continually added to the upper horizons, weathered parent material is continually added to the lower horizons, and vertical translocation of materials occurs through all horizons. Complete characterization of the geophysical and geochemical attributes of soils would require that attention be given to all of the horizons above the parent material (often designated as the A, B, C, and higher alphabetically designated horizons). We don't have the luxury of space to provide a comprehensive discussion of the properties and dynamics of all horizons, and given the emphasis in this book on surface-atmosphere interactions, most processes of interest to us will occur near the upper surface. Accordingly, we will focus on soil organic matter and its contributions to the O (for organic) and A (uppermost mineral) horizons.

Soil organic matter (SOM) originates from the dead leaf and root "litter" shed by plants, the exudation and deposition of organic material from both leaves and roots, and the dead biomass of soil animals and microorganisms. Once these organic materials are mixed into the soil a series of biological, chemical, and photochemical processes begin to decompose them to progressively smaller fragments and molecules, and transform them into alternative chemical forms. Much of this breakdown and transformation is due to the biological activity of soil organisms interacting with one another through various food webs, and extracting the fundamental elements (C, N, and P) and energy with which to construct and maintain their biomass. In general terms, *decomposition* refers to the breakdown of dead (or in some cases dying) biomass and its incorporation into various forms of SOM. Decomposition is an ecological "recycling" process that (1) converts organically bound elements into forms capable of re-assimilation by plants and microbes, (2) extracts and transforms energy from the chemical bonds that hold the organic compounds together, and (3) ultimately releases CO_2 back to the atmosphere (Paul 1984, Brussaard *et al.* 1997). Most of the nutrients taken up by plants exist as inorganic compounds; although some of the smaller organic molecules can also be assimilated. The freeing of soil elements from organic molecules through the oxidative processes of decomposition, and conversion to inorganic forms accessible to plants is referred to as *mineralization*. The reverse of mineralization, the conversion of inorganic nutrients to organic compounds, through assimilation and transformation by soil microbes is referred to as *immobilization*. Thus, immobilization is the process by which microbes compete with plants for available soil nutrients. During mineralization, some volatile compounds escape from the sequence of transformations spanning the organic-to-inorganic transition (e.g., nitrogen oxides) and are lost to the atmosphere, creating an important component of soil-atmosphere mass exchange. During immobilization a portion of available nutrients will become incorporated into stable fractions of SOM, and may be rendered inaccessible to further decomposition. An additional process important to soil-atmosphere mass exchange and involving the biogeochemical cycling of carbon occurs in anaerobic soils, *methanogenesis*. During methanogenesis, soil microorganisms utilize carbon in inorganic form (as CO_2) or organic form (as small organic molecules) as terminal electron acceptors (in the absence of O_2), and in the process produce methane (CH_4).

Decomposition is a complex process that occurs faster or slower for various compounds depending on their structure, chemical composition, and association with mineral components of the soil. Thus, SOM is composed of compounds with a spectrum of ages and turnover rates – some having appeared in the recent past and turning over in days to months, and some having persisted for millennia. As freshly deposited litter is processed within the soil, its decomposition rate slows; the more labile compounds, with faster turnover times, are lost relatively quickly, leaving behind more recalcitrant compounds, with slower turnover times. The decomposition rate of litter can change by over four orders of magnitude during its progression from freshly shed material to processed and stabilized SOM (Berg 2000), and this progression toward greater stability can be justified on thermodynamic principles (Bosatta and Ågren 1999). It is the most recalcitrant compounds, with the slowest turnover times, that render soils such effective systems for the sequestration and long-term storage of atmospheric CO_2 (Jenny 1980); the storage of carbon in all global soils has been estimated at 2000–3000 Pg of C (Jobbágy and Jackson 2000, Sabine *et al.* 2003), which is 3–4 times the carbon stored in the atmosphere. Radiocarbon dating of bulk SOM has revealed a mean age ranging from 1000–10 000 years (Paul and Clark 1996).

Recently, it has become apparent that in some ecosystems, especially in arid regions, decomposition is initiated while tissues are still attached to plants, or when they exist as dead surface litter that has not yet been mixed into the soil. This abiotic decomposition is due to the photodegradative action of ultraviolet light (Box 16.1). *Photodegradation* is becoming increasingly recognized as an important source of trace gas emissions to the atmosphere, including the emissions of CO and CO_2. Furthermore, photodegradation may be an important process rendering the polymeric chemical complexes that make up plant cell walls (such as lignin and lignocellulose), which are normally highly resistant to decomposition, significantly less resistant.

Not all organic compounds enter the soil as litter. Roots exude many low molecular weight compounds, particularly soluble sugars and amino acids, which have relatively short turnover times due to rapid immobilization by microorganisms. The exudation of compounds from roots is often referred to as *rhizodeposition*, and it has an important role in maintaining symbiotic associations between microbes and roots within the plant rhizosphere (Jones *et al.* 2009, Lambers *et al.* 2009). It has been estimated that rhizodeposition can cost up to 40% of the net photosynthetic gain by a plant (Doornbos *et al.* 2012). Rhizodeposition provides organic substrates for both energy and biomass production in mycorrhizal fungi and rhizospheric bacteria that live in close association with roots. Rhizodeposited compounds can become incorporated into more recalcitrant forms of SOM (Rasse *et al.* 2005), and they can stimulate the rate of decomposition of SOM, a process referred to as *priming* (Kuzyakov 2002). Substrate priming stimulates decomposition through increased activities of microbes as they pursue nutrient acquisition. However, there is debate as to whether microbes have an active role in the priming, or whether the rhizodeposited compounds interact directly with the soil organic matter to chemically alter decomposability of the compounds (Kemmitt *et al.* 2008).

Following the initial stages of biomass breakdown, most decomposition occurs at the scale of macromolecules or macromolecular complexes. Much of the molecular decomposition is catalyzed by extracellular enzymes (also called *exoenzymes*) produced by microorganisms, but which catalyze reactions external to the cells that produce them (Sinsabaugh 1994). Exoenzymes can be secreted into the soil, remain attached to the

| Box 16.1 | Photodegradation and decomposition |

Plant litter exposed to bright sunshine often exhibits higher rates of mass loss compared to litter protected from sunlight. This so-called *photodegradation* has been attributed to the action of UV-B solar radiation (280–320 nm), although other wavebands can also be involved (Anesio *et al.* 1999, Brandt *et al.* 2009). Most of the lost mass is due to degradation of lignin, which is resistant to microbial decomposition (Henry *et al.* 2008). Photodegradation is the source of direct emissions of carbon monoxide (CO) and CO_2 to the atmosphere (Tarr *et al.* 1995, Schade and Crutzen 1999, Brandt *et al.* 2009), and it has been shown to render lignin more susceptible to microbial decomposition once the sun-exposed litter mixes with the general pool of soil organic matter (Figure B16.1). Photodegradation has been shown to be especially important in arid ecosystems, where solar UV-B fluxes are high, and litter often remains on plants or the soil

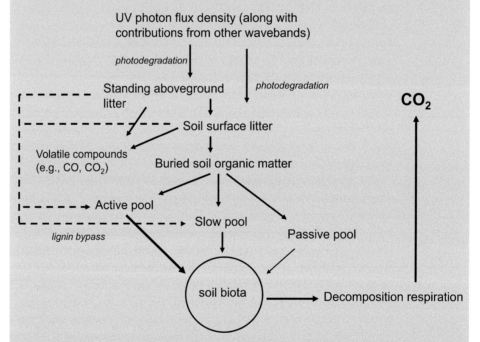

| Figure B16.1 | Relations among processes and pools involved in the photodegradation of lignin and its subsequent decomposition as a component of soil organic matter. Photodegradation causes chemical alteration of litter, resulting in the production of volatile products such as CO and CO_2, and it chemically transforms the litter in a way that enhances biotic decomposition from the various soil organic matter pools. The chemical transformation of litter is likely due to changes in lignin and/or to the priming of organisms capable of decomposing lignin, producing a "lignin bypass" that may allow a larger fraction of lignin to enter the active and slow pools of soil organic matter and thus decompose more quickly. |

surface for long periods of time (e.g., Austin and Vivanco 2006, Brandt *et al.* 2009, Rutledge *et al.* 2010). There is potential for species-specific differences in susceptibility to photodegradation, and this may be related to litter lignin content or other chemical attributes. There is still considerable work to be done to determine the principal controls and mechanisms that underlie photodegradation.

One issue that is likely to prove important involves potential direct and indirect interactions between photodegradation and microbial decomposition. As mentioned above, there is some evidence that the photochemical processing of lignin through photodegradation makes it more accessible for microbial decomposition once plant litter gets buried beneath the soil surface (Henry *et al.* 2008). However, any increase in the decomposability of lignin due to photodegradation would be balanced by the inhibitory effect of UV-B radiation on microbial biomass; thus, decreasing the decomposition potential of surface soils. These latter relations, in addition to the overall magnitude of photodegradation on soil carbon fluxes, remain to be elucidated.

external matrices of cell membranes or walls, or be contained within the space between membranes or walls (the periplasmic space). If secreted into the soil, most exoenzymes will catalyze their respective reactions while adsorbed to soil particles (Wallenstein and Weintraub 2008). Exoenzymes have roles in the breakdown of a broad diversity of macromolecules, including cellulose and proteins, and are ubiquitous to most soils. Many of the breakdown products of exoenzyme activity are taken up by the microorganisms that produce them, by other microorganisms, or by plant roots. For example, protease exoenzymes are known to break down larger soil proteins to short peptide chains capable of assimilation by many plants and fungi (Lipson and Näsholm 2001). Individual peptidase exoenzymes cleave specific amino acid residues from the ends of protein chains, progressively freeing amino acids for further metabolism or direct incorporation into plant or microbial biomass. Various exoenzyme cellulases are known to work in sequence to break down cellulose; endocellulases catalyze the cleavage of internal bonds in large, crystalline, multi-chain complexes, followed by exocellulases that catalyze the cleavage of short disaccharide and tetrasaccharide chains; hemicellulases hydrolyze the polysaccharide hemicelluloses that associate with pectin and cellulose in the cell walls of plants. In soils, ascomycete fungi (sometimes referred to collectively as the "cellulose decomposers") are the principal source of cellulose exoenzymes.

Lignin is among the most common recalcitrant compounds found in SOM, especially in forest soils. Lignin is a principal component of wood, forming the structural integrity of thickened, secondary cell walls in tree stems and branches. Lignin-degrading enzymes are less abundant in soils compared to most other exoenzymes. Lignin is structurally and chemically complex and lignin-degrading enzymes have poor catalytic access to the numerous cross-linked chains that compose its polymeric structure. The potential energy that can be extracted from lignin during microbial decomposition is less than that for other, more labile compounds, rendering it less favored as a growth substrate. In woody biomass, lignin is often combined with cellulose and hemicellulose to form *lignocellulose*. Lignocellulosic biomass decomposes more quickly than lignin alone, presumably because cellulose and hemicellulose have higher amounts of potential energy that can be extracted by microbial decomposers. As lignocellulose is decomposed, however, the most accessible cellulosic and hemicellulosic components disappear first, leaving behind less accessible components shielded by lignin.

The progressive processing of SOM through decomposition, whereby the more labile components are lost first and the more recalcitrant components are left behind, eventually leads to the formation of a particularly stable, long-lived fraction of SOM, referred to as *humus*. Humus is produced through a combination of biotic and abiotic processes, resulting in long chains of aromatic ring compounds linked together by amino acids, short peptides, and aliphatic compounds. Numerous exposed carboxylic acid groups provide humus with an acidic nature. Humus is highly processed and has lost all visible features that might be used to trace it to its original litter source. Soil scientists categorize humus as a mixture of polymers designated as fulvic acid, humic acid, or humin depending on acid/base solubility (of the three types, fulvic acids show the highest solubility in water solutions at low pH). Fulvic acids compose the greatest fraction of humus in forest ecosystems, whereas humic acids dominate the humus of grassland ecosystems (Stevenson 1982). Fulvic acids are smaller in molecular weight, contain lower C:N ratios and produce more acidity in soil solutions, compared to humic acids, and humins. These fractions of humus have long lifetimes in the soil (the lifetime of fulvic acids tends to be in the range of several hundred years, whereas that for humic acids and humin can be in the range of several thousand years). Because of its exceptional stability and long lifetime, humus formation is a principal process involved in long-term carbon sequestration by soils.

Soil carbon sequestration rates are highly dependent on ecosystem type, and they are especially susceptible to perturbation due to land-use change. The clearing of forests for agricultural use can cause significant loss of soil carbon accompanied by increased rates of soil respiration (Oades 1988). Afforestation of previously cultivated lands generally increases the rate of soil carbon sequestration, especially in the O and upper A horizons, due to greater rates of litter deposition; however, carbon is often lost from deeper mineral horizons, due to greater rates of humus oxidation (Richter *et al.* 1999, Thuille and Schulze 2005), although these trends can be quite variable depending on site and climate. Using a 62-year chronosequence of sites in Europe, Thuille *et al.* (2000) showed that the growth of Norway spruce forest on land previously occupied by grassland meadows, through natural secondary succession, caused the overall sequestration of carbon in the O and upper A horizons, with no consistent trend in losses from deeper mineral horizons. In observations utilizing chronosequences from several different European sites and covering 100 years, again involving the successional re-growth of Norway spruce forests on grasslands, Thuille and Schulze (2005) showed similar rates of gain in soil carbon in the organic horizon, but losses between 20–25% in the mineral horizons. In a study of secondary tropical forest re-growth using an 80-year chronosequence on former pasture lands in Puerto Rico, Marin-Spiotta *et al.* (2009) observed no change in total soil carbon concentrations; increased litterfall during forest re-growth was compensated by increased rates of carbon loss from SOM that had built up in the pastures prior to abandonment.

16.2 Control by substrate over soil respiration rate

Soil respiration reflects the contributions from a diversity of organisms, including plants and microbes. The contributions of these organisms are often grouped into two categories based on whether respiratory substrates originate from: (1) recently produced photosynthate that is oxidized during root respiration or during microbial respiration within the root rhizosphere; or (2) organic compounds that comprise SOM and are oxidized during microbial decomposition. These separate components of soil respiration are referred to by different names (e.g., autotrophic versus heterotrophic or root versus microbial, respectively); in this book they will be referred to as *rhizospheric respiration* versus *decomposition* (Figure 16.1). Rhizospheric respiration is due to mycorrhizal fungi, rhizospheric bacteria, and plant roots, all of which share the property of being dependent on substrates recently assimilated by photosynthesis. The decomposition fraction is due to the respiration of heterotrophic microorganisms as they decompose SOM. Some past studies have used experimental approaches to partition and quantify contributions from the rhizospheric and decomposition fractions (Kuzyakov 2006, Kuzyakov and Gavrichkova 2010). In one innovative approach, plants are "girdled" by cutting a swath of phloem tissue away from the stem, which eliminates the downward transport of photosynthate; any remaining soil CO_2 loss is assumed to be due to the decomposition fraction. Experiments of this type have revealed that the rhizospheric fraction represents 25–60% of the total soil respiration rate in several forest ecosystems (Högberg and Read 2006). The transfers of photoassimilate through the phloem, and losses as rhizospheric soil respiration, are coupled on relatively short time scales. While the exact time of the entire

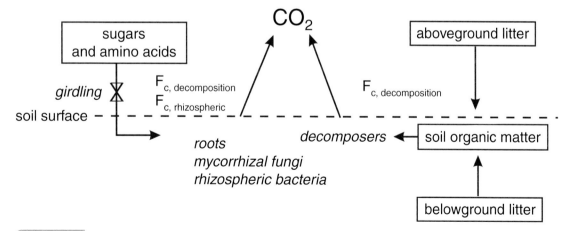

Figure 16.1 Partitioning of soil respiration into CO_2 fluxes due to rhizosphere-dependent organisms ($F_{c, rhizospheric}$) and decomposers ($F_{c, decomposition}$). The rhizosphere-dependent flux is supported by the downward transport of sugars and other substrates from the shoots of plants (often called rhizodeposition); this fraction can be eliminated by girdling the plants. The decomposition flux is supported by pools of soil organic matter derived primarily from litter inputs.

transfer will vary with environmental conditions and tree species, it appears to be a matter of one to a few days between the initial assimilation of CO_2 by leaves and the appearance of that assimilated carbon as rhizospheric respiration (Högberg *et al.* 2008).

Decomposition respiration can be predicted, to some extent, on the basis of chemical properties of the SOM, such as the lignin:N ratio (Parton *et al.* 1987, Wallenstein *et al.* 2010). In coniferous forests, biomass with a high lignin:N ratio undergoes slower decomposition, supports lower soil respiration rates and has a greater tendency for conversion to the more recalcitrant SOM pools, compared to decomposing biomass in broad-leaved forests with its lower lignin:N ratio (Stump and Binkley 1993, Osono and Takeda 2006, Nave *et al.* 2009). The needles of coniferous forests also contain greater amounts of resins and waxes, which are more resistant to decomposition. The high fungal biomass supported by mycorrhizal relations in coniferous forests may also be more resistant to decomposition than the bulk of litter deposited in broad-leaved forests. When the effects of litter chemical composition and climate are combined, it has been estimated from models that boreal forest litter loses, on average, 32% of its mass during the first two years of decomposition, compared to 85% in litter from broad-leaved tropical forests (Tuomi *et al.* 2009).

Decomposition respiration is dependent on interactions among microbial communities, or interactions among exoenzymes expressed differentially by the same communities, the nature and amounts of SOM substrates that they utilize, and the diffusive mobility of those substrates. For convenience, SOM substrates are generally designated according to three categories, based on the rates at which they are oxidized: active, slow, and passive (Figure 16.2). The *active* SOM pool has a turnover time ranging from several days to a

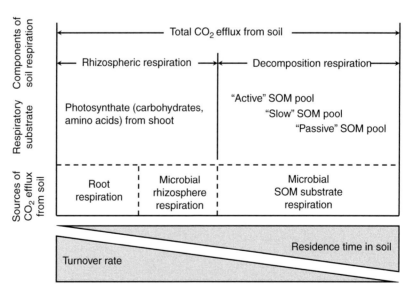

Figure 16.2   Scheme showing the components of soil respiration in relation to substrate type and CO_2 source. The transport of photosynthate from the shoot supports rhizospheric respiration with compounds that have a high turnover rate and short residence time in the soil. Decomposition respiration is derived from soil organic matter pools (SOM) that have lower turnover rates and longer residence time. Modified from a concept presented in Kuzyakov (2006), with permission from Elsevier B.V. Publishers.

few years, the *slow* pool has a turnover time of 20–50 years, and the *passive* pool has turnover times of 400 to several thousand years (Davidson *et al.* 2000a, Trumbore 2000). (It is important to keep in mind that these "pools" are not real biogeochemical entities; guilds of microbial organisms utilize distinct chemical substrates, not "pools"; the concept of pools is intended as an organizational convenience to facilitate taxonomic characterization and dynamic modeling.) Differences in turnover times among these pools are due to differences in chemical composition and accessibility to microbes and their associated exoenzymes (von Lutzöw *et al.* 2006). The chemical composition of different SOM pools is complex and varies among different ecosystems. However, to provide at least some context for discussion, we can state that the active pool often includes compounds such as disaccharides, simple sugars, amino acids, and short peptides, the slow pool includes structural compounds such as cellulose and hemicellulose, and the passive pool includes compounds such as lignin and the lignocelluloses. Efforts to model substrate effects on soil respiration rate focused, in the earliest models, on the general chemical characteristics of different substrate pools (e.g., lignin:N ratio), and on pool-specific decay coefficients that could be modified by climate (Parton *et al.* 1987, McGill 1996). More recently, modeling efforts have attempted to bring microbial processes, including control over exoenzyme activities and compound accessibility, to the forefront; providing for greater insight into the differential roles of specific microbial "guilds" (Appendix 16.1). As molecular identification techniques continue to be applied to the soil microbial communities of different ecosystems, we are likely to learn even more about specific substrate-microbe interactions and their control over soil respiration rate. Even yet to come in the mechanistic modeling of soil respiration is integration of enzyme kinetics with the transport and spatial dispersion of substrates and absorption/desorption of substrates to the inorganic mineral matrix of the soil.

16.3 Control by climate over soil respiration rate

The Q$_{10}$ for soil respiration rate is a principal parameter in the analysis and modeling of ecosystem and global carbon budgets (Raich and Schlesinger 1992, Raich and Potter 1995). Q$_{10}$ values ranging from 1.5 to 3.5 have been used to describe the temperature-dependence of soil respiration and, in general, the models that have been applied to plant tissue respiration also work well for describing soil respiration (Lloyd and Taylor 1994, Fang and Moncrieff 2001). There is evidence, however, from a broad range of site specific studies that the temperature response of soil respiration rate is controlled by complex interactions involving processes such as root and rhizospheric microbial metabolism that are coupled to temperature dynamics at the scale of hours-to-days, and processes involving the heterotrophic decomposition of different SOM pools and climate influences on net primary production and litter production that are coupled to temperature at the scale of months-to-years (Kirschbaum 1995). These complex interactions contribute to a combined influence on soil respiration, and some may dominate temperature sensitivities assessed over short time scales, whereas others may dominate assessments over long time scales. Thus, it has been difficult to derive exact values for the Q$_{10}$ of soil respiration. In one analysis, high-frequency flux data (averaged to 30 minute observation periods) collected across multiple years and sites was used to extract the "fast" and "slow" components of the temperature sensitivity of whole ecosystem respiration to reveal convergence toward an instantaneous, "metabolic" Q_{10} of ~ 1.4 (Mahecha *et al.* 2010). It must be emphasized that this refers to "ecosystem respiration" and not to "soil respiration," considered alone. However, it provides intriguing evidence that supports universal convergence toward fundamental temperature sensitivities in the respiratory metabolism of ecosystems, a result that was also derived from spectral analysis of ecosystem respiration (Stoy *et al.* 2009). If supported by future studies, these analyses suggest that previously observed spatial and temporal dynamics in soil respiration rate may be due principally to variation in the longer-term heterotrophic decomposition processes, especially those associated with different co-existing SOM pools. In the next few paragraphs we focus on these longer-term processes.

In a meta-analysis of studies on soil respiration distributed across a broad latitudinal gradient, Giardina and Ryan (2000) concluded that variation in substrate quality, rather than temperature, explained most of the variance in soil respiration rate. This conclusion was in contrast to those drawn from other studies in which temperature, as a principal component of climate, was identified as a dominant control over geographic variation in soil respiration rate (Trumbore *et al.* 1996, Davidson *et al.* 2000a, Trumbore 2000). In research intended to reconcile these differences, Knorr *et al.* (2005) showed that variation in SOM substrate quality and temperature could interact to cause variation in soil respiration rate. To illustrate this interaction, consider two hypothetical sites, one with a warmer climate compared to the other. If the SOM pool at the warmer site is smaller, then experimental incubations of soil (which often last for months) from each site at their respective mean temperatures will potentially yield similar amounts of respired CO_2. This is because CO_2 will be lost rapidly from the SOM pool at the warmer site, but then decrease over time as the most labile components of that pool are depleted; in contrast, CO_2 will be lost at a slower, but more sustained rate from the SOM pool at the cooler site (Figure 16.3). This observation may lead to the conclusion that soil respiration rate is insensitive to temperature, when comparing the two sites. In reality, it is the interaction between temperature and substrate availability that constitutes the dominant control. Several past studies have shown that older, more recalcitrant SOM pools exhibit higher temperature sensitivity, compared to younger, more labile pools (Couteaux *et al.* 2001, Dalias *et al.* 2001, Fierer *et al.* 2005, Knorr *et al.* 2005, Vanhala *et al.* 2007, Conant *et al.* 2008, Hartley and Ineson 2008, Craine *et al.* 2010) (Figure 16.3). A case has been made to support higher temperature sensitivity as recalcitrance increases on the basis of theoretical linkages between the activation energy of enzyme-substrate reactions and temperature sensitivity predicted from the Maxwell–Boltzmann relation (Bosatta and Ågren 1999). There is debate on this matter as other studies have shown that temperature sensitivity is similar among different SOM pools (Giardina and Ryan 2000, Fang *et al.* 2005, Conen *et al.* 2006, Czimczik and Trumbore 2007, Hopkins *et al.* 2012), or that greater temperature sensitivity exists in "young," labile pools, compared to older recalcitrant pools (Yuste *et al.* 2004, Rey and Jarvis 2006). The temperature dependence of respiration using different SOM pools is also complicated by the fact that acclimation can occur to seasonal changes in temperature, and potentially across even longer time scales (Luo 2007). Furthermore, the temperature sensitivities of substrate transport and absorption/desorption kinetics involving substrate binding to

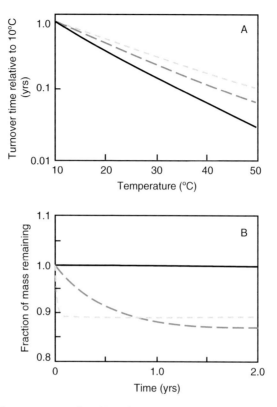

Figure 16.3 **A.** Modeled responses of turnover time in three SOM substrate pools: active (shown in light gray broken line), slow (shown in dark gray broken line), and passive (shown in black solid line) as a function of temperature. The results show that the passive pool is more sensitive to an increase in temperature than the slow and active pools. **B.** The fraction of mass lost due to decomposition of each SOM substrate pool as a function of time following a 2 °C step-wise increase in temperature. Redrawn from Knorr *et al.* (2005), with permission from Macmillan Publishers Ltd. Nature.

the inorganic soil matrix are poorly understood. Until our understanding of these complex interactions improves, conclusions about how climate might exert control over soil respiration rate, across both space and time, must carry significant uncertainty.

Soil moisture content also exerts an important control over soil respiration rate. Bacteria and fungi are dependent on the thin film of water that surrounds soil particles to facilitate the diffusion of substrates through the soil medium (Borken and Matzner 2009). Soil drying can force changes in microbial metabolism and accelerate cell death. Microbial death during drought releases organic compounds that can persist for weeks to months and serve as the substrate for growth of a new generation of microbes that arises upon re-wetting. Soil moisture content can also influence the physical nature of soil. For example, in many soils, particularly in coniferous forests, the organic horizon tends to be hydrophobic; intact leaf fragments and microbially processed polymers from the initial phases of decomposition are often composed of waxes, resins, and other aliphatic, hydrophobic compounds (Doerr *et al.* 2007). Hydrophobicity of the upper soil horizon often increases as partially decomposed

litter dries. Thus, when precipitation returns, following a prolonged drought, initial penetration of water into the soil may be limited. This pattern by which soil drying creates a
condition that sustains the dry state, even in the face of intermittent but small precipitation
events, is an example of a local positive feedback with potential to significantly influence the
coupling between soil respiration rate and seasonal climate variation.

Precipitation can also exert a purely physical influence on the apparent soil respiration
rate. Upon penetration into the soil, rain water displaces air in soil pores, flushing stored CO_2
to the atmosphere; this is often observed as a short pulse of emitted CO_2 following brief rain
events (Liu *et al.* 2002). In that case, observed soil-atmosphere CO_2 exchange is decoupled
in time from the respiratory processes that produce it. Following the initial flush of CO_2, soil
microorganisms can respond quickly and positively to the moisture pulse, causing a
sustained increase in observed CO_2 fluxes. This longer-term influence (hours-to-weeks) is
driven by microbial decomposition and mineralization, a phenomenon that was originally
described by H. F. Birch, a soil chemist working on agricultural and forest soils in Africa
in the 1950s and 1960s; this influence has often been called the *Birch effect*, and it has been
observed more recently in the patterns of ecosystem-atmosphere exchange following rain
pulses into the otherwise dry soils of arid and semi-arid regions (Huxman *et al.* 2004, Jarvis
et al. 2007, Vargas *et al.* 2010).

16.4 Coupling of soil respiration to net primary production and implications for carbon cycling in the face of global change

At steady state, the net CO_2 flux from an ecosystem's total SOM pool will reflect the
difference between decomposition (converting SOM to CO_2) and net primary production
(NPP) (converting CO_2 to autotrophic biomass). The net CO_2 flux from an ecosystem's
autotrophic biomass (due to both growth and maintenance respiration) will reflect the
difference between gross primary productivity (GPP) (assimilating CO_2 from the atmosphere) and NPP (committing a fraction of that assimilated CO_2 to biomass). Thus, NPP sets
an important constraint on both the decomposition and autotrophic components of ecosystem respiration (Kirschbaum 1995, 2000) (Figure 16.4). At the global scale, Raich and
Schlesinger (1992) observed a positive correlation between mean annual NPP for various
ecosystems and soil respiration rate (R_s), as expected by theory. However, mean annual R_s
was, on average, 24% higher than mean annual NPP across all ecosystems. Assuming that
NPP should place an upper limit on decomposition, in the steady state, the higher values for
R_s can be used as a minimum estimate of the contribution of root and rhizospheric
respiration to R_s. Respiration from many ecosystems does not exist at steady state, especially
given the recent influences of climate change, land-use change, and nitrogen deposition;
causing the need for a new generation of models with the potential to cover a broader set
of dynamic, forcing variables (Chapin *et al.* 2009).

Changes in climate have the potential to disturb the steady-state balance between NPP and
decomposition, as each responds differentially to variables such as temperature and precipitation. Perturbations will increase or decrease R_s, depending on these differential responses

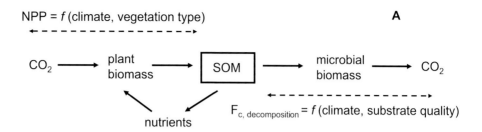

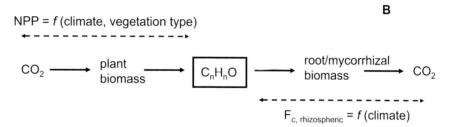

Figure 16.4 **A.** Conceptual model of heterotrophic soil respiration models recognizing that the heterotrophic component of the soil respiratory CO$_2$ flux (F$_{c,\ decomposition}$) is constrained by the input of litter from vegetation which is transformed into SOM from which soil microbes produce CO$_2$. **B.** General logic of autotrophic soil respiraton models recognizing that the rhizospheric component of the soil respiratory CO$_2$ flux (F$_{c,\ rhizospheric}$) is constrained by the availability of carbohydrate substrate (C$_n$H$_n$O) in the roots which serves as substrate for the respiratory metabolism of roots and mycorrhizal fungi.

(Kirschbaum 2000). To illustrate this point, imagine a C$_3$ ecosystem in a moderate-to-warm climate zone with SOM pools at steady state. If we now impose an increase in mean annual temperature on the system, we can predict on the basis of the first principles of metabolism that the rate of microbial SOM decomposition will increase, thus increasing R$_s$. However, it is also likely that NPP will decrease due to higher rates of leaf photorespiration and concomitantly lower rates of net CO$_2$ assimilation. If allowed to achieve a new steady state, in the presence of this temperature increase, the SOM pool should eventually stabilize at a smaller size. Now, imagine that the imposed temperature increase is accompanied by an increase in atmospheric CO$_2$ concentration. The higher CO$_2$ concentration should alleviate, to some extent, the effect of the higher temperature on photorespiration, and thus offset the decrease in NPP. This CO$_2$ effect will be most effective in warm climates, as this is where high photorespiration rates will also occur. If SOM substrate supply is in the range that limits R$_s$, then the higher NPP will potentially cause a shift toward higher R$_s$. If substrate supply is high enough not to limit R$_s$, but rather climate exhibits the dominant limiting control, then the increase in NPP will cause an increase in substrate pool size, but may cause little or no increase in R$_s$. Consideration of these complex interactions is a good exercise for students to consider, as it puts into practice some of the fundamental principles in metabolism and mass balance that we have discussed in previous chapters. However, it also illustrates the importance of quantitative models as tools to resolve the complexities imposed by feedback on the interactions between ecosystem processes and climate variation.

16.5 Methane emissions from soils

Methane (CH_4) is emitted at relatively high rates from wetland soils, the digestive tracts of ruminant animals, and human activities that extract and transfer natural gas (Wuebbles and Hayhoe 2002, IPCC 2007). Global emissions of CH_4 have increased from ~ 200 Tg yr^{-1} prior to the Industrial Revolution, to ~ 600 Tg yr^{-1} at the beginning of the twenty-first century (Dlugocencky *et al.* 1998, Houweling *et al.* 2000, Le Mer and Roger 2001). The current globally averaged tropospheric CH_4 mole fraction is ~ 1.8 µmol mol^{-1}. The annual mean growth rate of CH_4 in the atmosphere has varied since the early 1980s, from -4 to 15 nmol mol^{-1} yr^{-1}, but in all but three of the past 25 years the growth rate has been positive with a mean at ~ 7 nmol mol^{-1} yr^{-1} (Dlugocencky *et al.* 2011). Globally, soils are the source of 100–200 Tg yr^{-1} of CH_4 emissions (Le Mer and Roger 2001). Soils also take up CH_4 at the global rate of 20–40 Tg yr^{-1} (Dutaur and Verchot 2007). The production and consumption of CH_4 in soils is due to microbial metabolism. *Methanogenic* archaea (also called methanogens) inhabit anaerobic soils and produce CH_4 after transferring energy-rich electrons from SOM to CO_2, which they use as a terminal electron acceptor (Section 3.3). *Methanotrophic* bacteria inhabit aerobic or anaerobic soils and oxidize CH_4, which they use as a principal carbon and energy source.

The use of SOM by archaea to produce CH_4 follows the initial oxidation of some compounds to produce suitable methanogenic substrates. The anaerobic metabolism of alcohols, fatty acids, amino acids, and various aromatic compounds contained in SOM by hydrogenic and acetogenic bacteria produces the H_2 and CH_3COOH, respectively, that are used as an initial substrate in methanogenesis. Methanogens use acetate (CH_3COOH) and molecular hydrogen (H_2) and produce CH_4 according to:

$$CH_3COOH \rightarrow CH_4 + CO_2 \quad \Delta G^0 = -36\,kJ/mole, \tag{16.1}$$

$$CO_2 + 4H_2 \rightarrow CH_4 + 2H_2O \quad \Delta G^0 = -32.7\,kJ/mole. \tag{16.2}$$

Under standard conditions, the antecedent metabolism that produces CH_3COOH and H_2, as substrates, is endergonic (positive $\Delta G'^0$), whereas the methanogenic oxidation of the substrates to CH_4 is exergonic (negative $\Delta G'^0$). Thus, as a multi-species serial pathway, methanogenesis is ultimately dependent on the methanogenic archaea and their capacity to oxidize CH_3COOH and H_2 at a rate that maintains disequilibrium and allows the antecedent metabolism to progress according to mass action, despite its endergonic standard free energy. This forces the various microbial assemblages that produce and consume CH_3COOH and H_2 into thermodynamically dependent, obligatory, cooperative associations. The role of methanogens in these associations was nicely described by Schink (1999): "although (methanogens) are among the last actors in the (methanogenic) play, they conduct the whole process to a maximum energetic efficiency."

At the other end of the energetic spectrum, methanotrophic bacteria use CH_4 as an initial substrate, rather than produce it as a final product. Methanotrophs can grow aerobically or anaerobically with CH_4 as their only source of energy and carbon. In aerobic methanotrophy CH_4 is oxidized in an initial step to methanol according to:

$$CH_4 + NAD(P)H_2 + O_2 \rightarrow CH_3OH + NAD(P)^+ + H_2O. \qquad (16.3)$$

where either NADP$^+$ or NAD$^+$ can be used as electron carriers. The methanol produced in reaction (16.3) is progressively oxidized to formaldehyde, formate, and eventually, CO$_2$. The most common types of methanotrophs belong to the gram-negative eubacteria. Because of their specialization on CH$_4$ as a respiratory substrate, methanotrophs are common in the upper (aerobic) layers of wetland soils where they gain access to CH$_4$ released by methanogens living in the lower (anaerobic) layers (Figure 16.5A).

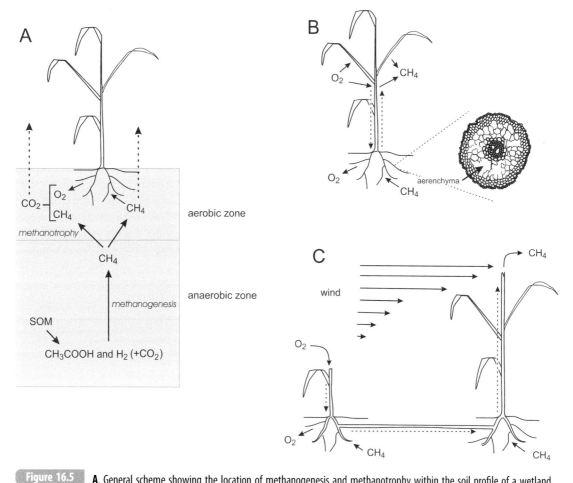

Figure 16.5   **A**. General scheme showing the location of methanogenesis and methanotrophy within the soil profile of a wetland system. **B**. Diffusion of O$_2$ (downward) and CH$_4$ (upward) through aerenchyma tissue in roots, stems, and leaves of a wetland grass, Spartina alterniflora. The diffusive fluxes of these gases will follow their respective concentration gradients between the atmosphere and water-logged sediments. **C**. Venturi-type advection of O$_2$ (downward) and CH$_4$ (upward) through broken stems of some wetland plants at different heights. Differential wind speed blowing across the canopy will create a pressure gradient that drives mass flow of air through connecting rhizomes between culms, thus creating a transport flow for O$_2$ and CH$_4$. Redrawn from a concept presented in Grosse et al. (1996).

The magnitude of CH_4 emissions, and the processes that transport it to the atmosphere, have become of special concern in recent years because of the contribution of increasing atmospheric CH_4 concentrations to climate warming. Methane is approximately twenty times more potent as a greenhouse gas, compared to CO_2, and contributes directly to approximately 0.5 W m^{-2} of the current atmospheric forcing that underlies rising global temperatures. The higher global warming potential of CH_4, compared to CO_2, is due to the lower concentration of CH_4 in the atmosphere, and thus the potential for each unit of emitted CH_4 to cause a greater proportional increase in the optical thickness of the lower atmosphere at those wavelengths absorbed by CH_4, compared to those absorbed by CO_2. Methane also exerts a dominant control over the oxidizing power of the troposphere because it is reactive with hydroxyl radical ($\cdot OH$), which oxidizes CH_4 to CO, and ultimately CO to CO_2. Estimates of tropospheric lifetime of methane have converged to the range of 8.6–8.9 years (Patra *et al.* 2009, Huijnen *et al.* 2010).

16.5.1 Methane transport mechanisms

Globally, most soil CH_4 production occurs in wetlands, particularly those at northern high latitudes. Transport of CH_4 from submerged soils to the atmosphere occurs through molecular diffusion (through water or plants), ebullition (bubbling), and advective transport through plants with pressure gradients sustained by a variety of mechanisms (Holzapfel-Pschorn *et al.* 1985, Holzapfel-Pschorn and Seiler 1986, Walter *et al.* 2007). Ebullition of methane occurs when methane production in submerged sediments exceeds the potential for solubility in the overlying water column, causing the CH_4 to come out of solution and partition into gaseous bubbles that subsequently rise through the column and deposit CH_4 directly to the atmosphere. In shallow-water wetlands, transport through aquatic macrophytic plants is significant, including submerged, floating-leaved, and emergent species. The transport of CH_4 through plants is facilitated by *aerenchyma*, a tissue in stems, roots, and leaves that is specialized for gas transport, especially in wetland plants. Aerenchyma can form from the genetically programmed separation of cells, creating air spaces between them, or from the death of cells. Aerenchyma forms continuous air passages through roots, stems, and leaves, allowing O_2 to diffuse downward into water-logged soils and CH_4 to diffuse upward to the atmosphere (Figure 16.5B). The formation of aerenchyma is promoted by anaerobiosis in soils.

Aerenchyma can also provide the conduits for pressure-driven, viscous flows through the extensive network of rhizomatous connections among emergent stems and leaves. Various mechanisms have been hypothesized as causes of pressure gradients in wetland plants, including thermal- and humidity-induced diffusion (Armstrong *et al.* 1996a). These mechanisms are dependent on the presence of extremely small pores in the epidermis of leaves and stems; either smaller-than-normal stomata or specialized pores. Pores of such a small size would sustain thermal- or humidity-induced pressure gradients between air spaces within the leaf and the outside atmosphere, and at the same time resist viscous flows from occurring directly across leaf and stem surfaces (Section 6.2.3). If, however, pressure gradients develop between two emergent stems connected by aerenchymatous rhizomes, CH_4 can diffuse into the system and be carried by viscous flows, down the pressure gradient

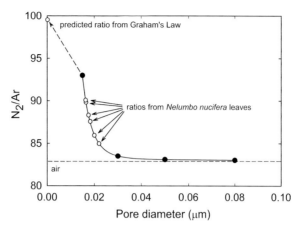

Figure 16.6 Relation between pore size and observed nitrogen (N$_2$) to argon (Ar) ratios collected after air diffused through nucleopore membranes (solid circles) or the upper surface of leaves of the wetland species *Nelumbo nucifera* (open circles). The open circle on the y-axis represents the ratio (99.2) predicted by Graham's Law assuming diffusion through a porous membrane. As pore size decreases, the ratio of Knudsen diffusion to bulk (viscous) flow through the pores increases and mass fractionation between the lighter compound Ar (mass 18) and the heavier compound N$_2$ (mass 28) increases. Air naturally contains approximately 78% N$_2$ and 0.94% Ar. Gas collected from inside the leaves has N$_2$/Ar ratios that are enriched in Ar relative to air, indicating that pore diameters in the leaf are in the range 0.015–0.03 μm, and thus likely to support mostly Knudsen diffusion. These pores may facilitate pressurization of the internal leaf air spaces by humidity-induced or thermal-induced diffusion. Redrawn from Dacey (1987).

and out into the atmosphere. In certain wetland reed canopies (e.g., *Phragmites australis*), aerenchymatous conduits connect living shoots with dead shoots that have been broken, leaving the conduits open to the atmosphere. This mechanism of viscous flow presumably accounts for relatively high CH$_4$ emission rates from some reed-dominated ecosystems (Kim *et al.* 1998). A single dead and broken shoot can serve as the common exit point for the viscous flows from numerous connected living shoots (Afreen *et al.* 2007).

There is evidence that some wetland plants have novel pores in the epidermis that are smaller than stomata (Dacey 1987). Observations of N$_2$/Ar ratios taken from gas inside leaves of the wetland plant *Nelumbo nucifera*, a species with these unique small pores, revealed mass fractionation as a result of air diffusion across the leaf surfaces (Figure 16.6). The fractionation was less than that predicted by Graham's Law, indicating that transport was due to a combination of diffusion and viscous flow. When calibrated against the fractionation of N$_2$ and Ar diffusing across synthetic nucleopore membranes, it was estimated that the leaf pores must be in the size range of 0.015–0.03 μm, considerably smaller than that for fully opened stomata. The structural nature of the pores is not known. Nonetheless, the observations provide evidence for the existence of pores capable of supporting Knudsen flow and the development of internal leaf pressure gradients.

The viscous flow of CH$_4$ through aquatic plants can also be driven by pressure gradients created through vertical wind speed gradients within a stand of aquatic grasses or reeds. The differential pressure forces that develop in these stands have an origin similar to that for pressure gradients in Venturi tubes (Armstrong *et al.* 1996b, Kim *et al.* 1998). The Venturi effect is typically discussed with regard to narrowing in a pipe and its influence on fluid flow; as fluid flows through the narrowed section of a pipe, its velocity must increase to satisfy continuity. The increase in velocity is accompanied by a decrease in pressure; kinetic energy that is transferred to acceleration of the fluid comes at the expense of pressure within the fluid. In some emergent aquatic plant species, such as those in the genera *Phragmites* and *Typhus*, broken stems, or culms are connected to other broken stems of different heights; this condition can produce Venturi-type pressure gradients. Wind speed decreases sharply as the ground is approached. Reduced wind speed at lower heights must be balanced by higher pressure (momentum must be conserved); thus, a pressure gradient exists between the two openings at different heights. If the stems are connected by aerenchymatous conduits, then viscous flow can be facilitated within the system (Figure 16.5C).

The use of stable isotopes (especially $^{13}C/^{12}C$ ratios) has facilitated studies of the processes underlying CH_4 emissions to the atmosphere. Molecules of the same compound, with different isotopic compositions, transported purely by diffusion will fractionate according to Graham's Law; molecules transported purely by viscous flow arising from pressure gradients, will not. The transport of CH_4 by diffusion alone will cause divergence in its $^{13}C/^{12}C$ ratio, which can be quantified after collecting CH_4 from the air spaces of leaves. As diffusion progresses, the CH_4 that is left behind will become progressively more enriched in ^{13}C. This empirical approach has been used to demonstrate the importance of both diffusive and viscous flows to the overall CH_4 flux. In the case of rice, diffusion appears to be the dominant transport mechanism (Chanton *et al.* 1997).

16.6 The fluxes of nitrogen oxides from soils

The emission of nitrogen oxides from soils is important to our study of the earth system in two important ways. First, nitrous oxide (N_2O) is a radiatively active, greenhouse gas. Second, the reactive nitrogen oxide compounds NO and NO_2 (collectively referred to as NO_x) support an active form of photochemistry that affects concentrations of atmospheric oxidants, such as ozone (O_3) and peroxyacyl nitrate (PAN). The NO_x compound that is emitted from soils in greatest abundance is nitric oxide (NO). The global emission rates of N_2O and NO have increased, along with the emission rates of CO_2 and CH_4, over the past century.

Most N_2O and NO originate from microbial nitrification and denitrification. *Nitrification* is an aerobic process in which bacteria use ammonia (typically present in soils as the ammonium ion, NH_4^+) as an electron source and require O_2 as the ultimate electron acceptor. In soils, nitrification occurs through the serial activity of two groups of bacteria, the ammonia-oxidizing bacteria, principally in the genus *Nitrosomonas*, and nitrite-oxidizing bacteria, principally in the genus, *Nitrobacter*. The energetics of ammonium-to-nitrite oxidation, the first stage of serial nitrification is represented as:

$$NH_4^+ + 1.5\,O_2 \rightarrow NO_2^- + 2H^+ + H_2O$$
$$\Delta G'^0 = -276\,kJ \text{ and } E^0 = 343\ mV. \tag{16.4}$$

The second stage of nitrification involves the oxidation of NO$_2^-$ to NO$_3^-$:

$$NO_2^- + 0.5\,O_2 \rightarrow NO_3^-$$
$$\Delta G'^0 = -75\ kJ \text{ and } E^o = 434mV. \tag{16.5}$$

The negative free energy change ($\Delta G'^0$) and positive reduction potential (E^o) are indicative of exergonic redox reactions (Section 3.2.1). The energy-yielding electrons that nitrifying bacteria strip from NH$_4^+$ are used to reduce CO$_2$, reflecting a form of autotrophic metabolism. During nitrification, some NO and N$_2$O is produced and emitted to the atmosphere, and the production of these products increases as O$_2$ availability in soils decreases. These fluxes are small, however, compared to the fluxes that occur during denitrification.

Microbial *denitrification* occurs under anaerobic conditions. Denitrifying bacteria replace O$_2$, as the final respiratory electron acceptor, with NO$_3^-$. The overall energetic balance of oxidizing glucose with NO$_3^-$ as the final electron acceptor, compared to O$_2$, is similar:

$$Glucose + 6\,O_2 \rightarrow 6\,CO_2 + 6\,H_2O$$
$$\Delta G^0 = -2870\ kJ, \tag{16.6}$$

$$Glucose + 4.8\,NO_3^- + 4.8H^+ \rightarrow 6\,CO_2 + 2.4\,N_2 + 8.4\,H_2O$$
$$\Delta G^0 = -2669\ kJ. \tag{16.7}$$

Denitrification occurs in a series of steps (Figure 16.7), with each step potentially carried out by the same or different specialist groups of bacteria or archaea. When considered to completion, the steps of denitrification convert soil NO$_3^-$ to N$_2$, which is emitted to the atmosphere.

During the steps of denitrification and nitrification some molecules of NO and N$_2$O are known to "leak" from molecule pools in the pathway. The amount of leakage has been shown to correlate positively with the overall availability of nitrogen and thus the overall activity of the nitrification and denitrification pathways. The ratio of NO to N$_2$O in the leaked

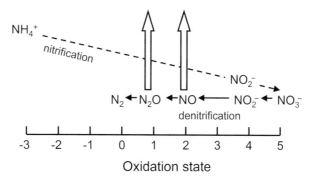

Figure 16.7 Schematic diagram showing steps of nitrification and denitrification in relation to the oxidation state of nitrogen. The intermediate compounds NO and N$_2$O potentially leak from both pathways and enter the atmosphere.

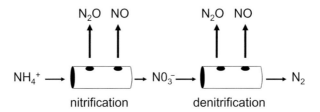

Figure 16.8 Hole-in-the-pipe conceptual model describing NO and N$_2$O fluxes from the nitrification and denitrification pathways. The flux rates are controlled by the amount of N compounds flowing through the pathways (pipes), and soil water content, which determines soil aeration. Based on a concept presented originally in Firestone and Davidson (1989), but also see Davidson and Verchot (2000).

compounds has been negatively correlated with soil water content, which in turn influences soil O$_2$ content and compound diffusion rates (Davidson and Verchot 2000). From these results it was hypothesized that at high soil water content, the diffusion rate of NO from the nitrification pathway is reduced and the ratio of nitrification to denitrification shifts to favor denitrification. Both of these changes may force more NO through the microbial steps that favor its reduction to N$_2$O (Davidson *et al.* 2000b). A conceptual model, known as the *hole-in-the-pipe model*, illustrates the controls over NO and N$_2$O emission rates, with the diameter of the pipe representing overall N flux through the pathways, and the diameter of the holes representing control by soil water content (Figure 16.8). The hole-in-the-pipe model has been important for organizing hypotheses about the controls over NO and N$_2$O emissions from soils. However, it is not quantitatively mechanistic and only loosely grounded in first-principles, such that it is limited as a tool for quantitative prediction.

Appendix 16.1 Derivation of first-order litter decomposition kinetics

The foundational assumption of litter decomposition models is that decomposition is proportional to microbial biomass and growth rate. As a general starting point, we can state that decomposition is a first-order decay process:

$$\frac{d\mathrm{c}}{dt} = -k\mathrm{c}, \tag{16.8}$$

where $d\mathrm{c}/dt$ is the molar loss of organic matter (per unit volume of soil), and k is a rate coefficient, or turnover coefficient, with units of reciprocal time. After integration, Eq. (16.8) can be used to calculate the amount of litter remaining after a given time interval (c_t) given an original concentration (c_0):

$$\mathrm{c}_t = \mathrm{c}_0 e^{-kt}. \tag{16.9}$$

Past laboratory studies have estimated k as 0.2 day^{-1} for proteins, 0.08 day^{-1} for cellulose and hemicellulose and 0.01 day^{-1} for lignin (Paul and Clark 1996). (For simplicity in developing theoretical principles, from this point forward, we will consider decomposition in a single SOM pool assuming a mean uniform turnover coefficient for all carbon in that pool. Models for multiple pools with different turnover coefficients can be developed along similar theoretical lines; see Amundson (2001).)

Deeper insight can be added to decomposition models if turnover is defined in terms of enzyme-substrate interactions and microbial diversity. Here, we will focus on the model developed by Moorhead and Sinsabaugh (2006), commonly called the Guild Decomposition Model (GDM), which in turn was influenced by theory developed in Sinsabaugh *et al.* (1991) and Schimel and Weintraub (2003). We start with a general analogy:

$$\frac{d\text{c}}{dt} = -k\text{c} = -\frac{\text{V}_{\max}}{K_m + \text{c}}\text{c}. \tag{16.10}$$

In this case, the decay coefficient (k) is defined as $\text{V}_{\max}/(K_m + \text{c})$, which places turnover time within the context of Michaelis–Menten kinetics, and $\text{V}_{\max}$ is defined with units mol (per unit volume soil) day^{-1}. In the original model it was assumed that the principal control over decomposition is the activity of extracellular enzymes produced by microorganisms and that the concentration of litter and associated SOM (i.e., substrate) is large compared to the concentration of enzymes (E). Under this condition, we can write:

$$\frac{d\text{c}}{dt} = -\frac{\text{V}_{\max}}{K_m + \text{E}}\text{E}. \tag{16.11}$$

Equation (16.11) can be expressed in terms of microbial biomass by assuming a fixed fraction of enzyme pool (a) to microbial biomass concentration (B), such that E = a/B, and defining K_m in terms of microbial biomass as $K_B = K_m/\text{a}$. Thus:

$$\frac{d\text{c}}{dt} = -\frac{\text{V}_{\max}}{K_B + \text{B}}\text{B}. \tag{16.12}$$

Assuming that $\text{V}_{\max}$ and K_B are constant for a particular soil with a stable community of microbial species, the dynamic variable in Eq. (16.12) is B; variations in B drive variations in decomposition rate. If we state the condition of extremely high, non-limiting B, then decomposition rate will be first-order with respect to c. Under this condition K_B is small compared to B, and we can write:

$$\frac{d\text{c}}{dt} = -k_c\text{c} \approx -\text{V}_{\max}, \tag{16.13}$$

where the decay coefficient (k_c) is now defined explicitly with respect to organic matter concentration. Alternatively, if we allow organic matter concentration (c) to increase to extremely high values, we can write:

$$\frac{d\text{c}}{dt} = -k_B\text{B} \approx -\text{V}_{\max}, \tag{16.14}$$

where the decay coefficient (k_B, also called r in some past model descriptions; Moorhead *et al.* 2006) is defined with respect to microbial biomass concentration. Thus, when $B < k_c c / k_B$, we can write:

$$\frac{dc}{dt} = -\frac{k_c c B}{K_B + B},$$

(16.15)

and when $c \ll k_B B / k_c$, we can write:

$$\frac{dc}{dt} = -\frac{k_B B c}{K_C + c},$$

(16.16)

where K_C is the Michaelis–Menten half-saturation coefficient with respect to organic matter concentration. This treatment allows for the assumption that both microbial biomass and SOM substrate have the potential to limit the rate of decomposition.

Moorhead and Sinsabaugh (2006) developed three microbial "guilds" or functional groups according to: (1) opportunists that specialize on labile, water/ethanol soluble substrates principally through intracellular metabolism (highest growth rates) (with concentration of substrates represented as c_1); (2) cellulose/lignocellulose decomposers that use extracellular enzymes to degrade the largest fraction of litter biomass (moderate-to-slow growth rates) (with concentration of substrates represented as c_2); and (3) recalcitrant litter decomposers that degrade the most complex compounds in SOM (slow growth rates) (with concentration of substrates represented as c_3). Thus, under the assumption that the size of each carbon pool limits decomposition we can develop separate equations for each guild (i) and each set of compounds (j) with the generic form:

$$\frac{dc_j}{dt} = U_{ij} = \frac{k_{cij} B_i c_j}{K_{Cij} + c_j},$$

(16.17)

where U_{ij} is the rate of carbon uptake by guild i from compound j. Moorhead and Sinsabaugh (2006) took the modeling even further by considering (1) negative feedback on the decomposition of cellulose by lignin, and (2) perturbations to decomposition by nitrogen deposition from the atmosphere (which tends to stimulate the early stages of decomposition and inhibit the latter stages of decomposition).

17 Fluxes of biogenic volatile compounds between plants and the atmosphere

> The nose can reveal much qualitative information about the release of organic volatiles by plants, but since it is preferentially sensitive to certain terpenes ... and rather insensitive to others ... analyses with a gas chromatograph are needed to obtain a quantitative picture of the volatile organics present in the air at all times. Thus we easily detect the aromaticity of a deciduous forest in autumn, and especially the sweet odor of the leaf litter on the forest floor, and we can tell a coniferous forest at a distance. But we are unprepared for the fact that an oak forest produces virtually as many aromatics as a pine forest, only of a lower odor level.
>
> Rasmussen and Went (1965)

Plants not only exchange inorganic C with the atmosphere, but also organic C in the form of a broad range of biogenic volatile organic compounds (BVOCs). The magnitude of the global BVOC flux is small compared to the global photosynthetic CO_2 flux: ~ 2 Pg of C contained in annual BVOC emissions (including CH_4 and CO) compared to global gross primary productivity, which is ~ 120 Pg of CO_2. However, the BVOC flux is critical to understanding chemical reactions that occur in the atmosphere, especially those that produce important oxidant compounds, such as ozone, those that determine the oxidation rate of important greenhouse gases, such as methane, and those that determine the production of organic aerosol particles, which affect the earth's radiation budget. Scientists did not recognize that the emission of compounds from vegetation could have such an important impact on atmospheric chemistry until the mid 1960s when researchers such as Rei Rasmussen and Fritz Went began collecting air samples in remote locations, far from the influences of urban pollution. This effort to describe the volatile chemical "fingerprints" of natural ecosystems revealed the presence of a large outward flux of reactive compounds that had previously been unidentified. For example, as reflected in the quote above, careful analysis using gas chromatography revealed the presence of reactive isoprene and other terpenes in an oak forest, which would have been undiscovered if left to detection by the human nose alone. Once these observations started to accumulate, it became clear that biogenic sources of reactive compounds were even more important than anthropogenic sources in their potential to catalyze oxidative photochemistry and, in many cases, control the overall oxidative capacity of the troposphere.

In this chapter we take up the topic of BVOC emissions beginning with the metabolic processes that produce the compounds and ending with their photochemical fate. The diversity of BVOCs capable of fueling atmospheric chemistry is immense, including alkenes, alkanes, alcohols, aldehydes, ketones, and organic acids. These compounds exhibit a broad range of atmospheric lifetimes, ranging from a few seconds to several years. Ultimately, they all represent sources of electrons that exist at higher potential energy levels than those in the inorganic products to which they are oxidized, e.g., CO and CO_2. Thus, thermodynamic free energy gradients favor the spontaneous reactions that are caused by BVOC emissions. In many ways, emitted BVOCs can be viewed as "fuel" powering the oxidative chemistry of the troposphere; as with all biologically relevant energetics on the earth's surface, the chemical power contained in BVOCs can be traced to the photosynthetic capture of solar photons. Until recently, most research on the chemical fate of BVOCs focused on gas phase reactions, driven by the hydroxyl radical ($\cdot$OH); commonly referred to as the "detergent of the atmosphere." Now, however, we recognize that some of the most important chemistry occurs in both the gas and liquid phases of the atmosphere, and is revealed in the production of organic aerosol particles that scatter the solar photon flux. The products that result from all of these reactions have important consequences for both the "health" of ecosystems and global climate. By considering these perspectives on ecosystem-atmosphere interactions, we will be able to push our understanding of connections within the earth system beyond those that form the traditional foci on CO_2 and H_2O exchanges, and begin consideration of processes within the realm of reactive photochemistry.

17.1 The chemical diversity of biogenic volatile organic compounds (BVOCs)

Plants emit a broad range of biogenic volatile organic compound (BVOC) types and they have an equally broad impact on atmospheric chemistry (Monson and Holland 2001, Laothawornkitkul *et al.* 2009). The famous plant physiologist, Fritz Went (1960), first described the importance of plants as sources of BVOCs when he used the mass of leaf oils in shrubs from the Western United States to estimate global terpene emissions as 175 Tg C yr^{-1} (1 Tg = 10^{12} g). Since that seminal study, estimates of global emissions (including CH_4 and all other types of BVOCs) have been refined through broader inventories of plant species and better maps of global vegetation and are now estimated to be $\sim$ 1 Pg C yr^{-1} (1 Pg = 10^{15} g) (Guenther *et al.* 2012). Current estimates of only global terpene emissions are $\sim$ 600 Tg C yr^{-1} (Arneth *et al.* 2008); three times higher than Went's original estimate. Terpenes represent one of the most reactive classes of emitted BVOCs. Terpenes consist of carbon chains (generally 5–15 carbons in length) containing one or more C-to-C double bonds, thus characterizing them as *alkenes* (often referred to by organic chemists as *olefins*). The C-to-C double bond consists of a *sigma bond* (the same type of bond formed between two nuclei in a single C-to-C bond) plus a *pi bond* (a bond involving electron orbital space above and below the plane of the C-to-C bond). Of the two bonds, the pi bond is weaker. Electrons contributing to the pi bond are more likely to leave their orbitals and form new bonding relations with other (reactant) atoms; thus, pi-bonded electrons provide the source for much of the reactive photochemistry that occurs between BVOCs and oxidative compounds in the atmosphere. The most common terpenes emitted from plants include: hemiterpenes (C_5H_n), monoterpenes ($C_{10}H_n$), and sesquiterpenes ($C_{15}H_n$). The biogenic production of these compounds has evolved for a variety of reasons, including protection from oxidative and thermal stresses in the chloroplast, protection from herbivory, and the maintenance of metabolic homeostasis (Harley *et al.* 1999, Sharkey and Yeh 2001, Monson *et al.* 2007, Harrison *et al.* 2013).

In terms of atmospheric chemistry the most important terpene emitted from plants is *isoprene* (2-methyl-1,3-butadiene), a branched-chain hemiterpene. Global isoprene emissions are likely to be in the range 500–550 Tg C yr^{-1} (Arneth *et al.* 2008). Isoprene is highly reactive, with an atmospheric lifetime of only 1–2 hours. The emissions of *monoterpenes* are also important contributors to photochemical reactions in the atmosphere. It is the monoterpenes, which often form ringed alkene structures. One type of monoterpene is camphor, an oxygenated compound that composes the leaf oils of sagebrush, the species used in Fritz Went's original estimate of global VOC emissions. Estimates of global monoterpene emissions have been less well constrained in global models, compared to isoprene, and thus emission estimates vary considerably; most have been in the range 75–120 Tg C yr^{-1} (Arneth *et al.* 2008). The chemical products that result from the oxidation of monoterpenes have lower boiling points than the emitted compounds; that is, the products from monoterpene oxidation exhibit lower volatility and a greater tendency to partition into the liquid phase of the atmosphere, compared to emitted monoterpenes. The oxidation of monoterpenes is an important source for the growth of atmospheric particles called *secondary organic aerosols* (Cahill *et al.* 2006). Secondary organic aerosols are small enough (with diameters in the sub-micron range), that they remain suspended in the atmosphere. Monoterpene lifetimes are in the range of 3–6 hours, a bit longer than isoprene, but nonetheless short enough to classify them also as "highly reactive." A third important class of plant-emitted terpenes is the *sesquiterpenes*, which are common components of plant defenses against herbivores. Sesquiterpenes are among the most reactive of all biogenic terpenes. Some of the more commonly emitted sesquiterpenes (e.g., β-caryophyllene) react so quickly after being emitted that they have atmospheric lifetimes of only 1–2 minutes. Like monoterpenes, the oxidation products of sesquiterpenes have an important role in the formation and growth of secondary organic aerosols.

In addition to terpenes, a variety of other BVOCs are emitted from plants. Recently, the *oxygenated BVOCs* have been recognized as an important class of compounds (Fall 2003). Oxygenated BVOC emissions are especially high from coniferous forest trees (acetone and methylbutenol), deciduous leaves undergoing expansion (methanol), and freshly cut turf grasses and harvested crops (hexenal, hexenol, and hexenal acetate). Oxygenated BVOCs tend to have longer atmospheric lifetimes than the terpenes, and this has contributed to their atmospheric persistence (Singh *et al.* 2001). Global ecosystems are also a source of atmospheric *acetone*, $(CH_3)_2CO$, producing ~ 95 Tg yr^{-1}. The principal sources of atmospheric acetone are direct emissions from terrestrial vegetation (33 Tg yr^{-1}), photochemical degradation of organic matter in oceans (27 Tg yr^{-1}), and the atmospheric oxidation of isoalkane compounds (principally propane) (21 Tg yr^{-1}) (Jacob *et al.* 2002). Acetone has an atmospheric lifetime of 15–60 days (Jacob *et al.* 2002) and is a source of peroxyacetyl radicals, which react with oxidized forms of nitrogen (NO_x) to form peroxyacetylnitrate (PAN), an important oxidant in the lower atmosphere. Global ecosystems are also a source of atmospheric *methanol* (CH_3OH), producing 240 Tg yr^{-1} (Jacob *et al.* 2005). Approximately 128 Tg methanol yr^{-1} is emitted directly from vegetation, with an important source being the expanding cell walls of leaves. *Methylbutenol* (2-methyl-3-buten-2-ol) is an oxidized hemiterpene, similar in structure and biochemical origin to isoprene and emitted at relatively high rates from North American pine forests (Guenther *et al.* 2000).

Acetaldehyde (CH₃CHO) and short-chain organic acids, including formic and acetic acids, are emitted from many plants and soil microbes (Seco *et al.* 2007). The estimated atmospheric lifetimes of many of the compounds emitted by vegetation and the oxidants responsible for their photochemical breakdown are presented in Table 17.1.

Although we have focused in this book on the exchanges of mass and energy between *terrestrial* ecosystems and the atmosphere, because of its potential importance to global climate, it is worth spending some time discussing the emission of sulfur compounds from *marine* ecosystems. The most common types of emitted compounds are dimethyl sulfide (CH₃SCH₃), carbonyl sulfide (COS), hydrogen sulfide (H₂S), and carbon disulfide (CS₂).

Table 17.1 Lifetimes (τ) for some of the important species in the oxidative chemical reaction sequences that occur in the lower atmosphere. From Monson and Holland (2001)

Chemical species	Chemical formula	Atmospheric lifetime
Hydroxyl radical	·OH	0.2–1 seconds[a]
Peroxy radical	RO_2·	5–900 seconds[b]
Nitrate radical	NO_3·	5–6 seconds (daytime)[c]
		> 1000 seconds (nighttime)
Nitrogen dioxide	NO_2	143 seconds (daytime)[d]
		7 hours (nighttime)
Isoprene	C_5H_8	0.06–1.5 day[e]
Monoterpenes	$C_{10}H_n$	0.06–20 days[f]
Sesquiterpenes	$C_{15}H_n$	2 min–3 hours[g]
Acetone	$(CH_3)_2CO$	15–61 days (·OH)[g]
Methanol	CH_3OH	12 days (·OH)[g]
Ozone	O_3	5–300 days[h]

[a] Calculated with respect to global average reaction with CH_4 (at total global content of 5000 Tg) and CO (at total global content of 360 Tg).

[b] Calculated for the methyl peroxy radical when reaction is with NO or HO_2·; reaction coefficients were taken from Finlayson-Pitts and Pitts (2000), for NO + CH_3O_2· 10 pptv NO, τ = 540 s; 1 ppbv NO, τ = 5.4 s for HO_2· + CH_3O_2· 8 pptv HO_2·, τ = 891 s

[c] Daytime zenith angle = 0°; nighttime τ calculated for reactions with NO_2 and VOCs in an unpolluted atmosphere, from Brasseur *et al.* (1999).

[d] Reaction coefficients were taken from Finlayson-Pitts and Pitts (2000). daytime, zenith angle = 50°, τ = 143 s; nighttime, 50 ppb O_3, τ = 7 h

[e] From: Atkinson and Arey (2003) with isoprene lifetimes for individual oxidants as: τ (·OH) = 1.4 h; τ (O_3) = 1.3 day; τ (NO_3·) = 1.6 h

[f] From: Atkinson and Arey (2003) with representative monoterpene lifetimes for individual oxidants as: α-pinene: τ (·OH) = 2.6 h; τ (O_3) = 4.6 h; τ (NO_3) = 11 min β-pinene: τ (·OH) = 1.8 h; τ (O_3) = 1.1 day; τ (NO_3·) = 27 min

[g] From: Atkinson and Arey (2003) and Jacob *et al.* (2002; in the case of acetone) and Millet *et al.* (2008; in the case of methanol).

[h] Photolysis and chemical reaction (surface deposition will cause lifetime to be less); from Brasseur *et al.* (1999): lower troposphere: 0–3 km τ = 5–8 days (summer); τ = 17–100 days (winter).upper troposphere: 6–10 km τ = 30–40 days (summer); τ = 90–300 days (winter).

Among these compounds, *dimethyl sulfide* (DMS) is the most important in terms of overall emission rates and potential to influence atmospheric chemistry and climate. DMS is produced by marine algae and bacteria through the breakdown of dimethyl sulfoniopropionate (DMSP), an organic metabolite that may have several physiological roles, including regulation of the osmotic potential of cells. Most DMSP is released from the broken cells of phytoplankton during zooplankton grazing or following viral infection. Once released, DMSP is converted to DMS by a combination of extracellular enzymes bound to algal cell walls, enzymes released from phytoplankton cells, and marine bacteria that are associated with phytoplankton blooms (Gonzalez *et al.* 2000, Pinhassi *et al.* 2005). In the latter case, communities of DMSP-cleaving bacteria are often closely associated with "ocean blooms" of phytoplankton and their associated zooplankton grazers, forming localized zones of high DMS production.

In addition to DMS, oceans are important sources of other sulfur compounds. Surface waters in the North Atlantic and Pacific Oceans are supersaturated with *carbon disulfide* (CS_2) and are assumed to be a net source of emissions to the atmosphere (Kim and Andreae 1987, Xie and Moore 1999). Emitted CS_2 is produced by microbial and phytoplankton processes, as well as photochemical degradation of dissolved organic matter. *Carbonyl sulfide* (COS) is produced by oxidation of CS_2 and DMS, and directly by photochemical breakdown of dissolved organic matter. Globally, the COS emission rate is ~ 500 $Gg\ yr^{-1}$ (1 Gg = 10^9 g) (Suntharalingam *et al.* 2008), much of which is subsequently assimilated by terrestrial plants.

Through a combination of gas-phase and surface-phase chemistry, DMS and other emitted sulfur compounds are ultimately oxidized to H_2SO_4, which forms hydrophilic aerosol particles, some of which function as cloud condensation nuclei. One of the classic feedback hypotheses involving biosphere-atmosphere interactions is based on the assumption that marine DMS emissions, which are positively correlated to sea-surface temperature, can enhance cloud formation, which then reduces solar radiation, cools sea-surface temperature, and thus functions as a negative feedback, reducing further DMS emissions (Charlson *et al.* 1987, Ayers and Cainey 2007, Vallina and Simo 2007).

17.2 The biochemical production of BVOCs

BVOCs that are emitted from vegetation are the primary or secondary products of plant metabolism. Most models that have been designed to predict BVOC emission rates are based on known interactions among biochemical processes. For the case of isoprene, the rate of emission is tightly coupled to instantaneous chloroplast metabolism, including the CO_2 assimilation rate (Figure 17.1). In fact, most of the isoprene emitted by leaves is constructed in the chloroplast using carbon substrates produced from recent photosynthesis, as well as ATP and NADPH produced from photosynthetic electron transport. Using these known connections to photosynthesis, researchers have been able to develop models of the temperature-, PPFD-, and CO_2-dependencies of isoprene emission rate (Monson *et al.*

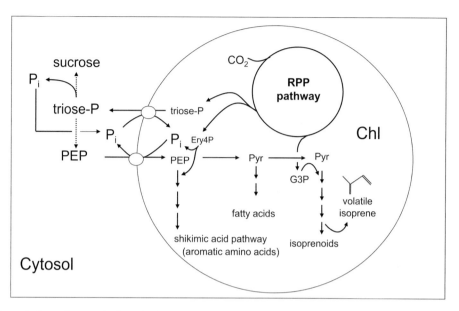

Figure 17.1 General scheme showing the flow of carbon from the photosynthetic RPP pathway in the chloroplast (Chl) to volatile isoprene. Two substrates are required for the 1-deoxy-D-xyulose-5-phosphate (DOXP) pathway that is responsible for the synthesis of isoprene: pyruvate (Pyr) which is derived from phosphoenolpyruvate phosphate (PEP) imported from the cytosol and the triose phosphate compound, glyceraldehyde 3-phosphate (G3P), which is a product of the RPP pathway. P_i = inorganic phosphate.

2012). The functions that connect isoprene emission rate (I_s) to temperature, PPFD, and CO_2 take the following general forms:

$$I_{sT} = f[I_{sb}, \exp^{c_1/RT_L}],$$

$$I_{sPPFD} = f\left[I_{sb}, \frac{\alpha c_2 I}{\sqrt{1 + \frac{\alpha^2 I^2}{c_3}}}\right],$$

$$I_{sCO_2} = f\left[I_{sb}, 1 - \frac{I_{smax} \; c_{ic}{}^h}{K_m + c_{ic}{}^h}\right], \tag{17.1}$$

where each of the equations is normalized to the *basal isoprene emission rate* (I_{sb}) which represents the rate of isoprene emission under a standard set of measurement conditions (typically $T_L = 30$ °C, PPFD = 1000 μmol m^{-2} s^{-1}, and $c_{ac} = 400$ μmol mol^{-1}). In the relations of Eq. (17.1), c_1, c_2, and c_3 represent scaling coefficients, I_{smax} represents the maximum isoprene emission rate, α represents the initial slope of the PPFD-dependence curve, I represents incident PPFD, and T_L represents leaf temperature. The basal isoprene emission rate is influenced by longer-term dynamics in the environment (e.g., those that occur at time scales of days-to-weeks).

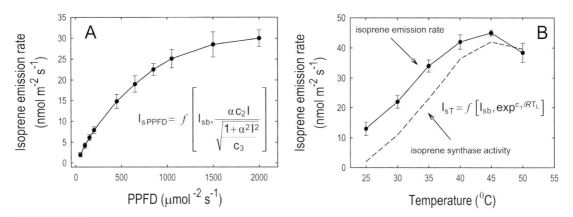

Figure 17.2 **A**. Relation between leaf isoprene emission rate and photosynthetic photon flux density (PPFD) for leaves of aspen trees. The modeled response is shown in general form as describing a rectangular hyperbola dependent on the base isoprene emission (I_{sb}), the initial slope of the light response (α), and the incident PPFD (I). **B**. Relation between leaf isoprene emission rate, the activity of the enzyme isoprene synthase, and temperature. The modeled response is shown in general form as describing an exponential (Arrhenius-type) relation dependent on I_{sb} and leaf temperature (T_L).

The temperature dependence of I_s exhibits a form similar to the Arrhenius equation, and it has been presumed to reflect the temperature dependence of isoprene synthase, the enzyme that catalyzes the final step in isoprene biosynthesis, except at leaf temperatures above 35–40 °C where substrate limitations (e.g., the rate of photosynthetic sugar production) begin to occur (Figure 17.2). The PPFD dependence reflects a rectangular hyperbola, similar to that developed for photosynthetic electron transport rate. Given the dependence of isoprene biosynthesis rate on the sugar products of photosynthesis, as well as NADPH and ATP generated from the photosynthetic electron transport system, it is not surprising that I_s reflects a dependence on PPFD that is similar to that for CO_2 assimilation rate. The one surprising relation, at least with regard to connections to photosynthesis, is the CO_2 response, which shows that I_s will decrease as c_{ic} increases; this is opposite to the relation between CO_2 assimilation rate and c_{ic}. The inhibition of I_s by c_{ic} is potentially due to CO_2 dependence in the transport of pyruvate equivalents into the chloroplast, which is required to produce substrates for the isoprene biosynthetic pathway (Rosenstiel *et al.* 2003).

The relationships presented in Eq. (17.1) have been developed independently of each other, and as a result provide no connection to integrated aspects of cellular carbon metabolism. Some researchers have developed models that provide a more synthetic framework for modeling isoprene emission, and these models have had some success in predicting the longer-term patterns that are difficult to evaluate using the relations in Eq. (17.1) (Niinemets *et al.* 1999, Grote *et al.* 2006, Harrison *et al.* 2012). However, it is currently not possible to validate the accuracy of these models because gaps persist in our knowledge of the specific biochemical interactions among cytosol and chloroplast processes.

Our ability to accurately model the emissions of monoterpenes and sesquiterpenes is even more limited than that for isoprene. Monoterpene emissions from broad-leaved trees tends to be light-dependent and linked to instantaneous metabolism, although monoterpenes can also be emitted at relatively high rates from stored pools contained in surface trichomes. In the case of Mediterranean oak trees, monoterpene emissions occur from chloroplasts in a manner similar to that for isoprene; in that case, the same models described in Eq. (17.1) for isoprene can be used to predict monoterpene emissions. Monoterpene emissions from coniferous trees often occur from stored reservoirs in resin ducts or surface "blisters." In these cases, the emission rate can be modeled on the basis of temperature and Henry's Law (see Section 4.2.2), the combination of which control the vapor pressure of the compound within the storage structure (Lerdau *et al.* 1997). There is some evidence that, contrary to conventional wisdom, even in coniferous trees a significant fraction of the emitted mono-terpenes are produced from the sugar products of recent photosynthesis, rather than from "older" stored carbon alone (Ghirardo *et al.* 2010); although exact control over the parti-tioning of substrates from "older" or "younger" pools of assimilated carbon has yet to be established.

Biochemical insight into the production of methanol and acetaldehyde by plants is also available, though we are not yet to a point of being able to accurately incorporate that insight into mathematical equations. Methanol is emitted in high amounts from expanding leaves and most likely originates from the demethylation of pectins (Figure 17.3). Pectins are a family of complex polysaccharides that are used in building the structural integrity of the primary cell walls of plants. Demethylation of pectins is required to produce the rigid cell wall matrix that includes embedded cellulose fibrils. The released methanol will partition

homogalacturonan

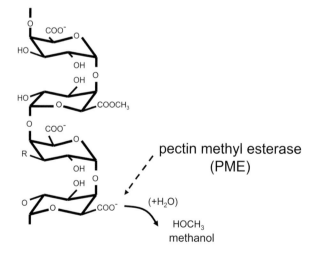

Figure 17.3 The role of pectin methyl esterase (PME) enzyme in catalyzing the release of methanol after hydrolysis of a methyl group from homogalacturonan, a complex pectin polysaccharide that is a principal component of the primary cell walls of plants.

into the aqueous environment of the cell wall water, and then equilibrate with air in the intercellular spaces of the leaf according to the prevailing leaf temperature and the relations of Henry's Law (see Section 4.2.2). Acetaldehyde is well-known as a product of fermentation during anaerobic conditions, and it has been shown to be transported from water-logged roots to leaves, followed by emission to the atmosphere (Kreuzwieser *et al.* 2004). However, acetaldehyde is also emitted from the leaves of non-flooded plants, especially during environmental transitions, such as from high to low light. In this case, imbalances in the pool sizes of respiratory metabolites can result in the emission of aerobic acetaldehyde (Karl *et al.* 2002). Finally, we know that when plants are exposed to abiotic and biotic stresses, including exposure to ozone, pathogen infection, or herbivory, key enzymes involved in cellular repair can produce a variety of oxygenated BVOCs (Heiden *et al.* 2003). Lipoxygenase enzymes catalyze the conversion of long-chain fatty acids found in damaged membranes into a variety of products, including six-carbon aldehyde compounds such as normal hexanal (n-hexanal) and cis-3-hexenal. Many of these six-carbon aldehyde compounds provide the "sweet" odor to injured vegetation, such as the odor detected from freshly mown grass. Thus, the emission of these compounds is the result of cellular repair through replacement of damaged biochemical constituents.

17.3 Emission of metabolic NH_3 and NO_2 from plants

Human activities, such as meat production, crop fertilization, and biomass burning, as well as direct emission from leaves, cause the emission of NH_3 and NO_2 to the atmosphere (Monson and Holland 2001). The production and assimilation of NH_3 and NO_2 in leaf cells is associated with photorespiration and nitrate assimilation. Whether leaves act as sources or sinks for atmospheric NH_3 and NO_2, is determined by the concentrations of each compound in the intercellular air spaces of the leaf (Sparks 2009). Thus, we can identify a leaf compensation point for these compounds, similar to the CO_2 compensation point (Lerdau *et al.* 2000, Sparks *et al.* 2001, Krupa 2003). As a general estimate, the intercellular air spaces of a C_3 leaf will contain NH_3 at ~ 2.5 nmol mol^{-1} during normal daytime metabolism (Farquhar *et al.* 1980a). General measurements of NH_3 concentrations in the atmosphere range from 0.2–15 nmol mol^{-1} (Roberts *et al.* 1988, Krupa 2003). Thus, a broad range of atmospheric conditions exist at which NH_3 would be assimilated by plants, and a relatively narrow range of conditions exist at which vegetation would act as a source for NH_3. In a detailed case study of a forest ecosystem, which receives air from natural forested landscapes with NH_3 concentrations of 0.2–0.3 nmol mol^{-1}, or nearby from agriculturally developed landscapes with NH_3 concentrations of 5–6 nmol mol^{-1}, Langford and Fehsenfeld (1992) found that the forest vegetation acted alternately as a source or sink for atmospheric NH_3 depending on wind direction; the forest-averaged NH_3 compensation point was ~ 1 nmol mol^{-1}.

Measurements of NO_2 uptake by isolated leaves have also revealed an apparent compensation point, with most estimates in the range 1–3 nmol mol^{-1} (Weber and Rennenberg 1996, Hereid and Monson 2001, Sparks *et al.* 2001). We might predict that vegetation is a source of NO_2 at sites with atmospheric concentrations below this range, and a sink at sites with concentrations above this range. Measurements of NO_2 exchange above forests, however, have revealed uptake of NO_2 in most atmospheric conditions, including some with concentrations as low as 0.2 nmol mol^{-1} (Jacob and Wofsy 1990, Rondon *et al.* 1993). This has led some to doubt the validity of the compensation points that have been measured in the past (Lerdau *et al.* 2000). Most models that are constrained by the need for mass balance in the atmospheric NO_x budget (which includes NO and NO_2) predict that plants should be an NO_2 sink at most places on earth.

17.4 Stomatal control over the emission of BVOCs from leaves

When BVOCs are produced and emitted from the mesophyll tissues of leaves, they exit the leaf through stomatal pores. In the case of some BVOCs which are produced within leaves, such as isoprene, however, there is no apparent change in the steady-state emission rate from the leaf, following a change in stomatal conductance (Figure 17.4). How can we account for this apparent paradox? In other words, if the isoprene molecules diffuse through the stomatal pores, how can we account for the observation that as the pores are forced to close, there is no change in the steady-state diffusion rate. An examination of a Fick's Law analog equation describing diffusive fluxes provides an answer: $F_{dI} = g_{sI}(c_{Ii} - c_{Ia})$ (see Section 6.1.4). In the case of the data in Figure 17.4, we know the response of two variables in this equation: stomatal conductance to isoprene diffusion (g_{sI}) must decrease in proportion to the observed stomatal conductance to water vapor (g_{sw}), but clearly the leaf isoprene flux (F_{dI}) remains constant. This means that the mole fraction gradient for isoprene from inside the leaf to the atmosphere (i.e., $c_{Ii} - c_{Ia}$) must increase in proportion to the decrease in g_{sI}. More directly, as the stomata close, and leaf mesophyll tissues continue to produce isoprene and release it to the intercellular air spaces of the leaf, the internal concentration of isoprene (c_{Ii}) increases, but the ambient concentration of isoprene (c_{Ia}) remains unchanged, and the cross-leaf gradient in isoprene concentration (Δc_I) increases. The internal buildup of isoprene will continue until a new steady state flux is reached. The upper limit on the flux will equal the metabolic rate of isoprene production in the leaf. *Thus, in the steady-state condition, and assuming no destruction of isoprene within the leaf, F_{dI} will always equal the metabolic rate of isoprene production, and g_{sI} and Δc_I will compensate for each other in proportional manner.*

Given this analysis, a logical question is why do the diffusive exchanges of CO_2 and H_2O decrease as g_s decreases? In other words, what is different about the exchanges of CO_2 and H_2O, compared to isoprene, that cause them to behave differently in their diffusive exchanges across leaf surfaces? In fact, there are other BVOCs, such as methanol, for which the diffusive flux from leaves also decreases in the face of decreasing g_s, similar to the case for CO_2 and H_2O and in contrast to the case for isoprene (Figure 17.4). One of two possible conditions can be invoked to explain a decrease in flux in response to a change in g_s: (1) the source or sink strength that ultimately determines the flux is affected by the change in g_s; or (2) the concentration gradient across the leaf (Δc) is incapable of responding to the change in g_s.

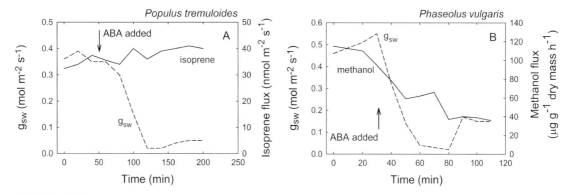

Figure 17.4 **A**. Responses of stomatal conductance to water vapor (g_{sw}) and isoprene flux from a leaf as a function of time following the addition of abscisic acid (ABA) to the leaf's xylem transpiration stream, which tends to force the stomata to close. **B**. Responses of g_{sw} and methanol flux from a leaf as a function of time following the addition of ABA to the leaf's xylem transpiration stream. Note the opposite responses, with isoprene flux not responding (or slightly increasing perhaps due to a small increase in leaf temperature), and methanol flux decreasing over time. Redrawn from Monson and Fall (1989) and Nemecek-Marshall *et al.* (1995).

For the case of a steady state, photosynthetic CO_2 flux, a decrease in g_s will cause CO_2 to diffuse across the leaf at a lower rate than the rate of metabolic assimilation of CO_2 from the intercellular air spaces. This will cause c_{ic} to decrease, and the metabolic sink strength for CO_2 will also decrease. The system will settle on a new, lower steady state CO_2 flux density, at a lower c_{ic}. Thus, in the case of the photosynthetic CO_2 flux, the first condition listed above explains the reduction in flux density as a function of reduced g_s.

For the cases of decreasing H_2O and methanol fluxes in response to a decrease in g_s, the second condition listed above provides an explanation. To understand this better, let us restate the conditions that allow isoprene flux to resist changes in g_s. In order for Δc_I to increase in the face of a decrease in g_{sI}, the concentration of isoprene in the intercellular air spaces of the leaf must exist below saturation. That is, "there must be an opportunity" for Δc_I to increase. Additionally, isoprene must be volatile enough that molecules will partition into the vapor phase of the intercellular air spaces after a perturbation to g_{sI}. If the air spaces were to be saturated with isoprene, holding as much as they could hold, then a reduction in g_{sI} could not be met with an increase in Δc_I, and F_{dI} would have to decrease; this is clearly not the case. Similarly, if the volatility of isoprene were low enough as to delay re-establishment of the steady state to well beyond the sampling frequency, then the system would be observed at a dynamic, non-steady state. In the latter case, the observer may falsely think that they are observing the system at the steady state, when in reality they are not. In the case for H_2O, the intercellular air spaces are saturated with respect to c_{iw}, and thus as long as leaf temperature remains constant, decreases in g_{sw} are accompanied by decreases in F_{dw}. In the case for methanol, the Henry's Law coefficient is so low that the system is effectively forced to a non-steady state during the sampling period of most flux observations, making it appear as though the methanol flux (F_{dm}), exhibits a steady-state decrease in response to a decrease in g_{sm} (at least in the short term). If given sufficient time to re-establish the steady state, the observed methanol flux should return to the same level as before perturbation to g_{sm}, and thus come back into balance with rate of synthesis. The observed responses of the diffusive flux for different BVOCs to changes in g_s should be reflected in their different Henry's Law coefficients (k_H); either responding like isoprene for those compounds with high k_H, or like methanol for those compounds with low k_H (Niinemets *et al.* 2004).

17.5 The fate of emitted BVOCs in the atmosphere

> Approximately 80% of our air pollution stems from hydrocarbons released by vegetation; so let's not go overboard in setting and reinforcing tough emission standards from man-made sources. (Ronald Reagan, Campaign Speech, 1980 *Sierra Magazine*, volume 65, page 7)

In the atmosphere, and in the presence of anthropogenic nitrogen oxide (NO_x) emissions, the oxidation of BVOCs leads to a variety of chemical products, including ozone and peroxyacetyl nitrate, two prominent tropospheric pollutants. In the past, this fact has been separated from its scientific context and used for political purposes, as evidenced in the quote above which was made by presidential candidate Ronald Reagan in 1980. In that case, he was using the importance of BVOCs in atmospheric chemistry to deflect potential governmental regulations against anthropogenic emissions of hydrocarbons; the implicit assumption being that natural processes are the source of most pollution, not people. Ronald Reagan was correct in noting the importance of BVOCs as a component in the production of tropospheric oxidants. However, he neglected to note that much of the NO_x that catalyzes the oxidation of hydrocarbons is indeed emitted from anthropogenic sources. People, and their associated economic enterprises, are indeed complicit in creating atmospheric pollution!

Once emitted to the atmosphere, BVOCs are oxidized through a complex series of gas- and liquid-phase reactions, which are capable of ultimately producing CO_2, if carried to completion (Figure 17.5). Many of the intermediate products produced during this chemistry, however, are lost through dry and wet deposition, or partitioned into aerosols, so that complete oxidation (to CO_2) only occurs for a fraction of the emitted BVOC mass. Some estimates place the percentage of emitted isoprene, for example, that is completely oxidized to CO_2, at 30% (Bergamaschi *et al.* 2000, Pfister *et al.* 2008). The gas-phase oxidation of BVOCs is initiated by reaction with hydroxyl radicals ($\cdot OH$). Hydroxyl radicals are formed from the photolysis of O_3 in the presence of solar radiation with wavelengths less than ~ 330 nm according to:

$$O_3 + hv(<330\ nm) \rightarrow O \cdot (^3P) + O_2, \qquad (17.2)$$

$$O_3 + hv(<330\ nm) \rightarrow O \cdot (^1D) + O_2. \qquad (17.3)$$

The products of O_3 photolysis are both *ground state singlet oxygen* ($O \cdot (^3P)$, reaction (17.2)) and *excited singlet oxygen* ($O \cdot (^1D)$, reaction (17.3)), although $\cdot OH$ is only formed from $O \cdot (^1D)$:

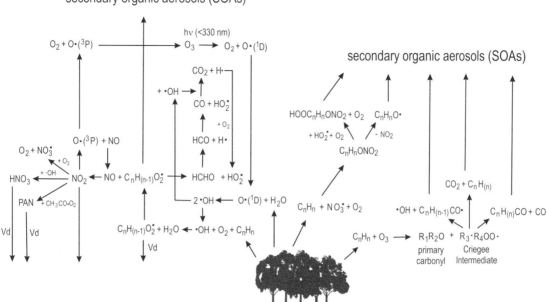

Figure 17.5 The complex chemical reactions that are associated with the oxidative photochemically driven breakdown of BVOCs emitted from vegetation. Principal products include O_3, organic nitrates, such as PAN, and secondary organic aerosols. Gas-phase chemistry involving hydroxyl radical is shown on the left-hand side of the figure. Oxidation by ozone and nitrate radical is shown on the right-hand side of the figure. From Monson (2010). V_d = deposition velocity.

$$O \cdot (^1D) + H_2O \rightarrow 2 \cdot OH. \qquad (17.4)$$

Reaction (17.4) provides the starting point for the gas-phase oxidation of BVOCs shown on the left-hand side of Figure 17.5.

Once formed, ·OH is capable of reacting with the most loosely associated electrons in organic compounds, such as the pi-bonded electrons of alkenes. The reaction is characterized by addition of ·OH to one of the double bonds, with the most favored addition being for terminal bonds. Following the addition of ·OH, O_2 is added, forming two peroxy radicals. The peroxy radicals, in turn are capable of reacting with nitric oxide (NO), if it is present at appropriate concentrations, producing nitrogen dioxide (NO_2). The NO_2 that is formed undergoes photolysis to form NO and ground-state singlet oxygen, $O\cdot(^3P)$. The $O\cdot(^3P)$ species has the potential to react with molecular oxygen (O_2) to form O_3; thus, in the presence of ·OH radicals and NO, the emission of BVOCs from ecosystems can sustain a catalytic cycle that generates significant quantities of O_3. In the absence of NO, peroxy radicals may react with other oxidants in the atmosphere (e.g., HO_2), in a catalytic process that generates additional ·OH. Thus, peroxy radicals may represent an important "buffering potential" in the atmosphere, allowing for the replenishment of ·OH following its initial consumption during the oxidation of alkenes (Taraborrelli *et al.* 2012).

In addition to reacting with ·OH, the double bonds of alkenes can be oxidized directly by O_3 through a reaction known as *ozonolysis*. Among terpenes, those with double bonds in cyclic rings, such as many of the monoterpenes, are most likely to participate in ozonolysis. During ozonolysis, O_3 reacts with pi-bonded electrons to produce a cyclic structure, known as a *Criegee Intermediate* (CI). The CI is capable of decomposing into a variety of products, many of which will be polar in their electrostatic character, and thus favored to dissolve into the liquid phase of aerosol particles. Isoprene is also capable of undergoing ozonolysis, although this reaction is considerably slower than that with ·OH (Atkinson and Carter 1984, Paulson *et al.* 1992). The rate coefficients for ozonolysis of alkenes, in general, tend to be low. However, tropospheric O_3 concentrations are high enough that, even with low rate coefficients, ozonolysis is a significant reaction mechanism in the gas-phase oxidation of BVOCs. One key difference between ozonolysis and reaction with ·OH is that ·OH is formed only during the day, when solar UV fluxes are high, and ·OH disappears relatively quickly after sunset. Ozone, however has a longer atmospheric lifetime, and its presence in the lower atmosphere is not dependent on diurnal transitions; thus, ozonolysis occurs during both the day and night. In addition to reacting with ·OH and O_3, alkenes are also capable of reacting with nitrate radicals (NO_3·). Nitrate radicals are formed from the reaction between O_3 and NO_2. NO_3 radicals are relatively unstable in the presence of UV radiation, and undergo rapid photodissociation during the day, so that they are most important as atmospheric oxidants at night (discussed in the next section).

The presence of plant canopies creates new complexities in understanding the oxidation of BVOCs. Canopies have the potential to modify the stoichiometries of BVOC-OH reactions because of the nature of the emitted BVOCs and the potential to create immense surfaces for the deposition of reaction intermediates. Vertical gradients of different species of reactive radicals within canopy air spaces can also create complexities in understanding photochemical dynamics. Some of the issues associated with reactive chemistry within the vicinity of canopies are presented in Box 17.1

17.6 Formation of organic secondary aerosol particles in the atmosphere

Liquid and solid particles ranging in size from a few nanometers to a few hundred micrometers are often small enough to remain suspended in the troposphere, and are often referred to as *aerosols*. Aerosols are large enough to scatter solar photons, and they therefore have an important influence on the earth's surface energy budget. Atmospheric aerosols can be introduced directly into the atmosphere (e.g., from the combustion of biomass or fossil fuels, or from windblown dust), in which case we refer to them as *primary aerosols*, or they can grow from smaller particles to form larger particles, in which case we refer to them as *secondary aerosols*. Secondary aerosols grow as compounds are photochemically processed in the troposphere, producing hydrophilic products that can partition into the liquid phase of the growing particles. The principal sources of secondary aerosol growth are the emission of sulfur compounds in marine ecosystems and the emission of BVOCs from terrestrial ecosystems. The potential for oxidized compounds to promote growth of existing aerosol particles is clear. It is less clear as to how effective such compounds are in nucleating particles. The chemical processing of compounds that partition into growing secondary aerosols does not cease once they are incorporated; heterogeneous reactions continue modifying compounds within the liquid phase of the aerosols themselves.

Compounds tend to partition into the gas or liquid phases of the atmosphere depending on temperature and volatility, according to Henry's Law. As BVOCs are progressively oxidized they become more polar (due to addition of oxygen), and less volatile. Oxidized BVOCs have a greater potential for entrainment into the liquid phase of growing secondary organic aerosol particles. The growth of secondary aerosol particles as emitted BVOCs are oxidized has been especially well-studied for the case of monoterpenes using *smog chambers*. Smog chambers are large vessels with minimally reactive surfaces into which compounds can be

---

Box 17.1   **Atmospheric chemistry within the canopy air space**

Within the air space of canopies, emitted BVOCs have the opportunity to react with atmospheric oxidants before being emitted into the lower atmosphere. Oxidation of emitted BVOCs within the canopy air space is catalyzed most often by $\cdot OH$, but can also be catalyzed by O_3 and $NO_3\cdot$. Observations above wet tropical forests have revealed that the chemical relations between emitted BVOCs and $\cdot OH$ are more complex than can be reconciled through conventional knowledge of O_x-HO_x-NO_x chemistry (Lelieveld *et al.* 2008, Taraborrelli *et al.* 2012). Conventional wisdom would suggest that in an unpolluted ecosystem, such as a tropical forest, BVOC emissions should reduce tropospheric $\cdot OH$ concentrations. However, observations by Lelieveld *et al.* (2008) showed that the reactions that occur between BVOCs and $\cdot OH$ have the potential to produce even more $\cdot OH$ than was present in the original pools of reactants. It is possible that reaction products, principally peroxy radicals, undergo secondary reactions, potentially with $HO_2\cdot$ that produce $\cdot OH$ at a greater rate than predicted for its initial consumption (Taraborrelli *et al.* 2012).

At forest sites in temperate latitudes, and with significant pollution by NO_x, the oxidation of BVOCs within and near the canopy can be driven to a greater extent by reaction with $NO_3\cdot$ radicals. For example, in an oak-hickory forest in Tennessee, which is impacted by NO_x deposition due to nearby automobile traffic, emitted terpene compounds are oxidized by $\cdot OH$ radicals above the forest canopy, where the daytime $\cdot OH$ production rate is high (Figure B17.1). However, deeper within the canopy, terpene losses due to oxidation by $NO_3\cdot$ increase. Even in this condition, $\cdot OH$ remains the primary oxidant reacting with emitted BVOCs, but the ratio of oxidation by $NO_3\cdot$ relative to $\cdot OH$ increases.

The chemistry that occurs within the air spaces of canopies has just begun to be explored and it is likely that new surprises will be revealed. The canopy air space provides unique opportunities for reactions to occur within "reaction systems" that are uncoupled from the more diffuse state of the well-mixed atmosphere. Canopy microenvironments create isolated reaction volumes, such as in areas of turbulent wake on the downwind side of tree boles (Crimaldi *et al.* 2008). The high degree of heterogeneity in the distribution of reactive species within canopies provides unique challenges in deciphering dynamics in the oxidative chemistry of plant-emitted BVOCs.

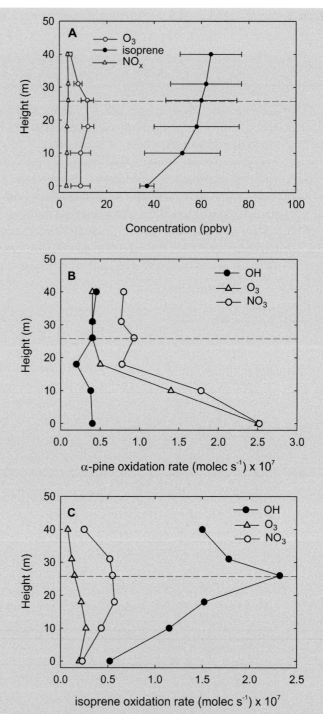

Figure B17.1 **A**. Vertical profiles of O_3, isoprene and NO_x concentrations measured within and above an oak-hickory forest. **B**. Vertical profile of modeled daytime oxidation rates of the monoterpene α-pinene (most likely emitted from pine needles) due to reaction with ·OH, O_3, or NO_3·. **C**. Vertical profile of modeled daytime oxidation rates of isoprene (most likely emitted from oak leaves) due to reaction with ·OH, O_3, or NO_3·. The horizontal dashed line indicates canopy height. Redrawn from Fuentes *et al.* (2007).

introduced at controlled rates, and from which air can be withdrawn for analysis of products. Some commonly emitted monoterpenes, such as the pinenes, can produce aerosol mass at up to 40% efficiency (mass of aerosol produced per unit mass of monoterpene oxidized). The capacity for isoprene, the BVOC emitted in highest amounts from global forests, to feed the growth of aerosol particles is not clear. Some smog chamber studies have suggested that the oxidation of isoprene can contribute to aerosol growth, but at low efficiencies (e.g., 3%; Kroll *et al.* 2006). Given that the total global isoprene emission flux is large, at ~ 500 Tg yr^{-1}, the annual growth of aerosol particles from isoprene oxidation is also large (~ 15 Tg yr^{-1}; Henze and Seinfeld 2006, Henze *et al.* 2008), despite low efficiencies. We emphasize that these estimates are based on studies from smog chambers. The efficiency by which isoprene oxidation can drive the growth of aerosol particles under natural conditions has been debated (Tsigaridis and Kanakidou 2007). The role of secondary organic aerosol formation in modifying the earth's radiation budget, and thus modeling of future climate trajectories is presented in Box 17.2.

Box 17.2   **The challenge of understanding global secondary organic aerosol production**

One of the greatest uncertainties in the current generation of global climate models is the role of atmospheric aerosols in modifying the surface radiation budget. Our lack of understanding in the formation of secondary organic aerosols is a key component of this uncertainty. In an inverse modeling study of global aerosol production, Goldstein and Galbally (2007) estimated that the annual increase in aerosol mass that would be required to accommodate photochemical oxidation rates using various models is 140–910 Tg yr^{-1}. When it was published, this estimate was considerably larger than those that had been reported from previous studies in which regional compound inventories were used to drive forward modeling studies; in this case, with most estimates of global aerosol mass production being in the range 10–70 Tg yr^{-1}. The gap between these estimate ranges was explained by Goldstein and Galbally with the assumption that BVOC emission inventories up to that time had grossly underestimated emission rates or poorly represented rates of certain key chemical reactions. The existence of this gap also had important implications for climate modeling. If we are forced to increase the rate of aerosol formation by 10-fold in earth energy budget models, then we must accommodate this increase by postulating higher sensitivity of surface warming to increases in greenhouse gas concentrations.

    The gap in the estimates of global aerosol production has not been closed, and it continues to represent an active area of research and debate. Our ability to represent in models the oxidative chemistry that drives aerosol growth has improved in recent years because of the continued study of gas- and liquid-phase chemistry under controlled smog chamber conditions (e.g., Varutbangkul *et al.* 2006, Lee *et al.* 2006). However, even in these studies, we have only made progress in our understanding of the immediate (first-generation) oxidation products that sustain aerosol growth. There exists a vast diversity of higher-generation reactions, beyond those involved in the initial oxidation, that have the potential to influence aerosol growth. Our understanding of these reactions is rudimentary, at best.

Appendix 17.1 Reactions leading to the oxidation of BVOCs to form tropospheric O_3

Photochemical oxidation of CH_4 and terpenes in the lower atmosphere is due to the reactive tendencies driven by O_x-HO_x-NO_x chemistry. The "x" in each of these compound formulas denotes that there are multiple forms capable of driving photochemical oxidation; the principal reactive compounds are O_2, O_3, $\cdot OH$, $HO_2\cdot$, NO, NO_2, and $NO_3\cdot$. For the purposes of this discussion, we will focus on oxidation driven by hydroxyl ($\cdot OH$) and hydroperoxy ($HO_2\cdot$) radicals in the unpolluted atmosphere, and by $\cdot OH$ and nitric oxide (NO) in the polluted atmosphere. As a starting point and in the interest of providing a complete description, we repeat the reactions that result in the formation of $\cdot OH$ radicals from the photolysis of O_3 in the presence of UV radiation and atmospheric humidity:

$$O_3 + h\nu(<330 \text{ nm}) \rightarrow O \cdot (^1D) + O_2, \tag{17.5}$$

$$O \cdot (^1D) + H_2O \rightarrow 2 \cdot OH, \tag{17.6}$$

where $h\nu$ is photon energy. The $O\cdot(^1D)$ species is an excited oxygen atom. Oxygen atoms have eight electrons, with six of those electrons located in orbital space within the valence energy shell. When two of those six electrons are unpaired in their orbitals, oxygen exists in the ground state, triplet form ($O\cdot(^3P)$). The potential energy contained in the unpaired electrons of $O\cdot(^3P)$ is thermodynamically insufficient to permit spontaneous reaction with H_2O. As a product of reaction (17.5), however, the unpaired electrons are forced into a momentary paired state, forming the unstable singlet oxygen species ($O\cdot(^1D)$). The potential energy level of $O\cdot(^1D)$ is higher than that of $O\cdot(^3P)$. Some of the additional energy in the $O\cdot(^1D)$ radical will be quenched when it collides with N_2 and O_2 atoms in the atmosphere, returning it to the more stable $O\cdot(^3P)$, ground state. Alternatively, some fraction of $O\cdot(^1D)$ will cause the formation of two $\cdot OH$ radicals if it collides with H_2O.

The principal gas-phase oxidation route for both alkanes, such as CH_4, and alkenes, such as isoprene, is through $\cdot OH$. In the case of CH_4 oxidation, $\cdot OH$ abstracts a carbon-bound H atom to form H_2O plus a methyl radical ($\cdot CH_3$) which reacts almost immediately with O_2 to form a methyl peroxy radical ($CH_3O_2\cdot$):

$$CH_4 + \cdot OH \rightarrow \cdot CH_3 + H_2O, \tag{17.7}$$

$$\cdot CH_3 + O_2 + M \rightarrow CH_3OO\cdot. \tag{17.8}$$

In the unpolluted atmosphere, the methyl peroxy radical is capable of reacting with the hydroperoxy radical, $HO_2\cdot$ to form methyl hydroperoxy (CH_3OOH) and molecular oxygen (O_2). Methyl hydroperoxy is capable of photolysis to form the methoxy radical ($CH_3O\cdot$) and $\cdot OH$, which in turn can be oxidized to produce formaldehyde (HCHO) and $HO_2\cdot$. Thus, this photo-oxidative sequence forms a complete catalytic cycle that begins and ends with $HO_2\cdot$, the hydroperoxy radical:

$$CH_3OO \cdot + HO_2 \cdot \rightarrow CH_3OOH + O_2, \qquad (17.9)$$

$$CH_3OOH + h\nu(<330 \text{ nm}) \rightarrow CH_3O \cdot + \cdot OH, \qquad (17.10)$$

$$CH_3O \cdot + O_2 \rightarrow HCHO + HO_2 \cdot. \qquad (17.11)$$

The formaldehyde formed from the oxidation of CH_4, can then be oxidized through further reaction with $\cdot OH$ or it can undergo photolysis when excited after absorption of photons at < 320 nm, in both cases producing carbon monoxide (CO):

$$HCHO + \cdot OH \rightarrow HCO + H_2O, \qquad (17.12)$$

$$HCHO + h\nu(<330 \text{ nm}) \rightarrow HCO + H, \qquad (17.13)$$

$$HCO + O_2 \rightarrow CO + HO_2 \cdot. \qquad (17.14)$$

The CO formed from the oxidation of formaldehyde can then be oxidized to form CO_2, thus completing the oxidation sequence that begins with CH_4 and ends with CO_2.

$$CO + \cdot OH \rightarrow CO_2 + \cdot H, \qquad (17.15)$$

$$\cdot H + O_2 \rightarrow HO_2 \cdot. \qquad (17.16)$$

Not all CH_4 will make it through this sequence; some of the intermediate compounds can be deposited back to the earth's surface, truncating their participation in further reactions.

In the polluted atmosphere, the oxidation of CH_4 can take a different path after the point of methyl peroxy radicals:

$$CH_3OO \cdot + NO \rightarrow CH_3O \cdot + NO_2, \qquad (17.17)$$

$$CH_3O \cdot + O_2 \rightarrow HCHO + HO_2 \cdot, \qquad (17.18)$$

$$HO_2 \cdot + NO \rightarrow NO_2 + \cdot OH. \qquad (17.19)$$

The series of reactions from (17.17) to (17.19) depends on the presence of NO at a critical concentration. The NO_2 formed from these reaction undergoes photolysis to form NO and ground-state singlet oxygen, $O \cdot (^3P)$. The $O \cdot (^3P)$ species has enough potential energy to react with molecular oxygen (O_2) to form O_3:

$$NO_2 + h\nu(<330 \text{ nm}) \rightarrow NO + O \cdot (^3P), \qquad (17.20)$$

$$O \cdot (^3P) + O_2 \rightarrow O_3. \qquad (17.21)$$

The NO produced in reaction (17.20) can then be recycled to react with additional methyl peroxy radicals (reaction (17.17)) to produce NO_2, thus re-initiating the reaction cycle. The $NO-NO_2$ cycle is capable of forming O_3, as long as the chemistry is sustained by available methyl peroxy radicals. Recognizing that the original reactant in this process was tropospheric O_3 (see reaction (17.5)), we can state: "it takes a little

O_3 to make a lot of O_3." We can construct a stoichiometric accounting of the potential to form O_3 as follows:

$$CH_4 + \cdot OH \, (+O_2) \rightarrow CH_3O_2 \cdot + H_2O, \tag{17.22}$$

$$CH_3O_2 \cdot + NO \rightarrow CH_3O \cdot + NO_2, \tag{17.23}$$

$$CH_3O \cdot + O_2 \rightarrow HCHO + HO_2 \cdot, \tag{17.24}$$

$$HO_2 \cdot + NO \rightarrow \cdot OH + NO_2, \tag{17.25}$$

$$2NO_2 + h\nu \rightarrow 2NO + 2O \cdot (^3P), \tag{17.26}$$

$$\underline{2O \cdot (^3P) + 2O_2 + M \rightarrow 2O_3 + M,} \tag{17.27}$$

$$\text{Net:} \quad CH_4 + 4O_2 + h\nu \rightarrow HCHO + 2O_3 + H_2O. \tag{17.28}$$

The summation given in Eq. (17.28) makes clear that hydrocarbons in the presence of NO drive a photochemically energized oxidation process capable of producing two O_3 molecules. Carrying the accounting through the oxidation of HCHO to CO_2 produces two more O_3 equivalents. Thus, the complete oxidation of CH_4 to CO_2 in the presence of sufficient NO has the potential to produce four O_3 molecules. Note that these processes do not result in the net production or loss of $\cdot OH$ or $HO_2 \cdot$; thus, this process is truly a "catalytic oxidation."

In the case of isoprene and other terpenes, the initial reaction with a hydroxyl radical is an *addition reaction*, which is fundamentally different than the *abstraction reaction* that occurs with CH_4. Using the case of isoprene as an example, $\cdot OH$ radical is added to a double bond, followed by the addition of O_2, to form an organic peroxy radical ($RO_2 \cdot$):

$$C_5H_8 + \cdot OH \rightarrow C \cdot C_4H_8OH, \tag{17.29}$$

$$C \cdot C_4H_8OH + O_2 \rightarrow COO \cdot \cdot C_4H_8OH. \tag{17.30}$$

In the unpolluted atmosphere, $RO_2 \cdot$ will react with $HO_2 \cdot$ in much the same manner as the methyl peroxy radical. Thus, during the oxidation of isoprene, formaldehyde will be formed, and some fraction will be ultimately oxidized to CO and CO_2, just as in the case for CH_4 oxidation. (Globally, most formaldehyde that is present in the atmosphere originates from the oxidation of non-methane BVOCs; a result of their greater cumulative emission rates and shorter lifetimes, compared to methane.) *In the polluted atmosphere*, the oxidation of organic peroxy radicals and organic alkoxy radicals will oxidize NO to form nitrogen dioxide (NO_2); once again, in a manner similar to CH_4 oxidation. The production of NO_2 can once again drive the formation of tropospheric O_3.

Stable isotope variants as tracers for studying biosphere-atmosphere exchange

> Rural air samples were all collected from the layer of air close to the ground … under circumstances where the metabolic activity of plants might be expected to influence the carbon dioxide composition of the air. This is so because plants exchange carbon dioxide with the atmosphere by means of respiration and assimilation and also because carbon dioxide is evolved from the ground through decay of organic material in the soil and respiration of plant roots … Thus the relationship between carbon isotope ratio and molar concentration observed for the carbon dioxide of rural air is explained if carbon dioxide is added to or subtracted from the atmosphere by plants or their decay products.
>
> Charles Keeling (1958), Scripp Institution of Oceanography

Of the 98 naturally occurring elements on earth, 18 are known to be radioactive, meaning that they exhibit time-dependent decay to lighter elements, and 80 are known to be stable. Of those 80 stable elements, 54 are known to exhibit isotopic variation, meaning that atoms within the same elemental category have different atomic masses due to variations in neutron number (Section 3.5). Given the dependency of diffusive flux on atomic mass, molecules of the same compound, but composed of different isotopes, will diffuse at different rates and thus segregate into isotopic fractions over time. Analysis of isotopic fractionation provides researchers with one of their most valuable tools for understanding rates of diffusive flux, enzyme-substrate interactions, interactions among metabolic pathways, and the sources and sinks of compounds used for various biogeochemical processes, even extending beyond those defined solely by diffusion. Of particular importance have been analyses of the isotopic composition of CO_2 and H_2O, given that C, H, and O are among those 54 elements that exhibit stable isotope variation.

Connections between the isotopic composition of atmospheric CO_2 and biogeochemical processes were recognized in studies conducted several decades ago by Charles Keeling, a geochemist at the California Institute of Technology. Keeling systematically collected air samples and used them to analyze changes in the isotopic state of the atmosphere above both terrestrial and marine ecosystems. Keeling not only pioneered the use of stable isotope analysis as a biogeochemical tool, but he made several discoveries about dynamics in the photosynthetic and respiratory activities of plants and soil in response to climate variation. As noted in the quote provided above from Keeling's early work, he detected a distinct influence of the metabolic activities of plants and soil microbes on the isotopic composition of atmospheric CO_2. These observations laid the foundation for our current understanding of biogeochemical dynamics in the carbon cycle – from processes at the scale of chloroplasts to budgets at the scale of the entire globe.

In this chapter we cover aspects of stable isotope fractionation by biogeochemical processes and discuss the application of stable isotope analysis to studies of CO_2 and H_2O

fluxes. We will begin our discussion with the biochemical and diffusive processes of photo-synthesis and their tendencies toward fractionation of the different isotopic forms of CO_2. Following its assimilation into the organic fraction of the biosphere, carbon can be released back to the atmosphere through respiration, which also has the potential for isotopic fractio-nation. Respiratory fractionation is not necessarily associated with enzyme active sites, and to understand its nature we need to also understand the potential for molecular structure and the systematic, but differential distribution of ^{13}C and ^{12}C in the organic molecules that serve as respiratory substrates. As the spatial scale at which we observe isotope ratios increases, the biochemical processes that influence isotope fractionation must be integrated into a broader perspective that includes atmospheric physics and micrometeorology. Thus, as we move beyond diffusive fractionation, we will consider the turbulent mixing of isotope fractions within the atmosphere and thus use the isotope fractions as tracers to understand how carbon and water move within the earth system. Finally, we will explicitly consider isotope fractio-nation during the evaporation of water from leaves and soil. While not quite as mature as the topic of carbon isotope fractionation, at least within the context of terrestrial ecosystems, analyses of plant-atmosphere exchanges of H_2O isotopes have more recently captured the attention of biogeochemists as another tool available for probing process interactions, espe-cially those involving coupling between the carbon and water cycles.

18.1 Stable isotope discrimination by Rubisco and at other points in plant carbon metabolism

Many years ago, Nier and Gulbransen (1939) reported the intriguing observation that the abundance of ^{13}C and ^{12}C varied in natural materials and Wickman (1952) reported variation of the $^{13}C/^{12}C$ ratio in different plant species. We now understand that much of the variation that was observed in these early studies was due to kinetic isotope effects that result from (1) the differential diffusion rates of $^{13}CO_2$ and $^{12}CO_2$ during photosynthetic CO_2 assimilation, *and* (2) the tendency for carboxylation enzymes to favor one isotopic form over the other as photosynthetic substrates. In C_3 plants, the carboxylation isotope effect is due to discrimination by the chloroplast enzyme Rubisco (see Section 3.5).

   The photosynthetic reaction catalyzed by Rubisco exhibits a relatively large kinetic isotope effect favoring the assimilation of $^{12}CO_2$ relative to $^{13}CO_2$, $\alpha \approx 1.030$ ($\Delta \approx 30‰$). The reaction sequence by which Rubisco catalyzes CO_2 assimilation is characterized by two sequential steps: (1) deprotonation of RuBP, one of the substrates used by Rubisco, to form the reactive enediolate; and (2) covalent bond formation between the enediolate and CO_2, the second substrate used by Rubisco. (Recall that the enediolate is an intermediate compound formed from RuBP within the active site of Rubisco, and which is highly reactive with CO_2 molecules that enter the active site; see Section 4.2.3.) The kinetic isotope effect, which favors the assimilation of $^{12}CO_2$, occurs during the second step in the reaction sequence; that involving the enediolate and CO_2 bond. The activation energy required for bond formation between $^{12}CO_2$ and the enediolate is lower than for $^{13}CO_2$; thus, reaction rates are faster for $^{12}CO_2$, leading to discrimination against $^{13}CO_2$. (It is common for lighter isotopic forms of compounds to be favored in the catalyzed formation of covalent bonds; see Section 3.5.) Variation has been observed among the Rubiscos of different organisms in the magnitude of this isotope effect. For example, the effect will be slightly larger if the covalent bond forms late in the process by which the enediolate is strained toward forming the transition state (Tcherkez *et al.* 2006); the dynamics of enediolate strain are determined to large extent by the amino acid composition and sequence in the active site of Rubisco. The range of discrimination for Rubisco observed among a diverse selection of photosynthetic organisms ranging from bacteria to vascular plants is as low as 20‰ and as high as 32‰ (Tcherkez *et al.* 2006). For vascular, C_3 plants, carbon isotope discrimination values by Rubisco extracted from leaves and observed in vitro have been reported to range from 27.4–30.3‰ (Christeller *et al.* 1976, Roeske and O'Leary 1984, Guy *et al.* 1993, McNevin *et al.* 2007). Many past treatments have used a value of 29‰ as typical of C_3 Rubisco carbon isotope discrimination; though Farquhar *et al.* (1989) argued that this value should be reduced by approximately 1.9‰ when placed into the context of a C_3 leaf because of the compensatory effects of carboxylation by the enzyme, PEP carboxylase, which is active at low levels in C_3 leaves. Thus, in most treatments, a value of 27‰ has been used for the biochemical discrimination in C_3 leaves. This can be considered the "maximum" leaf discrimination value.

The actual discrimination expressed by Rubisco in C_3 leaves (the in vivo discrimination) is considerably less than 27‰ because of diffusive limitations. In order for full discrimination to be expressed at the active site of Rubisco, any $^{13}CO_2$ that is subjected to discrimination must be free to diffuse away from the reaction site. The existence of a diffusive resistance that limits the potential for $^{13}CO_2$ to leave the reaction site will reduce biochemical discrimination. Furthermore, the diffusion of $^{13}CO_2$ and $^{12}CO_2$ into a leaf will increase fractionation, due to the fact that they diffuse at different rates. The combined effect of diffusive limitations into or out of a leaf will cause the actual discrimination observed during photosynthesis to be 14–20‰. A full mathematical treatment of the combined diffusive and biochemical fractionations involved in leaf photosynthesis will be covered in the next section.

Once assimilated by photosynthetic reactions, much of the carbon, now stored in sugars, is oxidized through respiration. Hexose phosphate sugars are oxidized initially in the reactions of glycolysis, and finally in the mitochondrial reactions of the TCA cycle. The product of several of these oxidative steps is CO_2. If we assume that the initial distribution of ^{13}C and ^{12}C among the various C atoms in photosynthetically produced sugars is random, *and* that there is no fractionation associated with respiratory processes, then there should be no difference in the $^{13}C/^{12}C$ ratio of sugars compared to respired CO_2. Thus, it was of interest when observations began to emerge showing that respired CO_2 was enriched in ^{13}C, relative to leaf sugars (Ghashghaie *et al.* 2003, Pataki 2005).

After careful study, it was discovered that the distribution of ^{13}C and ^{12}C in glucose molecules produced by photosynthesis is not random, as originally assumed. For example, the carbon at position 4 in the glucose molecule tends to be enriched in ^{13}C, relative to assimilated carbon, while the carbon at position 6 tends to be depleted in ^{13}C (Rossman *et al.* 1991, Tcherkez *et al.* 2004). The cause of this non-random distribution is likely discrimination that favors ^{13}C in the aldolase reaction of the RPP (C_3) pathway (Gleixner and

Schmidt 1997). It is the carbon at position 4 that is lost as CO_2 in the initial step of mitochondrial respiration (the conversion of pyruvate to acetyl-CoA), resulting in the release of ^{13}C-enriched CO_2; the two-carbon acetyl-CoA molecule that is left behind is depleted in ^{13}C. Following this initial step, acetyl-CoA, enters the TCA cycle, and is oxidized further. If all of the acetyl-CoA molecules were oxidized to CO_2, the isotope ratios of carbon entering glycolysis and respiration would balance, and there would be no overall enrichment in ^{13}C in respired CO_2. However, some of the acetyl-CoA is used in metabolic pathways that branch off from the TCA cycle, including those responsible for the biosynthesis of lipids and other complex compounds, such as lignin. The carbon used in these branched pathways will be depleted in ^{13}C, relative to respired CO_2. This means that overall there is an enrichment of respired CO_2 in ^{13}C and a depletion in compounds synthesized from acetyl-CoA. As we gain more knowledge about the $^{13}C/^{12}C$ fractionation patterns in the various photosynthetic and respiratory pathways of a leaf, we will be able to better constrain the metabolic models that we produce to describe carbon partitioning in C_3 plants.

18.2 Fractionation of stable isotopes in leaves during photosynthesis

The fractionation of $^{13}CO_2$ and $^{12}CO_2$ during photosynthetic CO_2 assimilation cannot be reconciled as the simple mathematical sum of serial fractionation by diffusion followed by enzyme catalysis. Rather, discrimination is weighted toward reflecting one or the other process depending on the degree to which each limits the rate of net CO_2 assimilation. A mathematical model that describes the relative contributions of both components in C_3 leaves is stated as (Farquhar et al. 1982):

$$^{13}\Delta = a + (b - a)c_{ic}/c_{ac}, \qquad (18.1)$$

where $^{13}\Delta$ is the leaf discrimination against $^{13}CO_2$, a is the discrimination component due to diffusive fractionation, and b is the discrimination component due to the catalytic assimilation of CO_2 through Rubisco. Equation (18.1) provides a weighting of diffusive and enzymatic fractionation scaled to the ratio of the intercellular CO_2 concentration (c_{ic}) to the atmospheric CO_2 concentration (c_{ac}). Thus, as c_{ic}/c_{ac} decreases, diffusional fractionation becomes more important in determining $^{13}\Delta$ and as c_{ic}/c_{ac} increases, enzymatic fractionation becomes more important. Equation (18.1) contains the implicit recognition that fractionation depends on the capacity for $^{13}CO_2$ molecules, which accumulate in the intercellular air spaces during photosynthesis, to diffuse back out of the leaf before being assimilated through carboxylation (Figure 18.1). As the capacity for this "back-diffusion" decreases, the capacity for Rubisco to discriminate one isotopic form from the other will also decrease. Stated another way, if g_{sc} is low relative to the rate of net CO_2 assimilation, then CO_2 uptake will be limited to a greater extent by diffusion, and catalysis by Rubisco will reflect less isotopic discrimination. As g_{sc} increases, c_{ic} will increase and move closer to c_{ac}, diffusive limitations to CO_2 uptake will be relaxed, $^{13}CO_2$ will diffuse away from the site of

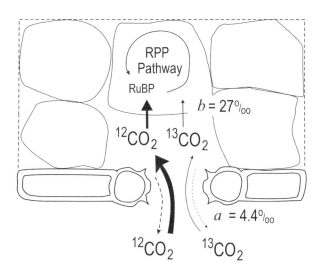

Figure 18.1 Representation of a leaf cross-section showing the diffusive and biochemical paths for the photosynthetic assimilation of $^{13}CO_2$ and $^{12}CO_2$. The diffusive and biochemical components of the leaf $^{13}CO_2$ discrimination are represented as a and b, respectively. Note that the c_{ic} and capacity for Rubisco to discriminate against $^{13}CO_2$ will depend on g_{sc} and the capacity for $^{13}CO_2$ that is subjected to discrimination to diffuse back out of the leaf prior to being assimilated by Rubisco. The ultimate carbon isotope ratio of the assimilated carbon will depend on the relative degrees to which diffusion (reflected in a) versus biochemistry (reflected in b) limit the rate of CO_2 assimilation.

discrimination at higher rates, and catalysis by Rubisco will reflect more discrimination. Thus, the degree to which the net CO_2 assimilation rate is limited by diffusion versus catalysis will be reflected in the c_{ic}/c_{ac} ratio of the leaf, which in turn is reflected in the $^{13}C/^{12}C$ ratio of biomass. General reviews of isotopic fractionation can be found in Rundel *et al.* (1989), Farquhar *et al.* (1989), O'Leary (1993), Lloyd and Farquhar (1994), and Bowling *et al.* (2008).

On the basis of this theory, we would expect correlations to exist among c_{ic}/c_{ac}, leaf carbon isotope discrimination and photosynthetic water-use efficiency (net CO_2 assimilation rate/transpiration rate; A/E_L) (Farquhar *et al.* 1989). Carbon isotope ratios are typically expressed in delta notation as:

$$\delta^{13}C_L = \delta^{13}C_a - a - (b-a)c_{ic}/c_{ac}, \tag{18.2}$$

where, in this case, $\delta^{13}C_L$ represents the $^{13}C/^{12}C$ ratio of the leaf biomass and $\delta^{13}C_a$ represents the $^{13}C/^{12}C$ ratio of the atmosphere, both relative to the commonly accepted Pee Dee Belemnite limestone standard. In leaves, a is estimated as 4.4‰, which reflects discrimination proportional to the ratio of the diffusivities of $^{13}CO_2$ and $^{12}CO_2$ in air, and b is taken as 27‰. The value of $\delta^{13}C_a$ is –8‰ for the well-mixed troposphere. The value of $\delta^{13}C_L$ typically varies between –20 and –30‰, for C_3 plant leaves with the less negative values reflecting lower c_{ic}/c_{ac} and higher diffusional limitations to CO_2 assimilation rate. The c_{ic}/c_{ac} ratio can be related to leaf photosynthetic water-use efficiency (A/E_L) according to:

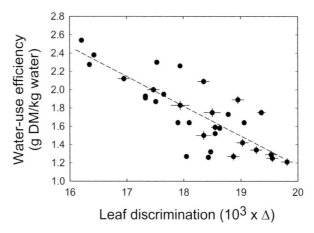

Figure 18.2 Observed relationship between the water-use efficiency of peanut plants, measured as the amount of dry mass (DM) gained per unit of water transpired, and the isotopic discrimination against $^{13}CO_2$ obtained from the $\delta^{13}C_L$. Data for different peanut genotypes have been pooled. Redrawn from Hubick *et al.* (1986), with permission from CSIRO Publishing.

$$\frac{c_{ic}}{c_{ac}} = 1 - \frac{1.58\, A\, \Delta c_w}{E_L c_{ac}}, \qquad (18.3)$$

where Δc_w represents the leaf-to-air water vapor mole fraction difference, A is the CO_2 assimilation rate, E_L is the leaf transpiration rate, and 1.58 represents the ratio of diffusivities for H_2O and CO_2 (i.e., K_{dw}/K_{dc}). In order to relate $\delta^{13}C_L$ to photosynthetic water-use efficiency we must recognize that $\delta^{13}C_L$ reflects the time-integrated assimilation of CO_2 over the life of the leaf (as well as any isotope effects due to the import of carbon from storage pools). According to Eqs. (18.2) and (18.3), leaves with higher lifetime photosynthetic water-use efficiencies should exhibit lower c_{ic}/c_{ac} ratios and correspondingly higher (less negative) $\delta^{13}C_L$. This prediction from theory has been supported by observations (Figure 18.2). Mathematical derivation of the carbon isotope fractionation models for leaves is available in Appendix 18.1.

18.3 Fractionation of the isotopic forms of H₂O during leaf transpiration

Water that moves through the soil-plant-atmosphere system is, for the most part, composed of the common isotopes of oxygen (^{16}O) and hydrogen (^{1}H); however, a small fraction is composed of different combinations of the heavier, but rarer, stable isotopes ^{18}O and ^{2}H. As water evaporates, two isotope effects will cause heavier molecules to leave the surfaces of a leaf more slowly and thus become concentrated in the liquid water that remains behind. A thermodynamic effect occurs whereby at equilibrium water will partition into the vapor phase at a lower $^{18}O/^{16}O$ ratio than exists in the liquid phase. A kinetic effect occurs because the binary diffusion rate through air of heavier water is less than that of lighter water. The combination of these effects has been incorporated into a theoretical framework known as the *Craig–Gordon model* (Craig and Gordon 1965):

$$R_E = \frac{1}{\alpha_k}\left(\frac{R_l}{\alpha^*}\frac{1}{(1-h)} - R_a\frac{h}{(1-h)}\right), \tag{18.4}$$

where R_E is the isotope ratio of evaporated water (e.g., $^2H_2O/H_2O$ or $H_2{}^{18}O/H_2{}^{16}O$); α_k represents the kinetic effect which is approximated as the ratio of diffusion coefficients K_{dl}/K_{dh} (where K_{dl} and K_{dh} are the diffusion coefficients for water containing the light versus heavy isotopes, respectively); α^* is thermodynamic fractionation, where $\alpha^* = R_l/R_v$, with R_l being the isotopic ratio (heavy/light) in liquid water and R_v being the ratio in water vapor; R_a is the isotopic ratio in water vapor immediately above the surface; and h is the relative humidity of the atmosphere above the surface. In Eq. (18.4), both α_k and α^* are greater than 1.

The Craig–Gordon model has been used to account for the enrichment of leaf water with regard to the heavier isotopes during transpiration. This is accomplished by algebraic rearrangement of Eq. (18.4) to solve for R_l, where R_l represents the isotopic ratio of liquid water at the site of evaporation. Recognizing that the mole fraction of water vapor in the leaf intercellular air spaces (c_{iw}) is saturated (i.e., $h = 1$), and assuming no difference in temperature between the leaf and the atmosphere, we can express the relative humidity of the air outside the boundary layer of the leaf as c_{aw}/c_{iw}. From this point, we can derive an expression analogous to Eq. (18.4), but with respect to mole fractions and solving for R_l as:

$$R_l = \alpha^*\left(\alpha_k\,R_E\left(\frac{c_{iw}-c_{aw}}{c_{iw}}\right) + R_a\left(\frac{c_{aw}}{c_{iw}}\right)\right). \tag{18.5}$$

Equation (18.5) has been further modified to account for the serial diffusive resistances provided by stomata and the leaf boundary layer (Flanagan *et al.* 1991) to provide:

$$R_l = \alpha^*\left(\alpha_k\,R_E\left(\frac{c_{iw}-c_{sw}}{c_{iw}}\right) + \alpha_{kh}\,R_E\left(\frac{c_{sw}-c_{aw}}{c_{iw}}\right) + R_a\left(\frac{c_{aw}}{c_{iw}}\right)\right), \tag{18.6}$$

where c_{iw}, c_{sw}, and c_{aw} are the mole fractions of water vapor in the leaf intercellular air spaces, at the outer surface of the leaf, and in the well-mixed atmosphere, respectively, and α_k and α_{kh} are the kinetic fractionation factors in still air and within the leaf boundary layer, respectively. (Kinetic fractionation in the turbulent boundary layer is slightly less than that in still air.) Mathematical derivation of the Craig–Gordon model is provided in Appendix 18.2.

18.4 Isotopic exchange of ^{18}O and ^{16}O between CO_2 and H_2O in leaves

Most of the CO_2 molecules that diffuse into a leaf are not assimilated by photosynthesis. Rather, they diffuse back out of the leaf and re-enter the atmosphere. This CO_2 flux is referred to as a "*retroflux*." Depending on the type of leaf and environmental conditions, the retroflux is between 50–70% of the inward flux (Gillon and Yakir 2001). Prior to leaving the leaf, retrofluxed CO_2 molecules may exchange an oxygen atom with H_2O in the leaf's chloroplasts, catalyzed by the enzyme carbonic anhydrase (Section 4.2.2). Each CO_2 molecule can undergo several sequential hydration reactions during its short residence in the leaf. During periods of evaporative fractionation, the leaf water that hydrates CO_2 will be enriched with ^{18}O compared to water that is conducted into the leaf from the xylem. As a result, retrofluxed CO_2 molecules are "tagged" with an $^{18}O/^{16}O$ ratio that reflects evaporative fractionation:

$$COO + H_2^{18}O \leftrightarrow CO^{18}O + H_2O. \tag{18.7}$$

In this reaction we recognize that the water and carbon fluxes of a plant intersect in a way that can be analyzed through the $^{18}O/^{16}O$ ratio of CO_2 molecules. As retrofluxed CO_2 molecules diffuse from the leaf they cause a local increase in the atmospheric concentration of $CO^{18}O$. This increase can be viewed as the result of a biophysical discrimination against $CO^{18}O$ that is dependent on, and proportional to, the net CO_2 assimilation rate. In order to develop this relation further, and in mathematical terms, we state the retroflux as:

$$F_r = g_{Lc} c_{cc}, \tag{18.8}$$

where F_r is the one-way flux of CO_2 from the chloroplast to the atmosphere (mol CO_2 m^{-2} s^{-1}), g_{Lc} is the total leaf conductance to CO_2 (mol air m^{-2} s^{-1}), and c_{cc} is the CO_2 mole fraction in the chloroplast (mol CO_2 mol^{-1} air). (It is important to note that the form of the diffusion equation for gross fluxes differs from that for net fluxes, in that the gross flux is driven not by the difference in concentration of molecular collisions between two points, but rather by the absolute concentration of molecular collisions at one point.) The net assimilation rate of CO_2 (A) can be expressed in terms of binary diffusion, derived from Fick's Law, as:

$$A = - g_{Lc} (c_{ac} - c_{cc}), \tag{18.9}$$

where c_{ac} is the CO_2 mole fraction in the well-mixed atmosphere outside the leaf. Using Eq. (18.9) to substitute for g_{Lc} in Eq. (18.8), we obtain an expression for the retroflux in terms of A:

$$F_r = - \frac{c_{cc}}{(c_{ac} - c_{cc})} A. \tag{18.10}$$

Plants with the C_3 and C_4 photosynthetic pathways differ in the degree to which retrofluxed CO_2 equilibrates with the isotopic composition of leaf water; this is due to differences in leaf carbonic anhydrase activity and in leaf residence time (which scales with A; see Gillon and Yakir 2000, 2001). In general, C_4 grasses exhibit less potential for CO_2/H_2O isotopic equilibrium than C_3 grasses. Given that the coverage by C_4 grasslands in subtropical latitudes is significant, C_3/C_4 differences in leaf-scale isotopic equilibrium can have large effects on global trends in the isotopic composition of atmospheric CO_2. The overall global mean fraction of leaf isotopic equilibrium between CO_2 and H_2O is estimated to be 78%

(Gillon and Yakir 2001), which is lower than the value of 100% assumed in some previous studies (Farquhar *et al.* 1993, Cias *et al.* 1997).

18.5 Assessing the isotopic signature of ecosystem respired CO$_2$ – the "Keeling plot"

In the late 1950s, Charles Keeling recognized that the mathematical relation between CO$_2$ concentration and the $^{13}C/^{12}C$ or $^{18}O/^{16}O$ ratios of atmospheric CO$_2$ is described by a reciprocal linear relation:

$$\delta^{13}C_a = I_{13C} + M \ (1/c_{ac}), \qquad (18.11)$$

where $\delta^{13}C_a$ is the isotope ratio of $^{13}C/^{12}C$ in atmospheric CO$_2$ written in delta notation (see Section 3.5), I_{13C} is an empirical coefficient that represents the *y*-intercept of the regression, M is an empirical coefficient that represents the slope of the regression, and c_{ac} is the atmospheric CO$_2$ mole fraction. At the time of his studies, the CO$_2$ mole fraction of the well-mixed atmosphere was 314 μmol mol^{-1} and Keeling measured a relatively constant, global value for $\delta^{13}C_a$ of −7‰. (It is informative for students to ask themselves why the value of $\delta^{13}C_a$ decreased from −7‰ to less than −8‰ during the period between 1958 and the present.) In contrast to the constancy that Keeling observed for $\delta^{13}C_a$ at 314 μmol mol^{-1}, he noted that the value for I_{13C} (i.e., as $1/c_{ac} \rightarrow 0$), was different for the air above different ecosystems and varied as a function of time of year. Keeling reasoned that the isotopic ratio of CO$_2$ in air measured at any instant in time and for any point in space ($\delta^{13}C_a$) must reflect an isotopic mixture concocted from two CO$_2$ sources; the well-mixed background (tropospheric) air ($\delta^{13}C_b$ at $\sim$ −7‰) and the CO$_2$ added from respiration sources within an ecosystem ($\delta^{13}C_s$ generally < -20‰).

We can formalize Keeling's conclusions according to the requirement for conservation of mass in the total CO$_2$ and isotopic mole fractions, respectively, as:

$$c_{ac} = c_{bc} + c_{sc}, \qquad (18.12)$$

$$\delta^{13}C_a c_{ac} = \delta^{13}C_b c_{bc} + \delta^{13}C_s c_{sc}, \qquad (18.13)$$

where c_{ac}, c_{bc}, and c_{sc} refer to the CO$_2$ mole fractions of the atmosphere observed within the vicinity of the ecosystem, the background CO$_2$ of the well-mixed troposphere, and the CO$_2$ added from an ecosystem respiration source, respectively (Pataki *et al.* 2003). Thus, as CO$_2$ is added to the background atmosphere from a respiratory source, it will change the isotopic composition of the air in a way that can be described by a linear mixing model:

$$\delta^{13}C_a = c_{bc}(\delta^{13}C_b - \delta^{13}C_s) \ (1/c_{ac}) + \delta^{13}C_s. \qquad (18.14)$$

With this theory in place, one can measure variations in the isotope composition and concentration of CO$_2$ near the surface of an ecosystem and determine the $\delta^{13}C$ of

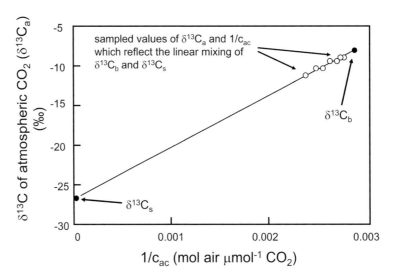

Figure 18.3   A sample nighttime Keeling plot showing the linear regression used to define the two end-member mixing pattern of $\delta^{13}C_s$ and $\delta^{13}C_b$ as a function of $1/c_{ac}$. The y-intercept ($\delta^{13}C_s$) represents the $^{13}C/^{12}C$ ratio of the CO_2 source at the lower boundary, and is typically accepted as the $\delta^{13}C$ of total ecosystem-respired CO_2. This intercept is often designated as $\delta^{13}C_R$, rather than $\delta^{13}C_s$.

ecosystem-respired CO_2; equivalent to the I_{13C} of Keeling's original analysis. This type of analysis is best conducted at night, when surface-atmosphere CO_2 exchange is restricted to respiration. The graphical regression analysis that characterizes CO_2 according to this theory is called a *Keeling plot* (Figure 18.3).

Use of the Keeling plot has allowed biogeochemists to understand the coupling of climate gradients to ecosystem carbon cycling. As discussed above, for the leaf scale, it has been established that increased moisture limitation, either through decreased soil moisture or decreased atmospheric humidity, causes decreased discrimination against $^{13}CO_2$ (Section 18.2). Across spatial and temporal gradients of precipitation, it has been shown that the lower c_{ic}/c_{ac} ratio expected for plants from the drier end of the gradient is correlated with a greater $\delta^{13}C$ value for the y-intercept of the Keeling plot; thus, reflecting less discrimination against $^{13}CO_2$ during the photosynthetic assimilation of carbon that subsequently supports ecosystem respiration (Bowling *et al.* 2002, Pataki *et al.* 2003). In this case, we are able to couple processes at the leaf scale with landscape-scale impacts on the local isotopic composition of atmospheric CO_2.

18.6 The influence of ecosystem CO_2 exchange on the isotopic composition of atmospheric CO_2

As is clear in the examples discussed above, dynamics in the isotope composition of atmospheric CO_2 have become valuable diagnostics by which to evaluate the influences

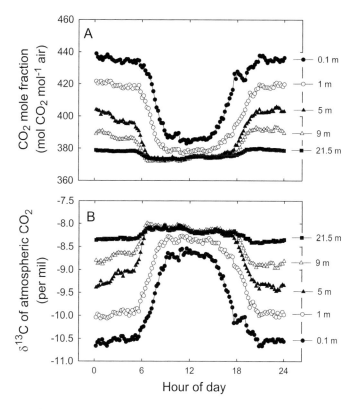

Figure 18.4 **A.** Diurnal variation in the CO_2 mole fraction in air at various heights above a subalpine forest ecosystem in Colorado. **B.** Diurnal variation in the isotopic ratio ($^{13}C/^{12}C$) expressed in delta notation, for various heights above a subalpine forest in Colorado. Redrawn from Bowling *et al.* (2005).

of climate on ecosystem carbon cycling. Immediately above a terrestrial ecosystem, and assuming the lack of isotopic influences from neighboring areas, photosynthetic CO_2 assimilation will cause enrichment in atmospheric $^{13}CO_2$ during the day, and respiratory CO_2 loss will cause depletion during the night (Figure 18.4). Theoretically, if the daily photosynthetic and respiratory CO_2 fluxes were tightly coupled, the mean diurnal $^{13}C/^{12}C$ ratio of air in the vicinity of the ecosystem $\left(\bar{\delta}^{13}C_a \right)$ should not change; enrichment with ^{13}C during the day should be exactly balanced by depletion with ^{13}C at night. This condition can be developed from the principle of continuity (see Section 6.3) as:

$$\frac{d\rho_m c_j}{dt} = \pm \frac{\partial S_{bj}}{\partial z}, \qquad (18.15)$$

where c_j is the mole fraction of atmospheric constituent j, ρ_m is the molar density of air (mol air m^{-3}), and S_{bj} represents biological sources or sinks of constituent j within a defined control volume. When integrated across the height of the control volume (to reference height z_r) we can write:

$$\int_0^{z_r} \frac{d\rho_m c_{ac}}{dt}\, dz = \pm \int_0^{z_r} \frac{\partial \overline{S}_{bj}}{\partial z}\, dz = \int_0^{z_r} \frac{\partial\, (\overline{A} + \overline{R}_E)}{\partial z}\, dz = \overline{F}_A + \overline{F}_R,$$

(18.16)

Term I   II   III   IV

where time-dependent divergence in the atmospheric CO_2 concentration is integrated between height 0 and z_r (Term I), and is balanced by the time-averaged biological sink or source activity (Term II), which in this case is equal to the vertically integrated and time-averaged net photosynthetic ($\overline{A}$) or ecosystem respiratory ($\overline{R}_E$) molar fluxes (Term III), which for convenience will be designated as $\overline{F}_A$ and $\overline{F}_R$, respectively (Term IV), where over-bars refer to time averaging. Thus:

$$\int_0^{z_r} \frac{d\rho_m c_{ac}}{dt}\, dz = \overline{F}_A + \overline{F}_R.$$

(18.17)

When expanded to account for the mole fractions of ^{13}C and ^{12}C in CO_2, and considering the case for integration with respect to time across a 24-hour cycle, we can write:

$$\int_0^{z_r}\int_0^{24h} \frac{d\rho_m c_{ac} \bar{\delta}^{13} C_a}{dt}\, dt\, dz = \bar{\delta}^{13} C_A\, \overline{F}_A + \bar{\delta}^{13} C_R \overline{F}_R$$

$$= \left(\bar{\delta}^{13} C_a - \bar{\Delta}_E\right)\overline{F}_A + \bar{\delta}^{13} C_R\, \overline{F}_R = 0,$$

(18.18)

where $\bar{\delta}^{13} C_a$ is the mean diurnal $^{13}C/^{12}C$ ratio of air in the vicinity of the ecosystem, $\bar{\delta}^{13} C_A$ represents the daily $^{13}C/^{12}C$ ratio of the photosynthetic products provided by the canopy, $\bar{\delta}^{13} C_R$ is the daily time-averaged $^{13}C/^{12}C$ ratio of the ecosystem respired CO_2, and $\bar{\Delta}_E$ is the time-averaged photosynthetic discrimination against ^{13}C provided by the ecosystem. Equation (18.18) states the condition of *isotopic equilibrium* between the 24-hour integrated photosynthetic and respiratory components of the ecosystem, which occurs when $\overline{F}_A = \overline{F}_R$ and $\bar{\delta}^{13} C_A = \bar{\delta}^{13} C_R$; and thus the mean diurnal change in the $^{13}C/^{12}C$ ratio of air in the vicinity of the ecosystem equals zero.

Isotopic equilibrium is not achieved in real ecosystems, even when $\overline{F}_A = \overline{F}_R$. This is due to the fact that each unit mass of CO_2 that is photosynthetically removed from the atmosphere at one isotope ratio is replaced by a unit mass of respired CO_2 that exists at a different isotope ratio; in other words, $\bar{\delta}^{13} C_A \neq \bar{\delta}^{13} C_R$. The result is *isotopic disequilibrium*. The isotopic ratio of CO_2 produced through soil respiration is affected by the age of soil organic matter (SOM) pools that serve as substrates and by isotopic fractionation that occurs during respiration. Some of the SOM that contributes to soil respiration is many decades old. Much of the carbon in this "old" SOM was assimilated prior to the Industrial Revolution; prior to the large-scale emission of fossilized carbon from coal and oil with their inherently low $^{13}C/^{12}C$ ratios (due to ancient photosynthetic discrimination). Because of these emissions, both CO_2 in the atmosphere and the SOM produced from autotrophic biomass have become isotopically "lighter," since the Industrial Revolution. Microbial use of older SOM substrates will result in respired CO_2

that is isotopically heavier than CO_2 produced from younger substrates. The degree of isotopic disequilibrium due to photosynthetic and respiratory activities will be influenced by the relative rates of microbial use of SOM pools of different ages. At disequilibrium the left-side term of Eq. (18.18) will not equal 0, and the resulting time-dependent change in the $^{13}CO_2/^{12}CO_2$ ratio is called the *disequilibrium flux*. The disequilibrium flux is considered for a 24-hour integral in Eq. (18.18), but it can also be defined for the longer time scales. The presence of an isotopic disequilibrium means that the relative concentrations of carbon isotopes in the CO_2 of the atmosphere could change, despite no change in overall CO_2 mole fraction. The effect of differential oxidation of SOM pools, and its influence on isotopic disequilibrium, must be taken into account for accurate quantification of the effect of soil respiration on the $^{13}CO_2/^{12}CO_2$ composition of the atmosphere (Fung *et al.* 1997). Past efforts have relied on models that provide turnover times for each SOM pool separately. Once resolved, differential turnover times can be coupled to predicted changes in the $^{13}CO_2/^{12}CO_2$ ratio of atmospheric CO_2, and thus estimation of the disequilibrium flux (Ciais *et al.* 1999, Scholze *et al.* 2008).

The concept of isotopic equilibrium is illustrated at the global scale in Figure 18.5. A decrease in the $\delta^{13}C$ ratio of the background atmosphere due to the combustion of fossil fuels over the past two centuries is commonly referred to as the *Suess effect* (Keeling 1979, Tans *et al.* 1979). Over this same time, the $\delta^{13}C$ ratio of photosynthate would have tracked that of the atmosphere, but with an offset due to photosynthetic discrimination of ~ 18‰. If isotopic equilibrium were to exist, then changes in the $\delta^{13}C$ of photosynthate and respiratory CO_2 should track one another in equal proportion. In the case of isotopic disequilibrium, the change in the $\delta^{13}C$ of respired CO_2 should be lagged by several years from the change in the $\delta^{13}C$ of photosynthate, depending on the globally averaged residence time of SOM. In the analysis of Figure 18.5, we have presented two alternative residence times for global SOM (τ): 18 years (after Randerson *et al.* 1999) and 27 years (after Yakir 2004). The effect of an increase in residence time is seen as causing an increase in the lagged effect of isotopic disequilibrium; the isotopic disequilibrium with $\tau = 18$ years (D_{18}) is estimated as 0.4‰, whereas with $\tau = 27$ years (D_{27}) it is estimated as 0.6‰. This small difference underscores the importance of accuracy in estimating the global disequilibrium flux; an estimate that is highly dependent on our understanding of global patterns and rates in SOM decomposition.

Isotopic disequilibrium is also affected by enrichment in ^{13}C that occurs in SOM during progressive mineralization; this is often observed as a 1–3‰ increase in $\delta^{13}C$ of SOM as a function of depth in forest soils (Nadelhoffer and Fry 1988) (Figure 18.6). Deeper SOM is also older, so some of the increase in $\delta^{13}C$ can be attributed to CO_2 assimilation in a pre-industrial atmosphere. However, this can only account for ~ 1.4‰ of observed increases. The residual difference may be due to: (1) the fact that plant biomass tends to be isotopically lighter than microbial biomass, SOM is composed of progressively higher fractions of microbial biomass as it is processed through time, and the mean age of SOM in deeper soil horizons is older than that for shallower horizons; and/or (2) microbes discriminate against ^{13}C during SOM oxidation (Ehleringer *et al.* 2000, Torn *et al.* 2002). These natural fractionations that occur during SOM mineralization cannot account for the Suess effect; such differential use would have occurred prior to, as well as after, the Industrial Revolution, so there is no basis to explain shifts in the $^{13}C/^{12}C$ ratio of atmospheric CO_2 during the past several decades. However, the

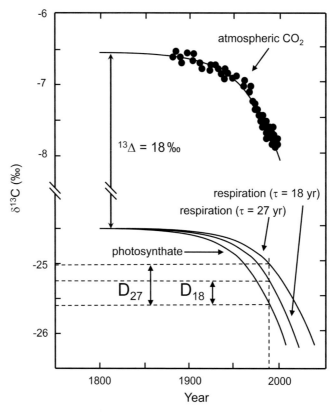

Figure 18.5 Graphic representation of the isotopic disequilibrium between the CO_2 assimilated by photosynthesis and emitted by respiration. The upper line and symbols (for the historical trend in the $\delta^{13}C$ of atmospheric CO_2) is taken from Francey *et al.* (1999). We assume a photosynthetic discrimination of 18‰, from which we obtain an offset curve labeled here as "photosynthate." When the photosynthate curve is offset by 18 years or 27 years, representing estimates of the global mean residence time of assimilated CO_2, we obtain the curves labeled "$\tau = 18$ yr" and "$\tau = 27$ yr." The global mean isotopic disequilibrium that occurs because respiration is offset by 18 or 27 years from photosynthesis is shown as D_{18} and D_{27}, respectively. Redrawn from a concept presented by Yakir (2004), with kind permission from Springer Science + Business Media.

presence of this SOM-dependent fractionation does make it more difficult to assess the Suess effect as it is carried forward in time. (By this point in the discussion students should have already recognized the answer to the question posed above as to why the $\delta^{13}C_a$ observed by Keeling in 1958 as –7‰ has decreased to its current value of less than –8‰.)

An additional isotopic tool that can be used to study the turnover rates of SOM involves the quantification of ^{14}C (Trumbore 2000). In SOM with residence times of centuries to millennia, age can be determined by the rate of radioactive decay of ^{14}C, which exhibits an isotopic half-life of ~ 5700 years. However, ^{14}C abundance can also be used in a slightly different way, permitting the estimation of turnover times for much younger SOM pools. During the period 1959–1963, ^{14}C was introduced into the atmosphere through above-ground atomic bomb testing. This pulse of ^{14}C was assimilated into the biosphere through

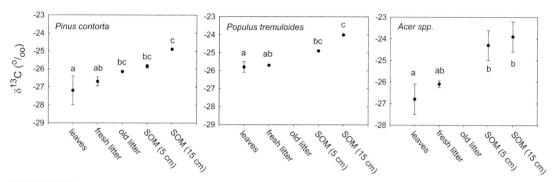

Figure 18.6   Enrichment in ^{13}C in soil litter and SOM at deeper layers of the soil profile in three ecosystems in northern Utah, USA. Vertical bars are $\pm$ 1 S.E. and points marked by different letters are statistically significant. Redrawn from Buchmann *et al.* (1997), with kind permission from Springer Science + Business Media.

photosynthesis and can be used as a tracer to estimate carbon compound turnover times as it moves through pools of leaves, litter, and SOM (Goh *et al.* 1976). In the soil of a deciduous oak-maple forest in the northeastern USA, for example, the ^{14}C content of various SOM pools was used to estimate turnover times of 2–5 years for leaf litter, 5–10 years for root litter, 40–100 years for low-density SOM, and > 100 years for mineral-associated SOM (Gaudinski *et al.* 2000). In the same study, ^{14}C was used to show that CO_2 derived from SOM that was older than one year represented ~ 40% of the respiration flux.

Appendix 18.1   Derivation of the Farquhar *et al.* (1982) model and augmentations for leaf carbon isotope discrimination

The following derivation of a model to relate the $^{13}CO_2/^{12}CO_2$ ratio of photosynthetically assimilated carbon to the time-averaged c_{ic}/c_{ac} ratio, and thus photosynthetic water-use efficiency, is based on that presented in Farquhar *et al.* (1982). The rate of $^{12}CO_2$ diffusion into a leaf can be represented as:

$$A = g_{sc}(c_{ac} - c_{ic}). \tag{18.19}$$

The rate of $^{12}CO_2$ carboxylation by Rubisco can be represented as:

$$A = k\, c_{ic}, \tag{18.20}$$

where k is a composite biochemical "transfer coefficient" that we will use for the moment to hold the processes true to form, but which will be eventually cancelled out. Substituting A/k for c_{ic} in Eq. (18.20), and using algebra, we obtain:

$$A = \frac{k\, g_{sc}}{k + g_{sc}} c_{ac}. \tag{18.21}$$

We can write the same equations for the diffusion of $^{13}CO_2$ into a leaf and subsequent carboxylation, but using "primes" to distinguish from $^{12}CO_2$:

$$A' = g'_{sc}(c'_{ac} - c'_{ic}); \qquad A' = k'c'_{ic}; \qquad A' = \frac{k'g'_{sc}}{k' + g'_{sc}}c'_{ac}. \tag{18.22}$$

Taking into account the discrimination against $^{13}CO_2$, relative to $^{12}CO_2$ by diffusion and Rubisco carboxylation, respectively, we can relate g'_{sc} and k' to g_{sc} and k according to:

$$g'_{sc} = g_{sc}(1 - a/1000), \tag{18.23}$$

$$k' = k(1 - b/1000), \tag{18.24}$$

where a is the steady-state fractionation of $^{13}CO_2$, relative to $^{12}CO_2$, due to diffusion through air, and b is the steady-state fractionation of $^{13}CO_2$, relative to $^{12}CO_2$, due to discrimination by the active site of Rubisco; both in units of parts per thousand. Using all of these relationships we can write:

$$\frac{A'/A}{c'_{ac}/c_{ac}} = \frac{k'(1 - a/1000) + g'_{sc}(1 - b/1000)}{k' + g'_{sc}}. \tag{18.25}$$

The overall discrimination against $^{13}CO_2$ relative to $^{12}CO_2$ ($^{13}\Delta$) can be written as:

$$^{13}\Delta = \left(1 - \frac{A'/A}{c'_{ac}/c_{ac}}\right) \times 1000, \tag{18.26}$$

and using Eq. (18.25) for substitution we can write:

$$^{13}\Delta = \frac{k'a + g'_{sc}b}{k' + g'_{sc}}. \tag{18.27}$$

Recognizing that $g'_{sc} = A'/(c'_{ac} - c'_{ic})$ and $A' = k'c'_{ic}$, we can write $g'_{sc} = k'c'_{ic}/(c'_{ac} - c'_{ic})$, and after substituting into Eq. (18.27), and recognizing that $c'_{ic}/c'_{ac} \approx c_{ic}/c_{ac}$, we obtain:

$$^{13}\Delta = a + (b - a)\, c_{ic}/c_{ac}. \tag{18.28}$$

The atmosphere has a $\delta^{13}C$ value that is approximately 8‰ more negative than the PDB standard that is typically used as the reference point for $^{13}C/^{12}C$ ratios, and so the isotopic ratio of the leaf, relative to the PDB standard is expressed as:

$$\delta^{13}C_L = \delta^{13}C_a - a - (b - a)c_{ic}/c_{ac}. \tag{18.29}$$

It should be recognized that the isotopic composition of the leaf is the result of assimilation processes over the entire life of the leaf, as well as any influences due to the import of carbon from other parts of the plant. Thus, the relevant c_{ic}/c_{ac} in Eq. (18.29) is the time-integrated value determined for the entire life of the leaf, and for other assimilatory organs that transport carbon into the leaf.

The relationship between c_{ic}/c_{ac} and photosynthetic water-use efficiency can be appreciated if we state:

$$A = g_{sc}(c_{ac} - c_{ic}),$$
$$E_L = 1.58\ g_{sc}\ (\Delta c_w). \tag{18.30}$$

Using the definition of photosynthetic water-use efficiency as A/E_L, we can state:

$$\frac{A}{E_L} = \frac{g_{sc}c_{ac} - g_{sc}c_{ic}}{1.58\ g_{sc}(\Delta c_w).} \tag{18.31}$$

Using algebra to rearrange Eq. (18.31), we can write:

$$1 - \frac{1.58\ A(\Delta c_w)}{c_{ac}E_L} = \frac{c_{ic}}{c_{ac}}. \tag{18.32}$$

We have simplified much of the theory to this point by focusing on the relation between $^{13}CO_2$ and $^{12}CO_2$ exchange between the ambient atmosphere and the air of the leaf intercellular air spaces. This form of the theory allows us to develop the concepts between $^{13}C/^{12}C$ fractionation and leaf water-use efficiency in a straightforward manner. A more detailed treatment of leaf isotopic fractionation would develop the transport pathway from the ambient atmosphere to the chloroplastic site of carboxylation, as well as accounting for other metabolic processes in the leaf that discriminate against ^{13}C. Following Lloyd and Farquhar (1994) we can write an augmented form of Eq. (18.28) as:

$$\begin{aligned}
^{13}\Delta = a_b \frac{c_{ac} - c_{sc}}{c_{ac}} &+ a\frac{c_{sc} - c_{ic}}{c_{ac}} + a\frac{c_{ic} - c_{wc}}{c_{ac}} + (e_s + a_s)\frac{c_{wc} - c_{cc}}{c_{ac}} \\
&+ b\frac{c_{cc}}{c_{ac}} - \frac{e\ R_d/(k + f\ \Gamma*)}{c_{ac}},
\end{aligned} \tag{18.33}$$

where a_b is the fractionation of $^{13}CO_2$ during diffusion across the leaf boundary layer (2.9‰); c_{ac}, c_{sc}, c_{ic}, c_{wc}, and c_{cc} are the average CO_2 mole fractions in the well-mixed atmosphere, at the surface of the leaf, in the leaf intercellular air spaces, at the wall of mesophyll cells, and in the chloroplast, respectively; e_s is the equilibrium fraction associated with CO_2 solubility into the aqueous phase of the cells; a_s is the fractionation during diffusion in the soluble phase; e and f are fractionations associated with "dark" respiration and photorespiration, respectively; R_d is the dark respiration rate from the tricarboxylic acid cycle; $\Gamma*$ is the photocompensation point (i.e., the CO_2 compensation point in the presence of CO_2 efflux from only photorespiration); and k is a biochemical "transfer coefficient," defined here as the carboxylation efficiency (according to $A = k\ c_{cc}$, or more formally as v_{cl}/c_{cc}, where v_{cl} is the RuBP carboxylation rate when not limited by RuBP substrate availability). In some cases, and in fact to account more accurately for biochemical processes, b is calculated as an "effective" biochemical fractionation combining several processes:

$$b = b_3 - \beta(b_3 - b_4), \tag{18.34}$$

where, by convention, b_3 refers to the carboxylation process that leads to a stable 3-C product (i.e., phosphoglyceric acid) and is the fractionation caused by the active site of Rubisco using soluble CO_2 as substrate (taken as 29‰), b_4 refers to the carboxylation process that leads to a stable 4-C product (i.e., oxaloacetic acid) and is the fractionation due to carboxylation catalyzed by the cytosolic enzyme phosphoenolpyruvate (PEP) carboxylase including the equilibrium fractionation associated with the reaction of CO_2 with H_2O (because PEP carboxylase uses HCO_3^- as substrate rather than CO_2) (taken as −5.7‰), and β is the ratio of PEP to RuBP carboxylation rates in C_3 leaves (typically less than 0.10). Using this approach a value for b can be approximated as 27‰ at 25°C.

The fractionations associated with mitochondrial respiration (e) and photorespiration (f) in the right-hand term of Eq. (18.33) are not well constrained compared to the other principal fractionations associated with leaf net CO_2 assimilation. It is known that mitochondrial respiration (R_d) during the night results in ^{13}C-enriched CO_2. However, during the day, mitochondrial respiration rates are down-regulated, and in fact recent observations have suggested that day-respired CO_2 (which includes photorespiration) is depleted in ^{13}C, not enriched (Tcherkez *et al.* 2010). Thus, the value of e during the day and night is uncertain. In the process of photorespiration, glycine decarboxylase, the enzyme-catalyzed step that releases CO_2, discriminates against ^{13}C. Both theoretical and empirical studies have estimated that the value of f is related to both the $^{13}C/^{12}C$ ratio of recently assimilated photo-assimilates (which enter photorespiration as oxygenated RuBP) and discrimination by glycine decarboxylase; estimates of f range between 7.5 and 13.7‰, with 11.5‰ being a "reasonable" first approximation (Lanigan *et al.* 2008).

Appendix 18.2 Derivation of the leaf form of the Craig–Gordon model

Following Flanagan *et al.* (1991), the total rate of transpiration from a leaf (E_L) and the rates of transpiration of the heavy (E'_L) and light (E''_L) isotopic forms of water can be expressed, respectively, as:

$$E_L = g_L \, (c_{iw} - c_{aw}), \tag{18.35}$$

$$E'_L = g'_L \, (R_{iv} c_{iw} - R_{av} c_{aw}), \tag{18.36}$$

$$E''_L = g''_L \left((1/R_{iv}) \, c_{iw} - (1/R_{av}) \, c_{aw} \right), \tag{18.37}$$

where g_L, g'_L, and g''_L are leaf conductances to total water vapor (heavy plus light), heavy, and light isotopic forms of water vapor, respectively; R_{iv} is the isotopic ratio (heavy/light) of water vapor in the intercellular air spaces of the leaf; and R_{av} is the isotopic ratio (heavy/light) of water vapor in the well mixed air outside the leaf. Given that $E''_L \approx E_L$, we can state that $E'_L/E''_L \approx E'_L/E_L$. Defining α^* as the thermodynamic fractionation factor (where $\alpha^* = R_l/R_v$, and R_l and R_v are the isotopic ratios in the liquid and vapor phases, respectively), and α_k as the kinetic fractionation factor (where $\alpha_k = g_L/g_L'$), we can write:

$$R_l \approx \frac{E_L'}{E_L} = \frac{1}{\alpha_k} \frac{(R_{iv}\, c_{iw} - R_{av}\, c_{aw})}{(c_{iw} - c_{aw})} = \frac{1}{\alpha_k} \frac{\left(\frac{R_l}{\alpha^*}\, c_{iw} - R_{av}\, c_{aw}\right)}{(c_{iw} - c_{aw})}. \tag{18.38}$$

During steady-state transpiration, and recognizing the constraint of mass balance, the isotopic composition of xylem water in the leaf (R_X) must equal that of the water transpired from the leaf; thus, $R_l = R_X$. Substituting R_X for R_l on the left-hand side of Eq. (18.5), and rearranging to solve for R_l on the right-hand side of Eq. (18.38), we can write:

$$R_l = \alpha^* \left(\alpha_k R_X \left(\frac{c_{iw} - c_{aw}}{c_{iw}} \right) + R_A \left(\frac{c_{aw}}{c_{iw}} \right) \right). \tag{18.39}$$

In still air, the values of α_k for the differential diffusion of water containing the heavy or light isotopes of hydrogen and oxygen, respectively, are $H_2O/^2H_2O = 1.025$ and $H_2^{16}O/H_2^{18}O = 1.0285$ (Merlivat 1978). Within the turbulent boundary layer of the leaf, the values for α_k (represented as α_{kh}) for both H_2O versus ^2H_2O and $H_2^{16}O$ versus $H_2^{18}O$ are reduced due to advective transfer of the isotopes, which is not subject to mass fractionation, and from the principles of fluid mechanics can be estimated as $H_2O/^2H_2O = 1.017$ and $H_2^{16}O/H_2^{18}O = 1.0189$ (see Flanagan *et al.* 1991). Thus, Eq. (18.39) can be further modified to account for the leaf boundary layer, as:

$$R_l = \alpha^* \left(\alpha_k R_X \left(\frac{c_{iw} - c_{sw}}{c_{iw}} \right) + \alpha_{kh} R_X \left(\frac{c_{sw} - c_{aw}}{c_{iw}} \right) + R_A \left(\frac{c_{aw}}{c_{iw}} \right) \right). \tag{18.40}$$

References

Ackerly, D. D. and Bazzaz, F. A. (1995) Leaf dynamics, self-shading and carbon gain in seedlings of a tropical pioneer tree. *Oecologia* **101**: 289–298.

Afreen, F., Zobayed, S. M. A., Armstrong, J. and Armstrong, W. (2007) Pressure gradients along whole culms and leaf sheaths, and other aspects of humidity-induced gas transport in Phragmites australis. *Journal of Experimental Botany* **58**: 1651–1662.

Albertson, J. D., Kustas, W. P. and Scanlon, T. M. (2001) Large-eddy simulation over heterogeneous terrain with remotely sensed land surface conditions. *Water Resources Research* **37**: 1939–1953.

Alton, P. (2009) A simple retrieval of ground albedo and vegetation absorptance from MODIS satellite data for parameterisation of global land-surface models. *Agricultural and Forest Meteorology* **149**: 1769–1775.

Amiro, B. D. (1990) Drag coefficients and turbulence spectra in three boreal forest canopies. *Boundary-Layer Meteorology* **52**: 227–246.

Amthor, J. S. (1994) Scaling CO_2-photosynthesis relationships from the leaf to canopy. *Photosynthesis Research* **39**: 321–350.

Amthor, J. S. (2000) The McCree-de Wit-Penning de Vries-Thornley respiration paradigms 30 years later. *Annals of Botany* **86**: 1–20.

Amthor, J. S., Goulden, M. L., Munger, J. W. and Wofsy, S. C. (1994) Testing a mechanistic model of forest-canopy mass and energy exchange using eddy correlation: Carbon dioxide and ozone uptake by a mixed-oak stand. *Australian Journal of Plant Physiology* **21**: 623–651.

Amundson, R. (2001) The carbon budget in soils. *Annual Review of Earth and Planetary Sciences* **29**: 535–562.

Anderson, M. C. (1966) Stand structure and light penetration. 2: A theoretical analysis. *Journal of Applied Ecology* **3**: 41–54.

Anderson, R. G., Canadell, J. G., Randerson, J. T., *et al*. (2011) Biophysical considerations in forestry for climate protection. *Frontiers in Ecology and the Environment* **9**: 174–182.

Anesio, A. M., Tranvik, L. J. and Granèli, W. (1999) Production of inorganic carbon from aquatic macrophytes by solar radiation. *Ecology* **80**: 1852–1859.

Anten, N. P. R. (2005) Optimal photosynthetic characteristics of individual plants in vegetation stands and implications for species coexistence. *Annals of Botany* **95**: 495–506.

Aphalo, P. J. and Jarvis, P. G. (1991) Do stomata respond to relative humidity? *Plant, Cell and Environment* **14**: 127–132.

Archontoulis, S. V., Yin, X., Vos, J., *et al*. (2012) Leaf photosynthesis and respiration of three bioenergy crops in relation to temperature and leaf nitrogen: How conserved are biochemical model parameters among crop species? *Journal of Experimental Botany* **63**: 895–911.

Armstrong, W., Armstrong, J. and Beckett, P. M. (1996a) Pressurised ventilation in emergent macrophytes: The mechanism and mathematical modelling of humidity-induced convection. *Aquatic Botany* **54**: 121–135.

Armstrong, J., Armstrong, W., Beckett, P. M., *et al.* (1996b) Pathways of aeration and the mechanisms and beneficial effects of humidity- and Venturi-induced convections in *Phragmites australis* (Cav) Trin ex Steud. *Aquatic Botany* **54**: 177–197.

Arneth, A., Monson, R. K., Schurgers, G., *et al.* (2008) Why are estimates of global isoprene emissions so similar (and why is this not so for monoterpenes)? *Atmospheric Chemistry and Physics* **8**: 4605–4620.

Arrhenius, S. (1908) *Das Werden der Welten*. Leipzig: Academic Publishing House.

Asner, G. P., Hughes, R. F., Mascaro, J., *et al.* (2011) High-resolution carbon mapping on the million-hectare island of Hawaii. *Frontiers in Ecology* **9**: 434–439.

Asner, G. P., Hughes, R. F., Vitousek, P. M., *et al.* (2008a) Invasive plants transform the three-dimensional structure of rain forests. *Proceedings of the National Academy of Sciences (USA)* **105**: 4519–4523.

Asner, G. P., Knapp, D. E., Kennedy-Bowdoin, T., *et al.* (2008b) Invasive species detection in Hawaiian rainforests using airborne imaging spectroscopy and LiDAR. *Remote Sensing of Environment* **112**: 1942–1955.

Asner, G. P., Scurlock, J. M. O., and Hicke, J. A. (2003) Global synthesis of leaf area index observations: Implications for ecological and remote sensing studies. *Global Ecology and Biogeography* **12**: 191–205.

Asrar, G., Myneni, R. B. and Kanemasu, E. T. (1989) Estimation of plant canopy attributes from spectral reflectance measurements. In: *Theory and Applications of Optical Remote Sensing* (Asrar, G., ed.). New York: John Wiley, pp. 252–292.

Atkin, O. K., Bruhn, D., Hurry, V. M. and Tjoelker, M. G. (2005) The hot and the cold: Unravelling the variable response of plant respiration to temperature. *Functional Plant Biology* **32**: 87–105.

Atkin, O. K. and Tjoelker, M. G. (2003) Thermal acclimation and the dynamic response of plant respiration to temperature. *Trends in Plant Science* **8**: 343–351.

Atkinson, R. and Arey, J. (2003) Atmospheric degradation of volatile organic compounds. *Chemical Reviews* **103**: 4605–4638.

Atkinson, R. and Carter, W. P. L. (1984) Kinetics and mechanisms of the gas-phase reactions of ozone with organic compounds under atmospheric conditions. *Chemical Reviews* **84**: 437–470.

Austin, A. T. and Vivanco, L. (2006) Plant litter decomposition in a semi-arid ecosystem controlled by photodegradation. *Nature* **442**: 555–558.

Ayers, G. P. and Cainey, J. M. (2007) The CLAW hypothesis: A review of the major developments. *Environmental Chemistry* **4**: 366–374.

Badger, M. R., Sharkey, T. D., and von Caemmerer, S. (1984) The relationship between steady-state gas exchange of bean leaves and the levels of carbon reduction cycle intermediates. *Planta* **160**: 305–313.

Baidya, S. and Avissar, R. (2000) Scales of response of the convective boundary layer to land-surface heterogeneity. *Geophysical Research Letters* **27**: 533–536.

Baldocchi, D. (1992) A Lagrangian random-walk model for simulating water vapor, CO_2, and sensible heat flux densities and scalar profiles over and within a soybean canopy. *Boundary-Layer Meteorology* **61**: 113–144.

Baldocchi, D. D. (2003) Assessing the eddy covariance technique for evaluating carbon dioxide exchange rates of ecosystems: Past, present and future. *Global Change Biology* **9**: 479–492.

Baldocchi, D. D. and Collineau, S. (1994) The physical nature of solar radiation in heterogeneous canopies: Spatial and temporal attributes. In: *Exploitation of Environmental Heterogeneity* (Pearcy, R. W. and Caldwell, M. M., eds.). San Diego, CA: Academic Press, pp. 21–71.

Baldocchi, D., Falge, E., Gu, L. H., *et al.* (2001) FLUXNET: A new tool to study the temporal and spatial variability of ecosystem-scale carbon dioxide, water vapor, and energy flux densities. *Bulletin of the American Meteorological Society* **82**: 2415–2434.

Baldocchi, D. D., Hutchinson, B. A., Matt, D. R. and McMillen, R. T. (1985) Canopy radiative transfer models for spherical and known leaf inclination distribution angles: A test in an oak-hickory forest. *Journal of Applied Ecology* **22**: 539–555.

Baldocchi, D. D. and Meyers, T. P. (1988) Turbulence structure in a deciduous forest. *Boundary-Layer Meteorology* **43**: 345–364.

Baldocchi, D. D. and Meyers, T. (1998) On using eco-physiological, micrometeorological and biogeochemical theory to evaluate carbon dioxide, water vapor and trace gas fluxes over vegetation: A perspective. *Agricultural and Forest Meteorology* **90**: 1–25.

Ball J. T. and Berry J. A. (1982) The Ci/Ca ratio: A basis for predicting stomatal control of photosynthesis. *Carnegie Institute of Washington Yearbook* **81**: 88–92.

Ball, J. T., Woodrow, I. E. and Berry, J. A. (1987) A model predicting stomatal conductance and its contribution to the control of photosynthesis under different environmental conditions. In: *Progress in Photosynthesis Research* (Biggens, J., ed.). Dordrecht: Martinus Nijhoff Publishers, pp. IV.5.221–IV.5.224.

Banta, R. M., Newsom, R. K., Lundquist, J. K., *et al.* (2002) Nocturnal low-level jet characteristics over Kansas during CASES-99. *Boundary Layer Meteorology* **105**: 221–252.

Berg, B. (2000) Litter decomposition and organic matter turnover in northern forest soils. *Forest Ecology and Management* **133**: 13–22.

Bergamaschi, P., Braunlich, M., Marik, T. and Brenninkmeijer, C. A. M. (2000) Measurements of the carbon and hydrogen isotopes of atmospheric methane at Izana, Tenerife: Seasonal cycles and synoptic-scale variations. *Journal of Geophysical Research–Atmospheres* **105**: 14531–14546.

Bernacchi, C. J., Pimentel, C. and Long, S. P. (2003) *In vivo* temperature response functions of parameters required to model RuBP-limited photosynthesis. *Plant, Cell and Environment* **26**: 1419–1430.

Bernacchi, C. J., Singsaas, E. L., Pimentel, C., *et al.* (2001) Improved temperature response functions for models of Rubisco-limited photosynthesis. *Plant, Cell and Environment* **24**: 253–259.

Berry, J. A. and Farquhar, G. D. (1978) The CO_2 concentrating function of C_4 plants: A biochemical model. In: *Proceedings of the Fourth International Congress on*

Photosynthesis (Hall, D., Coombs, J., and Goodwin, T., eds.). London: Biochemical Society of London, pp. 119–131.

Bittner, S., Janott, M. and Ritter, D., *et al.* (2012) Functional-structural water flow model reveals differences between diffuse- and ring-porous tree species. *Agricultural and Forest Meteorology* **158**: 80–89.

Björkman, O. (1975) Environmental and biological control of photosynthesis: Inaugural address. In: *Environmental and Biological Control of Photosynthesis* (Marcelle, R., ed.). The Hague: W. Junk, pp. 1–16.

Blackadar, A. K. (1957) Boundary layer wind maxima and their significance for the growth of nocturnal inversions. *Bulletin of the American Meteorological Society* **38**: 283–290.

Blackadar, A. K. (1997) *Turbulence and Diffusion in the Atmosphere*. Berlin: Springer-Verlag.

Blanken, P. D., Black, T. A., Yang, P. C., *et al.* (1997) Energy balance and canopy conductance of a boreal aspen forest: Partitioning overstory and understory components. *Journal of Geophysical Research – Atmospheres* **102**: 28915–28927.

Bohrer, G., Mourad, H., Laursen, T. A., *et al.* (2005) Finite element tree crown hydrodynamics model (FETCH) using porous media flow within branching elements: A new representation of tree hydrodynamics. *Water Resources Research* **41**: Article number W11404.

Bonan, G. B. (2008) Forests and climate change: Forcings, feedbacks, and the climate benefits of forests. *Science* **320**: 1444–1449.

Bonan, G. B., Lawrence, P. J., Oleson, K. W., *et al.* (2011) Improving canopy processes in the Community Land Model version 4 (CLM4) using global flux fields empirically inferred from FLUXNET data. *Journal of Geophysical Research – Biogeosciences* **116**: Article number G02014.

Bond-Lamberty, B., Wang, C. and Gower, S. T. (2004) Net primary production and net ecosystem production of a boreal black spruce wildfire chronosequence. *Global Change Biology* **10**: 473–487.

Borken, W. and Matzner, E. (2009) Reappraisal of drying and wetting effects on C and N mineralization and fluxes in soils. *Global Change Biology* **15**: 808–824.

Bosatta, E. and Ågren, G. I. (1999) Soil organic matter quality interpreted thermodynamically. *Soil Biology and Biochemistry* **31**: 1889–1891.

Bosatta, E. and Ågren, G. I. (2002) Quality and irreversibility: Constraints on ecosystem development. *Proceedings of the Royal Society of London, Series B* **269**: 203–210.

Bowling, D. R., Bowling, N. G., Bond, B. J., *et al.* (2002) ^{13}C content of ecosystem respiration is linked to precipitation and vapor pressure deficit. *Oecologia* **131**: 113–124.

Bowling, D. R., Burns, S. P., Conway, T. J., *et al.* (2005) Extensive observations of CO_2 carbon isotope content in and above a high-elevation, subalpine forest. *Global Biogeochemical Cycles* **19**: Article number GB 3023.

Bowling, D. R., Pataki, D. E. and Randerson, J. T. (2008) Carbon isotopes in terrestrial ecosystem pools and CO_2 fluxes. *New Phytologist* **178**: 24–40.

Brandt, L. A., Bohnet, C. and King, J. Y. (2009) Photochemically induced carbon dioxide production as a mechanism for carbon loss from plant litter in arid ecosystems. *Journal of Geophysical Research* **114**: Article number G02004.

Bridges, E. M. (1990) *Soil Horizon Designations*, Technical Paper 19, ISRIC, Wageningen.

Brooks, A. and Farquhar, G. D. (1985) Effect of temperature on the CO_2/O_2 specificity of ribulose-1,5-bisphosphate carboxylase/oxygenase and the rate of respiration in the light. *Planta* **165**: 397–406.

Brooks, R. H. and Corey, A. T. (1966) Properties of porous media affecting fluid flow. *Journal of Irrigation Drainage Division of the American Society of Civil Engineering* **92**: 61–87.

Brown, J. H. and West, G. B. (eds.) (2000) *Scaling in Biology*. New York: Oxford University Press.

Brown, H. and Escombe, F. (1900) Static diffusion of gases and liquids in relation to the assimilation of carbon and translocation in plants. *Philosophical Transactions of the Royal Society* **B193**: 223–291.

Brugnoli, E. and Björkman, O. (1992) Chloroplast movements in leaves: Influence on chlorophyll fluorescence and measurements of light-induced absorbance changes related to ΔpH and zeaxanthin formation. *Photosynthesis Research* **32**: 23–35.

Brunet, Y. and Irvine, M. R. (2000) The control of coherent eddies in vegetation canopies: Streamwise structure spacing, canopy shear scale and atmospheric stability. *Boundary-Layer Meteorology* **94**: 139–163.

Brussaard, L., Behan-Pelletier, V. M., Bignell, D. E., *et al.* (1997) Biodiversity and ecosystem functioning in soil. *Ambio* **26**: 563–570.

Buchmann, N., Kao, Y.-W. and Ehleringer, J. (1997) Influence of stand structure on carbon-13 of vegetation, soils, and canopy air within deciduous and evergreen forests in Utah, United States. *Oecologia* **110**: 109–119.

Buck, A. L. (1981) New equations for computing vapor pressure and enhancement factor. *Journal of Applied Meteorology* **20**: 1527–1532.

Buckingham, E. (1907) Studies on the movement of soil moisture. Bureau of Soils Bulletin, 38, US Department of Agriculture, Washington, DC.

Buckley, T. N., Mott, K. A. and Farquhar, G. D. (2003) A hydromechanical and biochemical model of stomatal conductance. *Plant, Cell and Environment* **26**: 1767–1785.

Buckley, T. N., Sack, L. and Gilbert, M. E. (2011) The role of bundle sheath extensions and life form in stomatal responses to leaf water status. *Plant Physiology* **156**: 962–973.

Budyko, M. I. (1974) *Climate and Life*. New York: Academic Press.

Cahill, T. M., Seaman, V. Y., Charles, M. J., *et al.* (2006) Secondary organic aerosols formed from oxidation of biogenic volatile organic compounds in the Sierra Nevada Mountains of California. *Journal of Geophysical Research – Atmospheres* **111**: Article number D16312.

Campbell, G. S. and Norman, J. M. (1998) *An Introduction to Environmental Biophysics*. New York: Springer-Verlag.

Canadell, J. G., Canadell, C., Raupach, M. R., *et al.* (2007) Contributions to accelerating atmospheric CO_2 growth from economic activity, carbon intensity, and efficiency of natural sinks. *Proceedings of the National Academy of Sciences (USA)* **104**: 18866–18870.

Canadell, J. G. and Raupach, M. R. (2008) Managing forests for climate change mitigation. *Science* **320**: 1456–1457.

Carter, G. A., Bahadur, R. and Norby, R. J. (2000) Effects of elevated atmospheric CO_2 and temperature on leaf optical properties in *Acer saccharum*. *Environmental and Experimental Botany* **43**: 267–273.

Cava, D. and Katul, G. G. (2008) Spectral short-circuiting and wake production within the canopy trunk space of an alpine hardwood forest. *Boundary-Layer Meteorology* **126**: 415–431.

Cerling, T. E., Harris, J. M., MacFadden, B. J., *et al.* (1997) Global vegetation change through the Miocene/Pliocene boundary. *Nature* **389**:153–158.

Cescatti, A. (1997) Modelling the radiative transfer in discontinuous canopies of asymmetric crowns.1: Model structure and algorithms. *Ecological Modelling* **101**: 263–274.

Chandrasekhar, S. (1950) *Radiative Transfer*. New York: Dover Publishers.

Chanton, J. P., Whiting, G. J., Blair, N. E., Lindau, C. W. and Bollich, P. K. (1997) Methane emission from rice: Stable isotopes, diurnal variations, and CO_2 exchange. *Global Biogeochemical Cycles* **11**: 15–27.

Chapin, F. S., McFarland, J., McGuire, A. D., *et al.* (2009) The changing global carbon cycle: Linking plant-soil carbon dynamics to global consequences. *Journal of Ecology* **97**: 840–850.

Charlson, R. J., Lovelock, J. E., Andreae, M. O. and Warren, G. (1987) Oceanic phytoplankton, atmospheric sulfur, cloud albedo and climate. *Nature* **326**: 655–661.

Chen, J. L., Reynolds, J. F., Harley, P. C. and Tenhunen, J. D. (1993) Coordination theory of leaf nitrogen distribution in a canopy. *Oecologia* **93**: 63–69.

Chen, J. M. (1996) Optically based methods for measuring seasonal variation in leaf area index in boreal conifer stands. *Agricultural and Forest Meteorology* **80**: 135–163.

Chen, J. M. and Black, T. A. (1992) Defining leaf-area index for non-flat leaves. *Plant, Cell and Environment* **15**: 421–429.

Chen, J. M. and Leblanc, S. G. (2001) Multiple-scattering scheme useful for geometric optical modeling. *IEEE Transactions on Geoscience and Remote Sensing* **39**: 1061–1071.

Chen, J. M., Rich, P. M., Gower, S. T., *et al.* (1997) Leaf area index of boreal forests: Theory, techniques, and measurements. *Journal of Geophysical Research* **102**(D24): 29429–29443.

Cheng, Y. G., Parlange, M. B. and Brutsaert, W. (2005) Pathology of Monin–Obukhov similarity in the stable boundary layer. *Journal of Geophysical Research, Atmospheres* **110**: Article number D06101.

Choat, B., Jansen, S., Brodribb, T. J., *et al.* (2012) Global convergence in the vulnerability of forests to drought. *Nature* **491**: 752–755.

Christeller, J. T., Laing, W. A. and Troughton, J. H. (1976) Isotope discrimination by ribulose 1,5-diphosphate carboxylase – no effect of temperature or HCO_3^- concentration. *Plant Physiology* **57**: 580–582.

Ciais, P., Denning, A. S., Tans, P. P., *et al.* (1997) A three dimensional synthesis study of $\delta^{18}O$ in atmospheric CO_2. Part 1: Surface fluxes. *Journal of Geophysical Research* **102**: 5873–5883.

Ciais, P., Friedlingstein, P., Schimel, D. S. and Tans, P. P. (1999) A global calculation of the $\delta^{13}C$ of soil respired carbon: Implications for the biospheric uptake of anthropogenic CO_2. *Global Biogeochemical Cycles* **13**: 519–530.

Cihlar, J., St-Laurent, L. and Dyer, J. A. (1991) Relation between the normalized vegetation index and ecological variables. *Remote Sensing of Environment* **35**: 279–298.

Cochard, H., Venisse, J. S., Barigah, T. S., *et al.* (2007) Putative role of aquaporins in variable hydraulic conductance of leaves in response to light. *Plant Physiology* **143**: 122–133.

Collatz, G. J., Ball, J. T., Grivet, C. and Berry, J. A. (1991) Regulation of stomatal conductance and transpiration: A physiological model of canopy processes. *Agricultural and Forest Meteorology* **54**: 107–136.

Collatz, G. J., Ribas-Carbo, M. and Berry, J. A. (1992) Coupled photosynthesis-stomatal conductance model for leaves of C_4 plants. *Australian Journal of Plant Physiology* **19**: 519–538.

Comstock, J. P. and Ehleringer, J. R. (1992) Correlating genetic variation in carbon isotopic composition with complex climatic gradients. *Proceedings of the National Academy of Sciences (USA)* **89**: 7747–7751.

Conant, R. T., Steinweg, J. M., Haddix, M. L., *et al.* (2008) Experimental warming shows that decomposition temperature sensitivity increases with soil organic matter recalcitrance. *Ecology* **89**: 2384–2391.

Conen, F., Leifeld, J., Seth, B. and Alewell, C. (2006) Warming mineralises young and old soil carbon equally. *Biogeosciences* **3**: 515–519.

Cooper, D. I., Leclerc, M. Y., Archuleta, J., *et al.* (2006) Mass exchange in the stable boundary layer by coherent structures. *Agricultural and Forest Meteorology* **136**: 114–131.

Couteaux, M. M., Bottner, P., Anderson, J. M., *et al.* (2001) Decomposition of ^{13}C-labelled standard plant material in a latitudinal transect of European coniferous forests: Differential impact of climate on the decomposition of soil organic matter compartments. *Biogeochemistry* **54**: 147–170.

Cowan, I. R. (1986) Stomatal function in relation to leaf metabolism and environment. In: *On the Economy of Plant Form and Function* (Givnish, T., ed.). Cambridge: Cambridge University Press, pp. 171–213.

Cowan, I. R. and Farquhar, G. D. (1977) Stomatal function in relation to leaf metabolism and environment. In: *Society for Experimental Biology Symposium, Integration of Activity in the Higher Plant Vol. 31* (Jennings, D. H., ed.). Cambridge: Society for Experimental Biology, pp. 471–505.

Craig, H. (1957) Isotopic standards for carbon and oxygen and correction factors for mass spectrometric analysis of carbon dioxide. *Geochimica and Cosmochimica Acta* **12**: 133–149.

Craig, H. and Gordon, L. I. (1965) Deuterium and oxygen 18 variations in the ocean and the marine atmosphere. In: *Proceedings of a Conference on Stable Isotopes in Oceanographic Studies and Paleotemperatures* (Tongiorgi, E., ed.). Pisa: Lischi and Figli Publishers, pp. 9–130.

Craine, J. M., Fierer, N. and McLauchlan, K. K. (2010) Widespread coupling between the rate and temperature sensitivity of organic matter decay. *Nature Geoscience* **3**: 854–857.

Cramer, W., Bondeau, A., Woodward, F. I., *et al.* (2001) Global response of terrestrial ecosystem structure and function to CO_2 and climate change: results from six dynamic global vegetation models. *Global Change Biology* **7**: 357–373.

Crimaldi, J. P., Cadwell, J. R. and Weiss, J. B. (2008) Reaction enhancement of isolated scalars by vortex stirring. *Physics of Fluids* **20**: Article number 073605.

Culf, A. D., Fisch, G. and Hodnett, M. G. (1995) The albedo of Amazonian forest and rangeland. *Journal of Climate* **8**: 1544–1554.

Cunningham, R. E. and Williams, R. J. J. (1980) *Diffusion in Gases and Porous Media*. New York: Plenum Press.

Czimczik, C. I. and Trumbore, S. E. (2007) Short-term controls on the age of microbial carbon sources in boreal forest soils. *Journal of Geophysical Research – Biogeosciences* **112**: Article number G03001.

Dacey, J. W. H. (1987) Knudsen-transitional flow and gas pressurization in leaves of Nelumbo. *Plant Physiology* **85**: 199–203.

Dalias, P., Anderson, J. M., Bottner, P. and Couteaux, M. M. (2001) Long-term effects of temperature on carbon mineralisation processes. *Soil Biology and Biochemistry* **33**: 1049–1057.

Dalton, J. (1802) Experimental essays on the constitution of mixed gases; on the force of steam or vapour from water and other liquids in different temperatures; both in a Torricellian vacuum and in air; on evaporation; and on the expansion of gases by heat. *Memoirs of the Proceedings of the Literature Philosophical Society of Manchester* Part 2, 535–602.

Davidson, E. A., Keller, M., Erickson, H. E., Verchot, L. V. and Veldkamp, E. (2000b) Testing a conceptual model of soil emissions of nitrous and nitric oxides. *Bioscience* **50**: 667–680.

Davidson, E. A., Trumbore, S. E. and Amundson, R. (2000a) Biogeochemistry – soil warming and organic carbon content. *Nature* **408**: 789–790.

Davidson, E. A. and Verchot, L. V. (2000) Testing the hole-in-the-pipe model of nitric and nitrous oxide emissions from soils using the TRAGNET database. *Global Biogeochemical Cycles* **14**: 1035–1043.

Davis, K. J., Lenschow, D. H., Oncley, S. P., *et al.* (1997) Role of entrainment in surface-atmosphere interactions over the boreal forest. *Journal of Geophysical Research – Atmospheres* **102**: 29219–29230.

De Bort, L. T. (1902) Variations of temperature of the free air in the zone between 8 km and 13 km of the altitude. *Comptes Rendus Hebdomadaires des Seances de l'Academie des Sciences* **134**: 987–989.

Denholm, J. V. (1981) The influence of penumbra on canopy photosynthesis: Theoretical considerations. *Agricultural Meteorology* **25**: 145–166.

Denmead, O. T. and Bradley, E. F. (1985) Flux-gradient relationships in a forest canopy. In: *The Forest-Atmosphere Interaction* (Hutchinson, B. A. and Hicks, B. B., eds.). Dordrecht: D. Reidel Publishing Co., pp. 421–442.

Denmead, O. T. and Bradley, E. F. (1987) On scalar transport in plant canopies. *Irrigation Science* **8**: 131–149.

Denmead, O. T., Raupach, M. R., Dunin, F. X., *et al.* (1996) Boundary layer budgets for regional estimates of scalar fluxes. *Global Change Biology* **2**: 255–264.

DePury, D. G. G. and Farquhar, G. D. (1997) Simple scaling of photosynthesis from leaves to canopies without the errors of big leaf models. *Plant, Cell and Environment* **20**: 537–557.

Dewar, R. C. (1992) Inverse modeling and the global carbon cycle. *Trends in Ecology and Evolution* **7**: 105–107.

Dewar, R. C. (1995) Interpretation of an empirical model for stomatal conductance in terms of guard cell function. *Plant, Cell and Environment* **18**: 365–372.

Dewar, R. C. (2002) The Ball-Berry-Leuning and Tardieu-Davies stomatal models: Synthesis and extension within a spatially aggregated picture of guard cell function. *Plant, Cell and Environment* **25**: 1383–1398.

Dewar, R. C., Tarvainen, L., Parker, K., *et al.* (2012) Why does leaf nitrogen decline within tree canopies less rapidly than light? An explanation from optimization subject to a lower bound on leaf mass per area. *Tree Physiology* **32**: 520–534.

Di Marco, G., Manes, F., Tricoli, D. and Vitale, E. (1990) Fluorescence parameters measured concurrently with net photosynthesis to investigate chloroplastic CO_2 concentration in leaves of *Quercus ilex* L. *Journal of Plant Physiology* **136**: 538–543.

Dickinson, R. E. (1983) Land surface processes and climate-surface albedos and energy balance. *Advances in Geophysics* **25**: 305–353.

Disney, M. I., Lewis, P. and North, P. R. J. (2000) Monte Carlo ray tracing in optical canopy reflectance modelling. *Remote Sensing Reviews* **18**: 163–196.

Dlugocencky, E. J., Masarie, K. A., Lang, P. M. and Tans, P. P. (1998) Continuing decline in the growth rate of the atmospheric methane burden. *Nature* **393**: 447–450.

Dlugokencky, E. J., Nisbet, E. G., Fisher, R. and Lowry, D. (2011) Global atmospheric methane: Budget, changes and dangers. *Philosophical Transactions of the Royal Society, Series A* **369**: 2058–2072.

Doerr, S. H., Ritsema, C. J., Dekker, L. W., *et al.* (2007) Water repellence of soils: New insights and emerging research needs. *Hydrological Processes* **21**: 2223–2228.

Doornbos, R. F., Loon, L. C. and Bakker, P. A. H. M. (2012) Impact of root exudates and plant defense signaling on bacterial communities in the rhizosphere: A review. *Agronomy for Sustainable Development* **32**: 227–243.

Dupont, S. and Patton, E. G. (2012) Momentum and scalar transport within a vegetation canopy following atmospheric stability and seasonal canopy changes: The CHATS experiment. *Atmospheric Chemistry and Physics* **12**: 5913–5935.

Dutaur, L. and Verchot, L. V. (2007) A global inventory of the soil CH_4 sink. *Global Biogeochemical Cycles* **21**: Article number GB 4013.

Dwyer, M. J., Patton, E. G. and Shaw, R. H. (1997) Turbulent kinetic energy budgets from a large-eddy simulation of airflow above and within a forest canopy. *Boundary-Layer Meteorology* **84**: 23–43.

Edwards, E. J., Osborne, C. P., Stromberg, C. A. E., *et al.* (2010) The origins of C_4 grasslands: Integrating evolutionary and ecosystem science. *Science* **328**: 587–591.

Ehleringer, J. R. and Björkman, O. (1978) Pubescence and leaf spectral characteristics in a desert shrub, *Encelia farinosa*. *Oecologia* **36**: 151–162.

Ehleringer, J. R. and Monson, R. K. (1993) Ecology and evolution of photosynthetic pathway variation. *Annual Review of Ecology and Systematics* **24**: 411–439.

Ehleringer, J. R., Buchmann, N. and Flanagan, L. B. (2000) Carbon isotope ratios in below-ground carbon cycle processes. *Ecological Applications* **10**: 412–422.

Einstein, A. (1905) On the motion – required by the molecular kinetic theory of heat – of small particles suspended in a stationary liquid. *Annalen der Physik* **17**: 549–560.

Evans, J. R., Sharkey, T. D., Berry, J. A. and Farquhar, G. D. (1986) Carbon isotope discrimination measured concurrently with gas exchange to investigate CO_2 diffusion in leaves of higher plants. *Australian Journal of Plant Physiology* **13**: 281–292.

Evans, J. R. and von Caemmerer, S. (1996) Carbon dioxide diffusion inside leaves. *Plant Physiology* **110**: 339–346.

Eyring, H. (1935) The activated complex in chemical reactions. *Journal of Chemical Physics* **3**: 107–115.

Fall, R. (2003) Abundant oxygenates in the atmosphere: A biochemical perspective. *Chemical Reviews* **103**: 4941–4951.

Fang, C. and Moncrieff, J. B. (2001) The dependence of soil CO_2 efflux on temperature. *Soil Biology and Biochemistry* **33**: 155–165.

Fang, C., Smith, P., Moncrieff, J. and Smith, J. U. (2005) Similar response of labile and resistant soil organic matter pools to changes in temperature. *Nature* **433**: 57–59.

Farquhar, G. D. (1989) Models of integrated photosynthesis of cells and leaves. *Philosophical Transactions of the Royal Society of London. Series B: Biological Sciences* **323**: 357–367.

Farquhar, G. D., Ehleringer, J. R. and Hubick, K. T. (1989) Carbon isotope discrimination and photosynthesis. *Annual Review of Plant Physiology and Molecular Biology* **40**: 503–537.

Farquhar, G. D., Firth, P. M., Wetselaar, R. and Weir, B. (1980a) On the gaseous exchange of ammonia between leaves and the environment: Determination of the ammonia compensation point. *Plant Physiology* **66**: 710–714.

Farquhar, G. D., Lloyd, J., Taylor, J. A., *et al.* (1993) Vegetation effects on the isotope composition of oxygen in atmospheric carbon dioxide. *Nature* **363**: 439–443.

Farquhar, G. D., O'Leary, M. H. and Berry, J. A. (1982) On the relationship between carbon isotope discrimination and the intercellular carbon dioxide concentration in leaves. *Australian Journal of Plant Physiology* **9**: 121–137.

Farquhar, G. D. and von Caemmerer, S. (1982) Modelling of photosynthetic response to environmental conditions. In: *Physiological Plant Ecology II. Water Relations and Carbon Assimilation* (Lange, O. L., Nobel, P. S., Osmond, C. B., and Ziegler, H., eds.). Berlin: Springer-Verlag, pp. 549–588.

Farquhar, G. D., von Caemmerer, S. and Berry, J. A. (1980b) A biochemical model of photosynthetic CO_2 assimilation in leaves of C_3 plants. *Planta* **149**: 78–90.

Farquhar, G. D., von Caemmerer, S. and Berry, J. A. (2001) Models of photosynthesis. *Plant Physiology* **125**: 42–45.

Farquhar, G. D. and Wong, S. C. (1984) An empirical model of stomatal conductance. *Australian Journal of Plant Physiology* **11**: 191–209.

Fassnacht, K. S., Gower, S. T., Norman, J. M. and McMurtrie, R. E. (1994) A comparison of optical and direct methods for estimating foliage surface area index in forests. *Agricultural and Forest Meteorology* **71**: 183–207.

Fell, D. A. (1992) Metabolic control analysis: A survey of its theoretical and experimental development. *Biochemical Journal* **286**: 313–330.

Fernàndez, N., Paruelo, J. M., Delibes, M., *et al.* (2010) Ecosystem functioning of protected and altered Mediterranean environments: A remote sensing classification in Donana, Spain. *Remote Sensing of Environment* **114**: 211–220.

Fernie, A. R., Carrari, F. and Sweetlove, L. J. (2004) Respiratory metabolism: Glycolysis, the TCA cycle and mitochondrial electron transport. *Current Opinion in Plant Biology* **7**: 254–261.

Fick, A. (1855) On liquid diffusion. *Philosophical Magazine and Journal of Science* **10**: 31–39.

Field, C. B. (1983) Allocating leaf nitrogen for the maximisation of carbon gain: Leaf age as a control on the allocation programme. *Oecologia* **56**: 341–347.

Field, C. B., Lobell, D. B., Peters, H. A. and Chiariello, N. R. (2007) Feedbacks of terrestrial ecosystems to climate change. *Annual Review of Environment and Resources* **32**: 1–29.

Fierer, N., Craine, J. M., McLauchlan, K. and Schimel, J. P. (2005) Litter quality and the temperature sensitivity of decomposition. *Ecology* **86**: 320–326.

Finlayson-Pitts, B. J. and Pitts, J. N., Jr. (2000) *Chemistry of the Upper and Lower Atmosphere*. San Diego, CA: Academic Press.

Finnigan, J. J. (2000) Turbulence in plant canopies. *Annual Review of Fluid Mechanics* **32**: 519–571.

Finnigan, J. J., Einaudi, F. and Fua, D. (1984) The interaction between an internal gravity wave and turbulence in the stably-stratified nocturnal boundary layer. *Journal of Atmospheric Science* **41**: 2409–2436.

Finnigan, J. J., Shaw, R. H. and Patton, E. G. (2009) Turbulence structure above a vegetation canopy. *Journal of Fluid Mechanics* **637**: 387–424.

Firestone, M. K. and Davidson, E. A. (1989) Microbiological basis of NO and N_2O production and consumption in soil. In: *Exchange of Trace Gases between Terrestrial Ecosystems and the Atmosphere* (Andreae, M. O. and Schimel, D. S., eds.). New York: John Wiley & Sons, pp. 7–21.

Fitzjarrald, D. R. and Moore, K. E. (1990) Mechanisms of nocturnal exchange between the rain forest and atmosphere. *Journal of Geophysical Research* **95**: 16839–16850.

FitzPatrick, E. A. (1993) Soil horizons. *Catena* **20**: 361–430.

Flanagan, L. B., Comstock, J. P. and Ehleringer, J. R. (1991) Comparison of modeled and observed environmental influences on the stable oxygen and hydrogen isotope composition of leaf water in *Phaseolus vulgaris* L. *Plant Physiology* **96**: 588–596.

Flexas, J., Barbour, M. M., Brendel, O., *et al.* (2012) Mesophyll diffusion conductance to CO_2: An unappreciated central player in photosynthesis. *Plant Science* **193–194**: 70–84.

Foken, T. (2008) The energy balance closure problem: An overview. *Ecological Applications* **18**: 1351–1367.

Foken, T. (2008) *Micrometeorology*. Heidelberg: Springer-Verlag.

Francey, R. J., Allison, C. E., Etheridge, D. M., *et al.* (1999) A 1000-year high precision record of $\delta^{13}C$ in atmospheric CO_2. *Tellus (B)* **51**: 170–193.

Franks, P. J., Cowan, I. R. and Farquhar, G. D. (1998) A study of stomatal mechanics using the cell pressure probe. *Plant, Cell and Environment* **21**: 94–100.

Franks, P. J., Cowan, I. R., Tyerman, S. D., *et al.* (1995) Guard-cell pressure aperture characteristics measured with the pressure probe. *Plant, Cell and Environment* **18**: 795–800.

Franks, P. J. and Farquhar, G. D. (2001) The effect of exogenous abscisic acid on stomatal development, stomatal mechanics, and leaf gas exchange in *Tradescantia virginiana*. *Plant Physiology* **125**: 935–942.

Frenzen, P. and Vogel, C. A. (1992) The turbulent kinetic energy budget in the atmospheric surface layer: A review and an experimental reexamination in the field. *Boundary-Layer Meteorology* **60**: 49–76.

Friend, A. D. (1991) Use of a model of photosynthesis and leaf microenvironment to predict optimal stomatal conductance and leaf nitrogen partitioning. *Plant, Cell and Environment* **14**: 895–905.

Friend, A. D., Woodward, F. I. and Switsur, V. R. (1989) Field measurements of photosynthesis, stomatal conductance, leaf nitrogen and $\delta^{13}C$ along altitudinal gradients in Scotland. *Functional Ecology* **3**: 117–122.

Fuentes, J. D., Wang, D., Bowling, D. R., *et al.* (2007) Biogenic hydrocarbon chemistry within and above a mixed deciduous forest. *Journal of Atmospheric Chemistry* **56**: 165–185.

Fuller, E. N., Schetter, P. D. and Giddings, J. C. (1966) A new method for prediction of binary gas-phase diffusion coefficients. *Industrial and Engineering Chemistry* **58**: 19–27.

Fung, I., Field, C. B., Berry, J. A., *et al.* (1997) Carbon 13 exchanges between the atmosphere and the biosphere. *Global Biogeochemical Cycles* **11**: 507–533.

Gale, J. (1973) Availability of carbon dioxide for photosynthesis at high altitudes: theoretical considerations. *Ecology* **53**: 494–497.

Gao, W., Shaw, R. H. and Paw, U. K. T. (1989) Observation of organized structure in turbulent flow within and above a forest canopy. *Boundary-Layer Meteorology* **47**: 349–377.

Gardiner, B. A. (1994) Wind and wind forces in a plantation spruce forest. *Boundary-Layer Meteorology* **67**: 161–186.

Gates, D. M. (1980) *Biophysical Ecology.* New York: Springer-Verlag.

Gates, D. M., Alderfer, R. and Taylor, E. (1968) Leaf temperatures of desert plants. *Science* **159**: 994–995.

Gaudinski, J. B., Trumbore, S. E., Davidson, E. A. and Zheng, S. H. (2000) Soil carbon cycling in a temperate forest: Radiocarbon-based estimates of residence times, sequestration rates and partitioning of fluxes. *Biogeochemistry* **51**: 33–69.

Gausman, H. W. (1985) *Plant Leaf Optical Properties in Visible and Near-Infrared Light.* Lubbock, TX: Texas Tech. Press.

Ghashghaie, J., Badeck, F.-W., Lanigan, G., *et al.* (2003) Carbon isotope fractionation during dark respiration and photorespiration in C_3 plants. *Phytochemistry Reviews* **2**: 145–161.

Ghirardo, A., Koch, K., Taipale, R., *et al.* (2010) Determination of de novo and pool emissions of terpenes from four common boreal/alpine trees by $^{13}CO_2$ labelling and PTR-MS analysis. *Plant, Cell and Environment* **33**: 781–792.

Giardina, C. P. and Ryan, M. G. (2000) Evidence that decomposition rates of organic carbon in mineral soil do not vary with temperature. *Nature* **404**: 858–861.

Gifford, R. M. (1994) The global carbon cycle: A viewpoint on the missing sink. *Australian Journal of Plant Physiology* **21**: 1–15.

Gillon, J. and Yakir, D. (2000) Internal conductance to CO_2 diffusion and $C^{18}OO$ discrimination in C_3 leaves. *Plant Physiology* **123**: 201–213.

Gillon, J. and Yakir, D. (2001) Influence of carbonic anhydrase activity in terrestrial vegetation on the ^{18}O content of atmospheric CO_2. *Science* **291**: 2584–2587.

Givnish, T. J. (1986) Optimal stomatal conductance, allocation of energy between leaves and roots, and the marginal cost of transpiration. In: *On the Economy of Plant Form and Function* (Givnish, T., ed.), pp. 171–213.

Gleixner, G. and Schmidt, H. L. (1997) Carbon isotope effects on the fructose-1,6-bisphosphate aldolase reaction, origin for non-statistical ^{13}C distributions in carbohydrates. *Journal of Biological Chemistry* **272**: 5382–5387.

Goh, K. M., Rafter, T. A., Stout, J. D. and Walker, T. W. (1976) Accumulation of soil organic matter and its carbon isotope content in a chronosequence of soils developed on aeolian sand in New Zealand. *Journal of Soil Science* **27**: 89–100.

Goldstein, A. H. and Galbally, I. E. (2007) Known and unexplored organic constituents in the earth's atmosphere. *Environmental Science and Technology* **41**: 1514–1521.

Gonzalez, J. M., Simo, R., Massana, R., *et al.* (2000) Bacterial community structure associated with a dimethyl sulfoniopropionate-producing North Atlantic algal bloom. *Applied Environmental Microbiology* **66**: 4237–4246.

Gorton, H. L., Herbert, S. K. and Vogelmann, T. C. (2003) Photoacoustic analysis indicates that chloroplast movement does not alter liquid-phase CO_2 diffusion in leaves of *Alocasia brisbanensis*. *Plant Physiology* **132**: 1529–1539.

Gorton, H. L., Williams, W. E. and Vogelmann, T. C. (1999) Chloroplast movement in *Alocasia macrorrhiza*. *Physiologia Plantarum* **106**: 421–428.

Gower, S. T. and Norman, J. M. (1990) Rapid estimation of leaf area index in forests using the LiCor LA1 2000. *Ecology* **72**: 1896–1900.

Grosse, W., Armstrong, J. and Armstrong, W. (1996) A history of pressurised gas-flow studies in plants. *Aquatic Botany* **54**: 87–100.

Grote, R., Mayrhofer, S., Fischbach, R. J., *et al.* (2006) Process-based modelling of isoprenoid emissions from evergreen leaves of *Quercus ilex* (L.). *Atmospheric Environment* **40**:152–165.

Guenther, A. B., Jiang, X., Heald, C. L., *et al.* (2012) The Model of Emissions of Gases and Aerosols from Nature version 2.1 (MEGAN2.1): An extended and updated framework for modeling biogenic emissions. *Geoscientific Model Development* **5**: 1471–1492.

Gutschick, V. P. (1991) Joining leaf photosynthesis models and canopy photon-transport models. In: *Photon-Vegetation Interaction: Applications in Optical Remote Sensing and Plant Ecology* (Myneni, R. B. and Ross, J., eds.). Berlin: Springer-Verlag, pp. 501–535.

Gutschick, V. P. and Simonneau, T. (2002) Modelling stomatal conductance of field-grown sunflower under varying soil water content and leaf environment: Comparison of three models of stomatal response to leaf environment and coupling with an abscisic acid-based model of stomatal response to soil drying. *Plant, Cell and Environment* **25**: 1423–1434.

Gutschick, V. P. and Weigel, F. W. (1984) Radiation transfer in vegetative canopies and layered media: Rapidly solvable exact integral equation not requiring Fourier resolution. *Journal of Quantitative Spectroscopy and Radiation Transfer* **31**: 71–82.

Gutteridge, S. and Pierce, J. (2006) A unified theory for the basis of the limitations of the primary reaction of photosynthetic CO_2 fixation: Was Dr. Pangloss right? *Proceedings of the National Academy of Science (USA)* **103**: 7203–7204.

Guy, R. D., Fogel, M. L. and Berry, J. A. (1993) Photosynthetic fractionation of the stable isotopes of oxygen and carbon. *Plant Physiology* **101**: 37–47.

Guyot, G., Scoffoni, C. and Sack, L. (2012) Combined impacts of irradiance and dehydration on leaf hydraulic conductance: Insights into vulnerability and stomatal control. *Plant, Cell and Environment* **35**: 857–871.

Haefner, J. W., Buckley, T. N. and Mott, K. A. (1997) A spatially explicit model of patchy stomatal responses to humidity. *Plant, Cell and Environment* **20**: 1087–1097.

Ham, J. M. and Heilman, J. L. (2003) Experimental test of density and energy-balance corrections on carbon dioxide flux as measured using open-path eddy covariance. *Agronomy Journal* **95**: 1393–1403.

Hansen, L. D., Hopkin, M. S., Rank, D. R., *et al.* (1994) The relation between plant growth and respiration: A thermodynamic model. *Planta* **194**: 77–85.

Harding, D. J., Lefsky, M. A., Parker, G. G. and Blair, J. B. (2001) Laser altimeter canopy height profiles: Methods and validation for closed-canopy, broadleaf forests. *Remote Sensing of Environment* **76**: 283–297.

Hari, P., Makela, A., Berninger, F. and Pohja, T. (1999) Field evidence for the optimality hypothesis of gas exchange in plants. *Australian Journal of Plant Physiology* **26**: 239–244.

Harley, P. C., Loreto, F., Di Marco, G. and Sharkey, T. D. (1992a) Theoretical considerations when estimating the mesophyll conductance to CO_2 flux by analysis of the response of photosynthesis to CO_2. *Plant Physiology* **98**: 1429–1436.

Harley, P. C., Monson, R. K. and Lerdau, M. T. (1999) Ecological and evolutionary aspects of isoprene emission from plants. *Oecologia* **118**: 109–123.

Harley, P. C. and Sharkey, T. D. (1991) An improved model of C_3 photosynthesis at high CO_2 – reversed O_2 sensitivity explained by lack of glycerate re-entry into the chloroplast. *Photosynthesis Research* **27**: 169–178.

Harley, P. C., Thomas, R. B., Reynolds, J. F. and Strain, B. R. (1992b) Modelling photosynthesis of cotton grown in elevated CO_2. *Plant, Cell and Environment* **15**: 271–282.

Harrison, S. P., Morfopoulos, C., Srikanta-Dani, S., *et al.* (2012) Volatile isoprenoid emissions from plastid to planet. *New Phytologist* **197**: 49–57.

Hartley, I. P. and Ineson, P. (2008) Substrate quality and the temperature sensitivity of soil organic matter decomposition. *Soil Biology and Biochemistry* **40**: 1567–1574.

Heiden, A. C., Kobel, K., Langebartels, C., *et al.* (2003) Emissions of oxygenated volatile organic compounds from plants – Part I: Emissions from lipoxygenase activity. *Journal of Atmospheric Chemistry* **45**: 143–172.

Heinrich, R. and Rapoport, T. A. (1974) A linear steady-state treatment of enzymatic chains: General properties, control and effector strength. *European Journal of Biochemistry* **42**: 107–120.

Heinsch, F. A., Zhao, M. S., Running, S. W., *et al.* (2006) Evaluation of remote sensing based terrestrial productivity from MODIS using regional tower eddy flux network observations. *IEEE Transactions on Geoscience and Remote Sensing* **44**: 1908–1925.

Henry, H. A. L., Brizgys, K. and Field, C. B. (2008) Litter decomposition in a California annual grassland: Interactions between photodegradation and litter layer thickness. *Ecosystems* **11**: 545–554.

Henze, D. K. and Seinfeld, J. H. (2006) Global secondary organic aerosol from isoprene oxidation. *Geophysical Research Letters* **33**: Article number L09812.

Henze, D. K., Seinfeld, J. H., Ng, N. L., *et al.* (2008) Global modeling of secondary organic aerosol formation from aromatic hydrocarbons: High- vs. low-yield pathways. *Atmospheric Chemistry and Physics* **8**: 2405–2420.

Hereid, D. P. and Monson, R. K. (2001) Nitrogen oxide fluxes between corn (*Zea mays* L.) leaves and the atmosphere. *Atmospheric Environment* **35**: 975–983.

Hirose, T. and Werger, M. J. A. (1987) Maximizing daily canopy photosynthesis with respect to the leaf nitrogen allocation pattern in the canopy. *Oecologia* **72**: 520–526.

Högberg, P., Högberg, M. N., Gottlicher, S. G., *et al.* (2008) High temporal resolution tracing of photosynthate carbon from the tree canopy to forest soil microorganisms. *New Phytologist* **177**: 220–228.

Högberg, P. and Read, D. J. (2006) Towards a more plant physiological perspective on soil ecology. *Trends in Ecology and Evolution* **21**: 548–554.

Hogstrom, U. (1988) Non-dimensional wind and temperature profiles in the atmospheric surface layer – a re-evaluation. *Boundary-Layer Meteorology* **42**: 55–78.

Holbrook, N. M. and Zwieniecki, M. A. (1999) Embolism repair and xylem tension: Do we need a miracle? *Plant Physiology* **120**: 7–10.

Hollinger, D. Y., Ollinger, S. V., Richardson, A. D., *et al.* (2010) Albedo estimates for land surface models and support for a new paradigm based on foliage nitrogen concentration. *Global Change Biology* **16**: 696–710.

Holtslag, A. A. M. and Nieuwstadt, F. T. M. (1986) Scaling the atmospheric boundary layer. *Boundary-Layer Meteorology* **36**: 201–209.

Holzapfel-Pschorn, A., Conrad, R. and Seiler, W. (1985) Production, oxidation and emission of methane in rice paddies. *FEMS Microbiology Letters* **31**: 343–351.

Holzapfel-Pschorn, A. and Seiler, W. (1986) Methane emission during a cultivation period from an Italian rice paddy. *Journal of Geophysical Research* **91**: 11803–11814.

Hopkins, F. M., Torn, M. S. and Trumbore, S. E. (2012) Warming accelerates decomposition of decades-old carbon in forest soils. *Proceedings of the National Academy of Sciences (USA)* **109**: E1753–E1761.

Houweling, S., Dentener, F. and Lelieveld, J. (2000) Simulation of preindustrial atmospheric methane to constrain the global source strength of natural wetlands. *Journal of Geophysical Research – Atmospheres* **105**:17243–17255.

Hrmova, M. and Fincher, G. B. (2001) Plant enzyme structure: Explaining substrate specificity and the evolution of function. *Plant Physiology* **125**: 54–57.

Hubick, K. T., Farquhar, G. D. and Shorter, R. (1896) Correlation between water-use efficiency and carbon isotope discrimination in diverse peanut (*Arachis*) germplasm. *Australian Journal of Plant Physiology* **13**: 803–816.

Hudak, A. T., Crookston, N. L., Evans, J. S., *et al.* (2008) Nearest neighbor imputation of species-level, plot-scale forest structure attributes from LiDAR data. *Remote Sensing of Ennvironment* **112**: 2232–2245.

Huijnen, V., Williams, J., van Weele, M., *et al.* (2010) The global chemistry transport model TM5: Description and evaluation of the tropospheric chemistry version 3.0. *Geoscientific Model Development* **3**: 445–473.

Hunt, J. C. R., Kaimal, J. C. and Gaynor, J. E. (1988) Eddy structure in the convective boundary layer – new measurements and new concepts. Quarterly *Journal of the Royal Meteorological Society* **114**: 827–858.

Huxman, T. E., Snyder, K. A., Tissue, D., *et al.* (2004) Precipitation pulses and carbon fluxes in semiarid and arid ecosystems. *Oecologia* **141**: 254–268.

IPCC (2001) *Intergovernmental Panel on Climate Change, Third Assessment Report.* Geneva: United Nations Environmental Program.

IPCC (2007) *Climate Change 2007: The Physical Science Basis. Fourth Assessment Report of the Intergovernmental Panel on Climate Change.* Cambridge: Cambridge University Press.

Ishii, H. T., Jennings, G. M., Sillett, S. C. and Koch, G. W. (2008) Hydrostatic constraints on morphological exploitation of light in tall Sequoia sempervirens trees. *Oecologia* **156**: 751–763.

Jacob, D., Field, B. D., Jin, E. M., *et al.* (2002) Atmospheric budget of acetone. *Journal of Geophysical Research – Atmospheres* **107**: Article number 4100.

Jacob, D., Field, B. D., Li, Q. B., *et al.* (2005) Global budget of methanol: constraints from atmospheric observations. *Journal of Geophysical Research – Atmospheres* **110**: Article number D08303.

Jacob, D. J. and Wofsy, S. C. (1990) Budgets of reactive nitrogen, hydrocarbons and ozone over the Amazon forest during the wet season. *Journal of Geophysical Research – Atmospheres* **95**: 16737–16754.

Janott, M., Gayler, S., Gessler, A., *et al.* (2011) A one-dimensional model of water flow in soil-plant systems based on plant architecture. *Plant and Soil* **341**: 233–256.

Jarman, P. D. (1974) The diffusion of carbon dioxide and water vapour through stomata. *Journal of Experimental Botany* **25**: 927–936.

Jarvis, P. G. (1971) The estimation of resistances to carbon dioxide transfer. In: *Plant Photosynthetic Production Manual of Methods* (Sestàk, J., Catsky, J. and Jarvis, P. G., eds.). The Hague: W. Junk, pp. 566–631.

Jarvis, P. G. (1995) Scaling processes and problems. *Plant, Cell and Environment* **18**: 1079–1089.

Jarvis, P. G. and Leverenz, J. (1983) Productivity of temperate, deciduous and evergreen forests. In: *Encyclopedia of Plant Physiology* (Lange, O. L., Nobel, P. S., Osmond, C. B. and Ziegler, H., eds.). Berlin: Springer-Verlag, pp. 233–280.

Jarvis, P. G. and McNaughton, K. G. (1986) Stomatal control of transpiration: Scaling up from leaf to region. *Advances in Ecological Research* **15**: 1–49.

Jarvis, P., Rey, A., Petsikos, C., *et al.* (2007) Drying and wetting of Mediterranean soils stimulates decomposition and carbon dioxide emission: The "Birch effect." *Tree Physiology* **27**: 929–940.

Jenny, H. (1980) *The Soil Resource: Origin and Behaviour.* New York: Springer-Verlag.

Jensen, H. W. (2001) *Realistic Image Synthesis Using Photon Mapping.* Wellesley, MA: AK Peters.

Jin, Z. H., Charlock, T. P., Smith, W. L., *et al.* (2004) A parameterization of ocean surface albedo. *Geophysical Research Letters* **31**: Article number L22301.

Jobbágy, E. G. and Jackson, R. B. (2000) The vertical distribution of soil organic carbon and its relation to climate and vegetation. *Ecological Applications* **10**: 423–436.

Joffre, S. M., Kangas, M., Heikinheimo, M. and Kitaigorodskii, S. A. (2001) Variability of the stable and unstable atmospheric boundary-layer height and its scales over a boreal forest. *Boundary-Layer Meteorology* **99**: 429–450.

Johnson, F. H., Eyring, H. and Williams, R. W. (1942) The nature of enzyme inhibitions in bacterial luminescence: Sulfanilamide, urethane, temperature and pressure. *Journal of Cellular and Comparative Physiology* **20**: 247–268.

Jones, D. L., Nguyen, C. and Finlay, R. D. (2009) Carbon flow in the rhizosphere: Carbon trading at the soil-root interface. *Plant and Soil* **321**: 5–33.

Jones, H. G. (1998) Stomatal control of photosynthesis and transpiration. *Journal of Experimental Botany* **49**: 387–398.

Jordan, D. B. and Ogren, W. L. (1981) Species variation in the specificity of ribulose-bisphosphate carboxlase-oxygenase. *Nature* **291**: 513–515.

Jordan, D. B. and Ogren, W. L. (1984) The CO_2/O_2 specificity of ribulose 1,5-bisphosphate carboxylase oxygenase – dependence on ribulose bisphosphate concentration, pH and temperature. *Planta* **161**: 308–313.

June, T., Evans, J. R. and Farquhar, G. D. (2004) A simple new equation for the reversible temperature dependence of photosynthetic electron transport: A study on soybean leaf. *Functional Plant Biology* **31**: 275–283.

Kacser, H. and Burns, J. A. (1973) The control of flux. *Society of Experimental Biology Symposium* **27**: 65–104.

Kaimal, J. C. and Finnigan, J. J. (1994) *Atmospheric Boundary Flows: Their Structure and Measurement*. New York: Oxford University Press.

Kaimal, J. C. and Wyngaard, J. C. (1990) The Kansas and Minnesota Experiments. *Boundary-Layer Meteorology* **50**: 31–47.

Kaimal, J. C., Wyngaard, J. C., Izumi, Y. and Coté, O. R. (1972) Spectral characteristics of surface layer turbulence. *Quarterly Journal of the Royal Meteorological Society* **98**: 563–589.

Karl, T., Curtis, A. J., Rosenstiel, T. N., *et al.* (2002) Transient releases of acetaldehyde from tree leaves – products of a pyruvate bypass mechanism? *Plant, Cell and Environment* **25**: 1121–1131.

Katul, G. G., Ellsworth, D. S. and Lai, C. T. (2000) Modelling assimilation and intercellular CO_2 from measured conductance: A synthesis of approaches. *Plant, Cell and Environment* **23**: 1313–1328.

Kattge, J., Knorr, W., Raddatz, T., *et al.* (2009) Quantifying photosynthetic capacity and its relationship to leaf nitrogen content for global-scale terrestrial biosphere models. *Global Change Biology* **15**: 976–991.

Keeley, J. E. and Rundel, P. W. (2005) Fire and the Miocene expansion of C_4 grasslands. *Ecology Letters* **8**: 683–690.

Keeling, C. D. (1958) The concentration and isotopic abundances of atmospheric carbon dioxide in rural areas. *Geochimica et Cosmochimica Acta* **13**: 322–334.

Keeling, C. D. (1961) The concentration and isotopic abundance of carbon dioxide in rural and marine air. *Geochimica et Cosmochimica Acta* **24**: 277–298.

Keeling, C. D. (1979) The Suess effect: [13]Carbon–[14]Carbon interrelations. *Environment International* **2**: 229–300.

Kemmitt, S. J., Lanyon, C. V., Waite, I. S., *et al.* (2008) Mineralization of native soil organic matter is not regulated by the size, activity or composition of the soil microbial biomass – a new perspective. *Soil Biology and Biochemistry* **40**: 61–73.

Kim, J., Verma, S. B., Billesbach, D. P. and Clement, R. J. (1998) Diel variation in methane emission from a midlatitude prairie wetland: Significance of convective through flow in *Phragmites australis*. *Journal of Geophysical Research – Atmospheres* **103**: 28029–28039.

Kim, K. -H., and Andreae, M. O. (1987) Carbon disulfide in seawater and the marine atmosphere over the North Atlantic. *Journal of Geophysical Research* **92**: 14733–14738.

Kirschbaum, M. U. F. (1995) The temperature dependence of soil organic matter decomposition, and the effect of global warming on soil organic C storage. *Soil Biology and Biochemistry* **27**: 753–760.

Kirschbaum, M. U. F. (2000) Will changes in soil organic carbon act as a positive or negative feedback on global warming? *Biogeochemistry* **48**: 21–51.

Knorr, W., Prentice, I. C., House, J. I. and Holland, E. A. (2005) Long-term sensitivity of soil carbon turnover to warming. *Nature* **433**: 298–301.

Kokhanovsky, A. and Schreier, M. (2009) The determination of snow specific surface area, albedo and effective grain size using AATSR space-borne measurements. *International Journal of Remote Sensing* **30**: 919–933.

Kolmogorov, A. (1941) The local structure of turbulence in incompressible viscous fluid for very large Reynolds' numbers. *Doklady Akademiia Nauk SSSR* **30**: 301–305.

Körner, C. and Diemer, M. (1987) In situ photosynthetic responses to light, temperature and carbon dioxide in herbaceous plants from low and high altitude. *Functional Ecology* **1**: 179–194.

Kraut, D. A., Carroll, K. S. and Herschlag, D. (2003) Challenges in enzyme mechanism and energetics. *Annual Review of Biochemistry* **72**: 517–571.

Kreim, M. and Giersch, C. (2007) Measuring in vivo elasticities of Calvin cycle enzymes: Network structure and patterns of modulations. *Phytochemistry* **68**: 2152–2162.

Kreuzwieser, J., Papadopoulou, E. and Rennenberg, H. (2004) Interaction of flooding with carbon metabolism of forest trees. *Plant Biology* **6**: 299–306.

Kroll, J. H., Ng, N. L., Murphy, S. M., *et al.* (2006) Secondary organic aerosol formation from isoprene photooxidation. *Environmental Science and Technology* **40**: 1869–1877.

Krupa, S. V. (2003) Effects of atmospheric ammonia (NH_3) on terrestrial vegetation: A review. *Environmental Pollution* **124**:179–221.

Kruse, J. and Adams, M. A. (2008) Sensitivity of respiratory metabolism and efficiency to foliar nitrogen during growth and maintenance. *Global Change Biology* **14**: 1233–1251.

Kuzyakov, Y. (2006) Sources of CO_2 efflux from soil and review of partitioning methods. *Soil Biology and Biochemistry* **38**: 425–448.

Kuzyakov, Y., Friedel, J. K. and Stahr, K. (2000) Review of mechanisms and quantification of priming effects. *Soil Biology and Biochemistry* **32**: 1485–1498.

Kuzyakov, Y. and Gavrichkova, O. (2010) Time lag between photosynthesis and carbon dioxide efflux from soil: A review of mechanisms and controls. *Global Change Biology* **16**: 3386–3406.

Laidler, K. J. and King, M. C. (1983) The development of transition state theory. *Journal of Physical Chemistry* **87**: 2657–2664.

Lambers, H., Mougel, C., Jaillard, B. and Hinsinger, P. (2009) Plant-microbe-soil interactions in the rhizosphere: An evolutionary perspective. *Plant and Soil* **321**: 83–115.

Langford, A. O. and Fehsenfeld, F. C. (1992) Natural vegetation as a source or sink for atmospheric ammonia – a case study. *Science* **255**: 581–583.

Lanigan, G. J., Betson, N., Griffiths, H. and Seibt, U. (2008) Carbon isotope fractionation during photorespiration and carboxylation in Senecio. *Plant Physiology* **148**: 2013–2020.

Laothawornkitkul, J., Taylor, J. E., Paul, N. D. and Hewitt, C. N. (2009) Biogenic volatile organic compounds in the Earth system. *New Phytologist* **183**: 27–51.

Le Mer, J. and Roger, P. (2001) Production, oxidation, emission and consumption of methane by soils: A review. *European Journal of Soil Biology* **37**: 25–50.

Leclerc, M. Y., Beissner, K. C., Shaw, R. H., *et al.* (1990) The influence of atmospheric stability on the budgets of the Reynolds stress and turblent kinetic energy within and above a deciduous forest. *Journal of Applied Meteorology* **29**: 916–933.

Leclerc, M. Y., Thurtell, G. W. and Kidd, G. E. (1988) Measurements and Langevin simulations of mean tracer concentration fields downwind from a circular line source inside an alfalfa canopy. *Boundary-Layer Meteorology* **43**: 287–308.

Lee, A., Goldstein, A. H., Kroll, J. H., *et al.* (2006) Gas-phase products and secondary aerosol yields from the photooxidation of 16 different terpenes. *Journal of Geophysical Research – Atmospheres* **111**: Article number D17305.

Lee, X., Goulden, M. L., Hollinger, D. Y., *et al.* (2011) Observed increase in local cooling effect of deforestation at higher latitudes. *Nature* **479**: 384–387.

Lee, X., Neumann, H. H., DenHartog, G., *et al.* (1997) Observation of gravity waves in a boreal forest. *Boundary-Layer Meteorology* **84**: 383–398.

Lefsky, M. A., Cohen, W. B., Parker, G. G. and Harding, D. J. (2002) Lidar remote sensing for ecosystem studies. *Bioscience* **52**: 19–30.

Legg, B. J. and Raupach, M. R. (1982) Markov chain simulation of particle dispersion in inhomogeneous flows – the mean drift velocity induced by a gradient in Eulerian velocity variance. *Boundary-Layer Meteorology* **24**: 3–13.

Lelieveld, J., Butler, T. M., Crowley, J. N., *et al.* (2008) Atmospheric oxidation capacity sustained by a tropical forest. *Nature* **452**: 737–740.

Lemeur, R. and Blad, B. L. (1974) Critical review of light models for estimating shortwave radiation regime of plant canopies. *Agricultural Meteorology* **14**: 255–286.

Lenschow, D. H. (1995) Micrometeorological techniques for measuring biosphere-atmosphere trace gas exchange. In: *Biogenic Trace Gases: Measuring Emissions from Soil and Water* (Matson, P. A. and Harriss, R. C., eds.). Oxford: Blackwell Scientific, pp. 126–163.

Lerdau, M., Guenther, A. and Monson, R. (1997) Plant production and emission of volatile organic compounds. *Bioscience* **47**: 373–383.

Lerdau, M. T., Munger, W. and Jacob, D. J. (2000) The NO_2 flux conundrum. *Science* **289**: 2291–2293.

Leuning, R. (1983) Transport of gases into leaves. *Plant, Cell and Environment* **6**: 181–194.

Leuning, R. (1995) A critical appraisal of a combined stomatal-photosynthesis model for C_3 plants. *Plant, Cell and Environment* **18**: 339–355.

Leuning, R., Kelliher, F. M., DePury, D. G. G. and Schulze, E. -D. (1995) Leaf nitrogen, photosynthesis, conductance and transpiration: scaling from leaves to canopies. *Plant, Cell and Environment* **18**: 1183–1200.

Li, J. and Dobbie, J. S. (1998) Four-stream isosector approximation for solar radiative transfer. *Journal of Atmospheric Science* **55**: 558–567.

Li, X. and Strahler, A. H. (1995) A hybrid geometric optical radiative transfer approach for modeling albedo and directional reflectance of discontinuous canopies. *IEEE Transactions on Geoscience and Remote Sensing* **33**: 466–480.

Liesack, W., Schnell, S. and Revsbech, N. P. (2000) Microbiology of flooded rice paddies. *FEMS Microbiology Reviews* **24**: 625–645.

Lipson, D. and Näsholm, T. (2001) The unexpected versatility of plants: Organic nitrogen use and availability in terrestrial ecosystems. *Oecologia* **128**: 305–316.

Liu, S. H., Liu, H. P., Xu, M., *et al.* (2001) Turbulence spectra and dissipation rates above and within a forest canopy. *Boundary-Layer Meteorology* **98**: 83–102.

Liu, X. Z., Wan, S. Q., Su, B., *et al.* (2002) Response of soil CO_2 efflux to water manipulation in a tallgrass prairie ecosystem. *Plant and Soil* **240**: 213–223.

Lloyd, J. (1991) Modelling stomatal responses to environment in *Macademia integrifolia* (L.) Batsch. *Australian Journal of Plant Physiology* **18**: 649–660.

Lloyd, J. and Farquhar, G. D. (1994) ^{13}C discrimination during CO_2 assimilation by the terrestrial biosphere. *Oecologia* **99**: 201–215.

Lloyd, J., Grace, J., Miranda, A. C., *et al.* (1995) A simple calibrated model of Amazon rainforest productivity based on leaf biochemical properties. *Plant, Cell and Environment* **18**: 1129–1145.

Lloyd, J., Patino, S., Paiva, R. Q., *et al.* (2010) Optimisation of photosynthetic carbon gain and within-canopy gradients of associated foliar traits for Amazon forest trees. *Biogeosciences* **7**: 1833–1859.

Lloyd, J. and Taylor, J. A. (1994) On the temperature dependence of soil respiration. *Functional Ecology* **8**: 315–323.

Lohammer, T., Larsson, S., Linder, S. and Falk, S. O. (1980) FAST – Simulation models of gaseous exchange in Scots pine. *Ecological Bulletin* **32**: 505–523.

Long, S. P., Ainsworth, E. A., Rogers, A. and Ort, D. R. (2004) Rising atmospheric carbon dioxide: Plants FACE the future. *Annual Review of Plant Biology* **55**: 591–628.

Long, S. P. and Bernacchi, C. J. (2003) Gas exchange measurements, what can they tell us about the underlying limitations of photosynthesis? Procedures and sources of error. *Journal of Experimental Botany* **54**: 2393–2401.

Long, S. P. and Drake, B. G. (1992) Photosynthetic CO_2 assimilation and rising atmospheric CO_2 concentrations. In: *Crop Photosynthesis: Spatial and Temporal Determinants* (Baker, N. and Thomas, H., eds.). Amsterdam: Elsevier Scientific, pp. 69–103.

Lothon, M., Lenschow, D. H. and Mayor, S. D. (2009) Doppler lidar measurements of vertical velocity spectra in the convective planetary boundary layer. *Boundary-Layer Meteorology* **132**: 205–226.

Lovelock, J. E. and Margulis, L. (1974) Atmospheric homeostasis by and for the biosphere: the Gaia hypothesis. *Tellus* **26**: 1–10.

Lumley, J. L. and Panofsky, H. A. (1964) *The Structure of Atmospheric Turbulence.* Monographs and Texts in Physics and Astronomy, Vol. XII. New York: John Wiley & Sons.

Luo, Y. Q. (2007) Terrestrial carbon-cycle feedback to climate warming. *Annual Review of Ecology and Systematics* **38**: 683–712.

Lusk, C. H., Reich, P. B., Montgomery, R. A., *et al.* (2008) Why are evergreen leaves so contrary about shade? *Trends in Ecology and Evolution* **23**: 299–303.

Mahecha, M. D., Reichstein, M., Carvalhais, N., *et al.* (2010) Global convergence in the temperature sensitivity of respiration at ecosystem level. *Science* **329**: 838–840.

Mahrt, L., Vickers, D., Nakamura, R., *et al.* (2001) Shallow drainage flows. *Boundary-Layer Meteorology* **101**: 243–260.

Majeau, N. and Coleman, J. R. (1996) Effect of CO_2 concentration on carbonic anhydrase and ribulose 1,5-bisphosphate carboxylase/oxygenase expression in pea. *Plant Physiology* **112**: 569–574.

Malcher, J. and Kraus, H. (1983) Low-level jet phenomena described by an integrated dynamical PBL model. *Boundary-Layer Meteorology* **27**: 327–343.

Marin-Spiotta, E., Silver, W. L., Swanston, C. W. and Ostertag, R. (2009) Soil organic matter dynamics during 80 years of reforestation of tropical pastures. *Global Change Biology* **15**: 1584–1597.

Massman, W. J. (1998) A review of the molecular diffusivities of H_2O, CO_2, CH_4, CO, O_3, SO_2, NH_3, N_2O, NO, and NO_2 in air, O_2 and N_2 near STP. *Atmospheric Environment* **32**: 1111–1127.

Massman, W. J. and Weil, J. C. (1999) An analytical one-dimensional second-order closure model of turbulence statistics and the Lagrangian time scale within and above plant canopies of arbitrary structure. *Boundary-Layer Meteorology* **94**: 81–107.

Matzner, S. and Comstock, J. (2001) The temperature dependence of shoot hydraulic resistance: implications for stomatal behaviour and hydraulic limitation. *Plant, Cell and Environment* **24**: 1299–1307.

Maurel, C., Verdoucq, L., Luu, D.-T. and Santoni, V. (2008) Plant aquaporins: membrane channels with multiple integrated functions. *Annual Review of Plant Biology* **59**: 595–624.

McCree, K. J. (1970) An equation for the rate of respiration of white clover plants grown under controlled conditions. In: *Prediction and Measurement of Photosynthetic Productivity* (Setlik, I., ed.). Wageningen: Pudoc Publishers, pp. 221–229.

McCulloh, K. A., Sperry, J. S. and Adler, F. R. (2003) Water transport in plants obeys Murray's law. *Nature* **421**: 939–942.

McCulloh, K., Sperry, J. S., Lachenbruch, B., *et al.* (2010) Moving water well: comparing hydraulic efficiency in twigs and trunks of coniferous, ring-porous, and diffuse-porous saplings from temperate and tropical forests. *New Phytologist* **186**: 439–450.

McGill, W. B. (1996) Review and classification of ten soil organic matter (SOM) models. In: *Evaluation of Soil Organic Matter Models* (Smith, U. J, ed.). Berlin: Springer-Verlag, pp. 111–132.

McNaughton, K. G. (1989) Regional interactions between canopies and the atmosphere. In: *Plant Canopies: their Growth, Form and Function* (Russell, G., Marshall, B. and Jarvis, P. G., eds.). Society for Experimental Biology Seminar Series 31. Cambridge: Cambridge University Press, pp. 63–81.

McNevin, D. B., Badger, M. R., Whitney, S. M., *et al.* (2007) Differences in carbon isotope discrimination of three variants of D-ribulose-1,5-bisphosphate carboxylase/oxygenase reflect differences in their catalytic mechanisms. *Journal of Biological Chemistry* **282**: 36068–36076.

Medlyn, B. E., Dreyer, E., Ellsworth, D., Forstreuter, M., Harley, P. C., Kirschbaum, M. U. F., Le Roux, X., Montpied, P., Strassemeyer, J., Walcroft, A., Wang, K. and Loustau, D. (2002) Temperature response of parameters of a biochemically based model of photosynthesis. II: A review of experimental data. *Plant, Cell and Environment* **25**: 1167–1179.

Meidner, H. and Mansfield, T. A. (1968) *Physiology of Stomata*. London: McGraw-Hill.

Meinzer, F. C., McCulloh, K. A., Lachenbruch, B., Woodruff, D. R. and Johnson, D. M. (2010) The blind men and the elephant: the impact of context and scale in evaluating conflicts between plant hydraulic safety and efficiency. *Oecologia* **164**: 287–296.

Meir, P., Kruijit, B., Broadmeadow, M., Barbosa, E., Kull, O., Carswell, F., Nobre, A. and Jarvis, P. G. (2002) Acclimation of photosynthetic capacity to irradiance in tree canopies in relation to leaf nitrogen concentration and leaf mass per unit area. *Plant, Cell and Environment* **25**: 343–357.

Merlivat, L. (1978) Molecular diffusivities of $H_2^{18}O$ in gases. *Journal of Chemistry and Physics* **69**: 2864–2871.

Meyers, T. P and Baldocchi, D. D. (1991) The budgets of turbulent kinetic energy and Reynolds stress within and above a deciduous forest. *Agricultural and Forest Meteorology* **53**: 207–222.

Miller, E. E. and Norman, J. M. (1971) Sunfleck theory for plant canopies. 2. Penumbra effect – intensity distributions along sunfleck segments. *Agronomy Journal* **63**: 739–743.

Millet, D. B., Jacob, D. J., Custer, T. G., *et al.* (2008) New constraints on terrestrial and oceanic sources of atmospheric methanol. *Atmospheric Chemistry and Physics* **8**: 6887–6905.

Monin, A. S. and Obukhov, A. M. (1954) Basic laws of turbulent mixing in the ground layer of the atmosphere. *Transactions of the Geophysical Institute of the Akademie Nauk, USSR* **151**: 163–187.

Monson, R. K. (2010) Reactions of biogenic volatile organic compounds in the atmosphere. In: *An introduction to the Chemistry and Biology of Volatiles* (Hermann, A., ed.). Chichester: John Wiley & Sons, pp. 363–388.

Monson, R. K., and Collatz, G. J. (2012) The ecophysiology and global biology of C_4 photosynthesis. In: *Terrestrial Photosynthesis in a Changing Environment: A Molecular, Physiological and Ecological Approach* (Flexas, J., Loreto, F. and Medrano, H., eds.). Cambridge: Cambridge University Press, pp. 54–70.

Monson, R. K. and Fall, R. (1989) Isoprene emission from aspen leaves. The influence of environment and relation to photosynthesis and photorespiration. *Plant Physiology* **90**: 267–274.

Monson, R. K. Grote, R., Niinemets, U. and Schnitzler, J. P. (2012) Modeling the isoprene emission rate from leaves. *New Phytologist* **195**: 541–559.

Monson, R. K. and Holland, E. A. (2001) Biospheric trace gas fluxes and their control over tropospheric chemistry. *Annual Review of Ecology and Systematics* **32**: 547–576.

Monson, R. K., Stidham, M. A., Williams, G. J., III, *et al.* (1982) Temperature dependence of photosynthesis in *Agropyron smithii* Rydb. I. Factors affecting net CO_2 uptake in intact leaves and contribution from ribulose 1,5-bisphosphate carboxylase measured *in vivo* and *in vitro*. *Plant Physiology* **69**: 921–928.

Monson, R. K., Trahan, N., Rosenstiel, T. N., *et al.* (2007) Isoprene emission from terrestrial ecosystems in response to global change: minding the gap between models and observations. *Philosophical Transactions of the Royal Society, Series A* **365**: 1677–1695.

Monteith, J. L. (1972) Solar radiation and productivity in tropical ecosystems. *Journal of Applied Ecology* **9**: 747–766.

Monteith, J. L. (1981) Evaporation and surface temperature. *Quarterly Journal of the Royal Meteorological Society* **107**: 1–27.

Monteith, J. L. (1995) A reinterpretation of stomatal responses to humidity. *Plant, Cell and Environment* **18**: 357–364.

Monteith, J. L. and Unsworth, M. H. (1990) *Principles of Environmental Physics*, 2nd edn. New York: Edward Arnold.

Montgomery, R. A. (2004) Effects of understory foliage on patterns of light attenuation near the forest floor. *Biotropica* **36**: 33–39.

Mooney, H. A., Gulmon, S. L., Rundel, P. W. and Ehleringer, J. (1980) Further observations of the water relations of *Prosopis tamarugo* of the northern Atacama Desert. *Oecologia* **44**: 177–180.

Moore, B. D., Cheng, S. -H., Rice, J. and Seemann, J. R. (1998) Sucrose cycling, Rubisco expression, and prediction of photosynthetic acclimation to elevated atmospheric CO_2. *Plant, Cell and Environment* **21**: 905–915.

Moore, B. D., Cheng, S. -H., Sims, D. and Seemann, J. R. (1999) The biochemical and molecular basis for photosynthetic acclimation to elevated atmospheric CO_2. *Plant, Cell and Environment* **22**: 567–582.

Moorhead, D. L. and Sinsabaugh, R. L. (2006) A theoretical model of litter decay and microbial interaction. *Ecological Monographs* **76**: 151–174.

Moorhead, D. L., Sinsabaugh, R. L., Linkins, A. E. and Reynolds, J. F. (1996) Decomposition processes: Modelling approaches and applications. *Science of the Total Environment* **183**: 137–149.

Morell, M. K., Paul, K., Kane, H. J. and Andrews, T. J. (1992) Rubisco – maladapted or misunderstood. *Australian Journal of Botany* **40**: 431–441.

Mori, S., Yamaji, K., Ishida, A., *et al.* (2010) Mixed-power scaling of whole-plant respiration from seedlings to giant trees. *Proceedings of the National Academy of Sciences (USA)* **107**: 1447–1451.

Mott, K. A. (1988) Do stomata respond to CO_2 concentrations other than intercellular? *Plant Physiology* **86**: 200–203.

Mott, K. A. (2009) Opinion: Stomatal responses to light and CO_2 depend on the mesophyll. *Plant, Cell and Environment* **32**: 1479–1486.

Mott, K. A. and Franks, P. J. (2001) The role of epidermal turgor in stomatal interactions following a local perturbation in humidity. *Plant, Cell and Environment* **24**: 657–662.

Mott, K. A. and Parkhurst, D. F. (1990) Stomatal responses to humidity in air and helox. *Plant, Cell and Environment* **14**. 509–515.

Mott, K. A. and Peak, D. (2010) Stomatal responses to humidity and temperature in darkness. *Plant, Cell and Environment* **33**: 1084–1090.

Mõttus, M. (2004) Measurement and modelling of the vertical distribution of sunflecks, penumbra and umbra in willow coppice. *Agricultural and Forest Meteorology* **121**: 79–91.

Mõttus, M. (2007) Photon recollision probability in discrete crown canopies. *Remote Sensing of Environment* **110**: 176–185.

Mu, Q., Heinsch, F. A., Zhao, M. and Running, S. W. (2007) Development of a global evapotranspiration algorithm based on MODIS and global meteorology data. *Remote Sensing of Environment* **111**: 519–536.

Myneni, R. B. (1991) Modeling radiative transfer and photosynthesis in three-dimensional vegetation canopies. *Agricultural and Forest Meteorology* **55**: 323–344.

Myneni, R. B., Hoffman, S., Knyazikhin, *et al.* (2002) Global products of vegetation leaf area and fraction absorbed PAR from year one of MODIS data. *Remote Sensing of Environment* **83**: 214–231.

Myneni, R. B., Maggion, S., Iaquinta, J., Privette, J. L., Gobron, N., Pinty, B., *et al.* (1995) Optical remote sensing of vegetation: Modeling, caveats, and algorithms. *Remote Sensing of Environment* **51**: 169–188.

Myneni, R. B., Ross, J. and Asrar, G. (1989) A review of the theory of photon transport in leaf canopies. *Agricultural and Forest Meteorology* **45**: 1–153.

Nadelhoffer, K. J. and Fry, B. (1988) Controls on natural ^{15}N and ^{13}C abundances in forest soil organic matter. *Soil Science Society of America Journal* **52**: 1633–1640.

Nave, L. E., Vance, E. D., Swanston, C. W. and Curtis, P. S. (2009) Impacts of elevated N inputs on north temperate forest soil C storage, C/N, and net N-mineralization. *Geoderma* **153**: 231–240.

Nemecek-Marshall, M., Macdonald, R. C., Franzen, J. J., Wojciechowski, C. and Fall, R. (1995) Methanol emission from leaves. Enzymatic detection of gas-phase methanol and relation of methanol fluxes to stomatal conductance and leaf development. *Plant Physiology* **108**: 1359–1368.

Neumann, R. B. and Cardon, Z. G. (2012) The magnitude of hydraulic redistribution by plant roots: A review and synthesis of empirical and modeling studies. *New Phytologist* **134**: 337–352.

Nie, G. Y., Hendrix, D. L., Webber, A. N., Kimball, B. A. and Long, S. P. (1995a) Increased accumulation of carbohydrates and decreased photosynthetic gene transcript levels in wheat grown at an elevated CO_2 concentration in the field. *Plant Physiology* **108**: 975–983.

Nie, G. Y., Long, S. P., Garcia, R. L., Kimball, B. A., Lamorte, R. A., Pinter, P. J., Wall, G. W. and Webber, A. N. (1995a) Effects of free air CO_2 enrichment on the development of the photosynthetic apparatus in wheat as indicated by changes in leaf proteins. *Plant, Cell and Environment* **18**: 855–864.

Nier, A. O. and Gulbransen, E. A. (1939) Variations in the relative abundance of the carbon isotopes. *Journal of the American Chemical Society* **61**: 697–698.

Niinemets, U. (2007) Photosynthesis and resource distribution through plant canopies. *Plant, Cell and Environment* **30**: 1052–1071.

Niinemets, U., Loreto, F. and Reichstein, M. (2004) Physiological and physicochemical controls on foliar volatile organic compound emissions. *Trends in Plant Science* **9**: 180–186.

Niinemets, U., Tenhunen, J. D., Harley, P. C. and Steinbrecher, R. (1999) A model of isoprene emission based on energetic requirements for isoprene synthesis and leaf photosynthetic properties for Liquidambar and Quercus. *Plant, Cell and Environment* **22**: 1319–1335.

Nilson, T. (1971) A theoretical analysis of the frequency of gaps in plant stands. *Agricultural Meteorology* **8**: 25–38.

Nisbet, E. G., Grassineau, N. V., Howe, C. J., Abell, P. I., Regelous, M. and Nisbet, R. E. R. (2007) The age of Rubisco: the evolution of oxygenic photosynthesis. *Geobiology* **5**: 311–335.

Nobel, P. S. (1999) *Physicochemical and Environmental Plant Physiology*, 2nd edn. San Diego, CA: Academic Press.

Nobel, P. S., Zaragoza, L. J. and Smith, W. K. (1975) Relation between mesophyll surface area, photosynthetic rate, and illumination of *Plectranthus parviflorus* Henckel. *Plant Physiology* **55**: 1067–1070.

Nonami, H. and Schulze, E. D. (1989) Cell water potential, osmotic potential and turgor in the epidermis and mesophyll of transpiring leaves – combined measurements with the cell pressure probe and nanoliter osmometer. *Planta* **177**: 35–46.

Norman, J. M. (1981) Interfacing leaf and canopy light interception models. In: *Predicting Photosynthesis for Ecosystem Models* (Hesketh, J. D. and Jones, J. W., eds.). Boca Raton, FL: CRC Press, pp. 49–67.

Norman, J. M. and Campbell, G. S. (1989) Canopy structure. In: *Plant Physiological Ecology: Field Methods and Instrumentation* (Pearcy, R. W., Ehleringer, J., Mooney, H. A. and Rundel, P. W., eds.). New York: Chapman and Hall, pp. 301–326.

Norman, J. M. and Welles, J. M. (1983) Radiative transfer in an array of canopies. *Agronomy Journal* **75**: 481–488.

Nowak, R. S., Ellsworth, D. S. and Smith, S. D. (2004) Functional responses of plants to elevated atmospheric CO_2 – do photosynthetic and productivity data from FACE experiments support early predictions? *New Phytologist* **162**: 253–280.

O'Leary, M. H. (1993) Biochemical basis of carbon isotope fractionation. In: *Isotopes and Plant Carbon-Water Relations* (Stable, J. R., Saranga, Y., Flash, I. and Yakir, D., eds.). New York: Academic Press, Inc., pp. 19–28.

Oades, J. M. (1988) The retention of organic matter in soils. *Biogeochemistry* **5**: 35–70.

Ögren, E. and Evans, J. R. (1993) Photosynthetic light response curves. I. The influence of CO_2 partial pressure and leaf inversion. *Planta* **189**: 180–190.

Oke, T. R. (1987) *Boundary Layer Climates*, 2nd edn. London: Routledge.

Oker-Blom, P. (1984) Penumbral effects of within-plant and between-plant shading on radiation distributions and leaf photosynthesis: a Monte-Carlo simulation. *Photosynthetica* **18**: 522–528.

Ollinger, S. V., Richardson, A. D., Martin, M. E., *et al.* (2008) Canopy nitrogen, carbon assimilation, and albedo in temperate and boreal forests: Functional relations and potential climate feedbacks. *Proceedings of the National Academy of Sciences (USA)* **105**: 19336–19341.

Osono, T. and Takeda, H. (2006) Fungal decomposition of Abies needle and Betula leaf litter. *Mycologia* **98**: 172–179.

Ow, L. F., Griffin, K. L., Whitehead, D., Walcroft, A. S. and Turnbull, M. H. (2008) Thermal acclimation of leaf respiration but not photosynthesis in *Populus deltoides* x *nigra*. *New Phytologist* **178**: 123–134.

Oya, V. and Laisk, A. (1976) Adaptation of the photosynthetic apparatus to the profile of light in the leaf. *Fiziologia Rastenii* **23**: 445–451.

Pacala, S. and Socolow, R. (2004) Stabilization wedges: solving the climate problem for the next 50 years with current technologies. *Science* **305**: 968–972.

Parkhurst, D. F. (1994) Diffusion of CO_2 and other gases inside leaves. *New Phytologist* **126**: 449–479.

Parkhurst, D. F., Wong, S.-C., Farquhar, G. D., and Cowan, I. R. (1988) Gradients of intercellular CO_2 levels across the leaf mesophyll. *Plant Physiology* **86**: 1032–1037.

Parton, W. J., Schimel, D. S., Cole, C. V. and Ojima, D. S. (1987) Analysis of factors controlling soil organic matter levels in Great Plains grasslands. *Soil Science Society of America Journal* **51**: 1173–1179.

Pataki, D. E. (2005) Emerging topics in stable isotope ecology: are there isotope effects in plant respiration? *New Phytologist* **167**: 321–323.

Pataki, D. E., Ehleringer, J. R., Flanagan, L. B., *et al.* (2003) The application and interpretation of Keeling plots in terrestrial carbon cycle research. *Global Biogeochemical Cycles* **17**: Article number 1022.

Patra, P. K., Takigawa, M., Ishijima, K., *et al.* (2009) Growth rate, seasonal, synoptic, diurnal variations and budget of methane in the lower atmosphere. *Journal of the Meteorological Society of Japan* **87**: 635–663.

Paul, E. A. (1984) Dynamics of organic matter in soils. *Plant and Soil* **76**: 275–285.

Paul, E. A. and Clark, F. E. (1996) *Soil Microbiology and Biochemistry*, 2nd edn. San Diego, CA: Academic Press.

Paulson, S. E., Flagan, R. C. and Seinfeld, J. H. (1992) Atmospheric photooxidation of isoprene part II: The ozone-isoprene reaction. *International Journal of Chemical Kinetics* **24**: 103–125.

Peak, D. and Mott, K. A. (2011) A new, vapour-phase mechanism for stomatal responses to humidity and temperature. *Plant, Cell and Environment* **34**: 162–178.

Pearman, G. I., Tanner, C. B. and Weaver, H. L. (1972) Boundary-layer heat transfer coefficients under field conditions. *Agricultural Meteorology* **10**: 83–92.

Penman, H. L. (1948) Natural evaporation from open water, bare soil and grass. *Proceedings of the Royal Society of London A* **193**: 120–145.

Penning de Vries, F. W. T., (1975) The cost of maintenance processes in plant cells. *Annals of Botany* **39**: 77–92.

Pfister, G. G., Emmons, L. K., Hess, P. G., *et al.* (2008) Contribution of isoprene to chemical budgets: A model tracer study with the NCAR CTM MOZART-4. *Journal of Geophysical Research – Atmospheres* **113**: Article number D05308.

Philip, J. R. (1966) Plant water relations: some physical aspects. *Annual Review of Plant Physiology* **17**: 245–268.

Pielke, R. A., Avissar, R., Raupach, M., Dolman, A. J., Zeng, X., Denning, A. S. (1998) Interactions between the atmosphere and terrestrial ecosystems: influence on weather and climate. *Global Change Biology* **4**: 461–475.

Pielke, R. A., Dalu, G. A., Snook, J. S., Lee, T. J. and Kittel, T. G. F. (1991) Nonlinear influence of mesoscale land use on weather and climate. *Journal of Climate* **4**: 1053–1069.

Pierce, J., Lorimer, G. H. and Reddy, G. S. (1986) Kinetic mechanism of ribulosebisphosphate carboxylase: Evidence for sequential, ordered reaction. *Biochemistry* **25**: 1636–1644.

Pieruschka, R., Huber, G. and Berry, J. A. (2010) Control of transpiration by radiation. *Proceedings of the National Academy of Sciences (USA)* **107**: 13371–13377.

Pinhassi, J., Simo, R., Gonzalez, J. M., *et al.* (2005) Dimethylsulfoniopropionate turnover is linked to the composition and dynamics of the bacterioplankton assemblage during a microcosm phytoplankton bloom. *Applied and Environmental Microbiology* **71**: 7650–7660.

Pinty, B. and Verstraete, M. M. (1998) Modeling and scattering of light by homogeneous vegetation in optical remote sensing. *Journal of the Atmospheric Sciences* **55**: 137–150.

Pittermann, J., Sperry, J. S., Hacke, U. G., Wheeler, J. K. and Sikkema, E. H. (2005) Torus-margo pits help conifers compete with angiosperms. *Science* **310**: 1924.

Plaxton, W. C. (1996) The organization and regulation of plant glycolysis. *Annual Review of Plant Physiology and Plant Molecular Biology* **47**: 185–214.

Poggi, D. and Katul, G. G. (2006) Two-dimensional scalar spectra in the deeper layers of a dense and uniform model canopy. *Boundary-Layer Meteorology* **121**: 267–281.

Poggi, D., Porporato, A., Ridolfi, L., *et al.* (2004) The effect of vegetation density on canopy sub-layer turbulence. *Boundary-Layer Meteorology* **111**: 565–587.

Poolman, M. G., Fell, D. A. and Thomas, S. (2000) Modelling photosynthesis and its control. *Journal of Experimental Botany* **51**: 319–328.

Poorter, L., Oberbauer, S. F. and Clark, D. B. (1995) Leaf optical properties along a vertical gradient in a tropical rain forest canopy in Costa Rica. *American Journal of Botany* **82**: 1257–1263.

Portis, A. R. and Parry, M. A. J. (2007) Discoveries in Rubisco (Ribulose 1,5-bisphosphate carboxylase/oxygenase): A historical perspective. *Photosynthesis Research* **94**: 121–143.

Prandtl, L. (1925) A report on testing for built-up turbulence. *Zeitschrift für Angewandte Mathematik und Mechanik* **5**: 136–139.

Priestley, C. H. B. and Taylor, R. J. (1972) Assessment of surface heat flux and evaporation using large-scale parameters. *Monthly Weather Review* **100**: 81–92.

Raich, J. W. and Potter, C. S. (1995) Global patterns of carbon dioxide emissions from soils. *Global Biogeochemical Cycles* **9**: 23–36.

Raich, J. W. and Schlesinger, W. H. (1992) The global carbon dioxide flux in soil respiration and its relationship to vegetation and climate. *Tellus, Series B* **44**: 81–99.

Ramier, D., Boulain, N., Cappelaere, B., *et al.* (2009) Towards an understanding of coupled physical and biological processes in the cultivated Sahel – 1: Energy and water. *Journal of Hydrology* **375**: 204–216.

Randerson, J. T., Thompson, M. V. and Field, C. B. (1999) Linking ^{13}C-based estimates of land and ocean sinks with predictions of carbon storage from CO_2 fertilization of plant growth. *Tellus, Series B* **51**: 668–678.

Raschke, K. (1960) Heat transfer between the plant and the environment. *Annual Review of Plant Physiology* **11**: 111–126.

Raschke, K. (1976) How stomata resolve the dilemma of opposing priorities. *Philosophical Transactions of the Royal Society of London, Series B* **273**: 551–560.

Rasse, D. P, Rumpel, C. and Dignac, M. F. (2005) Is soil carbon mostly root carbon? Mechanisms for a specific stabilization. *Plant and Soil* **269**: 341–356.

Rassmussen, R. A. and Went, F. W. (1965) Volatile organic material of plant origin in atmosphere. *Proceedings of the National Academy of Sciences (USA)* **53**: 215–220.

Raupach, M. R. (1988) Canopy transport processes. In: *Flow and Transport in the Natural Environment: Advances and Applications* (Steffen, W. C. and Denmead, O. T., eds.). Berlin: Springer-Verlag, pp. 95–132.

Raupach, M. R. (1989a) Stand overstorey processes. *Philosphical Transactions of the Royal Society of London, Series B* **324**: 175–190.

Raupach, M. R. (1989b) Turbulent transfer in plant canopy. In: *Plant Canopies: Their Growth, Form and Function* (Russell, G., Marshall, B. and Jarvis, P. G., eds.). Cambridge: Cambridge University Press, pp. 41–62.

Raupach, M. R. (1991) Vegetation-atmosphere interaction in homogeneous and heterogeneous terrain: Some implications of mixed-layer dynamics. *Vegetatio* **91**: 105–120.

Raupach, M. R. (1992) Drag and drag partition on rough surfaces. *Boundary-Layer Meteorology* **60**: 375–395.

Raupach, M. R. (1994) Simplified expressions for vegetation roughness length and zero-plane displacement as functions of canopy height and area index. *Boundary-Layer Meteorology* **71**: 211–216.

Raupach, M. R. and Finnigan, J. J. (1988) "Single-layer models of evaporation from plant canopies are incorrect but useful, whereas multilayer models are correct but useless": Discuss. *Australian Journal of Plant Physiology* **15**: 705–716.

Raupach, M. R., Finnigan, J. J. and Brunet, Y. (1996) Coherent eddies and turbulence in vegetation canopies: the mixing-layer analogy. *Boundary-Layer Meteorology* **78**: 351–382.

Raupach, M. R. and Thom, A. S. (1981) Turbulence in and above plant canopies. *Annual Review of Fluid Mechanics* **13**: 97–129.

Rautiainen, M., Mottus, M., Stenberg, P. and Ervasti, S. (2008) Crown envelope shape measurements and models. *Silva Fennica* **42**: 19–33.

Reich, P. B., Tjoelker, M. G., Machado, J. L. and Oleksyn, J. (2006) Universal scaling of respiratory metabolism, size, and nitrogen in plants. *Nature* **439**: 457–461.

Reich, P. B., Tjoelker, M. G., Pregitzer, K. S., Wright, I. J., Oleksyn, J. and Machado, J. L. (2008) Scaling of respiration to nitrogen in leaves, stems and roots of higher land plants. *Ecology Letters* **11**: 793–801.

Reich, P. B., Walters, M. B. and Ellsworth, D. S. (1997) From tropics to tundra: Global convergence in plant functioning. *Proceedings of the National Academy of Science, USA* **94**: 13730–13734.

Reich, P. B., Wright, I. J., Cavender-Bares, J., Craine, J. M., Oleksyn, J., Westoby, M. and Walters, M. B. (2003) The evolution of plant functional variation: Traits, spectra, and strategies. *International Journal of Plant Sciences* **164**: S143–S164.

Reich, P. B., Wright, I. J. and Lusk, C. H. (2007) Predicting leaf physiology from simple plant and climate attributes: A global GLOPNET analysis. *Ecological Applications* **17**: 1982–1988.

Rey, A. and Jarvis, P. (2006) Modelling the effect of temperature on carbon mineralization rates across a network of European forest sites (FORCAST). *Global Change Biology* **12**: 1894–1908.

Richter, D. D., Markewitz, D., Trumbore, S. E. and Wells, C. G. (1999) Rapid accumulation and turnover of soil carbon in a re-establishing forest. *Nature* **400**: 56–58.

Roberts, D. A., Nelson, B. W., Adams, J. B. and Palmer, F. (1998) Spectral changes with leaf aging in Amazon caatinga. *Trees – Structure and Function* **12**: 315–325.

Roberts, J. M., Langford, A. O., Goldan, P. and Fehsenfeld, F. C. (1988) Ammonia measurements at Niwot Ridge, Colorado and Point Arena, California using the tungsten-oxide denuder tube technique. *Journal of Atmospheric Chemistry* **7**: 137–152.

Rocha, A. V. and Shaver, G. R. (2009) Advantages of a two band EVI calculated from solar and photosynthetically active radiation fluxes. *Agricultural and Forest Meteorology* **149**: 1560–1563.

Rodean, H. C. (1996) Stochastic Lagrangian models of turbulent diffusion. *Meteorological Monographs* **26**: 1–84.

Roden, J. S. and Pearcy, R. W. (1993) Effect of leaf flutter on the light environment of poplars. *Oecologia* **93**: 201–207.

Roeske, C. A. and O'Leary, M. H. (1984) Carbon isotope effects on the enzyme-catalyzed carboxylation of ribulose bisphosphate. *Biochemistry* **23**: 6275–6284.

Rogers, A. (2013) The use and misuse of V_{cmax} in earth system models. *Photosynthesis Research*, DOI 10.1007/s11120–013–9818–1.

Romàn, M. O., Schaaf, C. B., Woodcock, C. E., *et al.* (2009) The MODIS (Collection V005) BRDF/albedo product: Assessment of spatial representativeness over forested landscapes. *Remote Sensing of Environment* **113**: 2476–2498.

Rondon, A., Johansson, C. and Granat, L. (1993) Dry deposition of nitrogen dioxide and ozone to coniferous forests. *Journal of Geophysical Research – Atmospheres* **98**: 5159–5172.

Rosenstiel, T., Potosnak, M., Griffen, K., *et al.* (2003) Elevated CO_2 uncouples growth and isoprene emission in a model agriforest ecosystem. *Nature* **421**: 256–259.

Rosenthal, D. M., Locke, A. M., Khozaei, M., Raines, C. A., Long, S. P. and Ort, D. R. (2011) Over-expressing the C_3 photosynthesis cycle enzyme sedoheptulose-1–7 bisphosphatase improves photosynthetic carbon gain and yield under fully open air CO_2 fumigation (FACE). *BMC Plant Biology* **11**: Article number 123.

Ross, J. K. (1981) *The Radiation Regime and Architecture of the Plant Stand*. The Hague: W. Junk.

Ross, J. and Môttus, M. (2000) Statistical treatment of umbra length inside willow coppice. *Agricultural and Forest Meteorology* **100**: 89–102.

Ross, J. and Nilson, T. (1968) A mathematical model of radiation regime of plant cover. In: *Actinometry and Atmospheric Optics*. Tallin, Estonia: Valus Publications, pp. 263–281.

Rossmann, A., Butzenlechner, M. and Schmidt, H. -L. (1991) Evidence for a nonstatistical carbon isotope distribution in natural glucose. *Plant Physiology* **96**: 609–614.

Rotenberg, E. and Yakir, D. (2010) Contribution of semi-arid forests to the climate system. *Science* **327**: 451–454.

Ruel, J. J. and Ayers, M. P. (1999) Jensen's inequality predicts effects of environmental variation. *Trends in Ecology and Evolution* **14**: 361–366.

Rundel, P. W., Ehleringer, J. R. and Nagy, K. A. (eds.) (1989) *Stable Isotopes in Ecological Research*. New York: Springer-Verlag.

Running, S. W. and Nemani, R. R. (1988) Relating seasonal patterns of the AVHRR vegetation index to simulated photosynthesis and transpiration of forests in different climates. *Remote Sensing of Environment* **24**: 347–367.

Running, S. W., Nemani, R. R., Heinsch, F. A., *et al.* (2004) A continuous satellite-derived measure of global terrestrial primary production. *Bioscience* **54**: 547–560.

Rutledge, S., Campbell, D. I., Baldocchi, D. and Schipper, L. A. (2010) Photodegradation leads to increased carbon dioxide losses from terrestrial organic matter. *Global Change Biology* **16**: 3065–3074.

Ryan, M. G. (1995) Foliar maintenance respiration of subalpine and boreal trees and shrubs in relation to nitrogen content. *Plant, Cell and Environment* **18**: 765–772.

Sabine, C., Heimann, M., Artaxo, P. *et al.* (2003) Current status and past trends of the carbon cycle. In: *The Global Carbon Cycle: Integrating Humans, Climate and the Natural World* (Field, C. B. and Raupach, M. R., eds.). Washington, DC: Island Press, pp. 17–44.

Sack, L. and Holbrook, N. M. (2006) Leaf hydraulics. *Annual Review of Plant Biology* **57**: 361–381.

Sage, R. F., Christin, P. A. and Edwards, E. J. (2011) The C_4 plant lineages of planet earth. *Journal of Experimental Botany* **62**: 3155–3169.

Sage, R. F. and Reid, C. D. (1994) Photosynthetic response mechanisms to environmental change in C_3 plants. In: *Plant-Environment Interactions* (Wilkinson, R. E., ed.). New York: Marcel Dekker, pp. 413–499.

Sage, R. F., Sharkey, T. D. and Seemann, J. R. (1989) Acclimation of photosynthesis to elevated CO_2 in five C_3 species. *Plant Physiology* **89**: 590–596.

Sage, R. F., Way, D. A. and Kubien, D. S. (2008) Rubisco, Rubisco activase, and global climate change. *Journal of Experimental Botany* **59**: 1581–1595.

Santiago, L. S. and Wright, S. J. (2007) Leaf functional traits of tropical forest plants in relation to growth form. *Functional Ecology* **21**: 19–27.

Sayre, J. D. (1926) Physiology of stomata of *Rumex patientia*. *Ohio Journal of Science* **26**: 233–266.

Schade, G. W. and Crutzen, P. J. (1999) CO emissions from degrading plant matter. II: Estimate of a global source strength. *Tellus* **51**: 908–919.

Schäffner, A. R. (1998) Aquaporin function, structure, and expression: are there more surprises to surface in water relations? *Planta* **204**: 131–139.

Scheffer, M., Holmgren, M., Brovkin, V., *et al.* (2005) Synergy between small- and large-scale feedbacks of vegetation on the water cycle. *Global Change Biology* **11**: 1003–1012.

Schimel, J. P. and Weintraub, M. N. (2003) The implications of exoenzyme activity on microbial carbon and nitrogen limitation in soil: a theoretical model. *Soil Biology and Biochemistry* **35**: 549–563.

Schink, B. (1999) Prokaryotes and the biosphere. In: *Biology of the Prokaryotes* (Lengeler, J. W., Drews, G. and Schlegel, H. G., eds.). Stuttgart: Georg Thieme Verlag, pp. 1028–1049.

Schlichting, H. and Gersten, K. (2004) *Boundary-Layer Theory*, 8th edn. New York: McGraw-Hill.

Scholze, M., Ciais, P. and Heimann, M. (2008) Modeling terrestrial ^{13}C cycling: Climate, land use and fire. *Global Biogeochemical Cycles* **22**: Article number GB1009.

Schulze, E.-D., Kelliher, F. M., *et al.* (1994) Relationships among maximum stomatal conductance, ecosystem surface conductance, carbon assimilation rate, and plant nitrogen nutrition: A global ecology scaling exercise. *Annual Review of Ecology and Systematics* **25**: 629–660.

Seco, R., Penuelas, J. and Filella, I. (2007) Sort-chain oxygenated VOCs: emissions and uptake by plants and atmospheric sources, sinks and concentrations. *Atmospheric Environment* **41**: 2477–2499.

Sellers, P. J., Berry, J. A., Collatz, G. J., *et al.* (1992) Canopy reflectance, photosynthesis and transpiration. III: A reanalysis using improved leaf models and a new canopy integration scheme. *Remote Sensing of Environment* **42**: 187–216.

Sellers, P. J., Randall, D. A., *et al.* (1996) A revised land surface parameterization (SiB2) for atmospheric GCMs. Part 1: Model formulation. *Journal of Climate* **9**: 676–705.

Sharkey, T. D. (1985) Photosynthesis in intact leaves of C$_3$ plants: physics, physiology and rate limitations. *Botanical Review* **51**: 53–105.

Sharkey, T. D. and Yeh, S. S. (2001) Isoprene emission from plants. *Annual Review of Plant Physiology and Plant Molecular Biology* **52**: 407–436.

Sharpe, P. J. and De Michelle, D. W. (1977) Reaction kinetics of poikilotherm development. *Journal of Theoretical Biology* **64**: 649–670.

Shaw, R. H. and Patton, E. G. (2003) Canopy element influences on resolved- and subgrid-scale energy within a large-eddy simulation. *Agricultural and Forest Meteorology* **115**: 5–17.

Shaw, R. H. and Pereira, A. R. (1982) Aerodynamic roughness of a plant canopy: A numerical experiment. *Agricultural Meteorology* **26**: 51–65.

Shaw, R. H. and Schumann, U. (1992) Large-eddy simulation of turbulent flow above and within a forest. *Boundary-Layer Meteorology* **61**: 47–64.

Shen, S. H. and Leclerc, M. Y. (1997) Modelling the turbulence structure in the canopy layer. *Agricultural and Forest Meteorology* **87**: 3–25.

Shimazaki, K., Doi, M., Assmann, S. M. and Kinoshita, T. (2007) Light regulation of stomatal movement. *Annual Review of Plant Biology* **58**: 219–247.

Sinclair, T. R., Murphy, C. E. and Knoerr, K. R. (1976) Development and evaluation of simplified models for simulating canopy photosynthesis and transpiration. *Journal of Applied Ecology* **13**: 813–829.

Sinclair, T. R., Tanner, C. B. and Bennet, J. M. (1984) Water-use efficiency in crop production. *Bioscience* **34**: 36–40.

Singh, H. B., Chen, Y., Staudt, A., Jacob, D., Blake, D., Heikes, B. and Snow, J. (2001) Evidence from the Pacific troposphere for large global sources of oxygenated organic compounds. *Nature* **410**: 1078–1081.

Sinoquet, H., Stephan, J., Sonohat, G., *et al.* (2007) Simple equations to estimate light interception by isolated trees from canopy structure features: Assessment with three-dimensional digitized apple trees. *New Phytologist* **175**: 94–106.

Sinsabaugh, R. L. (1994) Enzymatic analysis of microbial pattern and process. *Biology and Fertility of Soils* **17**: 69–74.

Sinsabaugh, R. L., Antibus, R. K. and Linkins, A. E. (1991) An enzymatic approach to the analysis of microbial activity during plant litter decomposition. *Agriculture Ecosystems and Environment* **34**: 43–54.

Siqueira, M., Katul, G. and Porporato, A. (2009) Soil moisture feedbacks on convection triggers: The role of soil-plant hydrodynamics. *Journal of Hydrometeorology* **10**: 96–112.

Smith, W. K. (1978) Temperature of desert plants – another perspective on adaptability of leaf size. *Science* **201**: 614–616.

Smith, W. K., Knapp, A. K. and Reiners, W. A. (1989) Penumbral effects on sunlight penetration in plant communities. *Ecology* **70**: 1603–1609.

Smolander, S. and Stenberg, P. (2001) A method for estimating light interception by a conifer shoot. *Tree Physiology* **21**: 797–803.

Smolander, S. and Stenberg, P. (2003) A method to account for shoot scale clumping in coniferous canopy reflectance models. *Remote Sensing of Environment* **88**: 363–373.

Smoluchowski, M. (1906) Zur kinetischen Theorie der Brownschen Molekularbewegung und der Suspensionen. *Annalen der Physik* **21**: 756–780.

Sparks, J. P. (2009) Ecological ramifications of the direct foliar uptake of nitrogen. *Oecologia* **159**: 1–13.

Sparks, J. P., Monson, R. K., Sparks, K. L. and Lerdau, M. T. (2001) Leaf uptake of nitrogen dioxide (NO_2) in a tropical wet forest: implications for tropospheric chemistry. *Oecologia* **127**: 214–221.

Sperry, J. S. (2003) Evolution of water transport and xylem structure. *International Journal of Plant Sciences* **164**: S115–S127.

Sperry, J. S., Alder, N. N. and Eastlack, S. E. (1993) The effect of reduced hydraulic conductance on stomatal conductance and xylem cavitation. *Journal of Experimental Botany* **44**: 1075–1082.

Sperry, J. S., Meinzer, F. C. and McCulloh, K. A. (2008) Safety and efficiency conflicts in hydraulic architecture: Scaling from tissues to trees. *Plant Cell and Environment* **31**: 632–645.

Spreitzer, R. J. and Salvucci, M. E. (2002) Rubisco: Structure, regulatory interactions and possibilities for a better enzyme. *Annual Review of Plant Physiology and Plant Molecular Biology* **53**: 449–475.

Stenberg, P. (1995) Penumbra in within-shoot and between-shoot shading in conifers and its significance for photosynthesis. *Ecological Modelling* **77**: 215–231.

Stenberg, P. (1996) Correcting LAI-2000 estimates for the clumping of needles in shoots of conifers. *Agricultural and Forest Meteorology* **79**: 1–8.

Stenberg, P. (2007) Simple analytical formula for calculating average photon recollision probability in vegetation canopies. *Remote Sensing of Environment* **109**: 221–224.

Steudle, E. and Peterson, C. A. (1998) How does water get through roots? *Journal of Experimental Botany* **49**: 775–788.

Steuer, R. (2007) Computational approaches to the topology, stability and dynamics of metabolic networks. *Phytochemistry* **68**: 2139–2151.

Stevenson, R. J. (1982) *Humus Chemistry: Genesis, Composition, Reactions*. New York: John Wiley & Sons.

Still, C. J., Berry, J. A., Collatz, G. J. and DeFries, R. S. (2003) Global distribution of C_3 and C_4 vegetation: Carbon cycle implications. *Global Biogeochemical Cycles* **17**: Article number 1006.

Stitt, M. (1994) Flux control at the level of the pathway: studies with mutants and transgenic plants having a decreased activity of enzymes involved in photosynthesis partitioning. In: *Flux Control in Biological Systems* (Schulze, E.-D., ed.). New York: Academic Press, Inc., pp. 13–36.

Stitt, M. (1996) Metabolic regulation of photosynthesis. In: *Photosynthesis and the Environment* (Baker, N., ed.). Dordrecht: Kluwer Academic Publishers, pp. 151–190.

Stoy, P. C., Richardson, A. D., Baldocchi, D. D., *et al.* (2009) Biosphere-atmosphere exchange of CO_2 in relation to climate: a cross-biome analysis across multiple time scales. *Biogeosciences* **6**: 2297–2312.

Stull, R. B. (1988) *An Introduction to Boundary Layer Meteorology.* Dordrecht: Kluwer Academic Publishers.

Stump, L. M. and Binkley, D. (1993) relationships between litter quality and nitrogen availability in Rocky Mountain forests. *Canadian Journal of Forest Research* **23**: 492–502.

Su, H. B., Schmid, H. P., Grimmond, C. S. B., *et al.* (2004) Spectral characteristics and correction of long-term eddy-covariance measurements over two mixed hardwood forests in non-flat terrain. *Boundary-Layer Meteorology* **110**: 213–253.

Su, H. -B., Schmid, H. P., Vogel, C. S. and Curtis, P. S. (2008) Effects of canopy morphology and thermal stability on mean flow and turbulence statistics observed inside a mixed hardwood forest. *Agricultural and Forest Meteorology* **148**: 862–882.

Sun, J., Lenschow, D. H., Burns, S. P., *et al.* (2004) Atmospheric disturbances that generate intermittent turbulence in nocturnal boundary layers. *Boundary-Layer Meteorology* **110**: 255–279.

Sun, J. L., Lenschow, D. H., Mahrt, L., *et al.* (1997) Lake induced atmospheric circulations during BOREAS. *Journal of Geophysical Research* **102**: 29155–29166.

Suntharalingam, P., Kettle, A. J., Montzka, S. M. and Jacob, D. J. (2008) Global 3-D model analysis of the seasonal cycle of atmospheric carbonyl sulfide: Implications for terrestrial vegetation uptake. *Geophysical Research Letters* **35**: Article number L19801.

Sweetlove, L. J. and Fernie, A. R. (2005) Regulation of metabolic networks: Understanding metabolic complexity in the systems biology era. *New Phytologist* **168**: 9–24.

Taneda, S. (1965) Experimental investigation of vortex streets. *Journal of the Physical Society of Japan* **20**: 714–721.

Tans, P. P., de Jong, A. F. M. and Mook, W. G. (1979) Natural atmospheric ^{14}C variation and the Suess effect. *Nature* **280**: 826–828.

Taraborrelli, D., Lawrence, M. G., Crowley, J. N., *et al.* (2012) Hydroxyl radical buffered by isoprene oxidation over tropical forests. *Nature Geosciences* **5**: 190–193.

Tarr, M. A., Miller, W. L. and Zepp, R. G. (1995) Direct carbon monoxide production from plant matter. *Journal of Geophysical Research* **100**: 11403–11413.

Taylor, C. M., Gounou, A., Guichard, F., *et al.* (2011) Frequency of Sahelian storm initiation enhanced over mesoscale soil-moisture patterns. *Nature Geoscience* **4**: 430–433.

Taylor, G. I. (1938) The spectrum of turbulence. *Proceedings of the Royal Society of London. Series A, Mathematical and Physical Sciences* **164**: 476–490.

Tcherkez, G., Farquhar, G. D. and Andrews, T. J. (2006) Despite slow catalysis and confused substrate specificity, all ribulose bisphosphate carboxylases may be nearly perfectly optimized. *Proceedings of the National Academy of Sciences (USA)* **103**: 7246–7251.

Tcherkez, G., Farquhar, G., Badeck, F. and Ghashghaie, J. (2004) Theoretical considerations about carbon isotope distribution in glucose of C_3 plants. *Functional Plant Biology* **31**: 857–877.

Tcherkez, G., Schaufele, R., Nogues, S., *et al.* (2010) On the $^{13}C/^{12}C$ isotopic signal of day and night respiration at the mesocosm level. *Plant, Cell and Environment* **33**: 900–913.

Tenhunen, J. D., Hanano, R., Abril, M., *et al.* (1994) Above- and below-ground environmental influences on leaf conductance of *Ceanothus thyrsiflorus* growing in a chaparral environment: Drought responses and the role of abscisic acid. *Oecologia* **99**: 306–314.

Terashima, I. and Hikosaka, K. (1995) Comparative ecophysiology of leaf and canopy photosynthesis. *Plant, Cell and Environment* **18**: 1111–1128.

Terashima, I. and Inoue, Y. (1984) Comparative photosynthetic properties of palisade tissue chloroplasts and spongy tissue chloroplasts of *Camilla japonica* L.: Functional adjustment of the photosynthetic apparatus to light environment within a leaf. *Plant and Cell Physiology* **25**: 555–563.

Terashima, I., Masuzawa, T., Ohba, H. and Yokoi, Y. (1995) Is photosynthesis suppressed at higher elevations due to low CO_2 pressure? *Ecology* **76**: 2663–2668.

Terashima, I., Miyazawa, S. I. and Hanba, Y. T. (2001) Why are sun leaves thicker than shade leaves? Consideration based on analyses of CO_2 diffusion in the leaf. *Journal of Plant Research* **114**: 93–105.

Terashima, I. and Saeki, T. (1983) Light environment within a leaf. I. Optical properties of paradermal sections of *Camellia* leaves with special reference to differences in the optical properties of palisade and spongy tissues. *Plant and Cell Physiology* **24**: 1493–1501.

Terashima, I. and Saeki, T. (1985) A new model for leaf photosynthesis incorporating the gradients of light environment and of photosynthetic properties of chloroplasts within a leaf. *Annals of Botany* **56**: 489–499.

Terashima, I., Wong, S. C., Osmond, C. B. and Farquhar, G. D. (1988) Characterization of non-uniform photosynthesis induced by abscisic-acid in leaves having different mesophyll anatomies. *Plant and Cell Physiology* **29**: 385–394.

Tholen, D., Boom, C., Noguchi, K., *et al.* (2008) The chloroplast avoidance response decreases internal conductance to CO_2 diffusion in *Arabdopsis thaliana* leaves. *Plant, Cell and Environment* **31**: 1688–1700.

Thom, A. S., Stewart, J. B., Oliver, H. R. and Gash, J. H. C. (1975) Comparison of aerodynamic and energy budget estimates of fluxes over a pine forest. *Quarterly Journal of the Royal Meteorological Society* **101**: 93–105.

Thomas, C. and Foken, T. (2007) Organised motion in a tall spruce canopy: Temporal scales, structure spacing and terrain effects. *Boundary-Layer Meteorology* **122**: 123–147.

Thomson, D. J. (1987) Random walk models of atmospheric dispersion. *Meteorological Magazine* **116**: 142–150.

Thornley, J. H. M. (1970) Respiration, growth and maintenance in plants. *Nature* **227**: 304–305.

Thornley, J. H. M. (1971) Energy, respiration and growth in plants. *Annals of Botany* **35**: 721–728.

Thuille, A., Buchmann, N. and Schulze, E. D. (2000) Carbon stocks and soil respiration rates during deforestation, grassland use and subsequent Norway spruce afforestation in the Southern Alps, Italy. *Tree Physiology* **20**: 849–857.

Thuille, A. and Schulze, E. D. (2005) Carbon dynamics in successional and afforested spruce stands in Thuringia and the Alps. *Global Change Biology* **12**: 325–342.

Torn, M. S., Lapenis, A. G., Timofeev, A., *et al.* (2002) Organic carbon and carbon isotopes in modern and 100-year-old-soil archives of the Russian steppe. *Global Change Biology* **8**: 941–953.

Trumbore, S. (2000) Age of soil organic matter and soil respiration: Radiocarbon constraints on belowground C dynamics. *Ecological Applications* **10**: 399–411.

Trumbore, S. E., Chadwick, O. A. and Amundson, R. (1996) Rapid exchange between soil carbon and atmospheric carbon dioxide driven by temperature change. *Science* **272**: 393–396.

Tsigaridis, K. and Kanakidou, M. (2007) Secondary organic aerosol importance in the future atmosphere. *Atmospheric Environment* **41**: 4682–4692.

Tuomi, M., Thum, T., Jarvinen, H., *et al.* (2009) Leaf litter decomposition: Estimates of global variability based on Yasso07 model. *Ecological Modelling* **220**: 3362–3371.

Turner, D. P., Ritts, W. D., Cohen, W. B. *et al.* (2006) Evaluation of MODIS NPP and GPP products across multiple biomes. *Remote Sensing of Environment* **102**: 282–292.

Turnipseed, A. A., Anderson, D. E., Blanken, P. D. *et al.* (2003) Airflows and turbulent flux measurements in mountainous terrain. Part 1: Canopy and local effects. *Agricultural and Forest Meteorology* **119**: 1–21.

Turnipseed, A. A., Anderson, D. E., Burns, S., *et al.* (2004) Airflows and turbulent flux measurements in mountainous terrain Part 2: Mesoscale effects. *Agricultural and Forest Meteorology* **125**: 187–205.

Tyerman, S. D., Bohnert, H. J., Maurel, C., *et al.* (1999) Plant aquaporins: their molecular biology, biophysics and significance for plant water relations. *Journal of Experimental Botany* **50**: 1055–1071.

Tyree, M. T. and Ewers, F. (1991) Tansley Review: The hydraulic architecture of trees and other woody plants. *New Phytologist* **119**: 345–360.

Tyree, M. T. and Sperry, J. S. (1989) Vulnerability of xylem to cavitation and embolism. *Annual Review of Plant Physiology and Plant Molecular Biology* **40**: 19–38.

Vallina, S. M. and Simo, R. (2007) Strong relationship between DMS and the solar radiation dose over the global surface ocean. *Science* **315**: 506–508.

van Genuchten, M. T. (1980) A closed-form equation for predicting the hydraulic conductivity of unsaturated soils. *Soil Science Society of America Journal* **44**: 892–898.

van Oosten, J.-J., Wilkens, D. and Besford, R. T. (1994) Regulation of the expression of photosynthetic nuclear genes by high CO_2 is mimicked by carbohydrates: a mechanism for the acclimation of photosynthesis to high CO_2. *Plant, Cell and Environment* **17**: 913–923.

van't Hoff, J. H. (1884) *Études de Dynamique Chimique* [*Studies in Chemical Dynamics*]. Amsterdam: F. Muller & Co.

Vandenhurk, B. J. J. M. and McNaughton, K. G. (1995) Implementation of near-field dispersion in a simple 2-layer surface resistance model. *Journal of Hydrology* **166**: 293–311.

Vanhala, P., Karhu, K., Tuomi, M., *et al.* (2007) Old soil carbon is more temperature sensitive than young in an agricultural field. *Soil Biology and Biochemistry* **39**: 2967–2970.

Vargas, R., Detto, M., Baldocchi, D. D. and Allen, M. F. (2010) Multiscale analysis of temporal variability of soil CO_2 production as influenced by weather and vegetation. *Global Change Biology* **16**: 1589–1605.

Varutbangkul, V., Brechtel, F. J., Bahreini, R., *et al.* (2006) Hygroscopicity of secondary organic aerosols formed by oxidation of cycloalkenes, monoterpenes, sesquiterpenes, and related compounds. *Atmospheric Chemistry and Physics* **6**: 2367–2388.

Veneklaas, E. J. and Poorter, L. (1998) Growth and carbon partitioning of tropical tree seedlings in contrasting light environments. In: *Inherent Variation in Plant Growth: Physiological Mechanisms and Ecological Consequences* (Lambers, H., Poorter, H. and Van Vuuren, M. M. I., eds.). Leiden: Backhuys Publishers, pp. 337–361.

Vernadsky, V. (1938) Problems of biogeochemistry II: On the fundamental material-energetic distinction between living and non-living natural bodies of the biosphere. (Translated from Russian by Jonathan Tennenbaum and Rachel Douglas, *21st Century*, Winter 2000–2001, pp. 20–39.)

Vogelmann, T. C., Bornman, J. F. and Josserand, S. (1989) Photosynthetic light gradients and spectral regime within leaves of *Medicago sativa*. *Philosophical Transactions of the Royal Society of London, Series B* **323**: 411–421.

Vogelman, T. C. and Martin, G. (1993) The functional significance of palisade tissue – penetration of directional versus diffuse light. *Plant, Cell and Environment* **16**: 65–72.

von Caemmerer, S., Evans, J. R., Hudson, G. S. and Andrews, T. J. (1994) The kinetics of ribulose-1,5-bisphosphate carboxylase/oxygenase *in vivo* inferred from measurements of photosynthesis in leaves of transgenic tobacco. *Planta* **195**: 88–97.

von Caemmerer, S. and Farquhar, G. D. (1981) Some relationships between the biochemistry of photosynthesis and the gas exchange of leaves. *Planta* **153**: 376–387.

von Caemmerer, S., Lawson, T., Oxborough, K., *et al.* (2004) Stomatal conductance does not correlate with photosynthetic capacity in transgenic tobacco with reduced amounts of Rubisco. *Journal of Experimental Botany* **55**: 1157–1166.

von Lutzöw, M., Kögel-Knabner, I., Ekschmitt, K., *et al.* (2006) Stabilization of organic matter in temperate soils: Mechanisms and their relevance under different soil conditions – a review. *European Journal of Soil Science* **57**: 426–445.

Wagner, W., Hollaus, M., Briese, C. and Ducic, V. (2008) 3D vegetation mapping using small-footprint full-waveform airborne laser scanners. *International Journal of Remote Sensing* **29**: 1433–1452.

Wallenstein, M. D., Hess, A. M., Lewis, M. R., *et al.* (2010) Decomposition of aspen leaf litter results in unique metabolomes when decomposed under different tree species. *Soil Biology and Biochemistry* **42**: 484–490.

Wallenstein, M. D. and Weintraub, M. N. (2008) Emerging tools for measuring and modeling the in situ activity of soil extracellular enzymes. *Soil Biology and Biochemistry* **40**: 2098–2106.

Walter, K. M., Smith, L. C. and Chapin, F. S. (2007) Methane bubbling from northern lakes: Present and future contributions to the global methane budget. *Philosophical Transactions of the Royal Society, Series A* **365**: 1657–1676.

Walter-Shea, E. A., Norman, J. M. and Blad, B. L. (1989) Leaf bidirectional reflectance and transmittance in corn and soybean. *Remote Sensing of Environment* **29**: 161–174.

Wang, G. L., Kim, Y. and Wang, D. G. (2007) Quantifying the strength of soil moisture-precipitation coupling and its sensitivity to changes in surface water budget. *Journal of Hydrometeorology* **8**: 551–570.

Wang, Y. P. and Jarvis, P. G. (1990) Description and validation of an array model. *Agricultural and Forest Meteorology* **51**: 257–280.

Warland, J. S. and Thurtell, G. W. (2000) A Lagrangian solution to the relationship between a distributed source and concentration profile. *Boundary-Layer Meteorology* **96**: 453–471.

Warren, C. (2006) Estimating the internal conductance to CO_2 movement. *Functional Plant Biology* **33**: 431–442.

Warren, J. M., Brooks, J. R., Dragila, M. I. and Meinzer, F. C. (2011) In situ separation of root hydraulic redistribution of soil water from liquid and vapor transport. *Oecologia* **166**: 899–911.

Webb, E. K., Pearman, G. I. and Leuning, R. (1980) Correction of flux measurements for density effects due to heat and water vapor transfer. *Quarterly Journal of the Royal Meteorological Society* **106**: 85–100.

Weber, P. and Rennenberg, H. (1996) Dependency of nitrogen dioxide (NO_2) fluxes to wheat (*Triticum aestivum* L.) leaves on NO_2 concentration, light intensity, temperature and relative humidity determined from controlled dynamic chamber experiments. *Atmospheric Environment* **30**: 3001–3009.

Went, F. W. (1960) Blue hazes in the atmosphere. *Nature* **187**: 641–643.

Westgate, M. E. and Steudle, E. (1985) Water transport in the midrib tissue of maize leaves – direct measurement of the propagation of changes in cell turgor across a plant tissue. *Plant Physiology* **78**: 183–191.

Whitney, S. M., von Caemmerer, S., Hudson, G. S. and Andrews, T. J. (1999) Directed mutation of the Rubisco large subunit of tobacco influences photorespiration and growth. *Plant Physiology* **121**: 579–588.

Wickman, F. E. (1952) Variations in the relative abundance of the carbon isotopes in plants. *Geochimica et Cosmochimica Acta* **2**: 243–254.

Wieringa, J. (1993) Representative roughness parameters for homogeneous terrain. *Boundary-Layer Meteorology* **63**: 323–363.

Wigley, G. and Clark, J. A. (1974) Heat transfer coefficients for constant energy flux models of broad leaves. *Boundary-Layer Meteorology* **7**: 139–150.

Williams, W. E. (1983) Optimal water-use efficiency in a California shrub. *Plant, Cell and Environment* **6**: 145–151.

Wilson, K. B., Baldocchi, D. D., and Hanson, P. J. (2000) Quantifying stomatal and non-stomatal limitations to carbon assimilation resulting from leaf aging and drought in mature deciduous tree species. *Tree Physiology* **20**: 787–797.

Wilson, J. D., Thurtell, G. E. and Kidd, G. E. (1981) Numerical simulation of particle trajectories in inhomogeneous turbulence. 2: Systems with variable turbulent velocity scale. *Boundary-Layer Meteorology* **21**: 423–441.

Wong, S. -C., Cowan, I. R. and Farquhar, G. D. (1978) Leaf conductance in relation to assimilation in *Eucalyptus pauciflora* Sieb. Ex Spreng: Influence of irradiance and partial pressure of carbon dioxide. *Plant Physiology* **62**: 670–674.

Wong, S. -C., Cowan, I. R. and Farquhar, G. D. (1985) Leaf conductance in relation to rate of CO_2 assimilation. I. Influence of nitrogen nutrition, phosphorus nutrition, photon flux density, and ambient partial pressure of CO_2 during ontogeny. *Plant Physiology* **78**: 821–825.

Woodrow, I. E. (1986) Control of the rate of photosynthetic carbon dioxide fixation. *Biochemica et Biophysica Acta* **851**: 181–192.

Woodrow, I. E., Ball, J. T. and Berry, J. A. (1990) Control of photosynthetic carbon dioxide fixation by the boundary layer, stomata, and ribulose 1,5-bisphosphate carboxylase-oxygenase. *Plant, Cell and Environment* **13**: 339–347.

Woodrow, I. E. and Berry, J. A. (1988) Enzymatic regulation of photosynthetic CO_2 fixation in C_3 plants. *Annual Review of Plant Physiology and Plant Molecular Biology* **39**: 533–594.

Woodrow, I. E. and Mott, K. A. (1993) Modelling C_3 photosynthesis: a sensitivity analysis of the photosynthetic carbon-reduction cycle. *Planta* **191**: 421–432.

Woodward, F. I. (1987) *Climate and Plant Distribution*. Cambridge: Cambridge University Press.

Wright, I. J., Reich, P. B. and Westoby, M. (2001) Strategy-shifts in leaf physiology, structure and nutrient content between species of high and low rainfall, and high and low nutrient habitats. *Functional Ecology* **15**: 423–434.

Wright, I. J., Reich, P. B., Westoby, M., *et al.* (2004) The worldwide leaf economics spectrum. *Nature* **428**: 821–827.

Wright, I. J., Westoby, M. and Reich, P. B. (2002) Convergence towards higher leaf mass per area in dry and nutrient-poor habitats has different consequences for leaf life span. *Journal of Ecology* **90**: 534–543.

Wu, H., Sharpe, P. J. H. and Spence, R. D. (1985) Stomatal mechanics. 3: Geometric interpretation of the mechanical advantage. *Plant, Cell and Environment* **8**: 269–274.

Wuebbles, D. J. and Hayhoe, K. (2002) Atmospheric methane and global change. *Earth Science Reviews* **57**: 177–210.

Wullschleger, S. D. (1993) Biochemical limitations to carbon assimilation in C_3 plants: A retrospective analysis of the A/Ci curves from 109 species. *Journal of Experimental Botany* **44**: 907–920.

Wyngaard, J. C. (1990) Scalar fluxes in the planetary boundary layer: Theory, modeling and measurement. *Boundary-Layer Meteorology* **50**: 49–75.

Wyngaard, J. C. (1992) Atmospheric turbulence. *Annual Review of Fluid Mechanics* **24**: 205–233.

Wyngaard, J. C. and Coté, O. R. (1971) The budgets of turbulent kinetic energy and temperature variance in the atmospheric surface layer. *Journal of Atmospheric Science* **28**: 190–201.

Wyngaard, J. C. and Coté, O. R. (1972) Cospectral similarity in the atmospheric surface layer. *Quarterly Journal of the Royal Meteorological Society* **98**: 590–603.

Xie, H. X. and Moore, R. M. (1999) Carbon disulfide in the North Atlantic and Pacific Oceans. *Journal of Geophysical Research* **104**: 5393–5402.

Yakir, D. (2004) The stable isotopic composition of atmospheric CO_2. In: *Treatise on Geochemistry, Volume 4: The Atmosphere* (Keeling, R. F., ed.). Amsterdam: Elsevier, pp. 175–212.

Ye, Q., Holbrook, N. M. and Zwieniecki, M. A. (2008) Cell-to-cell pathway dominates xylem-epidermis hydraulic connection in *Tradescantia fluminensis* (Vell. Conc.) leaves. *Planta* **227**: 1311–1319.

Yin, X., Van Oijen, M. and Schapendonk, A. H. C. M. (2004) Extension of a biochemical model for the generalized stoichiometry of electron transport limited C_3 photosynthesis. *Plant, Cell and Environment* **27**: 1211–1222.

Yoshie, F. (1986) Intercellular CO_2 concentration and water-use efficiency of temperate plants with different life-forms and from different microhabitats. *Oecologia* **68**: 370–374.

Yuste, J. C., Janssens, I. A., Carrara, A. and Ceulemans, R. (2004) Annual Q_{10} of soil respiration reflects plant phenological patterns as well as temperature sensitivity. *Global Change Biology* **10**: 161–169.

Zeng, X. and Pielke, R. A. (1995a) Further study on the predictability of landscape-induced atmospheric flow. *Journal of the Atmospheric Sciences* **52**: 1680–1698.

Zeng, X. and Pielke, R. A. (1995b) Landscape-induced atmospheric flow and its parameterization in large-scale numerical models. *Journal of Climate* **8**: 1156–1177.

Zhao, M. S., Heinsch, F. A., Nemani, R. R. and Running, S. W. (2005) Improvements of the MODIS terrestrial gross and net primary production global data set. *Remote Sensing of Environment* **95**: 164–176.

Zhao, M. S., Running, S. W. and Nemani, R. R. (2006) Sensitivity of Moderate Resolution Imaging Spectroradiometer (MODIS) terrestrial primary production to the accuracy of meteorological reanalyses. *Journal of Geophysical Research – Biogeosciences* **111**: Article number: G01002.

Zhou, L., Dickinson, R. E., Tian, Y., *et al.* (2003) A sensitivity study of climate and energy balance simulations with use of satellite-derived emissivity data over Northern Africa and the Arabian Peninsula. *Journal of Geophysical Research – Atmospheres* **108**: Article number 4795.

Zwieniecki, M. A., Brodribb, T. J. and Holbrook, N. M. (2007) Hydraulic design of leaves: insights from rehydration kinetics. *Plant, Cell and Environment* **30**: 910–921.

Index

Printed in the United States
by Baker & Taylor Publisher Services